STATA BASE REFERENCE MANUAL
VOLUME 3
N–R
RELEASE 12

A Stata Press Publication
StataCorp LP
College Station, Texas

Published by Stata Press, 4905 Lakeway Drive, College Station, Texas 77845
Typeset in TEX
Printed in the United States of America

10 9 8 7 6 5 4 3 2 1

ISBN-10: 1-59718-098-X (volumes 1–4)
ISBN-10: 1-59718-099-8 (volume 1)
ISBN-10: 1-59718-100-5 (volume 2)
ISBN-10: 1-59718-101-3 (volume 3)
ISBN-10: 1-59718-102-1 (volume 4)
ISBN-13: 978-1-59718-098-6 (volumes 1–4)
ISBN-13: 978-1-59718-099-3 (volume 1)
ISBN-13: 978-1-59718-100-6 (volume 2)
ISBN-13: 978-1-59718-101-3 (volume 3)
ISBN-13: 978-1-59718-102-0 (volume 4)

The suggested citation for this software is

StataCorp. 2011. *Stata: Release 12*. Statistical Software. College Station, TX: StataCorp LP.

Title

nbreg — Negative binomial regression

Syntax

Negative binomial regression model

`nbreg` *depvar* [*indepvars*] [*if*] [*in*] [*weight*] [, *nbreg_options*]

Generalized negative binomial model

`gnbreg` *depvar* [*indepvars*] [*if*] [*in*] [*weight*] [, *gnbreg_options*]

nbreg_options	Description
Model	
`noconstant`	suppress constant term
`dispersion(mean)`	parameterization of dispersion; `dispersion(mean)` is the default
`dispersion(constant)`	constant dispersion for all observations
`exposure(`$varname_e$`)`	include ln($varname_e$) in model with coefficient constrained to 1
`offset(`$varname_o$`)`	include $varname_o$ in model with coefficient constrained to 1
`constraints(`*constraints*`)`	apply specified linear constraints
`collinear`	keep collinear variables
SE/Robust	
`vce(`*vcetype*`)`	*vcetype* may be `oim`, `robust`, `cluster` *clustvar*, `opg`, `bootstrap`, or `jackknife`
Reporting	
`level(`*#*`)`	set confidence level; default is `level(95)`
`nolrtest`	suppress likelihood-ratio test
`irr`	report incidence-rate ratios
`nocnsreport`	do not display constraints
display_options	control column formats, row spacing, line width, and display of omitted variables and base and empty cells
Maximization	
maximize_options	control the maximization process; seldom used
`coeflegend`	display legend instead of statistics

gnbreg_options	Description
Model	
noconstant	suppress constant term
lnalpha(*varlist*)	dispersion model variables
exposure(*varname*$_e$)	include ln(*varname*$_e$) in model with coefficient constrained to 1
offset(*varname*$_o$)	include *varname*$_o$ in model with coefficient constrained to 1
constraints(*constraints*)	apply specified linear constraints
collinear	keep collinear variables
SE/Robust	
vce(*vcetype*)	*vcetype* may be oim, robust, cluster *clustvar*, opg, bootstrap, or jackknife
Reporting	
level(#)	set confidence level; default is level(95)
irr	report incidence-rate ratios
nocnsreport	do not display constraints
display_options	control column formats, row spacing, line width, and display of omitted variables and base and empty cells
Maximization	
maximize_options	control the maximization process; seldom used
coeflegend	display legend instead of statistics

indepvars and *varlist* may contain factor variables; see [U] **11.4.3 Factor variables**.

depvar, *indepvars*, *varname*$_e$, and *varname*$_o$ may contain time-series operators (nbreg only); see [U] **11.4.4 Time-series varlists**.

bootstrap, by (nbreg only), fracpoly (nbreg only), jackknife, mfp (nbreg only), mi estimate, nestreg (nbreg only), rolling, statsby, stepwise, and svy are allowed; see [U] **11.1.10 Prefix commands**.

vce(bootstrap) and vce(jackknife) are not allowed with the mi estimate prefix; see [MI] **mi estimate**.

Weights are not allowed with the bootstrap prefix; see [R] **bootstrap**.

vce() and weights are not allowed with the svy prefix; see [SVY] **svy**.

fweights, iweights, and pweights are allowed; see [U] **11.1.6 weight**.

coeflegend does not appear in the dialog box.

See [U] **20 Estimation and postestimation commands** for more capabilities of estimation commands.

Menu

nbreg

Statistics > Count outcomes > Negative binomial regression

gnbreg

Statistics > Count outcomes > Generalized negative binomial regression

Description

nbreg fits a negative binomial regression model of *depvar* on *indepvars*, where *depvar* is a nonnegative count variable. In this model, the count variable is believed to be generated by a Poisson-like process, except that the variation is greater than that of a true Poisson. This extra variation is referred to as overdispersion. See [R] **poisson** before reading this entry.

gnbreg fits a generalization of the negative binomial mean-dispersion model; the shape parameter α may also be parameterized.

If you have panel data, see [XT] **xtnbreg**.

Options for nbreg

Model

noconstant; see [R] **estimation options**.

dispersion(mean | constant) specifies the parameterization of the model. dispersion(mean), the default, yields a model with dispersion equal to $1+\alpha\exp(\mathbf{x}_j\boldsymbol{\beta}+\text{offset}_j)$; that is, the dispersion is a function of the expected mean: $\exp(\mathbf{x}_j\boldsymbol{\beta}+\text{offset}_j)$. dispersion(constant) has dispersion equal to $1+\delta$; that is, it is a constant for all observations.

exposure(*varname*$_e$), offset(*varname*$_o$), constraints(*constraints*), collinear; see [R] **estimation options**.

SE/Robust

vce(*vcetype*) specifies the type of standard error reported, which includes types that are derived from asymptotic theory, that are robust to some kinds of misspecification, that allow for intragroup correlation, and that use bootstrap or jackknife methods; see [R] ***vce_option***.

Reporting

level(*#*); see [R] **estimation options**.

nolrtest suppresses fitting the Poisson model. Without this option, a comparison Poisson model is fit, and the likelihood is used in a likelihood-ratio test of the null hypothesis that the dispersion parameter is zero.

irr reports estimated coefficients transformed to incidence-rate ratios, that is, e^{β_i} rather than β_i. Standard errors and confidence intervals are similarly transformed. This option affects how results are displayed, not how they are estimated or stored. irr may be specified at estimation or when replaying previously estimated results.

nocnsreport; see [R] **estimation options**.

display_options: noomitted, vsquish, noemptycells, baselevels, allbaselevels, cformat(%*fmt*), pformat(%*fmt*), sformat(%*fmt*), and nolstretch; see [R] **estimation options**.

Maximization

maximize_options: difficult, technique(*algorithm_spec*), iterate(*#*), [no]log, trace, gradient, showstep, hessian, showtolerance, tolerance(*#*), ltolerance(*#*), nrtolerance(*#*), nonrtolerance, and from(*init_specs*); see [R] **maximize**. These options are seldom used.

Setting the optimization type to technique(bhhh) resets the default *vcetype* to vce(opg).

The following option is available with nbreg but is not shown in the dialog box:

coeflegend; see [R] **estimation options**.

Options for gnbreg

Model

noconstant; see [R] **estimation options**.

lnalpha(*varlist*) allows you to specify a linear equation for $\ln\alpha$. Specifying lnalpha(male old) means that $\ln\alpha = \gamma_0 + \gamma_1$male $+ \gamma_2$old, where γ_0, γ_1, and γ_2 are parameters to be estimated along with the other model coefficients. If this option is not specified, gnbreg and nbreg will produce the same results because the shape parameter will be parameterized as a constant.

exposure(*varname$_e$*), offset(*varname$_o$*), constraints(*constraints*), collinear; see [R] **estimation options**.

SE/Robust

vce(*vcetype*) specifies the type of standard error reported, which includes types that are derived from asymptotic theory, that are robust to some kinds of misspecification, that allow for intragroup correlation, and that use bootstrap or jackknife methods; see [R] ***vce_option***.

Reporting

level(*#*); see [R] **estimation options**.

irr reports estimated coefficients transformed to incidence-rate ratios, that is, e^{β_i} rather than β_i. Standard errors and confidence intervals are similarly transformed. This option affects how results are displayed, not how they are estimated or stored. irr may be specified at estimation or when replaying previously estimated results.

nocnsreport; see [R] **estimation options**.

display_options: noomitted, vsquish, noemptycells, baselevels, allbaselevels, cformat(*%fmt*), pformat(*%fmt*), sformat(*%fmt*), and nolstretch; see [R] **estimation options**.

Maximization

maximize_options: difficult, technique(*algorithm_spec*), iterate(*#*), [no]log, trace, gradient, showstep, hessian, showtolerance, tolerance(*#*), ltolerance(*#*), nrtolerance(*#*), nonrtolerance, and from(*init_specs*); see [R] **maximize**. These options are seldom used.

Setting the optimization type to technique(bhhh) resets the default *vcetype* to vce(opg).

The following option is available with gnbreg but is not shown in the dialog box:

coeflegend; see [R] **estimation options**.

Remarks

Remarks are presented under the following headings:

Introduction to negative binomial regression
nbreg
gnbreg

Introduction to negative binomial regression

Negative binomial regression models the number of occurrences (counts) of an event when the event has extra-Poisson variation, that is, when it has overdispersion. The Poisson regression model is

$$y_j \sim \text{Poisson}(\mu_j)$$

where

$$\mu_j = \exp(\mathbf{x}_j\boldsymbol{\beta} + \text{offset}_j)$$

for observed counts y_j with covariates $\mathbf{x}_j$ for the jth observation. One derivation of the negative binomial mean-dispersion model is that individual units follow a Poisson regression model, but there is an omitted variable ν_j, such that e^{ν_j} follows a gamma distribution with mean 1 and variance α:

$$y_j \sim \text{Poisson}(\mu_j^*)$$

where

$$\mu_j^* = \exp(\mathbf{x}_j\boldsymbol{\beta} + \text{offset}_j + \nu_j)$$

and

$$e^{\nu_j} \sim \text{Gamma}(1/\alpha, \alpha)$$

With this parameterization, a Gamma(a,b) distribution will have expectation ab and variance ab^2.

We refer to α as the overdispersion parameter. The larger α is, the greater the overdispersion. The Poisson model corresponds to $\alpha = 0$. `nbreg` parameterizes α as $\ln\alpha$. `gnbreg` allows $\ln\alpha$ to be modeled as $\ln\alpha_j = \mathbf{z}_j\boldsymbol{\gamma}$, a linear combination of covariates $\mathbf{z}_j$.

`nbreg` will fit two different parameterizations of the negative binomial model. The default, described above and also given by the `dispersion(mean)` option, has dispersion for the jth observation equal to $1 + \alpha\exp(\mathbf{x}_j\boldsymbol{\beta} + \text{offset}_j)$. This is seen by noting that the above implies that

$$\mu_j^* \sim \text{Gamma}(1/\alpha, \alpha\mu_j)$$

and thus

$$\begin{aligned}\text{Var}(y_j) &= E\left\{\text{Var}(y_j|\mu_j^*)\right\} + \text{Var}\left\{E(y_j|\mu_j^*)\right\}\\ &= E(\mu_j^*) + \text{Var}(\mu_j^*)\\ &= \mu_j(1 + \alpha\mu_j)\end{aligned}$$

The alternative parameterization, given by the `dispersion(constant)` option, has dispersion equal to $1 + \delta$; that is, it is constant for all observations. This is so because the constant-dispersion model assumes instead that

$$\mu_j^* \sim \text{Gamma}(\mu_j/\delta, \delta)$$

and thus $\text{Var}(y_j) = \mu_j(1 + \delta)$. The Poisson model corresponds to $\delta = 0$.

For detailed derivations of both models, see Cameron and Trivedi (1998, 70–77). In particular, note that the mean-dispersion model is known as the NB2 model in their terminology, whereas the constant-dispersion model is referred to as the NB1 model.

See Long and Freese (2006) and Cameron and Trivedi (2010, chap. 17) for a discussion of the negative binomial regression model with Stata examples and for a discussion of other regression models for count data.

Hilbe (2011) provides an exclusive review of the negative binomial model and its variations, using Stata examples.

nbreg

It is not uncommon to posit a Poisson regression model and observe a lack of model fit. The following data appeared in Rodríguez (1993):

```
. use http://www.stata-press.com/data/r12/rod93
. list

     +---------------------------------------+
     | cohort   age_mos   deaths    exposure |
     |---------------------------------------|
  1. |      1       0.5      168       278.4 |
  2. |      1       2.0       48       538.8 |
  3. |      1       4.5       63       794.4 |
  4. |      1       9.0       89     1,550.8 |
  5. |      1      18.0      102     3,006.0 |
     |---------------------------------------|
  6. |      1      42.0       81     8,743.5 |
  7. |      1      90.0       40    14,270.0 |
  8. |      2       0.5      197       403.2 |
  9. |      2       2.0       48       786.0 |
 10. |      2       4.5       62     1,165.3 |
     |---------------------------------------|
 11. |      2       9.0       81     2,294.8 |
 12. |      2      18.0       97     4,500.5 |
 13. |      2      42.0      103    13,201.5 |
 14. |      2      90.0       39    19,525.0 |
 15. |      3       0.5      195       495.3 |
     |---------------------------------------|
 16. |      3       2.0       55       956.7 |
 17. |      3       4.5       58     1,381.4 |
 18. |      3       9.0       85     2,604.5 |
 19. |      3      18.0       87     4,618.5 |
 20. |      3      42.0       70     9,814.5 |
     |---------------------------------------|
 21. |      3      90.0       10     5,802.5 |
     +---------------------------------------+

. generate logexp = ln(exposure)
. poisson deaths i.cohort, offset(logexp)
Iteration 0:   log likelihood = -2160.0544
Iteration 1:   log likelihood = -2159.5162
Iteration 2:   log likelihood = -2159.5159
Iteration 3:   log likelihood = -2159.5159
Poisson regression                                Number of obs   =         21
                                                  LR chi2(2)      =      49.16
                                                  Prob > chi2     =     0.0000
Log likelihood = -2159.5159                       Pseudo R2       =     0.0113

------------------------------------------------------------------------------
      deaths |      Coef.   Std. Err.      z    P>|z|     [95% Conf. Interval]
-------------+----------------------------------------------------------------
      cohort |
          2  |  -.3020405   .0573319    -5.27   0.000    -.4144089   -.1896721
          3  |   .0742143   .0589726     1.26   0.208    -.0413698    .1897983
             |
       _cons |  -3.899488   .0411345   -94.80   0.000     -3.98011   -3.818866
      logexp |          1  (offset)
------------------------------------------------------------------------------
```

```
. estat gof
         Deviance goodness-of-fit =  4190.689
         Prob > chi2(18)          =    0.0000
         Pearson goodness-of-fit  =  15387.67
         Prob > chi2(18)          =    0.0000
```

The extreme significance of the goodness-of-fit χ^2 indicates that the Poisson regression model is inappropriate, suggesting to us that we should try a negative binomial model:

```
. nbreg deaths i.cohort, offset(logexp) nolog
Negative binomial regression                      Number of obs   =         21
                                                  LR chi2(2)      =       0.40
Dispersion     = mean                             Prob > chi2     =     0.8171
Log likelihood = -131.3799                        Pseudo R2       =     0.0015

------------------------------------------------------------------------------
      deaths |      Coef.   Std. Err.      z    P>|z|     [95% Conf. Interval]
-------------+----------------------------------------------------------------
      cohort |
          2  |  -.2676187   .7237203    -0.37   0.712    -1.686084    1.150847
          3  |  -.4573957   .7236651    -0.63   0.527    -1.875753    .9609618
             |
       _cons |  -2.086731    .511856    -4.08   0.000     -3.08995   -1.083511
      logexp |          1  (offset)
-------------+----------------------------------------------------------------
    /lnalpha |   .5939963   .2583615                      .0876171    1.100376
-------------+----------------------------------------------------------------
       alpha |   1.811212   .4679475                       1.09157    3.005295
------------------------------------------------------------------------------
Likelihood-ratio test of alpha=0:  chibar2(01) = 4056.27 Prob>=chibar2 = 0.000
```

Our original Poisson model is a special case of the negative binomial—it corresponds to $\alpha = 0$. nbreg, however, estimates α indirectly, estimating instead $\ln\alpha$. In our model, $\ln\alpha = 0.594$, meaning that $\alpha = 1.81$ (nbreg undoes the transformation for us at the bottom of the output).

To test $\alpha = 0$ (equivalent to $\ln\alpha = -\infty$), nbreg performs a likelihood-ratio test. The staggering χ^2 value of 4,056 asserts that the probability that we would observe these data conditional on $\alpha = 0$ is virtually zero, that is, conditional on the process being Poisson. The data are not Poisson. It is not accidental that this χ^2 value is close to the goodness-of-fit statistic from the Poisson regression itself.

❑ Technical note

The usual Gaussian test of $\alpha = 0$ is omitted because this test occurs on the boundary, invalidating the usual theory associated with such tests. However, the likelihood-ratio test of $\alpha = 0$ has been modified to be valid on the boundary. In particular, the null distribution of the likelihood-ratio test statistic is not the usual χ^2_1, but rather a 50:50 mixture of a χ^2_0 (point mass at zero) and a χ^2_1, denoted as $\overline{\chi}^2_{01}$. See Gutierrez, Carter, and Drukker (2001) for more details.

❑

❑ Technical note

The negative binomial model deals with cases in which there is more variation than would be expected if the process were Poisson. The negative binomial model is not helpful if there is less than Poisson variation—if the variance of the count variable is less than its mean. However, underdispersion is uncommon. Poisson models arise because of independently generated events. Overdispersion comes about if some of the parameters (causes) of the Poisson processes are unknown. To obtain underdispersion, the sequence of events somehow would have to be regulated; that is, events would not be independent but controlled based on past occurrences.

❑

gnbreg

gnbreg is a generalization of nbreg, dispersion(mean). Whereas in nbreg, one $\ln\alpha$ is estimated, gnbreg allows $\ln\alpha$ to vary, observation by observation, as a linear combination of another set of covariates: $\ln\alpha_j = \mathbf{z}_j\gamma$.

We will assume that the number of deaths is a function of age, whereas the $\ln\alpha$ parameter is a function of cohort. To fit the model, we type

```
. gnbreg deaths age_mos, lnalpha(i.cohort) offset(logexp)
Fitting constant-only model:
Iteration 0:   log likelihood =   -187.067  (not concave)
Iteration 1:   log likelihood =  -137.4064
Iteration 2:   log likelihood = -134.07766
Iteration 3:   log likelihood = -131.60668
Iteration 4:   log likelihood = -131.57951
Iteration 5:   log likelihood = -131.57948
Iteration 6:   log likelihood = -131.57948
Fitting full model:
Iteration 0:   log likelihood = -124.34327
Iteration 1:   log likelihood = -117.70256
Iteration 2:   log likelihood = -117.56373
Iteration 3:   log likelihood = -117.56164
Iteration 4:   log likelihood = -117.56164
Generalized negative binomial regression          Number of obs   =         21
                                                  LR chi2(1)      =      28.04
                                                  Prob > chi2     =     0.0000
Log likelihood = -117.56164                       Pseudo R2       =     0.1065

------------------------------------------------------------------------------
      deaths |      Coef.   Std. Err.      z    P>|z|     [95% Conf. Interval]
-------------+----------------------------------------------------------------
deaths       |
     age_mos |  -.0516657   .0051747    -9.98   0.000     -.061808   -.0415233
       _cons |  -1.867225   .2227944    -8.38   0.000    -2.303894   -1.430556
      logexp |          1  (offset)
-------------+----------------------------------------------------------------
lnalpha      |
      cohort |
          2  |   .0939546   .7187747     0.13   0.896    -1.314818    1.502727
          3  |   .0815279   .7365476     0.11   0.912    -1.362079    1.525135
             |
       _cons |  -.4759581   .5156502    -0.92   0.356    -1.486614    .5346978
------------------------------------------------------------------------------
```

We find that age is a significant determinant of the number of deaths. The standard errors for the variables in the $\ln\alpha$ equation suggest that the overdispersion parameter does not vary across cohorts. We can test this assertion by typing

```
. test 2.cohort 3.cohort
 ( 1)  [lnalpha]2.cohort = 0
 ( 2)  [lnalpha]3.cohort = 0
           chi2(  2) =    0.02
         Prob > chi2 =    0.9904
```

There is no evidence of variation by cohort in these data.

❑ Technical note

Note the intentional absence of a likelihood-ratio test for $\alpha = 0$ in gnbreg. The test is affected by the same boundary condition that affects the comparison test in nbreg; however, when α is parameterized by more than a constant term, the null distribution becomes intractable. For this reason, we recommend using nbreg to test for overdispersion and, if you have reason to believe that overdispersion exists, only then modeling the overdispersion using gnbreg.

❑

Saved results

nbreg and gnbreg save the following in e():

Scalars

e(N)	number of observations
e(k)	number of parameters
e(k_aux)	number of auxiliary parameters
e(k_eq)	number of equations in e(b)
e(k_eq_model)	number of equations in overall model test
e(k_dv)	number of dependent variables
e(df_m)	model degrees of freedom
e(r2_p)	pseudo-R-squared
e(ll)	log likelihood
e(ll_0)	log likelihood, constant-only mode
e(ll_c)	log likelihood, comparison model
e(alpha)	value of alpha
e(delta)	value of delta
e(N_clust)	number of clusters
e(chi2)	χ^2
e(chi2_c)	χ^2 for comparison test
e(p)	significance
e(rank)	rank of e(V)
e(rank0)	rank of e(V) for constant-only model
e(ic)	number of iterations
e(rc)	return code
e(converged)	1 if converged, 0 otherwise

Macros

e(cmd)	nbreg or gnbreg
e(cmdline)	command as typed
e(depvar)	name of dependent variable
e(wtype)	weight type
e(wexp)	weight expression
e(title)	title in estimation output
e(clustvar)	name of cluster variable
e(offset)	linear offset variable (nbreg)
e(offset1)	linear offset variable (gnbreg)
e(chi2type)	Wald or LR; type of model χ^2 test
e(chi2_ct)	Wald or LR; type of model χ^2 test corresponding to e(chi2_c)
e(dispers)	mean or constant
e(vce)	*vcetype* specified in vce()
e(vcetype)	title used to label Std. Err.
e(opt)	type of optimization
e(which)	max or min; whether optimizer is to perform maximization or minimization
e(ml_method)	type of ml method
e(user)	name of likelihood-evaluator program
e(technique)	maximization technique
e(properties)	b V
e(predict)	program used to implement predict
e(asbalanced)	factor variables fvset as asbalanced
e(asobserved)	factor variables fvset as asobserved

Matrices

`e(b)`	coefficient vector
`e(Cns)`	constraints matrix
`e(ilog)`	iteration log (up to 20 iterations)
`e(gradient)`	gradient vector
`e(V)`	variance–covariance matrix of the estimators
`e(V_modelbased)`	model-based variance

Functions

`e(sample)`	marks estimation sample

Methods and formulas

`nbreg` and `gnbreg` are implemented as ado-files.

See [R] **poisson** and Johnson, Kemp, and Kotz (2005, chap. 4) for an introduction to the Poisson distribution.

Methods and formulas are presented under the following headings:

Mean-dispersion model
Constant-dispersion model

Mean-dispersion model

A negative binomial distribution can be regarded as a gamma mixture of Poisson random variables. The number of times something occurs, y_j, is distributed as Poisson$(\nu_j\mu_j)$. That is, its conditional likelihood is

$$f(y_j \mid \nu_j) = \frac{(\nu_j\mu_j)^{y_j} e^{-\nu_j\mu_j}}{\Gamma(y_j+1)}$$

where $\mu_j = \exp(\mathbf{x}_j\boldsymbol{\beta} + \text{offset}_j)$ and ν_j is an unobserved parameter with a Gamma$(1/\alpha, \alpha)$ density:

$$g(\nu) = \frac{\nu^{(1-\alpha)/\alpha} e^{-\nu/\alpha}}{\alpha^{1/\alpha}\Gamma(1/\alpha)}$$

This gamma distribution has mean 1 and variance α, where α is our ancillary parameter.

The unconditional likelihood for the jth observation is therefore

$$f(y_j) = \int_0^\infty f(y_j \mid \nu) g(\nu)\, d\nu = \frac{\Gamma(m+y_j)}{\Gamma(y_j+1)\Gamma(m)}\, p_j^m (1-p_j)^{y_j}$$

where $p_j = 1/(1+\alpha\mu_j)$ and $m = 1/\alpha$. Solutions for α are handled by searching for $\ln\alpha$ because α must be greater than zero.

The log likelihood (with weights w_j and offsets) is given by

$$m = 1/\alpha \qquad p_j = 1/(1+\alpha\mu_j) \qquad \mu_j = \exp(\mathbf{x}_j\boldsymbol{\beta} + \text{offset}_j)$$

$$\ln L = \sum_{j=1}^n w_j \Big[\ln\{\Gamma(m+y_j)\} - \ln\{\Gamma(y_j+1)\} - \ln\{\Gamma(m)\} + m\ln(p_j) + y_j \ln(1-p_j) \Big]$$

For `gnbreg`, α can vary across the observations according to the parameterization $\ln\alpha_j = \mathbf{z}_j\boldsymbol{\gamma}$.

Constant-dispersion model

The constant-dispersion model assumes that y_j is conditionally distributed as Poisson(μ_j^*), where $\mu_j^* \sim$ Gamma($\mu_j/\delta, \delta$) for some dispersion parameter δ (by contrast, the mean-dispersion model assumes that $\mu_j^* \sim$ Gamma($1/\alpha, \alpha\mu_j$)). The log likelihood is given by

$$m_j = \mu_j/\delta \qquad p = 1/(1+\delta)$$

$$\ln L = \sum_{j=1}^{n} w_j \Big[\ln\{\Gamma(m_j + y_j)\} - \ln\{\Gamma(y_j+1)\} \\ - \ln\{\Gamma(m_j)\} + m_j \ln(p) + y_j \ln(1-p) \Big]$$

with everything else defined as before in the calculations for the mean-dispersion model.

`nbreg` and `gnbreg` support the Huber/White/sandwich estimator of the variance and its clustered version using `vce(robust)` and `vce(cluster` *clustvar*`)`, respectively. See [P] **_robust**, particularly *Maximum likelihood estimators* and *Methods and formulas*.

These commands also support estimation with survey data. For details on VCEs with survey data, see [SVY] **variance estimation**.

References

Cameron, A. C., and P. K. Trivedi. 1998. *Regression Analysis of Count Data*. Cambridge: Cambridge University Press.

——. 2010. *Microeconometrics Using Stata*. Rev. ed. College Station, TX: Stata Press.

Deb, P., and P. K. Trivedi. 2006. Maximum simulated likelihood estimation of a negative binomial regression model with multinomial endogenous treatment. *Stata Journal* 6: 246–255.

Gutierrez, R. G., S. Carter, and D. M. Drukker. 2001. sg160: On boundary-value likelihood-ratio tests. *Stata Technical Bulletin* 60: 15–18. Reprinted in *Stata Technical Bulletin Reprints*, vol. 10, pp. 269–273. College Station, TX: Stata Press.

Hilbe, J. M. 1998. sg91: Robust variance estimators for MLE Poisson and negative binomial regression. *Stata Technical Bulletin* 45: 26–28. Reprinted in *Stata Technical Bulletin Reprints*, vol. 8, pp. 177–180. College Station, TX: Stata Press.

——. 1999. sg102: Zero-truncated Poisson and negative binomial regression. *Stata Technical Bulletin* 47: 37–40. Reprinted in *Stata Technical Bulletin Reprints*, vol. 8, pp. 233–236. College Station, TX: Stata Press.

——. 2011. *Negative Binomial Regression*. 2nd ed. New York: Cambridge University Press.

Johnson, N. L., A. W. Kemp, and S. Kotz. 2005. *Univariate Discrete Distributions*. 3rd ed. New York: Wiley.

Long, J. S. 1997. *Regression Models for Categorical and Limited Dependent Variables*. Thousand Oaks, CA: Sage.

Long, J. S., and J. Freese. 2001. Predicted probabilities for count models. *Stata Journal* 1: 51–57.

——. 2006. *Regression Models for Categorical Dependent Variables Using Stata*. 2nd ed. College Station, TX: Stata Press.

Miranda, A., and S. Rabe-Hesketh. 2006. Maximum likelihood estimation of endogenous switching and sample selection models for binary, ordinal, and count variables. *Stata Journal* 6: 285–308.

Rodríguez, G. 1993. sbe10: An improvement to poisson. *Stata Technical Bulletin* 11: 11–14. Reprinted in *Stata Technical Bulletin Reprints*, vol. 2, pp. 94–98. College Station, TX: Stata Press.

Rogers, W. H. 1991. sbe1: Poisson regression with rates. *Stata Technical Bulletin* 1: 11–12. Reprinted in *Stata Technical Bulletin Reprints*, vol. 1, pp. 62–64. College Station, TX: Stata Press.

——. 1993. sg16.4: Comparison of nbreg and glm for negative binomial. *Stata Technical Bulletin* 16: 7. Reprinted in *Stata Technical Bulletin Reprints*, vol. 3, pp. 82–84. College Station, TX: Stata Press.

Also see

[R] **nbreg postestimation** — Postestimation tools for nbreg and gnbreg

[R] **glm** — Generalized linear models

[R] **poisson** — Poisson regression

[R] **tnbreg** — Truncated negative binomial regression

[R] **zinb** — Zero-inflated negative binomial regression

[MI] **estimation** — Estimation commands for use with mi estimate

[SVY] **svy estimation** — Estimation commands for survey data

[XT] **xtnbreg** — Fixed-effects, random-effects, & population-averaged negative binomial models

[U] **20 Estimation and postestimation commands**

Title

nbreg postestimation — Postestimation tools for nbreg and gnbreg

Description

The following postestimation commands are available after `nbreg` and `gnbreg`:

Command	Description
`contrast`	contrasts and ANOVA-style joint tests of estimates
`estat`	AIC, BIC, VCE, and estimation sample summary
`estat` (svy)	postestimation statistics for survey data
`estimates`	cataloging estimation results
`lincom`	point estimates, standard errors, testing, and inference for linear combinations of coefficients
`linktest`	link test for model specification
`lrtest`[1]	likelihood-ratio test
`margins`	marginal means, predictive margins, marginal effects, and average marginal effects
`marginsplot`	graph the results from margins (profile plots, interaction plots, etc.)
`nlcom`	point estimates, standard errors, testing, and inference for nonlinear combinations of coefficients
`predict`	predictions, residuals, influence statistics, and other diagnostic measures
`predictnl`	point estimates, standard errors, testing, and inference for generalized predictions
`pwcompare`	pairwise comparisons of estimates
`suest`	seemingly unrelated estimation
`test`	Wald tests of simple and composite linear hypotheses
`testnl`	Wald tests of nonlinear hypotheses

[1] `lrtest` is not appropriate with `svy` estimation results.

See the corresponding entries in the *Base Reference Manual* for details, but see [SVY] **estat** for details about `estat` (svy).

Syntax for predict

`predict` [*type*] *newvar* [*if*] [*in*] [, *statistic* `nooffset`]

`predict` [*type*] { *stub** | $newvar_{\text{reg}}$ $newvar_{\text{disp}}$ } [*if*] [*in*] , `scores`

statistic	Description
Main	
`n`	number of events; the default
`ir`	incidence rate (equivalent to `predict ..., n nooffset`)
`pr(`*n*`)`	probability $\Pr(y_j = n)$
`pr(`*a*`,`*b*`)`	probability $\Pr(a \le y_j \le b)$
`xb`	linear prediction
`stdp`	standard error of the linear prediction

In addition, relevant only after gnbreg are the following:

statistic	Description
Main	
alpha	predicted values of α_j
lnalpha	predicted values of $\ln\alpha_j$
stdplna	standard error of predicted $\ln\alpha_j$

These statistics are available both in and out of sample; type predict ... if e(sample) ... if wanted only for the estimation sample.

Menu

Statistics > Postestimation > Predictions, residuals, etc.

Options for predict

Main

n, the default, calculates the predicted number of events, which is $\exp(\mathbf{x}_j\boldsymbol{\beta})$ if neither offset(*varname*$_o$) nor exposure(*varname*$_e$) was specified when the model was fit; $\exp(\mathbf{x}_j\boldsymbol{\beta} + \text{offset}_j)$ if offset() was specified; or $\exp(\mathbf{x}_j\boldsymbol{\beta}) \times \text{exposure}_j$ if exposure() was specified.

ir calculates the incidence rate $\exp(\mathbf{x}_j\boldsymbol{\beta})$, which is the predicted number of events when exposure is 1. This is equivalent to specifying both the n and the nooffset options.

pr(*n*) calculates the probability $\Pr(y_j = n)$, where *n* is a nonnegative integer that may be specified as a number or a variable.

pr(*a*,*b*) calculates the probability $\Pr(a \le y_j \le b)$, where *a* and *b* are nonnegative integers that may be specified as numbers or variables;

b missing ($b \geq .$) means $+\infty$;
pr(20,.) calculates $\Pr(y_j \geq 20)$;
pr(20,*b*) calculates $\Pr(y_j \geq 20)$ in observations for which $b \geq .$ and calculates $\Pr(20 \le y_j \le b)$ elsewhere.

pr(.,*b*) produces a syntax error. A missing value in an observation of the variable *a* causes a missing value in that observation for pr(*a*,*b*).

xb calculates the linear prediction, which is $\mathbf{x}_j\boldsymbol{\beta}$ if neither offset() nor exposure() was specified; $\mathbf{x}_j\boldsymbol{\beta} + \text{offset}_j$ if offset() was specified; or $\mathbf{x}_j\boldsymbol{\beta} + \ln(\text{exposure}_j)$ if exposure() was specified; see nooffset below.

stdp calculates the standard error of the linear prediction.

alpha, lnalpha, and stdplna are relevant after gnbreg estimation only; they produce the predicted values of α_j, $\ln\alpha_j$, and the standard error of the predicted $\ln\alpha_j$, respectively.

nooffset is relevant only if you specified offset() or exposure() when you fit the model. It modifies the calculations made by predict so that they ignore the offset or exposure variable; the linear prediction is treated as $\mathbf{x}_j\boldsymbol{\beta}$ rather than as $\mathbf{x}_j\boldsymbol{\beta} + \text{offset}_j$ or $\mathbf{x}_j\boldsymbol{\beta} + \ln(\text{exposure}_j)$. Specifying predict ..., nooffset is equivalent to specifying predict ..., ir.

scores calculates equation-level score variables.

The first new variable will contain $\partial \ln L/\partial(\mathbf{x}_j\boldsymbol{\beta})$.

The second new variable will contain $\partial \ln L/\partial(\ln\alpha_j)$ for dispersion(mean) and gnbreg.

The second new variable will contain $\partial \ln L/\partial(\ln\delta)$ for dispersion(constant).

Remarks

After nbreg and gnbreg, predict returns the expected number of deaths per cohort and the probability of observing the number of deaths recorded or fewer.

```
. use http://www.stata-press.com/data/r12/rod93
. nbreg deaths i.cohort, nolog
Negative binomial regression                      Number of obs   =         21
                                                  LR chi2(2)      =       0.14
Dispersion     = mean                             Prob > chi2     =     0.9307
Log likelihood = -108.48841                       Pseudo R2       =     0.0007

------------------------------------------------------------------------------
      deaths |      Coef.   Std. Err.      z    P>|z|     [95% Conf. Interval]
-------------+----------------------------------------------------------------
      cohort |
          2  |   .0591305   .2978419     0.20   0.843    -.5246289      .64289
          3  |  -.0538792   .2981621    -0.18   0.857    -.6382662    .5305077
             |
       _cons |   4.435906   .2107213    21.05   0.000       4.0229    4.848912
-------------+----------------------------------------------------------------
    /lnalpha |  -1.207379   .3108622                     -1.816657   -.5980999
-------------+----------------------------------------------------------------
       alpha |     .29898   .0929416                      .1625683    .5498555
------------------------------------------------------------------------------
Likelihood-ratio test of alpha=0:  chibar2(01) =  434.62 Prob>=chibar2 = 0.000
. predict count
(option n assumed; predicted number of events)
. predict p, pr(0, deaths)
. summarize deaths count p
    Variable |       Obs        Mean    Std. Dev.       Min        Max
-------------+--------------------------------------------------------
      deaths |        21    84.66667    48.84192         10        197
       count |        21    84.66667     4.00773         80   89.57143
           p |        21    .4991542    .2743702   .0070255   .9801285
```

The expected number of deaths ranges from 80 to 90. The probability $\Pr(y_i \leq \texttt{deaths})$ ranges from 0.007 to 0.98.

Methods and formulas

All postestimation commands listed above are implemented as ado-files.

In the following, we use the same notation as in [R] **nbreg**.

Methods and formulas are presented under the following headings:

Mean-dispersion model
Constant-dispersion model

Mean-dispersion model

The equation-level scores are given by

$$\text{score}(\mathbf{x}\boldsymbol{\beta})_j = p_j(y_j - \mu_j)$$

$$\text{score}(\tau)_j = -m\left\{\frac{\alpha_j(\mu_j - y_j)}{1+\alpha_j\mu_j} - \ln(1+\alpha_j\mu_j) + \psi(y_j+m) - \psi(m)\right\}$$

where $\tau_j = \ln\alpha_j$ and $\psi(z)$ is the digamma function.

Constant-dispersion model

The equation-level scores are given by

$$\text{score}(\mathbf{x}\boldsymbol{\beta})_j = m_j\left\{\psi(y_j+m_j) - \psi(m_j) + \ln(p)\right\}$$

$$\text{score}(\tau)_j = y_j - (y_j+m_j)(1-p) - \text{score}(\mathbf{x}\boldsymbol{\beta})_j$$

where $\tau_j = \ln\delta_j$.

Also see

[R] **nbreg** — Negative binomial regression

[U] **20 Estimation and postestimation commands**

Title

nestreg — Nested model statistics

Syntax

Standard estimation command syntax

`nestreg` [, *options*] : *command_name depvar* (*varlist*) [(*varlist*) ...]
[*if*] [*in*] [*weight*] [*command_options*]

Survey estimation command syntax

`nestreg` [, *options*] : `svy` [*vcetype*] [, *svy_options*] : *command_name depvar*
(*varlist*) [(*varlist*) ...] [*if*] [*in*] [, *command_options*]

options	Description
Reporting	
`waldtable`	report Wald test results; the default
`lrtable`	report likelihood-ratio test results
`quietly`	suppress any output from *command_name*
`store(`*stub*`)`	store nested estimation results in `_est_`*stub#*

`by` is allowed; see [U] **11.1.10 Prefix commands**.

Weights are allowed if *command_name* allows them; see [U] **11.1.6 weight**.

A *varlist* in parentheses indicates that this list of variables is to be considered as a block. Each variable in a *varlist* not bound in parentheses will be treated as its own block.

All postestimation commands behave as they would after *command_name* without the `nestreg` prefix; see the postestimation manual entry for *command_name*.

Menu

Statistics > Other > Nested model statistics

Description

`nestreg` fits nested models by sequentially adding blocks of variables and then reports comparison tests between the nested models.

Options

Reporting

`waldtable` specifies that the table of Wald test results be reported. `waldtable` is the default.

`lrtable` specifies that the table of likelihood-ratio tests be reported. This option is not allowed if `pweights`, the `vce(robust)` option, or the `vce(cluster` *clustvar*`)` option is specified. `lrtable` is also not allowed with the `svy` prefix.

`quietly` suppresses the display of any output from *command_name*.

`store(`*stub*`)` specifies that each model fit by `nestreg` be stored under the name `_est_`*stub#*, where *#* is the nesting order from first to last.

Remarks

Remarks are presented under the following headings:

Estimation commands
Wald tests
Likelihood-ratio tests
Programming for nestreg

Estimation commands

`nestreg` removes collinear predictors and observations with missing values from the estimation sample before calling *command_name*.

The following Stata commands are supported by `nestreg`:

```
clogit        nbreg       regress
cloglog       ologit      scobit
glm           oprobit     stcox
intreg        poisson     stcrreg
logistic      probit      streg
logit         qreg        tobit
```

You do not supply a *depvar* for `stcox`, `stcrreg`, or `streg`; otherwise, *depvar* is required. You must supply two *depvar*s for `intreg`.

Wald tests

Use `nestreg` to test the significance of blocks of predictors, building the regression model one block at a time. Using the data from example 1 of [R] **test**, we wish to test the significance of the following predictors of birth rate: `medage`, `medagesq`, and `region` (already partitioned into four indicator variables: `reg1`, `reg2`, `reg3`, and `reg4`).

```
. use http://www.stata-press.com/data/r12/census4
(birth rate, median age)

. nestreg: regress brate (medage) (medagesq) (reg2-reg4)
Block  1: medage

      Source |       SS       df       MS              Number of obs =      50
-------------+------------------------------           F(  1,    48) =  164.72
       Model |  32675.1044     1  32675.1044           Prob > F      =  0.0000
    Residual |  9521.71561    48  198.369075           R-squared     =  0.7743
-------------+------------------------------           Adj R-squared =  0.7696
       Total |    42196.82    49  861.159592           Root MSE      =  14.084

------------------------------------------------------------------------------
       brate |      Coef.   Std. Err.      t    P>|t|     [95% Conf. Interval]
-------------+----------------------------------------------------------------
      medage |  -15.24893   1.188141   -12.83   0.000    -17.63785   -12.86002
       _cons |   618.3935   35.15416    17.59   0.000     547.7113    689.0756
------------------------------------------------------------------------------
```

```
Block  2: medagesq

      Source |       SS       df       MS              Number of obs =      50
-------------+------------------------------           F(  2,    47) =  158.75
       Model |  36755.8524     2  18377.9262           Prob > F      =  0.0000
    Residual |  5440.96755    47  115.765267           R-squared     =  0.8711
-------------+------------------------------           Adj R-squared =  0.8656
       Total |    42196.82    49  861.159592           Root MSE      =  10.759

------------------------------------------------------------------------------
       brate |      Coef.   Std. Err.      t    P>|t|     [95% Conf. Interval]
-------------+----------------------------------------------------------------
      medage |  -109.8925   15.96663    -6.88   0.000    -142.0132    -77.7718
    medagesq |   1.607332   .2707228     5.94   0.000     1.062708    2.151956
       _cons |   2007.071   235.4316     8.53   0.000     1533.444    2480.698
------------------------------------------------------------------------------

Block  3: reg2 reg3 reg4

      Source |       SS       df       MS              Number of obs =      50
-------------+------------------------------           F(  5,    44) =  100.63
       Model |   38803.419     5  7760.68381           Prob > F      =  0.0000
    Residual |  3393.40095    44  77.1227489           R-squared     =  0.9196
-------------+------------------------------           Adj R-squared =  0.9104
       Total |    42196.82    49  861.159592           Root MSE      =   8.782

------------------------------------------------------------------------------
       brate |      Coef.   Std. Err.      t    P>|t|     [95% Conf. Interval]
-------------+----------------------------------------------------------------
      medage |  -109.0957   13.52452    -8.07   0.000    -136.3526   -81.83886
    medagesq |   1.635208   .2290536     7.14   0.000     1.173581    2.096835
        reg2 |   15.00284   4.252068     3.53   0.001     6.433365    23.57233
        reg3 |   7.366435   3.953336     1.86   0.069    -.6009898    15.33386
        reg4 |   21.39679   4.650602     4.60   0.000     12.02412    30.76946
       _cons |    1947.61   199.8405     9.75   0.000     1544.858    2350.362
------------------------------------------------------------------------------

  +-------------------------------------------------------------+
  |       |          Block  Residual                     Change |
  | Block |       F     df        df   Pr > F       R2    in R2 |
  |-------+-----------------------------------------------------|
  |     1 |  164.72      1        48   0.0000   0.7743          |
  |     2 |   35.25      1        47   0.0000   0.8711   0.0967 |
  |     3 |    8.85      3        44   0.0001   0.9196   0.0485 |
  +-------------------------------------------------------------+
```

This single call to `nestreg` ran `regress` three times, adding a block of predictors to the model for each run as in

```
. regress brate medage

      Source |       SS       df       MS              Number of obs =      50
-------------+------------------------------           F(  1,    48) =  164.72
       Model |  32675.1044     1  32675.1044           Prob > F      =  0.0000
    Residual |  9521.71561    48  198.369075           R-squared     =  0.7743
-------------+------------------------------           Adj R-squared =  0.7696
       Total |    42196.82    49  861.159592           Root MSE      =  14.084

------------------------------------------------------------------------------
       brate |      Coef.   Std. Err.      t    P>|t|     [95% Conf. Interval]
-------------+----------------------------------------------------------------
      medage |  -15.24893   1.188141   -12.83   0.000    -17.63785   -12.86002
       _cons |   618.3935   35.15416    17.59   0.000     547.7113    689.0756
------------------------------------------------------------------------------
```

```
. regress brate medage medagesq

      Source |       SS       df       MS              Number of obs =      50
-------------+------------------------------           F(  2,    47) =  158.75
       Model |  36755.8524     2  18377.9262           Prob > F      =  0.0000
    Residual |  5440.96755    47  115.765267           R-squared     =  0.8711
-------------+------------------------------           Adj R-squared =  0.8656
       Total |    42196.82    49  861.159592           Root MSE      =  10.759

------------------------------------------------------------------------------
       brate |      Coef.   Std. Err.      t    P>|t|     [95% Conf. Interval]
-------------+----------------------------------------------------------------
      medage |  -109.8925   15.96663    -6.88   0.000    -142.0132    -77.7718
    medagesq |   1.607332   .2707228     5.94   0.000     1.062708    2.151956
       _cons |   2007.071   235.4316     8.53   0.000     1533.444    2480.698
------------------------------------------------------------------------------

. regress brate medage medagesq reg2-reg4

      Source |       SS       df       MS              Number of obs =      50
-------------+------------------------------           F(  5,    44) =  100.63
       Model |   38803.419     5  7760.68381           Prob > F      =  0.0000
    Residual |  3393.40095    44  77.1227489           R-squared     =  0.9196
-------------+------------------------------           Adj R-squared =  0.9104
       Total |    42196.82    49  861.159592           Root MSE      =   8.782

------------------------------------------------------------------------------
       brate |      Coef.   Std. Err.      t    P>|t|     [95% Conf. Interval]
-------------+----------------------------------------------------------------
      medage |  -109.0957   13.52452    -8.07   0.000    -136.3526   -81.83886
    medagesq |   1.635208   .2290536     7.14   0.000     1.173581    2.096835
        reg2 |   15.00284   4.252068     3.53   0.001     6.433365    23.57233
        reg3 |   7.366435   3.953336     1.86   0.069    -.6009898    15.33386
        reg4 |   21.39679   4.650602     4.60   0.000     12.02412    30.76946
       _cons |    1947.61   199.8405     9.75   0.000     1544.858    2350.362
------------------------------------------------------------------------------
```

nestreg collected the F statistic for the corresponding block of predictors and the model R^2 statistic from each model fit.

The F statistic for the first block, 164.72, is for a test of the joint significance of the first block of variables; it is simply the F statistic from the regression of brate on medage. The F statistic for the second block, 35.25, is for a test of the joint significance of the second block of variables in a regression of both the first and second blocks of variables. In our example, it is an F test of medagesq in the regression of brate on medage and medagesq. Similarly, the third block's F statistic of 8.85 corresponds to a joint test of reg2, reg3, and reg4 in the final regression.

Likelihood-ratio tests

The nestreg command provides a simple syntax for performing likelihood-ratio tests for nested model specifications; also see lrtest. Using the data from example 1 of [R] **lrtest**, we wish to jointly test the significance of the following predictors of low birthweight: age, lwt, ptl, and ht.

```
. use http://www.stata-press.com/data/r12/lbw
(Hosmer & Lemeshow data)

. xi: nestreg, lr: logistic low (i.race smoke ui) (age lwt ptl ht)
i.race            _Irace_1-3          (naturally coded; _Irace_1 omitted)

Block  1: _Irace_2 _Irace_3 smoke ui

Logistic regression                               Number of obs   =        189
                                                  LR chi2(4)      =      18.80
                                                  Prob > chi2     =     0.0009
Log likelihood = -107.93404                       Pseudo R2       =     0.0801

------------------------------------------------------------------------------
         low | Odds Ratio   Std. Err.      z    P>|z|     [95% Conf. Interval]
-------------+----------------------------------------------------------------
    _Irace_2 |   3.052746   1.498087     2.27   0.023     1.166747    7.987382
    _Irace_3 |   2.922593   1.189229     2.64   0.008     1.316457    6.488285
       smoke |   2.945742   1.101838     2.89   0.004     1.415167    6.131715
          ui |   2.419131   1.047359     2.04   0.041     1.035459    5.651788
       _cons |   .1402209   .0512295    -5.38   0.000     .0685216    .2869447
------------------------------------------------------------------------------

Block  2: age lwt ptl ht

Logistic regression                               Number of obs   =        189
                                                  LR chi2(8)      =      33.22
                                                  Prob > chi2     =     0.0001
Log likelihood =   -100.724                       Pseudo R2       =     0.1416

------------------------------------------------------------------------------
         low | Odds Ratio   Std. Err.      z    P>|z|     [95% Conf. Interval]
-------------+----------------------------------------------------------------
    _Irace_2 |   3.534767   1.860737     2.40   0.016     1.259736    9.918406
    _Irace_3 |   2.368079   1.039949     1.96   0.050     1.001356    5.600207
       smoke |   2.517698    1.00916     2.30   0.021     1.147676    5.523162
          ui |     2.1351   .9808153     1.65   0.099     .8677528      5.2534
         age |   .9732636   .0354759    -0.74   0.457     .9061578    1.045339
         lwt |   .9849634   .0068217    -2.19   0.029     .9716834    .9984249
         ptl |   1.719161   .5952579     1.56   0.118     .8721455    3.388787
          ht |   6.249602   4.322408     2.65   0.008     1.611152    24.24199
       _cons |   1.586014   1.910496     0.38   0.702     .1496092     16.8134
------------------------------------------------------------------------------

  +----------------------------------------------------------------+
  | Block |        LL       LR     df  Pr > LR       AIC        BIC |
  |-------+--------------------------------------------------------|
  |     1 |  -107.934    18.80      4   0.0009  225.8681   242.0768 |
  |     2 |  -100.724    14.42      4   0.0061   219.448   248.6237 |
  +----------------------------------------------------------------+
```

The estimation results from the full model are left in `e()`, so we can later use `estat` and other postestimation commands.

```
. estat gof

Logistic model for low, goodness-of-fit test

       number of observations =       189
 number of covariate patterns =       182
            Pearson chi2(173) =       179.24
                  Prob > chi2 =         0.3567
```

Programming for nestreg

If you want your user-written command (*command_name*) to work with nestreg, it must follow standard Stata syntax and allow the if qualifier. Furthermore, *command_name* must have sw or swml as a program property; see [P] **program properties**. If *command_name* has swml as a property, *command_name* must save the log-likelihood value in e(ll) and the model degrees of freedom in e(df_m).

Saved results

nestreg saves the following in r():

Matrices

r(wald)	matrix corresponding to the Wald table
r(lr)	matrix corresponding to the likelihood-ratio table

Methods and formulas

nestreg is implemented as an ado-file.

Acknowledgment

We thank Paul H. Bern, Syracuse University, for developing the hierarchical regression command that inspired nestreg.

Reference

Acock, A. C. 2010. *A Gentle Introduction to Stata*. 3rd ed. College Station, TX: Stata Press.

Also see

[P] **program properties** — Properties of user-defined programs

Title

net — Install and manage user-written additions from the Internet

Syntax

Set current location for net

`net from` *directory_or_url*

Change to a different net directory

`net cd` *path_or_url*

Change to a different net site

`net link` *linkname*

Search for installed packages

`net search` (see [R] **net search**)

Report current net location

`net`

Describe a package

`net describe` *pkgname* [`, from(`*directory_or_url*`)`]

Set location where packages will be installed

`net set ado` *dirname*

Set location where ancillary files will be installed

`net set other` *dirname*

Report net 'from', 'ado', and 'other' settings

`net query`

Install ado-files and help files from a package

`net install` *pkgname* [`, all replace force from(`*directory_or_url*`)`]

Install ancillary files from a package

`net get` *pkgname* [`, all replace force from(`*directory_or_url*`)`]

Shortcut to access Stata Journal (SJ) net site

net sj *vol-issue* [*insert*]

Shortcut to access Stata Technical Bulletin (STB) net site

net stb *issue* [*insert*]

List installed packages

ado [, find(*string*) from(*dirname*)]

ado dir [*pkgid*] [, find(*string*) from(*dirname*)]

Describe installed packages

ado describe [*pkgid*] [, find(*string*) from(*dirname*)]

Uninstall an installed package

ado uninstall *pkgid* [, from(*dirname*)]

where

pkgname is	name of a package
pkgid is	name of a package
or	a number in square brackets: [#]
dirname is	a directory name
or	PLUS (default)
or	PERSONAL
or	SITE

Description

net downloads and installs additions to Stata. The additions can be obtained from the Internet or from physical media. The additions can be ado-files (new commands), help files, or even datasets. Collections of files are bound together into *packages*. For instance, the package named zz49 might add the xyz command to Stata. At a minimum, such a package would contain xyz.ado, the code to implement the new command, and xyz.sthlp, the online help to describe it. That the package contains two files is a detail: you use net to download the package zz49, regardless of the number of files.

ado manages the packages you have installed by using net. The ado command lets you list and uninstall previously installed packages.

You can also access the net and ado features by selecting **Help > SJ and User-written Programs**; this is the recommended method to find and install additions to Stata.

Options

all is used with net install and net get. Typing it with either one makes the command equivalent to typing net install followed by net get.

`replace` is for use with `net install` and `net get`. It specifies that the downloaded files replace existing files if any of the files already exists.

`force` specifies that the downloaded files replace existing files if any of the files already exists, even if Stata thinks all the files are the same. `force` implies `replace`.

`find(`*string*`)` is for use with `ado`, `ado dir`, and `ado describe`. It specifies that the descriptions of the packages installed on your computer be searched, and that the package descriptions containing *string* be listed.

`from(`*dirname*`)`, when used with `ado`, specifies where the packages are installed. The default is `from(PLUS)`. `PLUS` is a code word that Stata understands to correspond to a particular directory on your computer that was set at installation time. On Windows computers, `PLUS` probably means the directory `c:\ado\plus`, but it might mean something else. You can find out what it means by typing `sysdir`, but doing so is irrelevant if you use the defaults.

`from(`*directory_or_url*`)`, when used with `net`, specifies the directory or URL where installable packages may be found. The directory or URL is the same as the one that would have been specified with `net from`.

Remarks

For an introduction to using `net` and `ado`, see [U] **28 Using the Internet to keep up to date**. The purpose of this documentation is

- to briefly, but accurately, describe `net` and `ado` and all their features and
- to provide documentation to those who wish to set up their own sites to distribute additions to Stata.

Remarks are presented under the following headings:

Definition of a package

A *package* is a collection of files—typically, `.ado` and `.sthlp` files—that together provide a new feature in Stata. Packages contain additions that you wish had been part of Stata at the outset. We write such additions, and so do other users.

One source of these additions is the *Stata Journal*, a printed and electronic journal with corresponding software. If you want the journal, you must subscribe, but the software is available for free from our website.

The purpose of the net and ado commands

The net command makes it easy to distribute and install packages. The goal is to get you quickly to a package-description page that summarizes the addition, for example,

```
. net describe rte_stat, from(http://www.wemakeitupaswego.edu/faculty/sgazer/)
-------------------------------------------------------------------------------
package rte_stat from http://www.wemakeitupaswego.edu/faculty/sgazer/
-------------------------------------------------------------------------------
TITLE
      rte_stat.  The robust-to-everything statistic; update.
DESCRIPTION/AUTHOR(S)
      S. Gazer, Dept. of Applied Theoretical Mathematics, WMIUAWG Univ.
      Aleph-0 100% confidence intervals proved too conservative for some
      applications; Aleph-1 confidence intervals have been substituted.
      The new robust-to-everything supplants the previous robust-to-
      everything-conceivable statistic.  See "Inference in the absence
      of data" (forthcoming).  After installation, see help rte.
INSTALLATION FILES                                  (type net install rte_stat)
      rte.ado
      rte.sthlp
      nullset.ado
      random.ado
-------------------------------------------------------------------------------
```

If you decide that the addition might prove useful, net makes the installation easy:

```
. net install rte_stat
checking rte_stat consistency and verifying not already installed...
installing into c:\ado\plus\ ...
installation complete.
```

The ado command helps you manage packages installed with net. Perhaps you remember that you installed a package that calculates the robust-to-everything statistic, but you cannot remember the command's name. You could use ado to search what you have previously installed for the rte command,

```
. ado
[1] package sg145 from http://www.stata.com/stb/stb56
      STB-56 sg145.  Scalar measures of fit for regression models.
  (output omitted)
[15] package rte_stat from http://www.wemakeitupaswego.edu/faculty/sgazer
      rte_stat.  The robust-to-everything statistic; update.
  (output omitted)
[21] package st0119 from http://www.stata-journal.com/software/sj7-1
      SJ7-1 st0119.  Rasch analysis
```

or you might type

```
. ado, find("robust-to-everything")
[15] package rte_stat from http://www.wemakeitupaswego.edu/faculty/sgazer
      rte_stat.  The robust-to-everything statistic; update.
```

Perhaps you decide that rte, despite the author's claims, is not worth the disk space it occupies. You can use ado to erase it:

```
. ado uninstall rte_stat
package rte_stat from http://www.wemakeitupaswego.edu/faculty/sgazer
      rte_stat.  The robust-to-everything statistic; update.
(package uninstalled)
```

ado uninstall is easier than erasing the files by hand because ado uninstall erases every file associated with the package, and, moreover, ado knows where on your computer rte_stat is installed; you would have to hunt for these files.

Content pages

There are two types of pages displayed by net: content pages and package-description pages. When you type net from, net cd, net link, or net without arguments, Stata goes to the specified place and displays the content page:

```
. net from http://www.stata.com
---------------------------------------------------------------------------
http://www.stata.com/
StataCorp
---------------------------------------------------------------------------

Welcome to StataCorp.

Below we provide links to sites providing additions to Stata, including
the Stata Journal, STB, and Statalist.  These are NOT THE OFFICIAL UPDATES;
you fetch and install the official updates by typing -update-.

PLACES you could -net link- to:
    sj                    The Stata Journal

DIRECTORIES you could -net cd- to:
    stb                   materials published in the Stata Technical Bulletin
    users                 materials written by various people, including StataCorp
                          employees
    meetings              software packages from Stata Users Group meetings
    links                 links to other locations providing additions to Stata
---------------------------------------------------------------------------
```

A content page tells you about other content pages and package-description pages. The example above lists other content pages only. Below we follow one of the links for the *Stata Journal*:

```
. net link sj
---------------------------------------------------------------------------
http://www.stata-journal.com/
The Stata Journal
---------------------------------------------------------------------------

The Stata Journal is a refereed, quarterly journal containing articles
of interest to Stata users.  For more details and subscription information,
visit the Stata Journal website at http://www.stata-journal.com.

PLACES you could -net link- to:
    stata                 StataCorp website

DIRECTORIES you could -net cd- to:
    production            Files for authors of the Stata Journal
    software              Software associated with Stata Journal articles
---------------------------------------------------------------------------
```

```
. net cd software
```

```
http://www.stata-journal.com/software/
The Stata Journal

PLACES you could -net link- to:
    stata             StataCorp website
    stb               Stata Technical Bulletin (STB) software archive
DIRECTORIES you could -net cd- to:
  (output omitted)
    sj7-1             volume 7, issue 1
  (output omitted)
    sj1-1             volume 1, issue 1
```

```
. net cd sj7-1
```

```
http://www.stata-journal.com/software/sj7-1/
Stata Journal volume 7, issue 1

DIRECTORIES you could -net cd- to:
    ..                Other Stata Journals
PACKAGES you could -net describe-:
    dm0027            File filtering in Stata: handling complex data
                      formats and navigating log files efficiently
    st0119            Rasch analysis
    st0120            Multivariable regression spline models
    st0121            mhbounds - Sensitivity Analysis for Average
                      Treatment Effects
```

dm0027, st0119, ..., st0121 are links to package-description pages.

1. When you type `net from`, you follow that with a location to display the location's content page.
   a. The location could be a URL, such as http://www.stata.com. The content page at that location would then be listed.
   b. The location could be `e:` on a Windows computer or a mounted volume on a Mac computer. The content page on that source would be listed. That would work if you had special media obtained from StataCorp or special media prepared by another user.
   c. The location could even be a directory on your computer, but that would work only if that directory contained the right kind of files.
2. Once you have specified a location, typing `net cd` will take you into subdirectories of that location, if there are any. Typing

   ```
   . net from http://www.stata-journal.com
   . net cd software
   ```

   is equivalent to typing

   ```
   . net from http://www.stata-journal.com/software
   ```

   Typing `net cd` displays the content page from that location.
3. Typing `net` without arguments redisplays the current content page, which is the content page last displayed.

4. net link is similar to net cd in that the result is to change the location, but rather than changing to subdirectories of the current location, net link jumps to another location:

```
. net from http://www.stata-journal.com
```

```
http://www.stata-journal.com/
The Stata Journal

The Stata Journal is a refereed, quarterly journal containing articles
of interest to Stata users.  For more details and subscription information,
visit the Stata Journal website at
http://www.stata-journal.com.

PLACES you could -net link- to:
    stata              StataCorp website

DIRECTORIES you could -net cd- to:
    production         Files for authors of the Stata Journal
    software           Software associated with Stata Journal articles
```

Typing net link stata would jump to http://www.stata.com:

```
. net link stata
```

```
http://www.stata.com/
StataCorp

Welcome to StataCorp.
  (output omitted )
```

Package-description pages

Package-description pages describe what could be installed:

```
. net from http://www.stata-journal.com/software/sj7-1
```

```
http://www.stata-journal.com/software/sj7-1/
  (output omitted)
. net describe st0119
```

```
package st0119 from http://www.stata-journal.com/software/sj7-1

TITLE
      SJ7-1 st0119.  Rasch analysis
DESCRIPTION/AUTHOR(S)
      Rasch analysis
      by Jean-Benoit Hardouin, University of Nantes, France
      Support:  jean-benoit.hardouin@univ-nantes.fr
      After installation, type help gammasym, gausshermite,
          geekel2d, raschtest, and raschtestv7
INSTALLATION FILES                               (type net install st0119)
      st0119/raschtest.ado
      st0119/raschtest.hlp
      st0119/raschtestv7.ado
      st0119/raschtestv7.hlp
      st0119/gammasym.ado
      st0119/gammasym.hlp
      st0119/gausshermite.ado
      st0119/gausshermite.hlp
      st0119/geekel2d.ado
      st0119/geekel2d.hlp
ANCILLARY FILES                                  (type net get st0119)
      st0119/data.dta
      st0119/outrasch.do
```

A package-description page describes the package and tells you how to install the component files. Package-description pages potentially describe two types of files:

1. Installation files: files that you type `net install` to install and that are required to make the addition work.
2. Ancillary files: additional files that you might want to install—you type `net get` to install them—but that you can ignore. Ancillary files are typically datasets that are useful for demonstration purposes. Ancillary files are not really installed in the sense of being copied to an official place for use by Stata itself. They are merely copied into the current directory so that you may use them if you wish.

You install the official files by typing `net install` followed by the package name. For example, to install `st0119`, you would type

```
. net install st0119
checking st0119 consistency and verifying not already installed...
installing into c:\ado\plus\ ...
installation complete.
```

You get the ancillary files—if there are any and if you want them—by typing `net get` followed by the package name:

```
. net get st0119
checking st0119 consistency and verifying not already installed...
copying into current directory...
      copying  data.dta
      copying  outrasch.do
ancillary files successfully copied.
```

Most users ignore the ancillary files.

Once you have installed a package—by typing `net install`—use `ado` to redisplay the package-description page whenever you wish:

```
. ado describe st0119
________________________________________________________________________________
[1] package st0119 from http://www.stata-journal.com/software/sj7-1
________________________________________________________________________________
TITLE
      SJ7-1 st0119.  Rasch analysis
DESCRIPTION/AUTHOR(S)
      Rasch analysis
      by Jean-Benoit Hardouin, University of Nantes, France
      Support:  jean-benoit.hardouin@univ-nantes.fr
      After installation, type help gammasym, gausshermite,
          geekel2d, raschtest, and raschtestv7
INSTALLATION FILES
      r/raschtest.ado
      r/raschtest.hlp
      r/raschtestv7.ado
      r/raschtestv7.hlp
      g/gammasym.ado
      g/gammasym.hlp
      g/gausshermite.ado
      g/gausshermite.hlp
      g/geekel2d.ado
      g/geekel2d.hlp
INSTALLED ON
      24 Apr 2011
________________________________________________________________________________
```

The package-description page shown by `ado` includes the location from which we got the package and when we installed it. It does not mention the ancillary files that were originally part of this package because they are not tracked by `ado`.

Where packages are installed

Packages should be installed in `PLUS` or `SITE`, which are code words that Stata understands and that correspond to some real directories on your computer. Typing `sysdir` will tell you where these are, if you care.

```
. sysdir
   STATA:  C:\Program Files\Stata12\
 UPDATES:  C:\Program Files\Stata12\ado\updates\
    BASE:  C:\Program Files\Stata12\ado\base\
    SITE:  C:\Program Files\Stata12\ado\site\
    PLUS:  c:\ado\plus\
PERSONAL:  c:\ado\personal\
OLDPLACE:  c:\ado\
```

If you type `sysdir`, you may obtain different results.

By default, net installs in the PLUS directory, and ado tells you about what is installed there. If you are on a multiple-user system, you may wish to install some packages in the SITE directory. This way, they will be available to other Stata users. To do that, before using net install, type

```
. net set ado SITE
```

and when reviewing what is installed or removing packages, redirect ado to that directory:

```
. ado ..., from(SITE)
```

In both cases, you type SITE because Stata will understand that SITE means the site ado-directory as defined by sysdir. To install into SITE, you must have write access to that directory.

If you reset where net installs and then, in the same session, wish to install into your private ado-directory, type

```
. net set ado PLUS
```

That is how things were originally. If you are confused as to where you are, type net query.

A summary of the net command

The net command displays content pages and package-description pages. Such pages are provided over the Internet, and most users get them there. We recommend that you start at http://www.stata.com and work out from there. We also recommend using net search to find packages of interest to you; see [R] **net search**.

net from moves you to a location and displays the content page.

net cd and net link change from your current location to other locations. net cd enters subdirectories of the original location. net link jumps from one location to another, depending on the code on the content page.

net describe lists a package-description page. Packages are named, and you type net describe *pkgname*.

net install installs a package into your copy of Stata. net get copies any additional files (ancillary files) to your current directory.

net sj and net stb simplify loading files from the *Stata Journal* and its predecessor, the *Stata Technical Bulletin*.

net sj *vol-issue*

is a synonym for typing

net from http://www.stata-journal.com/software/sj*vol-issue*

whereas

net sj *vol-issue* *insert*

is a synonym for typing

net from http://www.stata-journal.com/software/sj*vol-issue*
net describe *insert*

net set controls where net installs files. By default, net installs in the PLUS directory; see [P] **sysdir**. net set ado SITE would cause subsequent net commands to install in the SITE directory. net set other sets where ancillary files, such as .dta files, are installed. The default is the current directory.

net query displays the current net from, net set ado, and net set other settings.

A summary of the ado command

The ado command lists the package descriptions of previously installed packages.

Typing ado without arguments is the same as typing ado dir. Both list the names and titles of the packages you have installed.

ado describe lists full package-description pages.

ado uninstall removes packages from your computer.

Because you can install packages from a variety of sources, the package names may not always be unique. Thus the packages installed on your computer are numbered sequentially, and you may refer to them by name or by number. For instance, say that you wanted to get rid of the robust-to-everything statistic command you installed. Type

```
. ado, find("robust-to-everything")
[15] package rte_stat from http://www.wemakeitupaswego.edu/faculty/sgazer
      rte_stat.  The robust-to-everything statistic; update.
```

You could then type

```
. ado uninstall rte_stat
```

or

```
. ado uninstall [15]
```

Typing ado uninstall rte_stat would work only if the name rte_stat were unique; otherwise, ado would refuse, and you would have to type the number.

The find() option is allowed with ado dir and ado describe. It searches the package description for the word or phrase you specify, ignoring case (alpha matches Alpha). The complete package description is searched, including the author's name and the name of the files. Thus if rte was the name of a command that you wanted to eliminate, but you could not remember the name of the package, you could type

```
. ado, find(rte)
[15] package rte_stat from http://www.wemakeitupaswego.edu/faculty/sgazer
      rte_stat.  The robust-to-everything statistic; update.
```

Relationship of net and ado to the point-and-click interface

Users may instead select **Help > SJ and User-written Programs**. There are advantages and disadvantages:

1. Flipping through content and package-description pages is easier; it is much like a browser. See [GS] **19 Updating and extending Stata—Internet functionality** (GSM, GSU, or GSW).
2. When browsing a product-description page, note that the .sthlp files are highlighted. You may click on .sthlp files to review them before installing the package.
3. You may not redirect from where ado searches for files.

Creating your own site

The rest of this entry concerns how to create your own site to distribute additions to Stata. The idea is that you have written additions for use with Stata—say, xyz.ado and xyz.sthlp—and you wish to put them out so that coworkers or researchers at other institutions can easily install them. Or, perhaps you just have a dataset that you and others want to share.

In any case, all you need is a webpage. You place the files that you want to distribute on your webpage (or in a subdirectory), and you add two more files—a content file and a package-description file—and you are done.

Format of content and package-description files

The content file describes the content page. It must be named stata.toc:

——————————————————— begin stata.toc ———————

```
OFF                                   (to make site unavailable temporarily)
* lines starting with * are comments; they are ignored

* blank lines are ignored, too

* v indicates version—specify v 3, which is the current version of .toc files
v 3

* d lines display description text
* the first d line is the title, and the remaining ones are text
* blank d lines display a blank line
d title
d text
d text
d
...

* l lines display links
l word-to-show path-or-url [description]
l word-to-show path-or-url [description]
...

* t lines display other directories within the site
t path [description]
t path [description]
...

* p lines display packages
p pkgname [description]
p pkgname [description]
...
```

——————————————————— end stata.toc ———————

Package files describe packages and are named *pkgname*.pkg:

begin *pkgname*.pkg

```
* lines starting with * are comments; they are ignored
* blank lines are ignored, too
* v indicates version—specify v 3, which is the current version of .toc files
v 3
* d lines display package description text
* the first d line is the title, and the remaining ones are text
* blank d lines display a blank line
d title
d text
d Distribution-Date: date
d text
d
...
* f identifies the component files
f [path/]filename [description]
f [path/]filename [description]
...
* e line is optional; it means stop reading
e
```

end *pkgname*.pkg

Note the Distribution-Date description line. This line is optional but recommended. Stata can look for updates to user-written programs with the `adoupdate` command if the package files from which those programs were installed contain a Distribution-Date description line.

Example 1

Say that we want the user to see the following:

```
. net from http://www.university.edu/~me
------------------------------------------------------------------------
http://www.university.edu/~me
Chris Farrar, Uni University
------------------------------------------------------------------------

PACKAGES you could -net describe-:
    xyz             interval-truncated survival

. net describe xyz
------------------------------------------------------------------------
package xyz from http://www.university.edu/~me
------------------------------------------------------------------------

TITLE
      xyz.  interval-truncated survival.

DESCRIPTION/AUTHOR(S)
      C. Farrar, Uni University.

INSTALLATION FILES                              (type net install xyz)
      xyz.ado
      xyz.sthlp
ANCILLARY FILES                                 (type net get xyz)
      sample.dta
------------------------------------------------------------------------
```

The files needed to do this would be

begin stata.toc

```
v 3
d Chris Farrar, Uni University
p xyz interval-truncated survival
```

end stata.toc

```
                                                        begin xyz.pkg
v 3
d xyz.  interval-truncated survival.
d C. Farrar, Uni University.
f xyz.ado
f xyz.sthlp
f sample.dta
                                                        end xyz.pkg
```

On his homepage, Chris would place the following files:

stata.toc	(shown above)
xyz.pkg	(shown above)
xyz.ado	file to be delivered (for use by net install)
xyz.sthlp	file to be delivered (for use by net install)
sample.dta	file to be delivered (for use by net get)

Chris does nothing to distinguish ancillary files from installation files.

Example 2

S. Gazer wants to create a more complex site:

```
. net from http://www.wemakeitupaswego.edu/faculty/sgazer
--------------------------------------------------------------------------------
http://www.wemakeitupaswego.edu/faculty/sgazer
Data-free inference materials
--------------------------------------------------------------------------------

S. Gazer, Department of Applied Theoretical Mathematics
Also see my homepage for the preprint of "Irrefutable inference".
PLACES you could -net link- to:
    stata             StataCorp website
DIRECTORIES you could -net cd- to:
    ir               irrefutable inference programs (work in progress)
PACKAGES you could -net describe-:
    rtec             Robust-to-everything-conceivable statistic
    rte              Robust-to-everything statistic
```

```
. net describe rte
-----------------------------------------------------------------------------
package rte from http://www.wemakeitupaswego.edu/faculty/sgazer/
-----------------------------------------------------------------------------

TITLE
      rte.  The robust-to-everything statistic; update.
DESCRIPTION/AUTHOR(S)
      S. Gazer, Dept. of Applied Theoretical Mathematics, WMIUAWG Univ.
      Aleph-0 100% confidence intervals proved too conservative for some
      applications; Aleph-1 confidence intervals have been substituted.
      The new robust-to-everything supplants the previous robust-to-
      everything-conceivable statistic.  See "Inference in the absence
      of data" (forthcoming).  After installation, see help rte.

      Distribution-Date: 20110420

      Support:  email sgazer@wemakeitupaswego.edu
INSTALLATION FILES                               (type net install rte_stat)
      rte.ado
      rte.sthlp
      nullset.ado
      random.ado
ANCILLARY FILES                                  (type net get rte_stat)
      empty.dta
-----------------------------------------------------------------------------
```

The files needed to do this would be

```
                                                    begin stata.toc
v 3
d Data-free inference materials
d S. Gazer, Department of Applied Theoretical Mathematics
d
d Also see my homepage for the preprint of "Irrefutable inference".
l stata http://www.stata.com
t ir irrefutable inference programs (work in progress)
p rtec Robust-to-everything-conceivable statistic
p rte  Robust-to-everything statistic
                                                    end stata.toc
```

```
                                                    begin rte.pkg
v 3
d rte.  The robust-to-everything statistic; update.
d {bf:S. Gazer, Dept. of Applied Theoretical Mathematics, WMIUAWG Univ.}
d Aleph-0 100% confidence intervals proved too conservative for some
d applications; Aleph-1 confidence intervals have been substituted.
d The new robust-to-everything supplants the previous robust-to-
d everything-conceivable statistic.  See "Inference in the absence
d of data" (forthcoming).  After installation, see help {bf:rte}.
d
d Distribution-Date: 20110420
d
d Support:  email sgazer@wemakeitupaswego.edu
f rte.ado
f rte.sthlp
f nullset.ado
f random.ado
f empty.dta
                                                    end rte.pkg
```

On his homepage, Mr. Gazer would place the following files:

`stata.toc`	(shown above)
`rte.pkg`	(shown above)
`rte.ado`	(file to be delivered)
`rte.sthlp`	(file to be delivered)
`nullset.ado`	(file to be delivered)
`random.ado`	(file to be delivered)
`empty.dta`	(file to be delivered)
`rtec.pkg`	the other package referred to in `stata.toc`
`rtec.ado`	the corresponding files to be delivered
`rtec.sthlp`	
`ir/stata.toc`	the contents file for when the user types `net cd ir`
`ir/...`	whatever other `.pkg` files are referred to
`ir/...`	whatever other files are to be delivered

If Mr. Gazer later updated the `rte` package, he could change the Distribution-Date description line in his package. Then, if someone who had previously installed the `rte` packaged wanted to obtain the latest version, that person could use the `adoupdate` command; see [R] **adoupdate**.

For complex sites, a different structure may prove more convenient:

`stata.toc`	(shown above)
`rte.pkg`	(shown above)
`rtec.pkg`	the other package referred to in `stata.toc`
`rte/`	directory containing rte files to be delivered:
`rte/rte.ado`	(file to be delivered)
`rte/rte.sthlp`	(file to be delivered)
`rte/nullset.ado`	(file to be delivered)
`rte/random.ado`	(file to be delivered)
`rte/empty.dta`	(file to be delivered)
`rtec/`	directory containing rtec files to be delivered:
`rtec/...`	(files to be delivered)
`ir/stata.toc`	the contents file for when the user types `net cd ir`
`ir/*.pkg`	whatever other package files are referred to
`ir/*/...`	whatever other files are to be delivered

If you prefer this structure, it is simply a matter of changing the bottom of the `rte.pkg` from

```
f rte.ado
f rte.sthlp
f nullset.ado
f random.ado
f empty.dta
```

to

```
f rte/rte.ado
f rte/rte.sthlp
f rte/nullset.ado
f rte/random.ado
f rte/empty.dta
```

In writing paths and files, the directory separator forward slash (/) is used, regardless of operating system, because this is what the Internet uses.

It does not matter whether the files you put out are in Windows, Mac, or Unix format (how lines end is recorded differently). When Stata reads the files over the Internet, it will figure out the file format on its own and will automatically translate the files to what is appropriate for the receiver.

Additional package directives

F *filename* is similar to f *filename*, except that, when the file is installed, it will always be copied to the system directories (and not the current directory).

With f *filename*, the file is installed into a directory according to the file's suffix. For instance, xyz.ado would be installed in the system directories, whereas xyz.dta would be installed in the current directory.

Coding F xyz.ado would have the same result as coding f xyz.ado.

Coding F xyz.dta, however, would state that xyz.dta is to be installed in the system directories.

g *platformname filename* is also a variation on f *filename*. It specifies that the file be installed only if the user's operating system is of type *platformname*; otherwise, the file is ignored. The platform names are WIN (32-bit x86) and WIN64A (64-bit x86-64) for Windows; MACINTEL (32-bit Intel, GUI), OSX.X86 (32-bit Intel, console), MACINTEL64 (64-bit Intel, GUI), OSX.X8664 (64-bit Intel, console), MAC (32-bit PowerPC), and OSX.PPC (32-bit PowerPC), for Mac; and LINUX (32-bit x86), LINUX64 (64-bit x86-64), SOL64, and SOLX8664 (64-bit x86-64) for Unix.

G *platformname filename* is a variation on F *filename*. The file, if not ignored, is to be installed in the system directories.

g *platformname filename1 filename2* is a more detailed version of g *platformname filename*. In this case, *filename1* is the name of the file on the server (the file to be copied), and *filename2* is to be the name of the file on the user's system; for example, you might code

```
g WIN mydll.forwin mydll.plugin
g LINUX mydll.forlinux mydll.plugin
```

When you specify one *filename*, the result is the same as specifying two identical *filenames*.

G *platformname filename1 filename2* is the install-in-system-directories version of g *platformname filename1 filename2*.

h *filename* asserts that *filename* must be loaded, or this package is not to be installed; for example, you might code

```
g WIN mydll.forwin mydll.plugin
g LINUX mydll.forlinux mydll.plugin
h mydll.plugin
```

if you were offering the plugin mydll.plugin for Windows and Linux only.

SMCL in content and package-description files

The text listed on the second and subsequent d lines in both stata.toc and *pkgname*.pkg may contain SMCL as long as you include v 3; see [P] **smcl**.

Thus, in rte.pkg, S. Gazer coded the third line as

```
d {bf:S. Gazer, Dept. of Applied Theoretical Mathematics, WMIUAWG Univ.}
```

Error-free file delivery

Most people transport files over the Internet and never worry about the file being corrupted in the process because corruption rarely occurs. If, however, the files must be delivered perfectly or not at all, you can include checksum files in the directory.

For instance, say that `big.dta` is included in your package and that it must be sent perfectly. First, use Stata to make the checksum file for `big.dta`

```
. checksum big.dta, save
```

That command creates a small file called `big.sum`; see [D] **checksum**. Then copy both `big.dta` and `big.sum` to your homepage. If `set checksum` is `on` (the default is `off`), whenever Stata reads *filename*.*whatever* over the net, it also looks for *filename*.`sum`. If it finds such a file, it uses the information recorded in it to verify that what was copied was error free.

If you do this, be cautious. If you put `big.dta` and `big.sum` on your homepage and then later change `big.dta` without changing `big.sum`, people will think that there are transmission errors when they try to download `big.dta`.

References

Baum, C. F., and N. J. Cox. 1999. ip29: Metadata for user-written contributions to the Stata programming language. *Stata Technical Bulletin* 52: 10–12. Reprinted in *Stata Technical Bulletin Reprints*, vol. 9, pp. 121–124. College Station, TX: Stata Press.

Cox, N. J., and C. F. Baum. 2000. ip29.1: Metadata for user-written contributions to the Stata programming language. *Stata Technical Bulletin* 54: 21–22. Reprinted in *Stata Technical Bulletin Reprints*, vol. 9, pp. 124–126. College Station, TX: Stata Press.

Also see

[R] **adoupdate** — Update user-written ado-files

[R] **net search** — Search the Internet for installable packages

[R] **search** — Search Stata documentation

[R] **sj** — Stata Journal and STB installation instructions

[R] **ssc** — Install and uninstall packages from SSC

[D] **checksum** — Calculate checksum of file

[P] **smcl** — Stata Markup and Control Language

[R] **update** — Update Stata

[GSM] **19 Updating and extending Stata—Internet functionality**

[GSU] **19 Updating and extending Stata—Internet functionality**

[GSW] **19 Updating and extending Stata—Internet functionality**

[U] **28 Using the Internet to keep up to date**

Title

net search — Search the Internet for installable packages

Syntax

`net search` *word* [*word* ...] [, *options*]

options	Description
`or`	list packages that contain any of the keywords; default is all
`nosj`	search non-SJ and non-STB sources
`tocpkg`	search both tables of contents and packages; the default
`toc`	search tables of contents only
`pkg`	search packages only
`everywhere`	search packages for match
`filenames`	search filenames associated with package for match
`errnone`	make return code 111 instead of 0 when no matches found

Description

`net search` searches the Internet for user-written additions to Stata, including, but not limited to, user-written additions published in the *Stata Journal* (SJ) and in the *Stata Technical Bulletin* (STB). `net search` lists the available additions that contain the specified keywords.

The user-written materials found are available for immediate download by using the `net` command or by clicking on the link.

In addition to typing `net search`, you may select **Help > Search...** and choose **Search net resources**. This is the recommended way to search for user-written additions to Stata.

Options

`or` is relevant only when multiple keywords are specified. By default, `net search` lists only packages that include all the keywords. `or` changes the command to list packages that contain any of the keywords.

`nosj` specifies that `net search` not list matches that were published in the SJ or in the STB.

`tocpkg`, `toc`, and `pkg` determine what is searched. `tocpkg` is the default, meaning that both tables of contents (tocs) and packages (pkgs) are searched. `toc` restricts the search to tables of contents. `pkg` restricts the search to packages.

`everywhere` and `filenames` determine where in packages `net search` looks for *keywords*. The default is `everywhere`. `filenames` restricts `net search` to search for matches only in the filenames associated with a package. Specifying `everywhere` implies `pkg`.

`errnone` is a programmer's option that causes the return code to be 111 instead of 0 when no matches are found.

Remarks

`net search` searches the Internet for user-written additions to Stata. If you want to search the Stata documentation for a particular topic, command, or author, see [R] **search**. `net search` *word* [*word* ...] (without options) is equivalent to typing `search` *word* [*word* ...]`, net`.

Remarks are presented under the following headings:

Topic searches
Author searches
Command searches
Where does net search look?
How does net search work?

Topic searches

Example: Find what is available about random effects

```
. net search random effect
```

Comments:

- It is best to search using the singular form of a word. `net search random effect` will find both "random effect" and "random effects".
- `net search random effect` will also find "random-effect" because `net search` performs a string search and not a word search.
- `net search random effect` lists all packages containing the words "random" and "effect", not necessarily used together.
- If you wanted all packages containing the word "random" or the word "effect", you would type `net search random effect, or`.

Author searches

Example: Find what is available by author Jeroen Weesie

```
. net search weesie
```

Comments:

- You could type `net search jeroen weesie`, but that might list fewer results because sometimes the last name is used without the first.
- You could type `net search Weesie`, but it would not matter. Capitalization is ignored in the search.

Example: Find what is available by Jeroen Weesie, excluding SJ and STB materials

```
. net search weesie, nosj
```

- The SJ and the STB tend to dominate search results because so much has been published in them. If you know that what you are looking for is not in the SJ or in the STB, specifying the `nosj` option will narrow the search.
- `net search weesie` lists everything that `net search weesie, nosj` lists, and more. If you just type `net search weesie`, look down the list. SJ and STB materials are listed first, and non-SJ and non-STB materials are listed last.

Command searches

Example: Find the user-written command kursus

```
. net search kursus, file
```

- You could just type `net search kursus`, and that will list everything `net search kursus, file` lists, and more. Because you know `kursus` is a command, however, there must be a `kursus.ado` file associated with the package. Typing `net search kursus, file` narrows the search.
- You could also type `net search kursus.ado, file` to narrow the search even more.

Where does net search look?

`net search` looks everywhere, not just at http://www.stata.com.

`net search` begins by looking at http://www.stata.com, but then follows every link, which takes it to other places, and then follows every link again, which takes it to even more places, and so on.

Authors: Please let us know if you have a site that we should include in our search by sending an email to webmaster@stata.com. We will then link to your site from ours to ensure that `net search` finds your materials. That is not strictly necessary, however, as long as your site is directly or indirectly linked from some site that is linked to ours.

How does net search work?

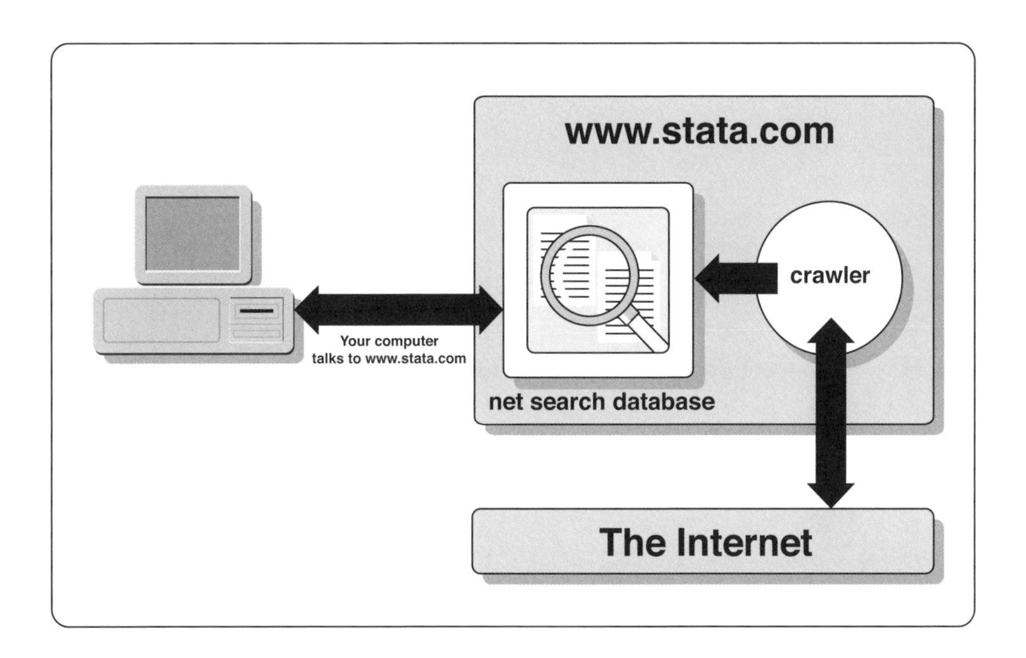

Our website maintains a database of Stata resources. When you use `net search`, it contacts http://www.stata.com with your request, http://www.stata.com searches its database, and Stata returns the results to you.

Another part of the system is called the crawler, which searches the web for new Stata resources to add to the `net search` database and verifies that the resources already found are still available. When a new resource becomes available, the crawler takes about 2 days to add it to the database, and, similarly, if a resource disappears, the crawler takes roughly 2 days to remove it from the database.

References

Baum, C. F., and N. J. Cox. 1999. ip29: Metadata for user-written contributions to the Stata programming language. *Stata Technical Bulletin* 52: 10–12. Reprinted in *Stata Technical Bulletin Reprints*, vol. 9, pp. 121–124. College Station, TX: Stata Press.

Cox, N. J., and C. F. Baum. 2000. ip29.1: Metadata for user-written contributions to the Stata programming language. *Stata Technical Bulletin* 54: 21–22. Reprinted in *Stata Technical Bulletin Reprints*, vol. 9, pp. 124–126. College Station, TX: Stata Press.

Gould, W. W., and A. R. Riley. 2000. stata55: Search web for installable packages. *Stata Technical Bulletin* 54: 4–6. Reprinted in *Stata Technical Bulletin Reprints*, vol. 9, pp. 10–13. College Station, TX: Stata Press.

Also see

[R] **net** — Install and manage user-written additions from the Internet

[R] **ssc** — Install and uninstall packages from SSC

[R] **sj** — Stata Journal and STB installation instructions

[R] **hsearch** — Search help files

[R] **search** — Search Stata documentation

[R] **adoupdate** — Update user-written ado-files

[R] **update** — Update Stata

Title

netio — Control Internet connections

Syntax

Turn on or off the use of a proxy server

`set httpproxy {on | off} [, init]`

Set proxy host name

`set httpproxyhost ["]`*name*`["]`

Set the proxy port number

`set httpproxyport #`

Turn on or off proxy authorization

`set httpproxyauth {on | off}`

Set proxy authorization user ID

`set httpproxyuser ["]`*name*`["]`

Set proxy authorization password

`set httpproxypw ["]`*password*`["]`

Set time limit for establishing initial connection

`set timeout1` *#seconds* `[, permanently]`

Set time limit for data transfer

`set timeout2` *#seconds* `[, permanently]`

Description

Several commands (for example, `net`, `news`, and `update`) are designed specifically for use over the Internet. Many other Stata commands that read a file (for example, `copy`, `type`, and `use`) can also read directly from a URL. All these commands will usually work without your ever needing to concern yourself with the `set` commands discussed here. These `set` commands provide control over network system parameters.

If you experience problems when using Stata's network features, ask your system administrator if your site uses a proxy. A proxy is a server between your computer and the rest of the Internet, and your computer may need to communicate with other computers on the Internet through this proxy. If your site uses a proxy, your system administrator can provide you with its host name and the port your computer can use to communicate with it. If your site's proxy requires you to log in to it before it will respond, your system administrator will provide you with a user ID and password.

set httpproxyhost sets the name of the host to be used as a proxy server. set httpproxyport sets the port number. set httpproxy turns on or off the use of a proxy server, leaving the proxy host name and port intact, even when not in use.

Under the Mac and Windows operating systems, when you set httpproxy on, Stata will attempt to obtain the values of httpproxyhost and httpproxyport from the operating system if they have not been previously set. set httpproxy on, init attempts to obtain these values from the operating system, even if they have been previously set.

If the proxy requires authorization (user ID and password), set authorization on via set httpproxyauth on. The proxy user and proxy password must also be set to the appropriate user ID and password by using set httpproxyuser and set httpproxypw.

Stata remembers the various proxy settings between sessions and does not need a permanently option.

set timeout1 changes the time limit in seconds that Stata imposes for establishing the initial connection with a remote host. The default value is 30. set timeout2 changes the time limit in seconds that Stata imposes for subsequent data transfer with the host. The default value is 180. If these time limits are exceeded, a "connection timed out" message and error code 2 are produced. You should seldom need to change these settings.

Options

init specifies that set httpproxy on attempts to initialize httpproxyhost and httpproxyport from the operating system (Mac and Windows only).

permanently specifies that, in addition to making the change right now, the timeout1 and timeout2 settings be remembered and become the default setting when you invoke Stata.

The various httpproxy settings do not have a permanently option because permanently is implied.

Remarks

If you receive an error message, see http://www.stata.com/support/faqs/web/ for the latest information.

1. remote connection failed r(677);

If you see

```
remote connection failed
r(677);
```

then you asked for something to be done over the web, and Stata tried but could not contact the specified host. Stata was able to talk over the network and look up the host but was not able to establish a connection to that host. Perhaps the host is down; try again later.

If all your web accesses result in this message, then perhaps your network connection is through a proxy server. If it is, then you must tell Stata.

Contact your system administrator. Ask for the name and port of the "HTTP proxy server". Say that you are told

HTTP proxy server: jupiter.myuni.edu
port number: 8080

In Stata, type

```
. set httpproxyhost jupiter.myuni.edu
. set httpproxyport 8080
. set httpproxy on
```

Your web accesses should then work.

2. connection timed out r(2);

If you see

```
connection timed out
r(2);
```

then an Internet connection has timed out. This can happen when

a. the connection between you and the host is slow, or

b. the connection between you and the host has disappeared, and so it eventually "timed out".

For (b), wait a while (say, 5 minutes) and try again (sometimes pieces of the Internet can break for up to a day, but that is rare). For (a), you can reset the limits for what constitutes "timed out". There are two numbers to set.

The time to establish the initial connection is **timeout1**. By default, Stata waits 30 seconds before declaring a timeout. You can change the limit:

```
. set timeout1 #seconds
```

You might try doubling the usual limit and specify 60; *#seconds* must be between 1 and 32,000.

The time to retrieve data from an open connection is **timeout2**. By default, Stata waits 180 seconds (3 minutes) before declaring a timeout. To change the limit, type

```
. set timeout2 #seconds
```

You might try doubling the usual limit and specify 360; *#seconds* must be between 1 and 32,000.

Also see

[R] **query** — Display system parameters

[P] **creturn** — Return c-class values

[U] **28 Using the Internet to keep up to date**

Title

news — Report Stata news

Syntax

`news`

Menu

Help > News

Description

`news` displays a brief listing of recent Stata news and information, which it obtains from Stata's website. `news` requires that your computer be connected to the Internet.

You may also execute `news` by selecting **Help > News**.

Remarks

`news` provides an easy way of displaying a brief list of the latest Stata news:

```
. news

  ___  ____  ____  ____  ____
 /__    /   ____/   /   ____/
___/   /   /___/   /   /___/  News        The latest from http://www.stata.com
--------------------------------------------------------------------------------

8 October 2011.  Official update available for download
    Click here (equivalent to pulling down Help and selecting
    Check for Updates) or type update from http://www.stata.com.

27 July 2011.  Stata 12 available
    Stata 12 -- structural equation models (SEM) -- multiple imputation
    using chained equations -- contrasts and pairwise comparisons --
    autoregressive fractionally integrated moving-average (ARFIMA) models --
    multivariate GARCH -- unobserved-components models -- time-series filters --
    receiver operating characteristic (ROC) regression -- contour plots --
    import Excel -- PDF export -- is now available.
    Visit http://www.stata.com/stata12/ for more information.

21 March 2011.  NetCourse schedule updated
    See http://www.stata.com/netcourse/ for more information.

    (output omitted)

<end>
```

Also see

[U] **28 Using the Internet to keep up to date**

Title

nl — Nonlinear least-squares estimation

Syntax

Interactive version

`nl` (*depvar* `=` *<sexp>*) [*if*] [*in*] [*weight*] [, *options*]

Programmed substitutable expression version

`nl` *sexp_prog* : *depvar* [*varlist*] [*if*] [*in*] [*weight*] [, *options*]

Function evaluator program version

`nl` *func_prog* `@` *depvar* [*varlist*] [*if*] [*in*] [*weight*] ,

{ `parameters(`*namelist*`)` | `nparameters(`*#*`)`} [*options*]

where

depvar is the dependent variable;

<sexp> is a substitutable expression;

sexp_prog is a substitutable expression program; and

func_prog is a function evaluator program.

options	Description
Model	
variables(*varlist*)	variables in model
initial(*initial_values*)	initial values for parameters
* parameters(*namelist*)	parameters in model (function evaluator program version only)
* nparameters(*#*)	number of parameters in model (function evaluator program version only)
sexp_options	options for substitutable expression program
func_options	options for function evaluator program
Model 2	
lnlsq(*#*)	use log least-squares where ln(*depvar* − *#*) is assumed to be normally distributed
noconstant	the model has no constant term; seldom used
hasconstant(*name*)	use *name* as constant term; seldom used
SE/Robust	
vce(*vcetype*)	*vcetype* may be gnr, robust, cluster *clustvar*, bootstrap, jackknife, hac *kernel*, hc2, or hc3
Reporting	
level(*#*)	set confidence level; default is level(95)
leave	create variables containing derivative of $E(y)$
title(*string*)	display *string* as title above the table of parameter estimates
title2(*string*)	display *string* as subtitle
display_options	control column formats and line width
Optimization	
optimization_options	control the optimization process; seldom used
eps(*#*)	specify # for convergence criterion; default is eps(1e-5)
delta(*#*)	specify # for computing derivatives; default is delta(4e-7)
coeflegend	display legend instead of statistics

* For function evaluator program version, you must specify parameters(*namelist*) or nparameters(*#*), or both.
bootstrap, by, jackknife, rolling, statsby, and svy are allowed; see [U] **11.1.10 Prefix commands**.
Weights are not allowed with the bootstrap prefix; see [R] **bootstrap**.
aweights are not allowed with the jackknife prefix; see [R] **jackknife**.
vce(), leave, and weights are not allowed with the svy prefix; see [SVY] **svy**.
aweights, fweights, and iweights are allowed; see [U] **11.1.6 weight**.
coeflegend does not appear in the dialog box.
See [U] **20 Estimation and postestimation commands** for more capabilities of estimation commands.

Menu

Statistics > Linear models and related > Nonlinear least squares

Description

`nl` fits an arbitrary nonlinear regression function by least squares. With the interactive version of the command, you enter the function directly on the command line or in the dialog box by using a *substitutable expression*. If you have a function that you use regularly, you can write a *substitutable expression program* and use the second syntax to avoid having to reenter the function every time. The function evaluator program version gives you the most flexibility in exchange for increased complexity; with this version, your program is given a vector of parameters and a variable list, and your program computes the regression function.

When you write a substitutable expression program or function evaluator program, the first two letters of the name must be `nl`. *sexp_prog* and *func_prog* refer to the name of the program without the first two letters. For example, if you wrote a function evaluator program named `nlregss`, you would type `nl regss @` ... to estimate the parameters.

Options

Model

`variables(`*varlist*`)` specifies the variables in the model. `nl` ignores observations for which any of these variables have missing values. If you do not specify `variables()`, then `nl` issues an error message with return code 480 if the estimation sample contains any missing values.

`initial(`*initial_values*`)` specifies the initial values to begin the estimation. You can specify a $1 \times k$ matrix, where k is the number of parameters in the model, or you can specify a parameter name, its initial value, another parameter name, its initial value, and so on. For example, to initialize `alpha` to 1.23 and `delta` to 4.57, you would type

```
nl ... , initial(alpha 1.23 delta 4.57) ...
```

Initial values declared using this option override any that are declared within substitutable expressions. If you specify a parameter that does not appear in your model, `nl` exits with error code 480. If you specify a matrix, the values must be in the same order that the parameters are declared in your model. `nl` ignores the row and column names of the matrix.

`parameters(`*namelist*`)` specifies the names of the parameters in the model. The names of the parameters must adhere to the naming conventions of Stata's variables; see [U] **11.3 Naming conventions**. If you specify both `parameters()` and `nparameters()`, the number of names in the former must match the number specified in the latter; if not, `nl` issues an error message with return code 198.

`nparameters(`*#*`)` specifies the number of parameters in the model. If you do not specify names with the `parameters()` option, `nl` names them `b1`, `b2`, ..., `b`*#*. If you specify both `parameters()` and `nparameters()`, the number of names in the former must match the number specified in the latter; if not, `nl` issues an error message with return code 198.

sexp_options refer to any options allowed by your *sexp_prog*.

func_options refer to any options allowed by your *func_prog*.

Model 2

`lnlsq(`*#*`)` fits the model by using log least-squares, which we define as least squares with shifted lognormal errors. In other words, ln(*depvar* − *#*) is assumed to be normally distributed. Sums of squares and deviance are adjusted to the same scale as *depvar*.

`noconstant` indicates that the function does not include a constant term. This option is generally not needed, even if there is no constant term in the model, unless the coefficient of variation (over observations) of the partial derivative of the function with respect to a parameter is less than `eps()` and that parameter is not a constant term.

`hasconstant(`*name*`)` indicates that parameter *name* be treated as the constant term in the model and that `nl` should not use its default algorithm to find a constant term. As with `noconstant`, this option is seldom used.

SE/Robust

`vce(`*vcetype*`)` specifies the type of standard error reported, which includes types that are derived from asymptotic theory, that are robust to some kinds of misspecification, that allow for intragroup correlation, and that use bootstrap or jackknife methods; see [R] ***vce_option***.

`vce(gnr)`, the default, uses the conventionally derived variance estimator for nonlinear models fit using Gauss–Newton regression.

`nl` also allows the following:

`vce(hac` *kernel* [#]`)` specifies that a heteroskedasticity- and autocorrelation-consistent (HAC) variance estimate be used. HAC refers to the general form for combining weighted matrices to form the variance estimate. There are three kernels available for `nl`:

`nwest` | `gallant` | `anderson`

specifies the number of lags. If # is not specified, $N-2$ is assumed.

`vce(hac` *kernel* [#]`)` is not allowed if weights are specified.

`vce(hc2)` and `vce(hc3)` specify alternative bias corrections for the robust variance calculation. `vce(hc2)` and `vce(hc3)` may not be specified with the `svy` prefix. By default, `vce(robust)` uses $\widehat{\sigma}_j^2 = \{n/(n-k)\}u_j^2$ as an estimate of the variance of the jth observation, where u_j is the calculated residual and $n/(n-k)$ is included to improve the overall estimate's small-sample properties.

`vce(hc2)` instead uses $u_j^2/(1-h_{jj})$ as the observation's variance estimate, where h_{jj} is the jth diagonal element of the hat (projection) matrix. This produces an unbiased estimate of the covariance matrix if the model is homoskedastic. `vce(hc2)` tends to produce slightly more conservative confidence intervals than `vce(robust)`.

`vce(hc3)` uses $u_j^2/(1-h_{jj})^2$ as suggested by Davidson and MacKinnon (1993 and 2004), who report that this often produces better results when the model is heteroskedastic. `vce(hc3)` produces confidence intervals that tend to be even more conservative.

See, in particular, Davidson and MacKinnon (2004, 239), who advocate the use of `vce(hc2)` or `vce(hc3)` instead of the plain robust estimator for nonlinear least squares.

Reporting

`level(`*#*`)`; see [R] **estimation options**.

`leave` leaves behind after estimation a set of new variables with the same names as the estimated parameters containing the derivatives of $E(y)$ with respect to the parameters. If the dataset contains an existing variable with the same name as a parameter, then using `leave` causes `nl` to issue an error message with return code 110.

`leave` may not be specified with `vce(cluster` *clustvar*`)` or the `svy` prefix.

`title(`*string*`)` specifies an optional title that will be displayed just above the table of parameter estimates.

`title2(`*string*`)` specifies an optional subtitle that will be displayed between the title specified in `title()` and the table of parameter estimates. If `title2()` is specified but `title()` is not, `title2()` has the same effect as `title()`.

display_options: `cformat(%`*fmt*`)`, `pformat(%`*fmt*`)`, `sformat(%`*fmt*`)`, and `nolstretch`; see [R] **estimation options**.

Optimization

optimization_options: `iterate(#)`, `[no]log`, `trace`. `iterate()` specifies the maximum number of iterations, `log`/`nolog` specifies whether to show the iteration log, and `trace` specifies that the iteration log should include the current parameter vector. These options are seldom used.

`eps(#)` specifies the convergence criterion for successive parameter estimates and for the residual sum of squares. The default is `eps(1e-5)`.

`delta(#)` specifies the relative change in a parameter to be used in computing the numeric derivatives. The derivative for parameter β_i is computed as $\{f(X, \beta_1, \beta_2, \ldots, \beta_i + d, \beta_{i+1}, \ldots) - f(X, \beta_1, \beta_2, \ldots, \beta_i, \beta_{i+1}, \ldots)\}/d$, where d is $\delta(\beta_i + \delta)$. The default is `delta(4e-7)`.

The following options are available with `nl` but are not shown in the dialog box:

`coeflegend`; see [R] **estimation options**.

Remarks

Remarks are presented under the following headings:

Substitutable expressions
Substitutable expression programs
Built-in functions
Lognormal errors
Other uses
Weights
Potential errors
General comments on fitting nonlinear models
Function evaluator programs

`nl` fits an arbitrary nonlinear function by least squares. The interactive version allows you to enter the function directly on the command line or dialog box using *substitutable expressions*. You can write a *substitutable expression program* for functions that you fit frequently to save yourself time. Finally, *function evaluator programs* give you the most flexibility in defining your nonlinear function, though they are more complicated to use.

The next section explains the substitutable expressions that are used to define the regression function, and the section thereafter explains how to write substitutable expression program files so that you do not need to type in commonly used functions over and over. Later sections highlight other features of `nl`.

The final section discusses function evaluator programs. If you find substitutable expressions adequate to define your nonlinear function, then you can skip that section entirely. Function evaluator programs are generally needed only for complicated problems, such as multistep estimators. The program receives a vector of parameters at which it is to compute the function and a variable into which the results are to be placed.

Substitutable expressions

You define the nonlinear function to be fit by nl by using a substitutable expression. Substitutable expressions are just like any other mathematical expressions involving scalars and variables, such as those you would use with Stata's generate command, except that the parameters to be estimated are bound in braces. See [U] **13.2 Operators** and [U] **13.3 Functions** for more information on expressions.

For example, suppose that you wish to fit the function

$$y_i = \beta_0(1 - e^{-\beta_1 x_i}) + \epsilon_i$$

where β_0 and β_1 are the parameters to be estimated and ϵ_i is an error term. You would simply type

```
. nl (y = {b0}*(1 - exp(-1*{b1}*x)))
```

You must enclose the entire equation in parentheses. Because b0 and b1 are enclosed in braces, nl knows that they are parameters in the model. nl will initialize b0 and b1 to zero by default. To request that nl initialize b0 to 1 and b1 to 0.25, you would type

```
. nl (y = {b0=1}*(1 - exp(-1*{b1=0.25}*x)))
```

That is, inside the braces denoting a parameter, you put the parameter name followed by an equal sign and the initial value. If a parameter appears in your function multiple times, you need only specify an initial value only once (or never, if you wish to set the initial value to zero). If you do specify more than one initial value for the same parameter, nl will use the *last* value given. Parameter names must follow the same conventions as variable names. See [U] **11.3 Naming conventions**.

Frequently, even nonlinear functions contain linear combinations of variables. As an example, suppose that you wish to fit the function

$$y_i = \beta_0 \left\{1 - e^{-(\beta_1 x_{1i} + \beta_2 x_{2i} + \beta_3 x_{3i})}\right\} + \epsilon_i$$

nl allows you to declare a linear combination of variables by using the shorthand notation

```
. nl (y = {b0=1}*(1 - exp(-1*{xb: x1 x2 x3})))
```

In the syntax {xb: x1 x2 x3}, you are telling nl that you are declaring a linear combination named xb that is a function of three variables, x1, x2, and x3. nl will create three parameters, named xb_x1, xb_x2, and xb_x3, and initialize them to zero. Instead of typing the previous command, you could have typed

```
. nl (y = {b0=1}*(1 - exp(-1*({xb_x1}*x1 + {xb_x2}*x2 + {xb_x3}*x3))))
```

and yielded the same result. You can refer to the parameters created by nl in the linear combination later in the function, though you must declare the linear combination first if you intend to do that. When creating linear combinations, nl ensures that the parameter names it chooses are unique and have not yet been used in the function.

In general, there are three rules to follow when defining substitutable expressions:

1. Parameters of the model are bound in braces: {b0}, {param}, etc.
2. Initial values for parameters are given by including an equal sign and the initial value inside the braces: {b0=1}, {param=3.571}, etc.
3. Linear combinations of variables can be included using the notation {*eqname*:*varlist*}, for example, {xb: mpg price weight}, {score: w x z}, etc. Parameters of linear combinations are initialized to zero.

If you specify initial values by using the `initial()` option, they override whatever initial values are given within the substitutable expression. Substitutable expressions are so named because, once values are assigned to the parameters, the resulting expression can be handled by `generate` and `replace`.

▷ Example 1

We wish to fit the CES production function

$$\ln Q_i = \beta_0 - \frac{1}{\rho} \ln \left\{ \delta K_i^{-\rho} + (1-\delta) L_i^{-\rho} \right\} + \epsilon_i \tag{1}$$

where $\ln Q_i$ is the log of output for firm i; K_i and L_i are firm i's capital and labor usage, respectively; and ϵ_i is a regression error term. Because ρ appears in the denominator of a fraction, zero is not a feasible initial value; for a CES production function, $\rho = 1$ is a reasonable choice. Setting $\delta = 0.5$ implies that labor and capital have equal impacts on output, which is also a reasonable choice for an initial value. We type

```
. use http://www.stata-press.com/data/r12/production
. nl (lnoutput = {b0} - 1/{rho=1}*ln({delta=0.5}*capital^(-1*{rho}) +
> (1 - {delta})*labor^(-1*{rho})))
(obs = 100)
Iteration 0:  residual SS =  29.38631
Iteration 1:  residual SS =  29.36637
Iteration 2:  residual SS =  29.36583
Iteration 3:  residual SS =  29.36581
Iteration 4:  residual SS =  29.36581
Iteration 5:  residual SS =  29.36581
Iteration 6:  residual SS =  29.36581
Iteration 7:  residual SS =  29.36581
      Source |       SS       df       MS
-------------+------------------------------         Number of obs =        100
       Model |  91.1449924     2  45.5724962         R-squared     =     0.7563
    Residual |  29.3658055    97  .302740263         Adj R-squared =     0.7513
-------------+------------------------------         Root MSE      =   .5502184
       Total |  120.510798    99  1.21728079         Res. dev.     =   161.2538

------------------------------------------------------------------------------
    lnoutput |      Coef.   Std. Err.      t    P>|t|     [95% Conf. Interval]
-------------+----------------------------------------------------------------
         /b0 |   3.792158    .099682    38.04   0.000     3.594316    3.989999
        /rho |   1.386993    .472584     2.93   0.004     .4490443    2.324941
      /delta |   .4823616   .0519791     9.28   0.000     .3791975    .5855258
------------------------------------------------------------------------------
  Parameter b0 taken as constant term in model & ANOVA table
```

`nl` will attempt to find a constant term in the model and, if one is found, mention it at the bottom of the output. `nl` found `b0` to be a constant because the partial derivative $\partial \ln Q_i / \partial$`b0` has a coefficient of variation less than `eps()` in the estimation sample.

The elasticity of substitution for the CES production function is $\sigma = 1/(1+\rho)$; and, having fit the model, we can use `nlcom` to estimate it:

```
. nlcom (1/(1 + _b[/rho]))
       _nl_1:  1/(1 + _b[/rho])

------------------------------------------------------------------------------
    lnoutput |      Coef.   Std. Err.      t    P>|t|     [95% Conf. Interval]
-------------+----------------------------------------------------------------
       _nl_1 |   .4189372   .0829424     5.05   0.000     .2543194     .583555
------------------------------------------------------------------------------
```

See [R] **nlcom** and [U] **13.5 Accessing coefficients and standard errors** for more information.

◁

nl's output closely mimics that of regress; see [R] **regress** for more information. The R^2, sums of squares, and similar statistics are calculated in the same way that regress calculates them. If no "constant" term is specified, the usual caveats apply to the interpretation of the R^2 statistic; see the comments and references in Goldstein (1992). Unlike regress, nl does not report a model F statistic, because a test of the joint significance of all the parameters except the constant term may not be relevant in a nonlinear model.

Substitutable expression programs

If you fit the same model often or if you want to write an estimator that will operate on whatever variables you specify, then you will want to write a substitutable expression program. That program will return a macro containing a substitutable expression that nl can then evaluate, and it may optionally calculate initial values as well. The name of the program must begin with the letters nl.

To illustrate, suppose that you use the CES production function often in your work. Instead of typing in the formula each time, you can write a program like this:

```
program nlces, rclass
        version 12
        syntax varlist(min=3 max=3) [if]
        local logout : word 1 of `varlist'
        local capital : word 2 of `varlist'
        local labor : word 3 of `varlist'
        // Initial value for b0 given delta=0.5 and rho=1
        tempvar y
        generate double `y' = `logout' + ln(0.5*`capital'^-1 + 0.5*`labor'^-1)
        summarize `y' `if', meanonly
        local b0val = r(mean)
        // Terms for substitutable expression
        local capterm "{delta=0.5}*`capital'^(-1*{rho})"
        local labterm "(1-{delta})*`labor'^(-1*{rho})"
        local term2   "1/{rho=1}*ln(`capterm' + `labterm')"
        // Return substitutable expression and title
        return local eq "`logout' = {b0=`b0val'} - `term2'"
        return local title "CES ftn., ln Q=`logout', K=`capital', L=`labor'"
end
```

The program accepts three variables for log output, capital, and labor, and it accepts an if *exp* qualifier to restrict the estimation sample. All programs that you write to use with nl must accept an if *exp* qualifier because, when nl calls the program, it passes a binary variable that marks the estimation sample (the variable equals one if the observation is in the sample and zero otherwise). When calculating initial values, you will want to restrict your computations to the estimation sample, and you can do so by using if with any commands that accept if *exp* qualifiers. Even if your program does not calculate initial values or otherwise use the if qualifier, the syntax statement must still allow it. See [P] **syntax** for more information on the syntax command and the use of if.

As in the previous example, reasonable initial values for δ and ρ are 0.5 and 1, respectively. Conditional on those values, (1) can be rewritten as

$$\beta_0 = \ln Q_i + \ln(0.5K_i^{-1} + 0.5L_i^{-1}) - \epsilon_i \qquad (2)$$

so a good initial value for β_0 is the mean of the right-hand side of (2) ignoring ϵ_i. Lines 7–10 of the function evaluator program calculate that mean and store it in a local macro. Notice the use of if in the summarize statement so that the mean is calculated only for the estimation sample.

The final part of the program returns two macros. The macro `title` is optional and defines a short description of the model that will be displayed in the output immediately above the table of parameter estimates. The macro `eq` is required and defines the substitutable expression that `nl` will use. If the expression is short, you can define it all at once. However, because the expression used here is somewhat lengthy, defining local macros and then building up the final expression from them is easier.

To verify that there are no errors in your program, you can call it directly and then use `return list`:

```
. use http://www.stata-press.com/data/r12/production

. nlces lnoutput capital labor
  (output omitted)

. return list

macros:
              r(title) : "CES ftn., ln Q=lnoutput, K=capital, L=labor"
                 r(eq) : "lnoutput = {b0=3.711606264663641} - 1/{rho=1}*ln({delt
> a=0.5}*capital^(-1*{rho}) + (1-{delta})*labor^(-1*{rho}))"
```

The macro `r(eq)` contains the same substitutable expression that we specified at the command line in the preceding example, except for the initial value for `b0`. In short, an `nl` substitutable expression program should return in `r(eq)` the same substitutable expression you would type at the command line. The only difference is that when writing a substitutable expression program, you do not bind the entire expression inside parentheses.

Having written the program, you can use it by typing

```
. nl ces: lnoutput capital labor
```

(There is a space between `nl` and `ces`.) The output is identical to that shown in example 1, save for the title defined in the function evaluator program that appears immediately above the table of parameter estimates.

❑ Technical note

You will want to store `nlces` as an ado-file called `nlces.ado`. The alternative is to type the code into Stata interactively or to place the code in a do-file. While those alternatives are adequate for occasional use, if you save the program as an ado-file, you can use the function anytime you use Stata without having to redefine the program. When `nl` attempts to execute `nlces`, if the program is not in Stata's memory, Stata will search the disk(s) for an ado-file of the same name and, if found, automatically load it. All you have to do is name the file with the `.ado` suffix and then place it in a directory where Stata will find it. You should put the file in the directory Stata reserves for user-written ado-files, which, depending on your operating system, is `c:\ado\personal` (Windows), `~/ado/personal` (Unix), or `~:ado:personal` (Mac). See [U] **17 Ado-files**. ❑

Sometimes you may want to pass additional options to the substitutable expression program. You can modify the `syntax` statement of your program to accept whatever options you wish. Then when you call `nl` with the syntax

. nl *func_prog*: *varlist*, *options*

any *options* that are not recognized by `nl` (see the table of options at the beginning of this entry) are passed on to your function evaluator program. The only other restriction is that your program cannot accept an option named `at` because `nl` uses that option with function evaluator programs.

Built-in functions

Some functions are used so often that `nl` has them built in so that you do not need to write them yourself. `nl` automatically chooses initial values for the parameters, though you can use the `initial(...)` option to override them.

Three alternatives are provided for exponential regression with one asymptote:

`exp3`	$y_i = \beta_0 + \beta_1 \beta_2^{x_i} + \epsilon_i$
`exp2`	$y_i = \beta_1 \beta_2^{x_i} + \epsilon_i$
`exp2a`	$y_i = \beta_1 \big(1 - \beta_2^{x_i}\big) + \epsilon_i$

For instance, typing `nl exp3: ras dvl` fits the three-parameter exponential model (parameters β_0, β_1, and β_2) using $y_i =$ `ras` and $x_i =$ `dvl`.

Two alternatives are provided for the logistic function (symmetric sigmoid shape; not to be confused with logistic regression):

`log4`	$y_i = \beta_0 + \beta_1 \Big/ \Big[1 + \exp\big\{-\beta_2(x_i - \beta_3)\big\}\Big] + \epsilon_i$
`log3`	$y_i = \beta_1 \Big/ \Big[1 + \exp\big\{-\beta_2(x_i - \beta_3)\big\}\Big] + \epsilon_i$

Finally, two alternatives are provided for the Gompertz function (asymmetric sigmoid shape):

`gom4`	$y_i = \beta_0 + \beta_1 \exp\Big[-\exp\big\{-\beta_2(x_i - \beta_3)\big\}\Big] + \epsilon_i$
`gom3`	$y_i = \beta_1 \exp\Big[-\exp\big\{-\beta_2(x_i - \beta_3)\big\}\Big] + \epsilon_i$

Lognormal errors

A nonlinear model with errors that are independent and identically distributed normal may be written

$$y_i = f(\mathbf{x}_i, \boldsymbol{\beta}) + u_i, \qquad u_i \sim N(0, \sigma^2) \tag{3}$$

for $i = 1, \dots, n$. If the y_i are thought to have a k-shifted lognormal instead of a normal distribution—that is, $\ln(y_i - k) \sim N(\zeta_i, \tau^2)$, and the systematic part $f(\mathbf{x}_i, \boldsymbol{\beta})$ of the original model is still thought appropriate for y_i—the model becomes

$$\ln(y_i - k) = \zeta_i + v_i = \ln\big\{f(\mathbf{x}_i, \boldsymbol{\beta}) - k\big\} + v_i, \quad v_i \sim N(0, \tau^2) \tag{4}$$

This model is fit if `lnlsq(`k`)` is specified.

If model (4) is correct, the variance of $(y_i - k)$ is proportional to $\big\{f(\mathbf{x}_i, \boldsymbol{\beta}) - k\big\}^2$. Probably the most common case is $k = 0$, sometimes called "proportional errors" because the standard error of y_i is proportional to its expectation, $f(\mathbf{x}_i, \boldsymbol{\beta})$. Assuming that the value of k is known, (4) is just another nonlinear model in $\boldsymbol{\beta}$, and it may be fit as usual. However, we may wish to compare the fit of (3) with that of (4) using the residual sum of squares (RSS) or the deviance D, $D = -2\times$ log-likelihood, from each model. To do so, we must allow for the change in scale introduced by the log transformation.

Assuming, then, the y_i to be normally distributed, Atkinson (1985, 85–87, 184), by considering the Jacobian $\prod |\partial \ln(y_i - k)/\partial y_i|$, showed that multiplying both sides of (4) by the geometric mean of $y_i - k$, $\dot{y}$, gives residuals on the same scale as those of y_i. The geometric mean is given by

$$\dot{y} = e^{n^{-1} \sum \ln(y_i - k)}$$

which is a constant for a given dataset. The residual deviance for (3) and for (4) may be expressed as

$$D(\widehat{\boldsymbol{\beta}}) = \Big\{1 + \ln(2\pi\widehat{\sigma}^2)\Big\} n \tag{5}$$

where $\widehat{\beta}$ is the maximum likelihood estimate (MLE) of β for each model and $n\widehat{\sigma}^2$ is the RSS from (3), or that from (4) multiplied by $\dot{y}^2$.

Because (3) and (4) are models with different error structures but the same functional form, the arithmetic difference in their RSS or deviances is not easily tested for statistical significance. However, if the deviance difference is large (>4, say), we would naturally prefer the model with the smaller deviance. Of course, the residuals for each model should be examined for departures from assumptions (nonconstant variance, nonnormality, serial correlations, etc.) in the usual way.

Alternatively, consider modeling

$$E(y_i) = 1/(C + Ae^{Bx_i}) \tag{6}$$

$$E(1/y_i) = E(y_i') = C + Ae^{Bx_i} \tag{7}$$

where C, A, and B are parameters to be estimated. Using the data $(y, x) = (0.04, 5)$, $(0.06, 12)$, $(0.08, 25)$, $(0.1, 35)$, $(0.15, 42)$, $(0.2, 48)$, $(0.25, 60)$, $(0.3, 75)$, and $(0.5, 120)$ (Danuso 1991), fitting the models yields

Model	C	A	B	RSS	Deviance
(6)	1.781	25.74	−0.03926	−0.001640	−51.95
(6) with `lnlsq(0)`	1.799	25.45	−0.04051	−0.001431	−53.18
(7)	1.781	25.74	−0.03926	8.197	24.70
(7) with `lnlsq(0)`	1.799	27.45	−0.04051	3.651	17.42

There is little to choose between the two versions of the logistic model (6), whereas for the exponential model (7), the fit using `lnlsq(0)` is much better (a deviance difference of 7.28). The reciprocal transformation has introduced heteroskedasticity into y_i', which is countered by the proportional errors property of the lognormal distribution implicit in `lnlsq(0)`. The deviances are not comparable between the logistic and exponential models because the change of scale has not been allowed for, although in principle it could be.

Other uses

Even if you are fitting linear regression models, you may find that `nl` can save you some typing. Because you specify the parameters of your model explicitly, you can impose constraints on them directly.

▷ Example 2

In example 2 of [R] **cnsreg**, we showed how to fit the model

$$\texttt{mpg} = \beta_0 + \beta_1\texttt{price} + \beta_2\texttt{weight} + \beta_3\texttt{displ} + \beta_4\texttt{gear\_ratio} + \beta_5\texttt{foreign} + \beta_6\texttt{length} + u$$

subject to the constraints

$$\beta_1 = \beta_2 = \beta_3 = \beta_6$$
$$\beta_4 = -\beta_5 = \beta_0/20$$

An alternative way is to use `nl`:

```
. use http://www.stata-press.com/data/r12/auto, clear
(1978 Automobile Data)
. nl (mpg = {b0} + {b1}*price + {b1}*weight + {b1}*displ +
> {b0}/20*gear_ratio - {b0}/20*foreign + {b1}*length)
(obs = 74)
Iteration 0:  residual SS =  1578.522
Iteration 1:  residual SS =  1578.522
      Source |       SS       df       MS
-------------+------------------------------         Number of obs =        74
       Model |  34429.4777     2  17214.7389         R-squared     =    0.9562
    Residual |  1578.52226    72  21.9239203         Adj R-squared =    0.9549
-------------+------------------------------         Root MSE      =  4.682299
       Total |       36008    74  486.594595         Res. dev.     =  436.4562

------------------------------------------------------------------------------
         mpg |      Coef.   Std. Err.      t    P>|t|     [95% Conf. Interval]
-------------+----------------------------------------------------------------
         /b0 |   26.52229   1.375178    19.29   0.000     23.78092    29.26365
         /b1 |   -.000923   .0001534    -6.02   0.000    -.0012288   -.0006172
------------------------------------------------------------------------------
```

The point estimates and standard errors for β_0 and β_1 are identical to those reported in example 2 of [R] **cnsreg**. To get the estimate for β_4, we can use `nlcom`:

```
. nlcom _b[/b0]/20
       _nl_1:  _b[/b0]/20

------------------------------------------------------------------------------
         mpg |      Coef.   Std. Err.      t    P>|t|     [95% Conf. Interval]
-------------+----------------------------------------------------------------
       _nl_1 |   1.326114   .0687589    19.29   0.000     1.189046    1.463183
------------------------------------------------------------------------------
```

The advantage to using `nl` is that we do not need to use the `constraint` command six times.

◁

`nl` is also a useful tool when doing exploratory data analysis. For example, you may want to run a regression of y on a function of x, though you have not decided whether to use sqrt(x) or ln(x). You can use `nl` to run both regressions without having first to generate two new variables:

```
. nl (y = {b0} + {b1}*ln(x))
. nl (y = {b0} + {b1}*sqrt(x))
```

Poi (2008) shows the advantages of using `nl` when marginal effects of transformed variables are desired as well.

Weights

Weights are specified in the usual way—analytic and frequency weights as well as `iweights` are supported; see [U] **20.22 Weighted estimation**. Use of analytic weights implies that the y_i have different variances. Therefore, model (3) may be rewritten as

$$y_i = f(\mathbf{x}_i, \boldsymbol{\beta}) + u_i, \qquad u_i \sim N(0, \sigma^2/w_i) \tag{3a}$$

where w_i are (positive) weights, assumed to be known and normalized such that their sum equals the number of observations. The residual deviance for $(3a)$ is

$$D(\widehat{\boldsymbol{\beta}}) = \{1 + \ln(2\pi\widehat{\sigma}^2)\}n - \sum \ln(w_i) \tag{5a}$$

[compare with (5)], where

$$n\widehat{\sigma}^2 = \text{RSS} = \sum w_i \{y_i - f(\mathbf{x}_i, \widehat{\boldsymbol{\beta}})\}^2$$

Defining and fitting a model equivalent to (4) when weights have been specified as in (3a) is not straightforward and has not been attempted. Thus deviances using and not using the `lnlsq()` option may not be strictly comparable when analytic weights (other than 0 and 1) are used.

You do not need to modify your substitutable expression in any way to use weights. If, however, you write a substitutable expression program, then you should account for weights when obtaining initial values. When `nl` calls your program, it passes whatever weight expression (if any) was specified by the user. Here is an outline of a substitutable expression program that accepts weights:

```
program nlname, rclass
        version 12
        syntax varlist [aw fw iw] if
        ...
        // Obtain initial values allowing weights
        // Use the syntax [`weight'`exp'].  For example,
        summarize varname [`weight'`exp'] `if'
        regress depvar varlist [`weight'`exp'] `if'
        ...
        // Return substitutable expression
        return local eq "substitutable expression"
        return local title "description of estimator"
end
```

For details on how the `syntax` command processes weight expressions, see [P] **syntax**.

Potential errors

`nl` is reasonably robust to the inability of your nonlinear function to be evaluated at some parameter values. `nl` does assume that your function can be evaluated at the initial values of the parameters. If your function cannot be evaluated at the initial values, an error message is issued with return code 480. Recall that if you do not specify an initial value for a parameter, then `nl` initializes it to zero. Many nonlinear functions cannot be evaluated when some parameters are zero, so in those cases specifying alternative initial values is crucial.

Thereafter, as `nl` changes the parameter values, it monitors your function for unexpected missing values. If these are detected, `nl` backs up. That is, `nl` finds a point between the previous, known-to-be-good parameter vector and the new, known-to-be-bad vector at which the function can be evaluated and continues its iterations from that point.

`nl` requires that once a parameter vector is found where the predictions can be calculated, small changes to the parameter vector be made to calculate numeric derivatives. If a boundary is encountered at this point, an error message is issued with return code 481.

When specifying `lnlsq()`, an attempt to take logarithms of $y_i - k$ when $y_i \le k$ results in an error message with return code 482.

If `iterate()` iterations are performed and estimates still have not converged, results are presented with a warning, and the return code is set to 430.

If you use the programmed substitutable expression version of `nl` with a function evaluator program, or vice versa, Stata issues an error message. Verify that you are using the syntax appropriate for the program you have.

General comments on fitting nonlinear models

Achieving convergence is often problematic. For example, a unique minimum of the sum-of-squares function may not exist. Much literature exists on different algorithms that have been used, on strategies for obtaining good initial parameter values, and on tricks for parameterizing the model to make its behavior as linear-like as possible. Selected references are Kennedy and Gentle (1980, chap. 10) for computational matters and Ross (1990) and Ratkowsky (1983) for all three aspects. Ratkowsky's book is particularly clear and approachable, with useful discussion on the meaning and practical implications of intrinsic and parameter-effects nonlinearity. An excellent text on nonlinear estimation is Gallant (1987). Also see Davidson and MacKinnon (1993 and 2004).

To enhance the success of `nl`, pay attention to the form of the model fit, along the lines of Ratkowsky and Ross. For example, Ratkowsky (1983, 49–59) analyzes three possible three-parameter yield-density models for plant growth:

$$E(y_i) = \begin{cases} (\alpha + \beta x_i)^{-1/\theta} \\ (\alpha + \beta x_i + \gamma x_i^2)^{-1} \\ (\alpha + \beta x_i^{\phi})^{-1} \end{cases}$$

All three models give similar fits. However, he shows that the second formulation is dramatically more linear-like than the other two and therefore has better convergence properties. In addition, the parameter estimates are virtually unbiased and normally distributed, and the asymptotic approximation to the standard errors, correlations, and confidence intervals is much more accurate than for the other models. Even within a given model, the way the parameters are expressed (for example, ϕ^{x_i} or $e^{\theta x_i}$) affects the degree of linearity and convergence behavior.

Function evaluator programs

Occasionally, a nonlinear function may be so complex that writing a substitutable expression for it is impractical. For example, there could be many parameters in the model. Alternatively, if you are implementing a two-step estimator, writing a substitutable expression may be altogether impossible. Function evaluator programs can be used in these situations.

`nl` will pass to your function evaluator program a list of variables, a weight expression, a variable marking the estimation sample, and a vector of parameters. Your program is to replace the dependent variable, which is the first variable in the variables list, with the values of the nonlinear function evaluated at those parameters. As with substitutable expression programs, the first two letters of the name must be `nl`.

To focus on the mechanics of the function evaluator program, again let's compare the CES production function to the previous examples. The function evaluator program is

```
program nlces2
        version 12
        syntax varlist(min=3 max=3) if, at(name)
        local logout : word 1 of `varlist'
        local capital : word 2 of `varlist'
        local labor : word 3 of `varlist'
        // Retrieve parameters out of at matrix
        tempname b0 rho delta
        scalar `b0' = `at'[1, 1]
        scalar `rho' = `at'[1, 2]
        scalar `delta' = `at'[1, 3]
        tempvar kterm lterm
        generate double `kterm' = `delta'*`capital'^(-1*`rho') `if'
        generate double `lterm' = (1-`delta')*`labor'^(-1*`rho') `if'
        // Fill in dependent variable
        replace `logout' = `b0' - 1/`rho'*ln(`kterm' + `lterm') `if'
end
```

Unlike the previous `nlces` program, this one is not declared to be r-class. The `syntax` statement again accepts three variables: one for log output, one for capital, and one for labor. An `if` *exp* is again required because `nl` will pass a binary variable marking the estimation sample. All function evaluator programs must accept an option named `at()` that takes a `name` as an argument—that is how `nl` passes the parameter vector to your program.

The next part of the program retrieves the output, labor, and capital variables from the variables list. It then breaks up the temporary matrix `at` and retrieves the parameters `b0`, `rho`, and `delta`. Pay careful attention to the order in which the parameters refer to the columns of the `at` matrix because that will affect the syntax you use with `nl`. The temporary names you use inside this program are immaterial, however.

The rest of the program computes the nonlinear function, using some temporary variables to hold intermediate results. The final line of the program then replaces the dependent variable with the values of the function. Notice the use of `` `if' `` to restrict attention to the estimation sample. `nl` makes a copy of your dependent variable so that when the command is finished your data are left unchanged.

To use the program and fit your model, you type

```
. use http://www.stata-press.com/data/r12/production, clear
. nl ces2 @ lnoutput capital labor, parameters(b0 rho delta)
> initial(b0 0 rho 1 delta 0.5)
```

The output is again identical to that shown in example 1. The order in which the parameters were specified in the `parameters()` option is the same in which they are retrieved from the `at` matrix in the program. To initialize them, you simply list the parameter name, a space, the initial value, and so on.

If you use the `nparameters()` option instead of the `parameters()` option, the parameters are named `b1`, `b2`, ..., `b`k, where k is the number of parameters. Thus you could have typed

```
. nl ces2 @ lnoutput capital labor, nparameters(3) initial(b1 0 b2 1 b3 0.5)
```

With that syntax, the parameters called `b0`, `rho`, and `delta` in the program will be labeled `b1`, `b2`, and `b3`, respectively. In programming situations or if there are many parameters, instead of listing the parameter names and initial values in the `initial()` option, you may find it more convenient to pass a column vector. In those cases, you could type

```
. matrix myvals = (0, 1, 0.5)
. nl ces2 @ lnoutput capital labor, nparameters(3) initial(myvals)
```

In summary, a function evaluator program receives a list of variables, the first of which is the dependent variable that you are to replace with the values of your nonlinear function. Additionally, it must accept an `if` *exp*, as well as an option named `at` that will contain the vector of parameters at which `nl` wants the function evaluated. You are then free to do whatever is necessary to evaluate your function and replace the dependent variable.

If you wish to use weights, your function evaluator program's `syntax` statement must accept them. If your program consists only of, for example, `generate` statements, you need not do anything with the weights passed to your program. However, if in calculating the nonlinear function you use commands such as `summarize` or `regress`, then you will want to use the weights with those commands.

As with substitutable expression programs, `nl` will pass to it any options specified that `nl` does not accept, providing you with a way to pass more information to your function.

❑ Technical note

Before version 9 of Stata, the `nl` command used a different syntax, which required you to write an *nlfcn* program, and it did not have a syntax for interactive use other than the seven functions that were built-in. The old syntax of `nl` still works, and you can still use those *nlfcn* programs. If `nl` does not see a colon, an at sign, or a set of parentheses surrounding the equation in your command, it assumes that the old syntax is being used.

The current version of `nl` uses scalars and matrices to store intermediate calculations instead of local and global macros as the old version did, so the current version produces more accurate results. In practice, however, any discrepancies are likely to be small.

❑

Saved results

nl saves the following in e():

Scalars

e(N)	number of observations
e(k)	number of parameters
e(k_eq_model)	number of equations in overall model test; always 0
e(df_m)	model degrees of freedom
e(df_r)	residual degrees of freedom
e(df_t)	total degrees of freedom
e(mss)	model sum of squares
e(rss)	residual sum of squares
e(tss)	total sum of squares
e(mms)	model mean square
e(msr)	residual mean square
e(ll)	log likelihood assuming i.i.d. normal errors
e(r2)	R-squared
e(r2_a)	adjusted R-squared
e(rmse)	root mean squared error
e(dev)	residual deviance
e(N_clust)	number of clusters
e(lnlsq)	value of lnlsq if specified
e(log_t)	1 if lnlsq specified, 0 otherwise
e(gm_2)	square of geometric mean of $(y-k)$ if lnlsq; 1 otherwise
e(cj)	position of constant in e(b) or 0 if no constant
e(delta)	relative change used to compute derivatives
e(rank)	rank of e(V)
e(ic)	number of iterations
e(converge)	1 if converged, 0 otherwise

Macros

e(cmd)	nl
e(cmdline)	command as typed
e(depvar)	name of dependent variable
e(wtype)	weight type
e(wexp)	weight expression
e(title)	title in estimation output
e(title_2)	secondary title in estimation output
e(clustvar)	name of cluster variable
e(hac_kernel)	HAC kernel
e(hac_lag)	HAC lag
e(vce)	*vcetype* specified in vce()
e(vcetype)	title used to label Std. Err.
e(type)	1 = interactively entered expression 2 = substitutable expression program 3 = function evaluator program
e(sexp)	substitutable expression
e(params)	names of parameters
e(funcprog)	function evaluator program
e(rhs)	contents of variables()
e(properties)	b V
e(predict)	program used to implement predict
e(marginsnotok)	predictions disallowed by margins

Matrices

e(b)	coefficient vector
e(init)	initial values vector
e(V)	variance–covariance matrix of the estimators

Functions

e(sample)	marks estimation sample

Methods and formulas

`nl` is implemented as an ado-file.

The derivation here is based on Davidson and MacKinnon (2004, chap. 6). Let $\boldsymbol{\beta}$ denote the $k \times 1$ vector of parameters, and write the regression function using matrix notation as $\mathbf{y} = \mathbf{f}(\mathbf{x}, \boldsymbol{\beta}) + \mathbf{u}$ so that the objective function can be written as

$$\text{SSR}(\boldsymbol{\beta}) = \{\mathbf{y} - \mathbf{f}(\mathbf{x}, \boldsymbol{\beta})\}' \mathbf{D} \{\mathbf{y} - \mathbf{f}(\mathbf{x}, \boldsymbol{\beta})\}$$

The $\mathbf{D}$ matrix contains the weights and is defined in [R] **regress**; if no weights are specified, then $\mathbf{D}$ is the $N \times N$ identity matrix. Taking a second-order Taylor series expansion centered at $\boldsymbol{\beta}_0$ yields

$$\text{SSR}(\boldsymbol{\beta}) \approx \text{SSR}(\boldsymbol{\beta}_0) + \mathbf{g}'(\boldsymbol{\beta}_0)(\boldsymbol{\beta} - \boldsymbol{\beta}_0) + \frac{1}{2}(\boldsymbol{\beta} - \boldsymbol{\beta}_0)'\mathbf{H}(\boldsymbol{\beta}_0)(\boldsymbol{\beta} - \boldsymbol{\beta}_0) \tag{8}$$

where $\mathbf{g}(\boldsymbol{\beta}_0)$ denotes the $k \times 1$ gradient of $\text{SSR}(\boldsymbol{\beta})$ evaluated at $\boldsymbol{\beta}_0$ and $\mathbf{H}(\boldsymbol{\beta}_0)$ denotes the $k \times k$ Hessian of $\text{SSR}(\boldsymbol{\beta})$ evaluated at $\boldsymbol{\beta}_0$. Letting $\mathbf{X}$ denote the $N \times k$ matrix of derivatives of $\mathbf{f}(\mathbf{x}, \boldsymbol{\beta})$ with respect to $\boldsymbol{\beta}$, the gradient $\mathbf{g}(\boldsymbol{\beta})$ is

$$\mathbf{g}(\boldsymbol{\beta}) = -2\mathbf{X}'\mathbf{D}\mathbf{u} \tag{9}$$

$\mathbf{X}$ and $\mathbf{u}$ are obviously functions of $\boldsymbol{\beta}$, though for notational simplicity that dependence is not shown explicitly. The (m, n) element of the Hessian can be written

$$H_{mn}(\boldsymbol{\beta}) = -2\sum_{i=1}^{i=N} d_{ii} \left[\frac{\partial^2 f_i}{\partial\beta_m \partial\beta_n} u_i - X_{im}X_{in}\right] \tag{10}$$

where d_{ii} is the ith diagonal element of $\mathbf{D}$. As discussed in Davidson and MacKinnon (2004, chap. 6), the first term inside the brackets of (10) has expectation zero, so the Hessian can be approximated as

$$\mathbf{H}(\boldsymbol{\beta}) = 2\mathbf{X}'\mathbf{D}\mathbf{X} \tag{11}$$

Differentiating the Taylor series expansion of $\text{SSR}(\boldsymbol{\beta})$ shown in (8) yields the first-order condition for a minimum

$$\mathbf{g}(\boldsymbol{\beta}_0) + \mathbf{H}(\boldsymbol{\beta}_0)(\boldsymbol{\beta} - \boldsymbol{\beta}_0) = \mathbf{0}$$

which suggests the iterative procedure

$$\boldsymbol{\beta}_{j+1} = \boldsymbol{\beta}_j - \alpha\mathbf{H}^{-1}(\boldsymbol{\beta}_j)\mathbf{g}(\boldsymbol{\beta}_j) \tag{12}$$

where α is a "step size" parameter chosen at each iteration to improve convergence. Using (9) and (11), we can write (12) as

$$\boldsymbol{\beta}_{j+1} = \boldsymbol{\beta}_j + \alpha(\mathbf{X}'\mathbf{D}\mathbf{X})^{-1}\mathbf{X}'\mathbf{D}\mathbf{u} \tag{13}$$

where $\mathbf{X}$ and $\mathbf{u}$ are evaluated at $\boldsymbol{\beta}_j$. Apart from the scalar α, the second term on the right-hand side of (13) can be computed via a (weighted) regression of the columns of $\mathbf{X}$ on the errors. `nl` computes the derivatives numerically and then calls `regress`. At each iteration, α is set to one, and a candidate value $\boldsymbol{\beta}^*_{j+1}$ is computed by (13). If $\text{SSR}(\boldsymbol{\beta}^*_{j+1}) < \text{SSR}(\boldsymbol{\beta}_j)$, then $\boldsymbol{\beta}_{j+1} = \boldsymbol{\beta}^*_{j+1}$ and the iteration is complete. Otherwise, α is halved, a new $\boldsymbol{\beta}^*_{j+1}$ is calculated, and the process is repeated. Convergence is declared when $\alpha|\beta_{j+1,m}| \leq \epsilon(|\beta_{jm}| + \tau)$ for all $m = 1, \ldots, k$. `nl` uses $\tau = 10^{-3}$ and, by default, $\epsilon = 10^{-5}$, though you can specify an alternative value of ϵ with the `eps()` option.

As derived, for example, in Davidson and MacKinnon (2004, chap. 6), an expedient way to obtain the covariance matrix is to compute $\mathbf{u}$ and the columns of $\mathbf{X}$ at the final estimate $\widehat{\boldsymbol{\beta}}$ and then regress that $\mathbf{u}$ on $\mathbf{X}$. The covariance matrix of the estimated parameters of that regression serves as an estimate of $\mathrm{Var}(\widehat{\boldsymbol{\beta}})$. If that regression employs a robust covariance matrix estimator, then the covariance matrix for the parameters of the nonlinear regression will also be robust.

All other statistics are calculated analogously to those in linear regression, except that the nonlinear function $f(\mathbf{x}_i, \boldsymbol{\beta})$ plays the role of the linear function $\mathbf{x}_i'\boldsymbol{\beta}$. See [R] **regress**.

This command supports estimation with survey data. For details on VCEs with survey data, see [SVY] **variance estimation**.

Acknowledgments

The original version of `nl` was written by Patrick Royston of the MRC Clinical Trials Unit, London, and published in Royston (1992). Francesco Danuso's menu-driven nonlinear regression program (1991) provided the inspiration.

References

Atkinson, A. C. 1985. *Plots, Transformations, and Regression: An Introduction to Graphical Methods of Diagnostic Regression Analysis*. Oxford: Oxford University Press.

Danuso, F. 1991. sg1: Nonlinear regression command. *Stata Technical Bulletin* 1: 17–19. Reprinted in *Stata Technical Bulletin Reprints*, vol. 1, pp. 96–98. College Station, TX: Stata Press.

Davidson, R., and J. G. MacKinnon. 1993. *Estimation and Inference in Econometrics*. New York: Oxford University Press.

——. 2004. *Econometric Theory and Methods*. New York: Oxford University Press.

Gallant, A. R. 1987. *Nonlinear Statistical Models*. New York: Wiley.

Goldstein, R. 1992. srd7: Adjusted summary statistics for logarithmic regressions. *Stata Technical Bulletin* 5: 17–21. Reprinted in *Stata Technical Bulletin Reprints*, vol. 1, pp. 178–183. College Station, TX: Stata Press.

Kennedy, W. J., Jr., and J. E. Gentle. 1980. *Statistical Computing*. New York: Dekker.

Poi, B. P. 2008. Stata tip 58: nl is not just for nonlinear models. *Stata Journal* 8: 139–141.

Ratkowsky, D. A. 1983. *Nonlinear Regression Modeling: A Unified Practical Approach*. New York: Dekker.

Ross, G. J. S. 1987. *MLP User Manual, Release 3.08*. Oxford: Numerical Algorithms Group.

——. 1990. *Nonlinear Estimation*. New York: Springer.

Royston, P. 1992. sg7: Centile estimation command. *Stata Technical Bulletin* 8: 12–15. Reprinted in *Stata Technical Bulletin Reprints*, vol. 2, pp. 122–125. College Station, TX: Stata Press.

——. 1993. sg1.4: Standard nonlinear curve fits. *Stata Technical Bulletin* 11: 17. Reprinted in *Stata Technical Bulletin Reprints*, vol. 2, p. 121. College Station, TX: Stata Press.

Also see

[R] **nl postestimation** — Postestimation tools for nl

[R] **gmm** — Generalized method of moments estimation

[R] **ml** — Maximum likelihood estimation

[R] **nlcom** — Nonlinear combinations of estimators

[R] **nlsur** — Estimation of nonlinear systems of equations

[R] **regress** — Linear regression

[SVY] **svy estimation** — Estimation commands for survey data

[U] **20 Estimation and postestimation commands**

Title

nl postestimation — Postestimation tools for nl

Description

The following postestimation commands are available after `nl`:

Command	Description
`estat`	AIC, BIC, VCE, and estimation sample summary
`estat` (svy)	postestimation statistics for survey data
`estimates`	cataloging estimation results
`lincom`	point estimates, standard errors, testing, and inference for linear combinations of coefficients
`lrtest`[1]	likelihood-ratio test
`margins`[2]	marginal means, predictive margins, marginal effects, and average marginal effects
`marginsplot`	graph the results from margins (profile plots, interaction plots, etc.)
`nlcom`	point estimates, standard errors, testing, and inference for nonlinear combinations of coefficients
`predict`	predictions and residuals
`predictnl`	point estimates, standard errors, testing, and inference for generalized predictions
`test`	Wald tests of simple and composite linear hypotheses
`testnl`	Wald tests of nonlinear hypotheses

[1] `lrtest` is not appropriate with `svy` estimation results.
[2] You must specify the `variables()` option with `nl`.

See the corresponding entries in the *Base Reference Manual* for details, but see [SVY] **estat** for details about `estat` (svy).

Syntax for predict

`predict` [*type*] *newvar* [*if*] [*in*] [, *statistic*]

`predict` [*type*] { *stub** | $newvar_1$ ... $newvar_k$ } [*if*] [*in*] , `scores`

where k is the number of parameters in the model.

statistic	Description
Main	
`yhat`	fitted values; the default
`residuals`	residuals
`pr(`a`,`b`)`	$\Pr(y_j \mid a < y_j < b)$
`e(`a`,`b`)`	$E(y_j \mid a < y_j < b)$
`ystar(`a`,`b`)`	$E(y_j^*)$, $y_j^* = \max\{a, \min(y_j, b)\}$

These statistics are available both in and out of sample; type `predict ... if e(sample) ...` if wanted only for the estimation sample.

Menu

Statistics > Postestimation > Predictions, residuals, etc.

Options for predict

Main

`yhat`, the default, calculates the fitted values.

`residuals` calculates the residuals.

`pr(`*a*`,`*b*`)` calculates $\Pr(a < \mathbf{x}_j\mathbf{b} + u_j < b)$, the probability that $y_j|\mathbf{x}_j$ would be observed in the interval (a, b).

a and *b* may be specified as numbers or variable names; *lb* and *ub* are variable names;
`pr(20,30)` calculates $\Pr(20 < \mathbf{x}_j\mathbf{b} + u_j < 30)$;
`pr(`*lb*`,`*ub*`)` calculates $\Pr(lb < \mathbf{x}_j\mathbf{b} + u_j < ub)$; and
`pr(20,`*ub*`)` calculates $\Pr(20 < \mathbf{x}_j\mathbf{b} + u_j < ub)$.

a missing ($a \geq .$) means $-\infty$; `pr(.,30)` calculates $\Pr(-\infty < \mathbf{x}_j\mathbf{b} + u_j < 30)$;
`pr(`*lb*`,30)` calculates $\Pr(-\infty < \mathbf{x}_j\mathbf{b} + u_j < 30)$ in observations for which $lb \geq .$
and calculates $\Pr(lb < \mathbf{x}_j\mathbf{b} + u_j < 30)$ elsewhere.

b missing ($b \geq .$) means $+\infty$; `pr(20,.)` calculates $\Pr(+\infty > \mathbf{x}_j\mathbf{b} + u_j > 20)$;
`pr(20,`*ub*`)` calculates $\Pr(+\infty > \mathbf{x}_j\mathbf{b} + u_j > 20)$ in observations for which $ub \geq .$
and calculates $\Pr(20 < \mathbf{x}_j\mathbf{b} + u_j < ub)$ elsewhere.

`e(`*a*`,`*b*`)` calculates $E(\mathbf{x}_j\mathbf{b} + u_j \mid a < \mathbf{x}_j\mathbf{b} + u_j < b)$, the expected value of $y_j|\mathbf{x}_j$ conditional on $y_j|\mathbf{x}_j$ being in the interval (a, b), meaning that $y_j|\mathbf{x}_j$ is truncated. *a* and *b* are specified as they are for `pr()`.

`ystar(`*a*`,`*b*`)` calculates $E(y_j^*)$, where $y_j^* = a$ if $\mathbf{x}_j\mathbf{b} + u_j \leq a$, $y_j^* = b$ if $\mathbf{x}_j\mathbf{b} + u_j \geq b$, and $y_j^* = \mathbf{x}_j\mathbf{b} + u_j$ otherwise, meaning that y_j^* is censored. *a* and *b* are specified as they are for `pr()`.

`scores` calculates the scores. The jth new variable created will contain the score for the jth parameter in `e(b)`.

Methods and formulas

All postestimation commands listed above are implemented as ado-files.

Also see

[R] **nl** — Nonlinear least-squares estimation

[U] **20 Estimation and postestimation commands**

Title

nlcom — Nonlinear combinations of estimators

Syntax

Nonlinear combination of estimators—one expression

`nlcom` [*name*:]*exp* [, *options*]

Nonlinear combinations of estimators—more than one expression

`nlcom` ([*name*:]*exp*) [([*name*:]*exp*)...] [, *options*]

options	Description
`level(#)`	set confidence level; default is `level(95)`
`iterate(#)`	maximum number of iterations
`post`	post estimation results
display_options	control column formats and line width
`noheader`	suppress output header

`noheader` does not appear in the dialog box.

The second syntax means that if more than one expression is specified, each must be surrounded by parentheses. The optional *name* is any valid Stata name and labels the transformations.

exp is a possibly nonlinear expression containing

`_b[`*coef*`]`
`_b[`*eqno*`:`*coef*`]`
`[`*eqno*`]`*coef*
`[`*eqno*`]_b[`*coef*`]`

eqno is

##
name

coef identifies a coefficient in the model. *coef* is typically a variable name, a level indicator, an interaction indicator, or an interaction involving continuous variables. Level indicators identify one level of a factor variable and interaction indicators identify one combination of levels of an interaction; see [U] **11.4.3 Factor variables**. *coef* may contain time-series operators; see [U] **11.4.4 Time-series varlists**.

Distinguish between `[]`, which are to be typed, and [ ], which indicate optional arguments.

Menu

Statistics > Postestimation > Nonlinear combinations of estimates

Description

nlcom computes point estimates, standard errors, test statistics, significance levels, and confidence intervals for (possibly) nonlinear combinations of parameter estimates after any Stata estimation command. Results are displayed in the usual table format used for displaying estimation results. Calculations are based on the "delta method", an approximation appropriate in large samples.

nlcom can be used with svy estimation results; see [SVY] **svy postestimation**.

Options

level(*#*) specifies the confidence level, as a percentage, for confidence intervals. The default is level(95) or as set by set level; see [U] **20.7 Specifying the width of confidence intervals**.

iterate(*#*) specifies the maximum number of iterations used to find the optimal step size in calculating numerical derivatives of the transformation(s) with respect to the original parameters. By default, the maximum number of iterations is 100, but convergence is usually achieved after only a few iterations. You should rarely have to use this option.

post causes nlcom to behave like a Stata estimation (eclass) command. When post is specified, nlcom will post the vector of transformed estimators and its estimated variance–covariance matrix to e(). This option, in essence, makes the transformation permanent. Thus you could, after posting, treat the transformed estimation results in the same way as you would treat results from other Stata estimation commands. For example, after posting, you could redisplay the results by typing nlcom without any arguments, or use test to perform simultaneous tests of hypotheses on linear combinations of the transformed estimators; see [R] **test**.

Specifying post clears out the previous estimation results, which can be recovered only by refitting the original model or by storing the estimation results before running nlcom and then restoring them; see [R] **estimates store**.

display_options: cformat(%*fmt*), pformat(%*fmt*), sformat(%*fmt*), and nolstretch; see [R] **estimation options**.

The following option is available with nlcom but is not shown in the dialog box:

noheader suppresses the output header.

Remarks

Remarks are presented under the following headings:

Introduction
Basics
Using the post option
Reparameterizing ML estimators for univariate data
nlcom versus eform

Introduction

nlcom and predictnl both use the delta method. They take nonlinear transformations of the estimated parameter vector from some fitted model and apply the delta method to calculate the variance, standard error, Wald test statistic, etc., of the transformations. nlcom is designed for functions of the parameters, and predictnl is designed for functions of the parameters and of the data, that is, for predictions.

nlcom generalizes lincom (see [R] **lincom**) in two ways. First, nlcom allows the transformations to be nonlinear. Second, nlcom can be used to simultaneously estimate many transformations (whether linear or nonlinear) and to obtain the estimated variance–covariance matrix of these transformations.

Basics

In [R] **lincom**, the following regression was performed:

```
. use http://www.stata-press.com/data/r12/regress
. regress y x1 x2 x3
      Source |       SS       df       MS              Number of obs =     148
-------------+------------------------------           F(  3,   144) =   96.12
       Model |  3259.3561     3  1086.45203           Prob > F      =  0.0000
    Residual |  1627.56282   144  11.3025196           R-squared     =  0.6670
-------------+------------------------------           Adj R-squared =  0.6600
       Total |  4886.91892   147  33.2443464           Root MSE      =  3.3619

------------------------------------------------------------------------------
           y |      Coef.   Std. Err.      t    P>|t|     [95% Conf. Interval]
-------------+----------------------------------------------------------------
          x1 |   1.457113    1.07461     1.36   0.177     -.666934    3.581161
          x2 |   2.221682   .8610358     2.58   0.011     .5197797    3.923583
          x3 |   -.006139   .0005543   -11.08   0.000    -.0072345   -.0050435
       _cons |   36.10135   4.382693     8.24   0.000     27.43863    44.76407
------------------------------------------------------------------------------
```

Then lincom was used to estimate the difference between the coefficients of x1 and x2:

```
. lincom _b[x2] - _b[x1]
 ( 1)  - x1 + x2 = 0

------------------------------------------------------------------------------
           y |      Coef.   Std. Err.      t    P>|t|     [95% Conf. Interval]
-------------+----------------------------------------------------------------
         (1) |   .7645682   .9950282     0.77   0.444     -1.20218    2.731316
------------------------------------------------------------------------------
```

It was noted, however, that nonlinear expressions are not allowed with lincom:

```
. lincom _b[x2]/_b[x1]
not possible with test
r(131);
```

Nonlinear transformations are instead estimated using nlcom:

```
. nlcom _b[x2]/_b[x1]
       _nl_1:  _b[x2]/_b[x1]

------------------------------------------------------------------------------
           y |      Coef.   Std. Err.      t    P>|t|     [95% Conf. Interval]
-------------+----------------------------------------------------------------
       _nl_1 |   1.524714   .9812848     1.55   0.122    -.4148688    3.464297
------------------------------------------------------------------------------
```

❑ Technical note

The notation _b[*name*] is the standard way in Stata to refer to regression coefficients; see [U] **13.5 Accessing coefficients and standard errors**. Some commands, such as lincom and test, allow you to drop the _b[] and just refer to the coefficients by *name*. nlcom, however, requires the full specification _b[*name*].

❑

Returning to our linear regression example, nlcom also allows simultaneous estimation of more than one combination:

```
. nlcom (_b[x2]/_b[x1]) (_b[x3]/_b[x1]) (_b[x3]/_b[x2])

       _nl_1:  _b[x2]/_b[x1]
       _nl_2:  _b[x3]/_b[x1]
       _nl_3:  _b[x3]/_b[x2]

------------------------------------------------------------------------------
           y |      Coef.   Std. Err.      t    P>|t|     [95% Conf. Interval]
-------------+----------------------------------------------------------------
       _nl_1 |   1.524714   .9812848     1.55   0.122    -.4148688    3.464297
       _nl_2 |  -.0042131   .0033483    -1.26   0.210    -.0108313     .002405
       _nl_3 |  -.0027632   .0010695    -2.58   0.011    -.0048772   -.0006493
------------------------------------------------------------------------------
```

We can also label the transformations to produce more informative names in the estimation table:

```
. nlcom (ratio21:_b[x2]/_b[x1]) (ratio31:_b[x3]/_b[x1]) (ratio32:_b[x3]/_b[x2])

     ratio21:  _b[x2]/_b[x1]
     ratio31:  _b[x3]/_b[x1]
     ratio32:  _b[x3]/_b[x2]

------------------------------------------------------------------------------
           y |      Coef.   Std. Err.      t    P>|t|     [95% Conf. Interval]
-------------+----------------------------------------------------------------
     ratio21 |   1.524714   .9812848     1.55   0.122    -.4148688    3.464297
     ratio31 |  -.0042131   .0033483    -1.26   0.210    -.0108313     .002405
     ratio32 |  -.0027632   .0010695    -2.58   0.011    -.0048772   -.0006493
------------------------------------------------------------------------------
```

nlcom saves the vector of estimated combinations and its estimated variance–covariance matrix in r().

```
. matrix list r(b)

r(b)[1,3]
       ratio21     ratio31     ratio32
c1   1.5247143  -.00421315  -.00276324

. matrix list r(V)

symmetric r(V)[3,3]
             ratio21     ratio31     ratio32
ratio21    .96291982
ratio31   -.00287781    .00001121
ratio32   -.00014234   2.137e-06   1.144e-06
```

Using the post option

When used with the post option, nlcom saves the estimation vector and variance–covariance matrix in e(), making the transformation permanent:

```
. quietly nlcom (ratio21:_b[x2]/_b[x1]) (ratio31:_b[x3]/_b[x1])
> (ratio32:_b[x3]/_b[x2]), post

. matrix list e(b)

e(b)[1,3]
        ratio21      ratio31      ratio32
y1    1.5247143  -.00421315  -.00276324

. matrix list e(V)

symmetric e(V)[3,3]
             ratio21      ratio31      ratio32
ratio21   .96291982
ratio31  -.00287781    .00001121
ratio32  -.00014234    2.137e-06    1.144e-06
```

After posting, we can proceed as if we had just run a Stata estimation (eclass) command. For instance, we can replay the results,

```
. nlcom
```

y	Coef.	Std. Err.	t	P>|t|	[95% Conf.	Interval]
ratio21	1.524714	.9812848	1.55	0.122	-.4148688	3.464297
ratio31	-.0042131	.0033483	-1.26	0.210	-.0108313	.002405
ratio32	-.0027632	.0010695	-2.58	0.011	-.0048772	-.0006493

or perform other postestimation tasks in the transformed metric, this time making reference to the new "coefficients":

```
. display _b[ratio31]
-.00421315

. estat vce, correlation

Correlation matrix of coefficients of nlcom model

        e(V) |  ratio21   ratio31   ratio32
-------------+------------------------------
     ratio21 |   1.0000
     ratio31 |  -0.8759    1.0000
     ratio32 |  -0.1356    0.5969    1.0000

. test _b[ratio21] = 1

 ( 1)  ratio21 = 1

       F(  1,   144) =    0.29
            Prob > F =    0.5937
```

We see that testing _b[ratio21]=1 in the transformed metric is equivalent to testing using testnl _b[x2]/_b[x1]=1 in the original metric:

```
. quietly reg y x1 x2 x3

. testnl _b[x2]/_b[x1] = 1

  (1)  _b[x2]/_b[x1] = 1

            F(1, 144) =        0.29
             Prob > F =        0.5937
```

We needed to refit the regression model to recover the original parameter estimates.

❑ Technical note

In a previous technical note, we mentioned that commands such as `lincom` and `test` permit reference to *name* instead of `_b[`*name*`]`. This is not the case when `lincom` and `test` are used after `nlcom, post`. In the above, we used

```
. test _b[ratio21] = 1
```

rather than

```
. test ratio21 = 1
```

which would have returned an error. Consider this a limitation of Stata. For the shorthand notation to work, you need a variable named *name* in the data. In `nlcom`, however, *name* is just a coefficient label that does not necessarily correspond to any variable in the data.

❑

Reparameterizing ML estimators for univariate data

When run using only a response and no covariates, Stata's maximum likelihood (ML) estimation commands will produce ML estimates of the parameters of some assumed univariate distribution for the response. The parameterization, however, is usually not one we are used to dealing with in a nonregression setting. In such cases, `nlcom` can be used to transform the estimation results from a regression model to those from a maximum likelihood estimation of the parameters of a univariate probability distribution in a more familiar metric.

▷ Example 1

Consider the following univariate data on Y = # of traffic accidents at a certain intersection in a given year:

```
. use http://www.stata-press.com/data/r12/trafint
. summarize accidents
```

Variable	Obs	Mean	Std. Dev.	Min	Max
accidents	12	13.83333	14.47778	0	41

A quick glance of the output from `summarize` leads us to quickly reject the assumption that Y is distributed as Poisson because the estimated variance of Y is much greater than the estimated mean of Y.

Instead, we choose to model the data as univariate negative binomial, of which a common parameterization is

$$\Pr(Y=y)=\frac{\Gamma(r+y)}{\Gamma(r)\Gamma(y+1)}p^r(1-p)^y \qquad 0\le p\le 1,\quad r>0,\quad y=0,1,\dots$$

with

$$E(Y)=\frac{r(1-p)}{p} \qquad \text{Var}(Y)=\frac{r(1-p)}{p^2}$$

There exist no closed-form solutions for the maximum likelihood estimates of p and r, yet they may be estimated by the iterative method of Newton–Raphson. One way to get these estimates would be to write our own Newton–Raphson program for the negative binomial. Another way would be to write our own ML evaluator; see [R] **ml**.

The easiest solution, however, would be to use Stata's existing negative binomial ML regression command, nbreg. The only problem with this solution is that nbreg estimates a different parameterization of the negative binomial, but we can worry about that later.

```
. nbreg accidents
Fitting Poisson model:
Iteration 0:   log likelihood = -105.05361
Iteration 1:   log likelihood = -105.05361
Fitting constant-only model:
Iteration 0:   log likelihood = -43.948619
Iteration 1:   log likelihood = -43.891483
Iteration 2:   log likelihood =  -43.89144
Iteration 3:   log likelihood =  -43.89144
Fitting full model:
Iteration 0:   log likelihood =  -43.89144
Iteration 1:   log likelihood =  -43.89144
Negative binomial regression                      Number of obs   =         12
                                                  LR chi2(0)      =       0.00
Dispersion     = mean                             Prob > chi2     =          .
Log likelihood = -43.89144                        Pseudo R2       =     0.0000
```

accidents	Coef.	Std. Err.	z	P>|z|	[95% Conf.	Interval]
_cons	2.627081	.3192233	8.23	0.000	2.001415	3.252747
/lnalpha	.1402425	.4187147			-.6804233	.9609083
alpha	1.150553	.4817534			.5064026	2.61407

```
Likelihood-ratio test of alpha=0:  chibar2(01) =  122.32 Prob>=chibar2 = 0.000
. nbreg, coeflegend
Negative binomial regression                      Number of obs   =         12
                                                  LR chi2(0)      =       0.00
Dispersion     = mean                             Prob > chi2     =          .
Log likelihood = -43.89144                        Pseudo R2       =     0.0000
```

accidents	Coef.	Legend
_cons	2.627081	_b[accidents:_cons]
/lnalpha	.1402425	_b[lnalpha:_cons]
alpha	1.150553	

```
Likelihood-ratio test of alpha=0:  chibar2(01) =  122.32 Prob>=chibar2 = 0.000
```

From this output, we see that, when used with univariate data, nbreg estimates a regression intercept, β_0, and the logarithm of some parameter α. This parameterization is useful in regression models: β_0 is the intercept meant to be augmented with other terms of the linear predictor, and α is an overdispersion parameter used for comparison with the Poisson regression model.

However, we need to transform $(\beta_0, \ln\alpha)$ to (p, r). Examining *Methods and formulas* of [R] **nbreg** reveals the transformation as

$$p = \{1 + \alpha \exp(\beta_0)\}^{-1} \qquad r = \alpha^{-1}$$

which we apply using nlcom:

```
. nlcom (p:1/(1 + exp([lnalpha]_b[_cons] + _b[_cons])))
> (r:exp(-[lnalpha]_b[_cons]))
           p:  1/(1 + exp([lnalpha]_b[_cons] + _b[_cons]))
           r:  exp(-[lnalpha]_b[_cons])
```

accidents	Coef.	Std. Err.	z	P>|z|	[95% Conf.	Interval]
p	.0591157	.0292857	2.02	0.044	.0017168	.1165146
r	.8691474	.3639248	2.39	0.017	.1558679	1.582427

Given the invariance of maximum likelihood estimators and the properties of the delta method, the above parameter estimates, standard errors, etc., are precisely those we would have obtained had we instead performed the Newton–Raphson optimization in the (p, r) metric.

◁

❏ Technical note

Note how we referred to the estimate of $\ln\alpha$ above as `[lnalpha]_b[_cons]`. This is not entirely evident from the output of `nbreg`, which is why we redisplayed the results using the `coeflegend` option so that we would know how to refer to the coefficients; [U] **13.5 Accessing coefficients and standard errors**.

❏

nlcom versus eform

Many Stata estimation commands allow you to display exponentiated regression coefficients, some by default, some optionally. Known as "`eform`" in Stata terminology, this reparameterization serves many uses: it gives odds ratios for logistic models, hazard ratios in survival models, incidence-rate ratios in Poisson models, and relative-risk ratios in multinomial logit models, to name a few.

For example, consider the following estimation taken directly from the technical note in [R] **poisson**:

```
. use http://www.stata-press.com/data/r12/airline
. gen lnN = ln(n)
. poisson injuries XYZowned lnN
Iteration 0:   log likelihood = -22.333875
Iteration 1:   log likelihood = -22.332276
Iteration 2:   log likelihood = -22.332276
Poisson regression                                Number of obs   =          9
                                                  LR chi2(2)      =      19.15
                                                  Prob > chi2     =     0.0001
Log likelihood = -22.332276                       Pseudo R2       =     0.3001
```

injuries	Coef.	Std. Err.	z	P>|z|	[95% Conf.	Interval]
XYZowned	.6840667	.3895877	1.76	0.079	-.0795111	1.447645
lnN	1.424169	.3725155	3.82	0.000	.6940517	2.154285
_cons	4.863891	.7090501	6.86	0.000	3.474178	6.253603

When we replay results and specify the `irr` (incidence-rate ratios) option,

```
. poisson, irr
Poisson regression                                Number of obs   =          9
                                                  LR chi2(2)      =      19.15
                                                  Prob > chi2     =     0.0001
Log likelihood = -22.332276                       Pseudo R2       =     0.3001
```

injuries	IRR	Std. Err.	z	P>|z|	[95% Conf.	Interval]
XYZowned	1.981921	.7721322	1.76	0.079	.9235678	4.253085
lnN	4.154402	1.547579	3.82	0.000	2.00181	8.621728
_cons	129.5272	91.84126	6.86	0.000	32.2713	519.8828

we obtain the exponentiated regression coefficients and their estimated standard errors.

Contrast this with what we obtain if we exponentiate the coefficients manually by using `nlcom`:

```
. nlcom (E_XYZowned:exp(_b[XYZowned])) (E_lnN:exp(_b[lnN]))
  E_XYZowned:  exp(_b[XYZowned])
       E_lnN:  exp(_b[lnN])
```

injuries	Coef.	Std. Err.	z	P>|z|	[95% Conf.	Interval]
E_XYZowned	1.981921	.7721322	2.57	0.010	.4685701	3.495273
E_lnN	4.154402	1.547579	2.68	0.007	1.121203	7.187602

There are three things to note when comparing `poisson, irr` (and `eform` in general) with `nlcom`:

1. The exponentiated coefficients and standard errors are identical. This is certainly good news.
2. The Wald test statistic (`z`) and level of significance are different. When using `poisson, irr` and other related `eform` options, the Wald test does not change from what you would have obtained without the `eform` option, and you can see this by comparing both versions of the `poisson` output given previously.

   When you use `eform`, Stata knows that what is usually desired is a test of

   $$H_0\colon \exp(\beta) = 1$$

   and not the uninformative-by-comparison

   $$H_0\colon \exp(\beta) = 0$$

   The test of $H_0\colon \exp(\beta) = 1$ is asymptotically equivalent to a test of $H_0\colon \beta = 0$, the Wald test in the original metric, but the latter has better small-sample properties. Thus if you specify `eform`, you get a test of $H_0\colon \beta = 0$.

   `nlcom`, however, is general. It does not attempt to infer the test of greatest interest for a given transformation, and so a test of

   $$H_0\colon \text{transformed coefficient} = 0$$

   is always given, regardless of the transformation.
3. You may be surprised to see that, even though the coefficients and standard errors are identical, the confidence intervals (both 95%) are different.

`eform` confidence intervals are standard confidence intervals with the endpoints transformed. For example, the confidence interval for the coefficient on `lnN` is $[0.694, 2.154]$, whereas the confidence interval for the incidence-rate ratio due to `lnN` is $[\exp(0.694), \exp(2.154)] = [2.002, 8.619]$, which, except for some roundoff error, is what we see from the output of `poisson, irr`. For exponentiated coefficients, confidence intervals based on transform-the-endpoints methodology generally have better small-sample properties than their asymptotically equivalent counterparts.

The transform-the-endpoints method, however, gives valid coverage only when the transformation is monotonic. `nlcom` uses a more general and asymptotically equivalent method for calculating confidence intervals, as described in *Methods and formulas*.

Saved results

`nlcom` saves the following in `r()`:

Scalars

`r(N)`	number of observations
`r(df_r)`	residual degrees of freedom

Matrices

`r(b)`	vector of transformed coefficients
`r(V)`	estimated variance–covariance matrix of the transformed coefficients

If `post` is specified, `nlcom` also saves the following in `e()`:

Scalars

`e(N)`	number of observations
`e(df_r)`	residual degrees of freedom
`e(N_strata)`	number of strata L, if used after `svy`
`e(N_psu)`	number of sampled PSUs n, if used after `svy`
`e(rank)`	rank of `e(V)`

Macros

`e(cmd)`	`nlcom`
`e(predict)`	program used to implement `predict`
`e(properties)`	`b V`

Matrices

`e(b)`	vector of transformed coefficients
`e(V)`	estimated variance–covariance matrix of the transformed coefficients
`e(V_srs)`	simple-random-sampling-without-replacement (co)variance $\widehat{V}_{\text{srswor}}$, if `svy`
`e(V_srswr)`	simple-random-sampling-with-replacement (co)variance $\widehat{V}_{\text{srswr}}$, if `svy` and `fpc()`
`e(V_msp)`	misspecification (co)variance $\widehat{V}_{\text{msp}}$, if `svy` and available

Functions

`e(sample)`	marks estimation sample

Methods and formulas

`nlcom` is implemented as an ado-file.

Given a $1 \times k$ vector of parameter estimates, $\widehat{\boldsymbol{\theta}} = (\widehat{\theta}_1, \dots, \widehat{\theta}_k)$, consider the estimated p-dimensional transformation

$$g(\widehat{\boldsymbol{\theta}}) = [g_1(\widehat{\boldsymbol{\theta}}), g_2(\widehat{\boldsymbol{\theta}}), \dots, g_p(\widehat{\boldsymbol{\theta}})]$$

The estimated variance–covariance of $g(\widehat{\boldsymbol{\theta}})$ is given by

$$\widehat{\text{Var}}\left\{g(\widehat{\boldsymbol{\theta}})\right\} = \mathbf{GVG}'$$

where $\mathbf{G}$ is the $p \times k$ matrix of derivatives for which

$$\mathbf{G}_{ij} = \left.\frac{\partial g_i(\boldsymbol{\theta})}{\partial \theta_j}\right|_{\boldsymbol{\theta}=\widehat{\boldsymbol{\theta}}} \qquad i = 1, \ldots, p \qquad j = 1, \ldots, k$$

and $\mathbf{V}$ is the estimated variance–covariance matrix of $\widehat{\boldsymbol{\theta}}$. Standard errors are obtained as the square roots of the variances.

The Wald test statistic for testing

$$H_0\colon g_i(\boldsymbol{\theta}) = 0$$

versus the two-sided alternative is given by

$$Z_i = \frac{g_i(\widehat{\boldsymbol{\theta}})}{\left[\widehat{\text{Var}}_{ii}\left\{g(\widehat{\boldsymbol{\theta}})\right\}\right]^{1/2}}$$

When the variance–covariance matrix of $\widehat{\boldsymbol{\theta}}$ is an asymptotic covariance matrix, Z_i is approximately distributed as Gaussian. For linear regression, Z_i is taken to be approximately distributed as $t_{1,r}$ where r is the residual degrees of freedom from the original fitted model.

A $(1-\alpha) \times 100\%$ confidence interval for $g_i(\boldsymbol{\theta})$ is given by

$$g_i(\widehat{\boldsymbol{\theta}}) \pm z_{\alpha/2}\left[\widehat{\text{Var}}_{ii}\left\{g(\widehat{\boldsymbol{\theta}})\right\}\right]^{1/2}$$

for those cases where Z_i is Gaussian and

$$g_i(\widehat{\boldsymbol{\theta}}) \pm t_{\alpha/2,r}\left[\widehat{\text{Var}}_{ii}\left\{g(\widehat{\boldsymbol{\theta}})\right\}\right]^{1/2}$$

for those cases where Z_i is t distributed. z_p is the $1-p$ quantile of the standard normal distribution, and $t_{p,r}$ is the $1-p$ quantile of the t distribution with r degrees of freedom.

References

Feiveson, A. H. 1999. FAQ: What is the delta method and how is it used to estimate the standard error of a transformed parameter? http://www.stata.com/support/faqs/stat/deltam.html.

Gould, W. W. 1996. crc43: Wald test of nonlinear hypotheses after model estimation. *Stata Technical Bulletin* 29: 2–4. Reprinted in *Stata Technical Bulletin Reprints*, vol. 5, pp. 15–18. College Station, TX: Stata Press.

Oehlert, G. W. 1992. A note on the delta method. *American Statistician* 46: 27–29.

Phillips, P. C. B., and J. Y. Park. 1988. On the formulation of Wald tests of nonlinear restrictions. *Econometrica* 56: 1065–1083.

Also see

[R] **lincom** — Linear combinations of estimators

[R] **predictnl** — Obtain nonlinear predictions, standard errors, etc., after estimation

[R] **test** — Test linear hypotheses after estimation

[R] **testnl** — Test nonlinear hypotheses after estimation

[U] **20 Estimation and postestimation commands**

Title

nlogit — Nested logit regression

Syntax

Nested logit regression

`nlogit` *depvar* [*indepvars*] [*if*] [*in*] [*weight*] [|| *lev1_equation*
[|| *lev2_equation* ...]] || *altvar*: [*byaltvarlist*] , `case(`*varname*`)` [*options*]

where the syntax of *lev#_equation* is

altvar: [*byaltvarlist*] [, `base(`*#* | *lbl*`)` `estconst`]

Create variable based on specification of branches

`nlogitgen` *newaltvar* = *altvar* (*branchlist*) [, `nolog`]

where *branchlist* is

branch, *branch* [, *branch* ...]

and *branch* is

[*label*:] *alternative* [| *alternative* [| *alternative* ...]]

Display tree structure

`nlogittree` *altvarlist* [*if*] [*in*] [*weight*] [, `choice(`*depvar*`)` `nolabel` `nobranches`]

options	Description
Model	
* case(*varname*)	use *varname* to identify cases
base(# | *lbl*)	use the specified level or label of *altvar* as the base alternative for the bottom level
noconstant	suppress the constant terms for the bottom-level alternatives
nonnormalized	use the nonnormalized parameterization
altwise	use alternativewise deletion instead of casewise deletion
constraints(*constraints*)	apply specified linear constraints
collinear	keep collinear variables
SE/Robust	
vce(*vcetype*)	*vcetype* may be oim, robust, cluster *clustvar*, bootstrap, or jackknife
Reporting	
level(#)	set confidence level; default is level(95)
notree	suppress display of tree-structure output; see also nolabel and nobranches
nocnsreport	do not display constraints
display_options	control column formats and line width
Maximization	
maximize_options	control the maximization process; seldom used

* case(*varname*) is required.

bootstrap, by, jackknife, statsby, and xi are allowed; see [U] **11.1.10 Prefix commands**.

Weights are not allowed with the bootstrap prefix; see [R] **bootstrap**.

fweights, iweights, and pweights are allowed with nlogit, and fweights are allowed with nlogittree; see [U] **11.1.6 weight**. Weights for nlogit must be constant within case.

See [U] **20 Estimation and postestimation commands** for more capabilities of estimation commands.

Menu

nlogit

Statistics > Categorical outcomes > Nested logit regression

nlogitgen

Statistics > Categorical outcomes > Setup for nested logit regression

nlogittree

Statistics > Categorical outcomes > Display nested logit tree structure

Description

nlogit performs full information maximum-likelihood estimation for nested logit models. These models relax the assumption of independently distributed errors and the independence of irrelevant alternatives inherent in conditional and multinomial logit models by clustering similar alternatives into nests.

By default, nlogit uses a parameterization that is consistent with random utility maximization (RUM). Before version 10 of Stata, a nonnormalized version of the nested logit model was fit, which you can request by specifying the nonnormalized option.

You must use nlogitgen to generate a new categorical variable to specify the branches of the decision tree before calling nlogit.

Options

Specification and options for lev#_equation

altvar is a variable identifying alternatives at this level of the hierarchy.

byaltvarlist specifies the variables to be used to compute the by-alternative regression coefficients for that level. For each variable specified in the variable list, there will be one regression coefficient for each alternative of that level of the hierarchy. If the variable is constant across each alternative (a case-specific variable), the regression coefficient associated with the base alternative is not identifiable. These regression coefficients are labeled as (base) in the regression table. If the variable varies among the alternatives, a regression coefficient is estimated for each alternative.

base(*#* | *lbl*) can be specified in each level equation where it identifies the base alternative to be used at that level. The default is the alternative that has the highest frequency.

If vce(bootstrap) or vce(jackknife) is specified, you must specify the base alternative for each level that has a *byaltvarlist* or if the constants will be estimated. Doing so ensures that the same model is fit with each call to nlogit.

estconst applies to all the level equations except the bottom-level equation. Specifying estconst requests that constants for each alternative (except the base alternative) be estimated. By default, no constant is estimated at these levels. Constants can be estimated in only one level of the tree hierarchy. If you specify estconst for one of the level equations, you must specify noconstant for the bottom-level equation.

Options for nlogit

Model

case(*varname*) specifies the variable that identifies each case. case() is required.

base(*#* | *lbl*) can be specified in each level equation where it identifies the base alternative to be used at that level. The default is the alternative that has the highest frequency.

If vce(bootstrap) or vce(jackknife) is specified, you must specify the base alternative for each level that has a *byaltvarlist* or if the constants will be estimated. Doing so ensures that the same model is fit with each call to nlogit.

noconstant applies only to the equation defining the bottom level of the hierarchy. By default, constants are estimated for each alternative of *altvar*, less the base alternative. To suppress the constant terms for this level, specify noconstant. If you do not specify noconstant, you cannot specify estconst for the higher-level equations.

nonnormalized requests a nonnormalized parameterization of the model that does not scale the inclusive values by the degree of dissimilarity of the alternatives within each nest. Use this option to replicate results from older versions of Stata. The default is to use the RUM–consistent parameterization.

`altwise` specifies that alternativewise deletion be used when marking out observations because of missing values in your variables. The default is to use casewise deletion. This option does not apply to observations that are marked out by the `if` or `in` qualifier or the `by` prefix.

`constraints(`*constraints*`)`; see [R] **estimation options**.

The inclusive-valued/dissimilarity parameters are parameterized as `ml` ancillary parameters. They are labeled as [*alternative*`_tau`]`_const`, where *alternative* is one of the alternatives defining a branch in the tree. To constrain the inclusive-valued/dissimilarity parameter for alternative `a1` to be, say, equal to alternative `a2`, you would use the following syntax:

```
. constraint 1 [a1_tau]_cons = [a2_tau]_cons
. nlogit ..., constraints(1)
```

`collinear` prevents collinear variables from being dropped. Use this option when you know that you have collinear variables and you are applying `constraints()` to handle the rank reduction. See [R] **estimation options** for details on using `collinear` with `constraints()`.

`nlogit` will not allow you to specify an independent variable in more than one level equation. Specifying the `collinear` option will allow execution to proceed in this case, but it is your responsibility to ensure that the parameters are identified.

SE/Robust

`vce(`*vcetype*`)` specifies the type of standard error reported, which includes types that are derived from asymptotic theory, that are robust to some kinds of misspecification, that allow for intragroup correlation, and that use bootstrap or jackknife methods; see [R] ***vce_option***.

If `vce(robust)` or `vce(cluster` *clustvar*`)` is specified, the likelihood-ratio test for the independence of irrelevant alternatives (IIA) is not computed.

Reporting

`level(`*#*`)`; see [R] **estimation options**.

`notree` specifies that the tree structure of the nested logit model not be displayed. See also `nolabel` and `nobranches` below for when `notree` is not specified.

`nocnsreport`; see [R] **estimation options**.

display_options: `cformat(%`*fmt*`)`, `pformat(%`*fmt*`)`, `sformat(%`*fmt*`)`, and `nolstretch`; see [R] **estimation options**.

Maximization

maximize_options: `difficult`, `technique(`*algorithm_spec*`)`, `iterate(`*#*`)`, [`no`]`log`, `trace`, `gradient`, `showstep`, `hessian`, `showtolerance`, `tolerance(`*#*`)`, `ltolerance(`*#*`)`, `nrtolerance(`*#*`)`, `nonrtolerance`, and `from(`*init_specs*`)`; see [R] **maximize**. These options are seldom used.

The `technique(bhhh)` option is not allowed.

Specification and options for nlogitgen

newaltvar and *altvar* are variables identifying alternatives at each level of the hierarchy.

label defines a label to associate with the branch. If no label is given, a numeric value is used.

alternative specifies an alternative, of *altvar* specified in the syntax, to be included in the branch. It is either a numeric value or the label associated with that value. An example of `nlogitgen` is

```
. nlogitgen type = restaurant(fast: 1 | 2,
> family: CafeEccell | LosNortenos | WingsNmore, fancy: 6 | 7)
```

`nolog` suppresses the display of the iteration log.

Specification and options for nlogittree

Main

altvarlist is a list of alternative variables that define the tree hierarchy. The first variable must define bottom-level alternatives, and the order continues to the variable defining the top-level alternatives.

`choice(`*depvar*`)` defines the choice indicator variable and forces `nlogittree` to compute and display choice frequencies for each bottom-level alternative.

`nolabel` forces `nlogittree` to suppress value labels in tree-structure output.

`nobranches` forces `nlogittree` to suppress drawing branches in the tree-structure output.

Remarks

Remarks are presented under the following headings:

Introduction

`nlogit` performs full information maximum-likelihood estimation for nested logit models. These models relax the assumption of independently distributed errors and the IIA inherent in conditional and multinomial logit models by clustering similar alternatives into nests. Because the nested logit model is a direct generalization of the alternative-specific conditional logit model (also known as McFadden's choice model), you may want to read [R] **asclogit** before continuing.

By default, `nlogit` uses a parameterization that is consistent with RUM. Before version 10 of Stata, a nonnormalized version of the nested logit model was fit, which you can request by specifying the `nonnormalized` option. We recommend using the RUM-consistent version of the model for new projects because it is based on a sound model of consumer behavior.

McFadden (1977, 1981) showed how this model can be derived from a rational choice framework. Amemiya (1985, chap. 9) contains a nice discussion of how this model can be derived under the assumption of utility maximization. Hensher, Rose, and Greene (2005) provide a lucid introduction to choice models including nested logit.

Throughout this entry, we consider a model of restaurant choice. We begin by introducing the data.

▷ Example 1

We have fictional data on 300 families and their choice of seven local restaurants. Freebirds and Mama's Pizza are fast food restaurants; Café Eccell, Los Norteños, and Wings 'N More are family restaurants; and Christopher's and Mad Cows are fancy restaurants. We want to model the decision of where to eat as a function of household income (`income`, in thousands of dollars), the number of children in the household (kids), the rating of the restaurant according to a local restaurant guide (rating, coded 0–5), the average meal cost per person (cost), and the distance between the household and the restaurant (distance, in miles). income and kids are attributes of the family, rating is an attribute of the alternative (the restaurant), and cost and distance are attributes of the alternative as perceived by the families—that is, each family has its own cost and distance for each restaurant.

We begin by loading the data and listing some of the variables for the first three families:

```
. use http://www.stata-press.com/data/r12/restaurant

. describe

Contains data from http://www.stata-press.com/data/r12/restaurant.dta
  obs:         2,100
 vars:             8                          10 Mar 2011 01:17
 size:        67,200
-------------------------------------------------------------------------------
              storage  display     value
variable name   type   format      label      variable label
-------------------------------------------------------------------------------
family_id       float  %9.0g                  family ID
restaurant      float  %12.0g      names      choices of restaurants
income          float  %9.0g                  household income
cost            float  %9.0g                  average meal cost per person
kids            float  %9.0g                  number of kids in the household
rating          float  %9.0g                  ratings in local restaurant
                                                guide
distance        float  %9.0g                  distance between home and
                                                restaurant
chosen          float  %9.0g                  0 no 1 yes
-------------------------------------------------------------------------------
Sorted by:  family_id
```

```
. list family_id restaurant chosen kids rating distance in 1/21, sepby(fam)
> abbrev(10)
```

	family_id	restaurant	chosen	kids	rating	distance
1.	1	Freebirds	1	1	0	1.245553
2.	1	MamasPizza	0	1	1	2.82493
3.	1	CafeEccell	0	1	2	4.21293
4.	1	LosNortenos	0	1	3	4.167634
5.	1	WingsNmore	0	1	2	6.330531
6.	1	Christophers	0	1	4	10.19829
7.	1	MadCows	0	1	5	5.601388
8.	2	Freebirds	0	3	0	4.162657
9.	2	MamasPizza	0	3	1	2.865081
10.	2	CafeEccell	0	3	2	5.337799
11.	2	LosNortenos	1	3	3	4.282864
12.	2	WingsNmore	0	3	2	8.133914
13.	2	Christophers	0	3	4	8.664631
14.	2	MadCows	0	3	5	9.119597
15.	3	Freebirds	1	3	0	2.112586
16.	3	MamasPizza	0	3	1	2.215329
17.	3	CafeEccell	0	3	2	6.978715
18.	3	LosNortenos	0	3	3	5.117877
19.	3	WingsNmore	0	3	2	5.312941
20.	3	Christophers	0	3	4	9.551273
21.	3	MadCows	0	3	5	5.539806

Because each family chose among seven restaurants, there are 7 observations in the dataset for each family. The variable `chosen` is coded 0/1, with 1 indicating the chosen restaurant and 0 otherwise.
◁

We could fit a conditional logit model to our data. Because `income` and `kids` are constant within each family, we would use the `asclogit` command instead of `clogit`. However, the conditional logit may be inappropriate. That model assumes that the random errors are independent, and as a result it forces the odds ratio of any two alternatives to be independent of the other alternatives, a property known as the IIA. We will discuss the IIA assumption in more detail later.

Assuming that unobserved shocks influencing a decision maker's attitude toward one alternative have no effect on his attitudes toward the other alternatives may seem innocuous, but often this assumption is too restrictive. Suppose that when a family was deciding which restaurant to visit, they were pressed for time because of plans to attend a movie later. The unobserved shock (being in a hurry) would raise the likelihood that the family goes to either fast food restaurant (Freebirds or Mama's Pizza). Similarly, another family might be choosing a restaurant to celebrate a birthday and therefore be inclined to attend a fancy restaurant (Christopher's or Mad Cows).

Nested logit models relax the independence assumption and allow us to group alternatives for which unobserved shocks may have concomitant effects. Here we suspect that restaurants should be grouped by type (fast, family, or fancy). The tree structure of a family's decision about where to eat might look like this:

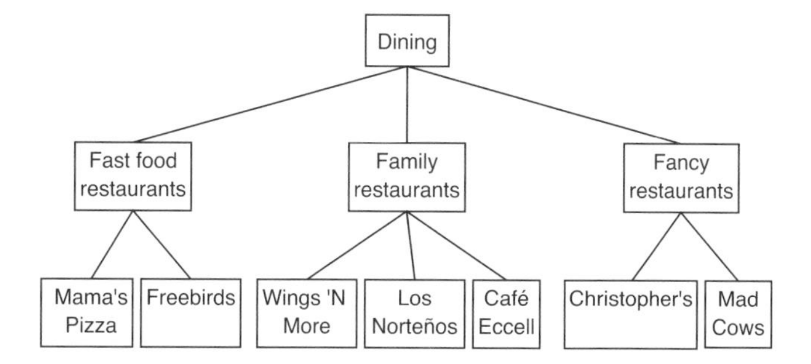

At the bottom of the tree are the individual restaurants, indicating that there are some random shocks that affect a family's decision to eat at each restaurant independently. Above the restaurants are the three types of restaurants, indicating that other random shocks affect the type of restaurant chosen. As is customary when drawing decision trees, at the top level is one box, representing the family making the decision.

We use the following terms to describe nested logit models.

level, or decision level, is the level or stage at which a decision is made. The example above has only two levels. In the first level, a type of restaurant is chosen—fast food, family, or fancy—and in the second level, a specific restaurant is chosen.

bottom level is the level where the final decision is made. In our example, this is when we choose a specific restaurant.

alternative set is the set of all possible alternatives at any given decision level.

bottom alternative set is the set of all possible alternatives at the bottom level. This concept is often referred to as the choice set in the economics-choice literature. In our example, the bottom alternative set is all seven of the specific restaurants.

alternative is a specific alternative within an alternative set. In the first level of our example, "fast food" is an alternative. In the second or bottom level, "Mad Cows" is an alternative. Not all alternatives within an alternative set are available to someone making a choice at a specific stage, only those that are nested within all higher-level decisions.

chosen alternative is the alternative from an alternative set that we observe someone having chosen.

❑ Technical note

Although decision trees in nested logit analysis are often interpreted as implying that the highest-level decisions are made first, followed by decisions at lower levels, and finally the decision among alternatives at the bottom level, no such temporal ordering is implied. See Hensher, Rose, and Greene (2005, chap. 13). In our example, we are not assuming that families first choose whether to attend a fast, family, or fancy restaurant and then choose the particular restaurant; we assume merely that they choose one of the seven restaurants.

❑

Data setup and the tree structure

To fit a nested logit model, you must first create a variable that defines the structure of your decision tree.

▷ Example 2

To run `nlogit`, we need to generate a categorical variable that identifies the first-level set of alternatives: fast food, family restaurants, or fancy restaurants. We can do so easily by using `nlogitgen`.

```
. nlogitgen type = restaurant(fast: Freebirds | MamasPizza,
> family: CafeEccell | LosNortenos| WingsNmore, fancy: Christophers | MadCows)
new variable type is generated with 3 groups
label list lb_type
lb_type:
           1 fast
           2 family
           3 fancy
. nlogittree restaurant type, choice(chosen)
tree structure specified for the nested logit model
 type    N       restaurant    N   k
─────────────────────────────────────
 fast    600 ─┬─ Freebirds     300  12
              └─ MamasPizza    300  15
 family  900 ─┬─ CafeEccell    300  78
              ├─ LosNortenos   300  75
              └─ WingsNmore    300  69
 fancy   600 ─┬─ Christophers  300  27
              └─ MadCows       300  24
─────────────────────────────────────
                      total  2100 300
k = number of times alternative is chosen
N = number of observations at each level
```

The new categorical variable is `type`, which takes on value 1 (fast) if `restaurant` is Freebirds or Mama's Pizza; value 2 (family) if `restaurant` is Café Eccell, Los Norteños, or Wings 'N More; and value 3 (fancy) otherwise. `nlogittree` displays the tree structure.

◁

❑ Technical note

We could also use values instead of value labels of `restaurant` in `nlogitgen`. Value labels are optional, and the default value labels for `type` are `type1`, `type2`, and `type3`. The vertical bar is also optional.

```
. use http://www.stata-press.com/data/r12/restaurant, clear
. nlogitgen type = restaurant(1 2, 3 4 5, 6 7)
new variable type is generated with 3 groups
label list lb_type
lb_type:
           1 type1
           2 type2
           3 type3
```

```
. nlogittree restaurant type
tree structure specified for the nested logit model
 type    N      restaurant    N
────────────────────────────────
 type1 600 ─┬─ Freebirds     300
            └─ MamasPizza    300
 type2 900 ─┬─ CafeEccell    300
            ├─ LosNortenos   300
            └─ WingsNmore    300
 type3 600 ─┬─ Christophers  300
            └─ MadCows       300
────────────────────────────────
                    total   2100
N = number of observations at each level
```

❑

In our dataset, every family was able to choose among all seven restaurants. However, in other applications some decision makers may not have been able to choose among all possible alternatives. For example, two cases may have choice hierarchies of

```
      case 1                           case 2
 type        restaurant          type        restaurant
───────────────────────          ───────────────────────
 fast      ─┬─ Freebirds         fast      ─┬─ Freebirds
            └─ MamasPizza                   └─ MamasPizza
 family    ─┬─ CafeEccell        family    ─┬─ LosNortenos
            ├─ LosNortenos                  └─ WingsNmore
            └─ WingsNmore
 fancy     ─┬─ Christophers      fancy     ─── Christophers
            └─ MadCows
```

where the second case does not have the restaurant alternatives Café Eccell or Mad Cows available to them. The only restriction is that the relationships between higher- and lower-level alternative sets be the same for all decision makers. In this two-level example, Freebirds and Mama's Pizza are classified as fast food restaurants for both cases; Café Eccell, Los Norteños, and Wings 'N More are family restaurants; and Christopher's and Mad Cows are fancy restaurants. `nlogit` requires only that hierarchy be maintained for all cases.

Estimation

▷ Example 3

With our `type` variable created that defines the three types of restaurants, we can now examine how the alternative-specific attributes (`cost`, `rating`, and `distance`) apply to the bottom alternative set (the seven restaurants) and how family-specific attributes (`income` and `kid`) apply to the alternative set at the first decision level (the three types of restaurants).

```
. nlogit chosen cost rating distance || type: income kids, base(family) ||
> restaurant:, noconstant case(family_id)
tree structure specified for the nested logit model
 type     N       restaurant     N   k
------------------------------------------
 fast    600 -+- Freebirds      300  12
              +- MamasPizza     300  15
 family  900 -+- CafeEccell     300  78
              |- LosNortenos    300  75
              +- WingsNmore     300  69
 fancy   600 -+- Christophers   300  27
              +- MadCows        300  24
------------------------------------------
                       total   2100 300
k = number of times alternative is chosen
N = number of observations at each level
Iteration 0:   log likelihood = -541.93581
  (output omitted)
Iteration 17:  log likelihood = -485.47331
RUM-consistent nested logit regression         Number of obs      =      2100
Case variable: family_id                       Number of cases    =       300
Alternative variable: restaurant               Alts per case: min =         7
                                                              avg =       7.0
                                                              max =         7
                                                  Wald chi2(7)    =     46.71
Log likelihood = -485.47331                       Prob > chi2     =    0.0000

------------------------------------------------------------------------------
      chosen |      Coef.   Std. Err.      z    P>|z|     [95% Conf. Interval]
-------------+----------------------------------------------------------------
restaurant   |
        cost |  -.1843847   .0933975    -1.97   0.048    -.3674404   -.0013289
      rating |    .463694   .3264935     1.42   0.156    -.1762215     1.10361
    distance |  -.3797474   .1003828    -3.78   0.000    -.5764941   -.1830007
------------------------------------------------------------------------------
type equations
-------------+----------------------------------------------------------------
fast         |
      income |  -.0266038   .0117306    -2.27   0.023    -.0495952   -.0036123
        kids |  -.0872584   .1385026    -0.63   0.529    -.3587184    .1842016
-------------+----------------------------------------------------------------
family       |
      income |          0  (base)
        kids |          0  (base)
-------------+----------------------------------------------------------------
fancy        |
      income |   .0461827   .0090936     5.08   0.000     .0283595    .0640059
        kids |  -.3959413   .1220356    -3.24   0.001    -.6351267   -.1567559
------------------------------------------------------------------------------
dissimilarity parameters
-------------+----------------------------------------------------------------
type         |
   /fast_tau |   1.712878    1.48685                     -1.201295    4.627051
 /family_tau |   2.505113   .9646351                       .614463    4.395763
  /fancy_tau |   4.099844   2.810123                     -1.407896    9.607583
------------------------------------------------------------------------------
LR test for IIA (tau = 1):           chi2(3) =     6.87   Prob > chi2 = 0.0762
------------------------------------------------------------------------------
```

First, let's examine how we called `nlogit`. The delimiters (`||`) separate equations. The first equation specifies the dependent variable, `chosen`, and three alternative-specific variables, `cost`, `rating`, and `distance`. We refer to these variables as alternative-specific because they vary among the bottom-level alternatives, the restaurants. We obtain one parameter estimate for each variable. These estimates are listed in the equation subtable labeled `restaurant`.

For the second equation, we specify the `type` variable. It identifies the first-level alternatives, the restaurant types. Following the colon after `type`, we specify two case-specific variables, `income` and `kids`. Here we obtain a parameter estimate for each variable for each alternative at this level. That is why we call these variable lists *by-alternative* variables. Because `income` and `kids` do not vary within each case, to identify the model one alternative's set of parameters must be set to zero. We specified the `base(family)` option with this equation to restrict the parameters for the `family` alternative.

The variable identifying the bottom-level alternatives, `restaurant`, is specified after the second equation delimiter. We do not specify any variables after the colon delimiter at this level. Had we specified variables here, we would have obtained an estimate for each variable in each equation. As we will see below, these variables parameterize the constant term in the utility equation for each bottom-level alternative. The `noconstant` option suppresses bottom-level alternative-specific constant terms.

Near the bottom of the output are the dissimilarity parameters, which measure the degree of correlation of random shocks within each of the three types of restaurants. Dissimilarity parameters greater than one imply that the model is inconsistent with RUM; Hensher, Rose, and Greene (2005, sec. 13.6) discuss this in detail. We will ignore the fact that all our dissimilarity parameters exceed one.

The conditional logit model is a special case of nested logit in which all the dissimilarity parameters are equal to one. At the bottom of the output, we find a likelihood-ratio test of this hypothesis. Here we have mixed evidence of the null hypothesis that all the parameters are one. Equivalently, the property known as the IIA imposed by the conditional logit model holds if and only if all dissimilarity parameters are equal to one. We discuss the IIA in more detail now.

◁

Testing for the IIA

The IIA is a property of the multinomial and conditional logit models that forces the odds of choosing one alternative over another to be independent of the other alternatives. For simplicity, suppose that a family was choosing only between Freebirds and Mama's Pizza, and the family was equally likely to choose either of the restaurants. The probability of going to each restaurant is 50%. Now suppose that Bill's Burritos opens up next door to Freebirds, which is also a burrito restaurant. If the IIA holds, then the probability of going to each restaurant must now be 33.33% so that the family remains equally likely to go to Mama's Pizza or Freebirds.

The IIA may sometimes be a plausible assumption. However, a more likely scenario would be for the probability of going to Mama's Pizza to remain at 50% and the probabilities of going to Freebirds and Bill's Burritos to be 25% each, because the two restaurants are next door to each other and serve the same food. Nested logit analysis would allow us to relax the IIA assumption of conditional logit. We could group Bill's Burritos and Freebirds into one nest that encompasses all burrito restaurants and create a second nest for pizzerias.

The IIA is a consequence of assuming that the errors are independent and identically distributed (i.i.d.). Because the errors are i.i.d., they cannot contain any alternative-specific unobserved information, and therefore adding a new alternative cannot affect the relationship between a pair of existing alternatives.

In the previous example, we saw that a joint test that the dissimilarity parameters were equal to one is one way to test for IIA. However, that test required us to specify a decision tree for the nested logit model, and different specifications could lead to conflicting results of the test. Hausman and McFadden (1984) suggest that if part of the choice set truly is irrelevant with respect to the other alternatives, omitting that subset from the conditional logit model will not lead to inconsistent estimates. Therefore, Hausman's (1978) specification test can be used to test for IIA, and this test will not be sensitive to the tree structure we specify for a nested logit model.

▷ Example 4

We want to test the IIA for the subset of family restaurants against the alternatives of fast food and fancy restaurants. To do so, we need to use Stata's `hausman` command; see [R] **hausman**.

We first run the estimation on the full bottom alternative set, save the results by using `estimates store`, and then run the estimation on the bottom alternative set, excluding the alternatives of family restaurants. We then run the `hausman` test.

```
. generate incFast = (type == 1) * income
. generate incFancy = (type == 3) * income
. generate kidFast = (type == 1) * kids
. generate kidFancy = (type == 3) * kids
. clogit chosen cost rating distance incFast incFancy kidFast kidFancy,
> group(family_id) nolog
Conditional (fixed-effects) logistic regression   Number of obs   =       2100
                                                  LR chi2(7)      =     189.73
                                                  Prob > chi2     =     0.0000
Log likelihood = -488.90834                       Pseudo R2       =     0.1625
```

chosen	Coef.	Std. Err.	z	P>|z|	[95% Conf.	Interval]
cost	-.1367799	.0358479	-3.82	0.000	-.2070404	-.0665193
rating	.3066622	.1418291	2.16	0.031	.0286823	.584642
distance	-.1977505	.0471653	-4.19	0.000	-.2901927	-.1053082
incFast	-.0390183	.0094018	-4.15	0.000	-.0574455	-.0205911
incFancy	.0407053	.0080405	5.06	0.000	.0249462	.0564644
kidFast	-.2398757	.1063674	-2.26	0.024	-.448352	-.0313994
kidFancy	-.3893862	.1143797	-3.40	0.001	-.6135662	-.1652061

```
. estimates store fullset
. clogit chosen cost rating distance incFast kidFast if type != 2,
> group(family_id) nolog
note: 222 groups (888 obs) dropped because of all positive or
      all negative outcomes.
Conditional (fixed-effects) logistic regression   Number of obs   =        312
                                                  LR chi2(5)      =      44.35
                                                  Prob > chi2     =     0.0000
Log likelihood = -85.955324                       Pseudo R2       =     0.2051
```

chosen	Coef.	Std. Err.	z	P>|z|	[95% Conf.	Interval]
cost	-.0616621	.067852	-0.91	0.363	-.1946496	.0713254
rating	.1659001	.2832041	0.59	0.558	-.3891698	.72097
distance	-.244396	.0995056	-2.46	0.014	-.4394234	-.0493687
incFast	-.0737506	.0177444	-4.16	0.000	-.108529	-.0389721
kidFast	.4105386	.2137051	1.92	0.055	-.0083157	.8293928

```
. hausman . fullset

                 ---- Coefficients ----
             |      (b)          (B)            (b-B)     sqrt(diag(V_b-V_B))
             |       .          fullset       Difference          S.E.
-------------+----------------------------------------------------------------
        cost |   -.0616621    -.1367799        .0751178        .0576092
      rating |    .1659001     .3066622       -.1407621        .2451308
    distance |    -.244396    -.1977505       -.0466456        .0876173
     incFast |   -.0737506    -.0390183       -.0347323         .015049
     kidFast |    .4105386    -.2398757        .6504143        .1853533
------------------------------------------------------------------------------
                           b = consistent under Ho and Ha; obtained from clogit
            B = inconsistent under Ha, efficient under Ho; obtained from clogit

    Test:  Ho:  difference in coefficients not systematic

                  chi2(5) = (b-B)'[(V_b-V_B)^(-1)](b-B)
                          =        10.70
                Prob>chi2 =      0.0577
                (V_b-V_B is not positive definite)
```

Similar to our findings in example 3, the results of the test of the IIA are mixed. We cannot reject the IIA at the commonly used 5% significance level, but we could at the 10% level. Substantively, a significant test result suggests that the odds of going to one of the fancy restaurants versus going to one of the fast food restaurants changes if we include the family restaurants in the alternative set and that a nested logit specification may be warranted.

◁

Nonnormalized model

Previous versions of Stata fit a nonnormalized nested logit model that is available via the `nonnormalized` option. The nonnormalized version is presented in, for example, Greene (2012, 768–770). Here we outline the differences between the RUM-consistent and nonnormalized models. Our discussion follows Heiss (2002) and assumes the decision tree has two levels, with M alternatives at the upper level and a total of J alternatives at the bottom level.

In a RUM framework, by consuming alternative j, decision maker i obtains utility

$$U_{ij} = V_{ij} + \epsilon_{ij} = \alpha_j + \mathbf{x}_{ij}\boldsymbol{\beta}_j + \mathbf{z}_i\boldsymbol{\gamma}_j + \epsilon_{ij}$$

where V_{ij} is the deterministic part of utility and ϵ_{ij} is the random part. $\mathbf{x}_{ij}$ are alternative-specific variables and $\mathbf{z}_i$ are case-specific variables. The set of errors $\epsilon_{i1}, \dots, \epsilon_{iJ}$ are assumed to follow the generalized extreme-value (GEV) distribution, which is a generalization of the type 1 extreme-value distribution that allows for alternatives within nests of the tree structure to be correlated. Let ρ_m denote the correlation in nest m, and define the dissimilarity parameter $\tau_m = \sqrt{1-\rho_m}$. $\tau_m = 0$ implies that the alternatives in nest m are perfectly correlated, whereas $\tau_m = 1$ implies independence.

The *inclusive value* for the mth nest corresponds to the expected value of the utility that decision maker i obtains by consuming an alternative in nest m. Denote this value by IV_m:

$$\mathrm{IV}_m = \ln \sum_{j \in B_m} \exp\left(V_k/\tau_m\right) \tag{1}$$

where B_m denotes the set of alternatives in nest m. Given the inclusive values, we can show that the probability that random-utility–maximizing decision maker i chooses alternative j is

$$\Pr_j = \frac{\exp\{V_j/\tau(j)\}}{\exp\{\mathrm{IV}(j)\}} \frac{\exp\{\tau(j)\mathrm{IV}(j)\}}{\sum_m \exp\left(\tau_m \mathrm{IV}_m\right)}$$

where $\tau(j)$ and $\text{IV}(j)$ are the dissimilarity parameter and inclusive value for the nest in which alternative j lies.

In contrast, for the nonnormalized model, we have a latent variable

$$\widetilde{V}_{i,j} = \widetilde{\alpha}_j + \mathbf{x}_{i,j}\widetilde{\boldsymbol{\beta}}_j + \mathbf{z}_i\widetilde{\boldsymbol{\gamma}}_j$$

and corresponding inclusive values

$$\widetilde{\text{IV}}_m = \ln \sum_{j \in B_m} \exp\left(\widetilde{V}_k\right) \tag{2}$$

The probability of choosing alternative j is

$$\Pr_j = \frac{\exp\left(\widetilde{V}_j\right)}{\exp\left\{\widetilde{\text{IV}}(j)\right\}} \frac{\exp\left\{\tau(j)\widetilde{\text{IV}}(j)\right\}}{\sum_m \exp\left(\tau_m \widetilde{\text{IV}}_m\right)}$$

Equations (1) and (2) represent the key difference between the RUM-consistent and nonnormalized models. By scaling the V_{ij} within each nest, the RUM-consistent model allows utilities to be compared across nests. Without the rescaling, utilities can be compared only for goods within the same nest. Moreover, adding a constant to each V_{ij} for consumer i will not affect the probabilities of the RUM-consistent model, but adding a constant to each $\widetilde{V}_{ij}$ will affect the probabilities from the nonnormalized model. Decisions based on utility maximization can depend only on utility differences and not the scale or zero point of the utility function because utility is an ordinal concept, so the nonnormalized model cannot be consistent with utility maximization.

Heiss (2002) showed that the nonnormalized model can be RUM consistent in the special case where all the variables are specified in the bottom-level equation. Then multiplying the nonnormalized coefficients by the respective dissimilarity parameters results in the RUM-consistent coefficients.

❑ Technical note

Degenerate nests occur when there is only one alternative in a branch of the tree hierarchy. The associated dissimilarity parameter of the RUM model is not defined. The inclusive-valued parameter of the nonnormalized model will be identifiable if there are alternative-specific variables specified in (1) of the model specification (the *indepvars* in the model syntax). Numerically, you can skirt the issue of nonidentifiable/undefined parameters by setting constraints on them. For the RUM model constraint, set the dissimilarity parameter to 1. See the description of `constraints()` in *Options* for details on setting constraints on the dissimilarity parameters.

❑

Saved results

nlogit saves the following in e():

Scalars

e(N)	number of observations
e(N_case)	number of cases
e(k_eq)	number of equations in e(b)
e(k_eq_model)	number of equations in overall model test
e(k_alt)	number of alternatives for bottom level
e(k_alt*j*)	number of alternatives for *j*th level
e(k_indvars)	number of independent variables
e(k_ind2vars)	number of by-alternative variables for bottom level
e(k_ind2vars*j*)	number of by-alternative variables for *j*th level
e(df_m)	model degrees of freedom
e(df_c)	clogit model degrees of freedom
e(ll)	log likelihood
e(ll_c)	clogit model log likelihood
e(N_clust)	number of clusters
e(chi2)	χ^2
e(chi2_c)	likelihood-ratio test for IIA
e(p)	p-value for model Wald test
e(p_c)	p-value for IIA test
e(i_base)	base index for bottom level
e(i_base*j*)	base index for *j*th level
e(levels)	number of levels
e(alt_min)	minimum number of alternatives
e(alt_avg)	average number of alternatives
e(alt_max)	maximum number of alternatives
e(const)	constant indicator for bottom level
e(const*j*)	constant indicator for *j*th level
e(rum)	1 if RUM model, 0 otherwise
e(rank)	rank of e(V)
e(ic)	number of iterations
e(rc)	return code
e(converged)	1 if converged, 0 otherwise

Macros

`e(cmd)`	`nlogit`
`e(cmdline)`	command as typed
`e(depvar)`	name of dependent variable
`e(indvars)`	name of independent variables
`e(ind2vars)`	by-alternative variables for bottom level
`e(ind2vars`*j*`)`	by-alternative variables for *j*th level
`e(case)`	variable defining cases
`e(altvar)`	alternative variable for bottom level
`e(altvar`*j*`)`	alternative variable for *j*th level
`e(alteqs)`	equation names for bottom level
`e(alteqs`*j*`)`	equation names for *j*th level
`e(alt`*i*`)`	*i*th alternative for bottom level
`e(alt`*j*`_`*i*`)`	*i*th alternative for *j*th level
`e(wtype)`	weight type
`e(wexp)`	weight expression
`e(title)`	title in estimation output
`e(clustvar)`	name of cluster variable
`e(chi2type)`	`Wald`, type of model χ^2 test
`e(vce)`	*vcetype* specified in `vce()`
`e(vcetype)`	title used to label Std. Err.
`e(opt)`	type of optimization
`e(which)`	`max` or `min`; whether optimizer is to perform maximization or minimization
`e(ml_method)`	type of `ml` method
`e(user)`	name of likelihood-evaluator program
`e(technique)`	maximization technique
`e(datasignature)`	the checksum
`e(datasignaturevars)`	variables used in calculation of checksum
`e(properties)`	`b V`
`e(estat_cmd)`	program used to implement `estat`
`e(predict)`	program used to implement `predict`
`e(marginsnotok)`	predictions disallowed by `margins`

Matrices

`e(b)`	coefficient vector
`e(Cns)`	constraints matrix
`e(k_altern)`	number of alternatives at each level
`e(k_branch`*j*`)`	number of branches at each alternative of *j*th level
`e(stats)`	alternative statistics for bottom level
`e(stats`*j*`)`	alternative statistics for *j*th level
`e(altidx`*j*`)`	alternative indices for *j*th level
`e(alt_ind2vars)`	indicators for bottom level estimated by-alternative variable—`e(k_alt)`$\times$`e(k_ind2vars)`
`e(alt_ind2vars`*j*`)`	indicators for *j*th level estimated by-alternative variable—`e(k_alt`*j*`)`$\times$`e(k_ind2vars`*j*`)`
`e(ilog)`	iteration log (up to 20 iterations)
`e(gradient)`	gradient vector
`e(V)`	variance–covariance matrix of the estimators
`e(V_modelbased)`	model-based variance

Functions

`e(sample)`	marks estimation sample

Methods and formulas

`nlogit`, `nlogitgen`, and `nlogittree` are implemented as ado-files.

Methods and formulas are presented under the following headings:

Two-level nested logit model
Three-level nested logit model

Two-level nested logit model

Consider our two-level nested logit model for restaurant choice. We define $T = \{1,2,3\}$ to be the set of indices denoting the three restaurant types and $R_1 = \{1,2\}$, $R_2 = \{3,4,5\}$, and $R_3 = \{6,7\}$ to be the set of indices representing each restaurant within type $t \in T$. Let C_1 and C_2 be the random variables that represent the choices made for the first level, restaurant type, and second level, restaurant, of the hierarchy, where we observe the choices $C_1 = t, t \in T$, and $C_2 = j, j \in R_t$. Let $\mathbf{z}_t$ and $\mathbf{x}_{tj}$, for $t \in T$ and $j \in R_t$, refer to the row vectors of explanatory variables for the first-level alternatives and bottom-level alternatives for one case, respectively. We write the utilities (latent variables) as $U_{tj} = \mathbf{z}_t \boldsymbol{\alpha}_t + \mathbf{x}_{tj}\boldsymbol{\beta}_j + \epsilon_{tj} = \boldsymbol{\eta}_{tj} + \epsilon_{tj}$, where $\boldsymbol{\alpha}_t$ and $\boldsymbol{\beta}_j$ are column vectors and the ϵ_{tj} are random disturbances. When the $\mathbf{x}_{tj}$ are alternative specific, we can drop the indices from $\boldsymbol{\beta}$, where we estimate one coefficient for each alternative in R_t, $t \in T$. These variables are specified in the first equation of the `nlogit` syntax (see example 3).

When the random-utility framework is used to describe the choice behavior, the alternative that is chosen is the alternative that has the highest utility. Assume for our restaurant example that we choose restaurant type $t \in T$. For the RUM parameterization of `nlogit`, the conditional distribution of ϵ_{tj} given choice of restaurant type t is a multivariate version of Gumbel's extreme-value distribution,

$$F_{R|T}(\boldsymbol{\epsilon} \,|\, t) = \exp\left[-\left\{ \sum_{m \in R_t} \exp(\epsilon_{tm}/\tau_t) \right\}^{\tau_t} \right] \tag{3}$$

where it has been shown that the ϵ_{tj}, $j \in R_t$, are exchangeable with correlation $1 - \tau_t^2$, for $\tau_t \in (0,1]$ (Kotz and Nadarajah 2000). For example, the probability of choosing Christopher's, $j = 6$ given type $t = 3$, is

$$\begin{aligned}
\Pr(C_2 = 6 \,|\, C_1 = 3) &= \Pr\left(U_{36} - U_{37} > 0\right) \\
&= \Pr\left(\epsilon_{37} \le \epsilon_{36} + \eta_{36} - \eta_{37}\right) \\
&= \int_{-\infty}^{\infty} \left\{ \int_{-\infty}^{\epsilon_{36}+\eta_{36}-\eta_{37}} f_{R|T}\left(\epsilon_{36}, \epsilon_{37}\right) d\epsilon_{37} \right\} d\epsilon_{36}
\end{aligned}$$

where $f = \dfrac{\partial F}{\partial \epsilon_{36} \partial \epsilon_{37}}$ is the joint density function of $\boldsymbol{\epsilon}$ given t. U_{37} is the utility of eating at Mad Cows, the other fancy ($t = 3$) restaurant. Amemiya (1985) demonstrates that this integral evaluates to the logistic function

$$\begin{aligned}
\Pr(C_2 = 6 \,|\, C_1 = 3) &= \frac{\exp(\boldsymbol{\eta}_{36}/\tau_3)}{\exp(\boldsymbol{\eta}_{36}/\tau_3) + \exp(\boldsymbol{\eta}_{37}/\tau_3)} \\
&= \frac{\exp(\mathbf{x}_{36}\boldsymbol{\beta}_6/\tau_3)}{\exp(\mathbf{x}_{36}\boldsymbol{\beta}_6/\tau_3) + \exp(\mathbf{x}_{37}\boldsymbol{\beta}_7/\tau_3)}
\end{aligned}$$

and in general

$$\Pr(C_2 = j \,|\, C_1 = t) = \frac{\exp(\mathbf{x}_{tj}\boldsymbol{\beta}_j/\tau_t)}{\sum_{m \in R_t} \exp(\mathbf{x}_{tm}\boldsymbol{\beta}_m/\tau_t)} \tag{4}$$

Letting $\tau_t = 1$ in (3) reduces to the product of independent extreme-value distributions, and (4) reduces to the multinomial logistic function.

For the logistic function in (4), we scale the linear predictors by the dissimilarity parameters. Another formulation of the conditional probability of choosing alternative $j \in R_t$ given choice $t \in T$ is the logistic function without this normalization:

$$\Pr(C_2 = j \,|\, C_1 = t) = \frac{\exp(\mathbf{x}_{tj}\boldsymbol{\beta}_j)}{\sum_{m\in R_t}\exp(\mathbf{x}_{tm}\boldsymbol{\beta}_m)}$$

and this is what is used in `nlogit`'s nonnormalized parameterization.

Amemiya (1985) defines the general form for the joint distribution of the ϵ's as

$$F_{T,R}(\boldsymbol{\epsilon}) = \exp\left\{-\sum_{k\in T}\theta_k\left(\sum_{m\in R_k}\exp(-\epsilon_{km}/\tau_k)\right)^{\tau_k}\right\}$$

from which the probability of choice $t, t \in T$ can be derived as

$$\Pr(C_1 = t) = \frac{\theta_t\left\{\sum_{m\in R_t}\exp(\boldsymbol{\eta}_{tm}/\tau_t)\right\}^{\tau_t}}{\sum_{k\in T}\theta_k\left\{\sum_{m\in R_k}\exp(\boldsymbol{\eta}_{km}/\tau_k)\right\}^{\tau_k}} \tag{5}$$

`nlogit` sets $\theta_t = 1$. Noting that

$$\begin{aligned}\left\{\sum_{m\in R_t}\exp(\boldsymbol{\eta}_{tm}/\tau_t)\right\}^{\tau_t} &= \left\{\sum_{m\in R_t}\exp\left(\frac{\mathbf{z}_t\boldsymbol{\alpha}_t + \mathbf{x}_{tm}\boldsymbol{\beta}_m}{\tau_t}\right)\right\}^{\tau_t}\\ &= \exp(\mathbf{z}_t\boldsymbol{\alpha}_t)\left\{\sum_{m\in R_t}\exp\left(\mathbf{x}_{tm}\boldsymbol{\beta}_m/\tau_t\right)\right\}^{\tau_t}\\ &= \exp(\mathbf{z}_t\boldsymbol{\alpha}_t + \tau_t I_t)\end{aligned}$$

we define the inclusive values I_t as

$$I_t = \ln\left\{\sum_{m\in R_t}\exp(\mathbf{x}_{tm}\boldsymbol{\beta}_m/\tau_t)\right\}$$

and we can view

$$\exp(\tau_t I_t) = \left\{\sum_{m\in R_t}\exp(x_{tm}\boldsymbol{\beta}_m)^{1/\tau_t}\right\}^{\tau_t}$$

as a weighted average of the $\exp(x_{tm}\boldsymbol{\beta}_m)$, for $m \in R_t$. For the `nlogit` RUM parameterization, we can express (5) as

$$\Pr(C_1 = t) = \frac{\exp(\mathbf{z}_t\boldsymbol{\alpha}_t + \tau_t I_t)}{\sum_{k\in T}\exp(\mathbf{z}_k\boldsymbol{\alpha}_k + \tau_k I_k)}$$

Next we define inclusive values for the nonnormalized model to be

$$\widetilde{I}_t = \ln\left\{\sum_{m\in R_t}\exp(\mathbf{x}_{tm}\boldsymbol{\beta}_m)\right\}$$

and we express $\Pr(C_1 = t)$ as

$$\Pr(C_1 = t) = \frac{\exp(\mathbf{z}_t\boldsymbol{\alpha}_t + \tau_t\widetilde{I}_t)}{\sum_{k\in T}\exp(\mathbf{z}_k\boldsymbol{\alpha}_k + \tau_k\widetilde{I}_k)} \tag{6}$$

Equation (5) is consistent with (6) only when $\boldsymbol{\eta}_{ij} = \mathbf{x}_{ij}\boldsymbol{\beta}_j$, so in general the `nlogit` nonnormalized model is not consistent with the RUM model.

Now assume that we have N cases where we add a third subscript, i, to denote case i, $i = 1, \dots, N$. Denote y_{itj} to be a binary variable indicating the choice made by case i so that for each i only one y_{itj} is 1 and the rest are 0 for all $t \in T$ and $j \in R_t$. The log likelihood for the two-level RUM-consistent model is

$$\log \ell = \sum_{i=1}^{N} \sum_{k \in T} \sum_{m \in R_k} y_{ikm} \log \{ \Pr(C_{i1} = i) \Pr(C_{i2} = m | C_{i1} = i) \}$$

$$= \sum_{i=1}^{N} \sum_{k \in T} \sum_{m \in R_k} y_{ikm} \left[\mathbf{z}_{ik}\boldsymbol{\alpha}_k + \tau_k I_{ik} - \log \left\{ \sum_{l \in T} \exp(\mathbf{z}_{il}\boldsymbol{\alpha}_l + \tau_l I_{il}) \right\} + \right.$$

$$\left. \mathbf{x}_{ikm}\boldsymbol{\beta}_m / \tau_k - \log \left\{ \sum_{l \in R_k} \exp(\mathbf{x}_{ikl}\boldsymbol{\beta}_l / \tau_k) \right\} \right]$$

The likelihood for the nonnormalized model has a similar form, replacing I with $\widetilde{I}$ and by not scaling $\mathbf{x}_{ikj}\boldsymbol{\beta}_j$ by τ_k.

Three-level nested logit model

Here we define a three-level nested logit model that can be generalized to the four-level and higher models. As before, let the integer set T be the indices for the first level of choices. Let sets S_t, $t \in T$, be mutually exclusive sets of integers representing the choices of the second level of the hierarchy. Finally, let R_j, $j \in S_t$, be the bottom-level choices. Let $U_{tjk} = \eta_{tjk} + \epsilon_{tjk}$, $k \in R_j$, and the distribution of ϵ_{tjk} be Gumbel's multivariate extreme value of the form

$$F(\boldsymbol{\epsilon}) = \exp\left(- \sum_{t \in T} \left[\sum_{j \in S_t} \left\{ \sum_{k \in R_j} \exp(-\eta_{tjk}/\tau_j) \right\}^{\tau_j/\upsilon_t} \right]^{\upsilon_j} \right)$$

Let C_1, C_2, and C_3 represent the choice random variables for levels 1, 2, and the bottom, respectively. Then the set of conditional probabilities is

$$\Pr(C_3 = k \,|\, C_1 = t, C_2 = j) = \frac{\exp(\eta_{tjk}/\tau_j)}{\sum_{l \in R_j} \exp(\eta_{tjl}/\tau_j)}$$

$$\Pr(C_2 = j \,|\, C_1 = t) = \frac{\left\{ \sum_{k \in R_j} \exp(\eta_{tjk}/\tau_j) \right\}^{\tau_j/\upsilon_t}}{\sum_{l \in S_t} \left\{ \sum_{k \in R_l} \exp(\eta_{tlk}/\tau_l) \right\}^{\tau_l/\upsilon_t}}$$

$$\Pr(C_1 = t) = \frac{\left[\sum_{j \in S_t} \left\{ \sum_{k \in R_j} \exp(\eta_{tjk}/\tau_j) \right\}^{\tau_j/\upsilon_t} \right]^{\upsilon_t}}{\sum_{l \in T} \left[\sum_{j \in S_l} \left\{ \sum_{k \in R_j} \exp(\eta_{ljk}/\tau_j) \right\}^{\tau_j/\upsilon_l} \right]^{\upsilon_l}}$$

Assume that we can decompose the linear predictor as $\boldsymbol{\eta}_{tjk} = \mathbf{z}_t\boldsymbol{\alpha}_t + \mathbf{u}_{tj}\boldsymbol{\gamma}_j + \mathbf{x}_{tjk}\boldsymbol{\beta}_k$. Here $\mathbf{z}_t$, $\mathbf{u}_{tj}$, and $\mathbf{x}_{tjk}$ are the row vectors of explanatory variables for the first, second, and bottom levels of the hierarchy, respectively, and $\boldsymbol{\alpha}_t$, $\boldsymbol{\gamma}_j$, and $\boldsymbol{\beta}_k$ are the corresponding column vectors of regression coefficients for $t \in T$, $j \in S_t$, and $k \in R_j$. We then can define the inclusive values for the first and second levels as

$$I_{tj} = \log \sum_{k\in R_j} \exp(\mathbf{x}_{tjk}\boldsymbol{\beta}_k/\tau_j)$$

$$J_t = \log \sum_{j\in S_t} \exp(\mathbf{u}_{tj}\boldsymbol{\gamma}_j/\upsilon_t + \frac{\tau_j}{\upsilon_t} I_{tj})$$

and rewrite the probabilities

$$\Pr(C_3 = k \,|\, C_1 = t, C_2 = j) = \frac{\exp(\mathbf{x}_{tjk}\boldsymbol{\beta}_k/\tau_j)}{\sum_{l\in R_j} \exp(\mathbf{x}_{tjl}\boldsymbol{\beta}_l/\tau_j)}$$

$$\Pr(C_2 = j \,|\, C_1 = t) = \frac{\exp(\mathbf{u}_{tj}\boldsymbol{\gamma}_j/\upsilon_t + \frac{\tau_j}{\upsilon_t} I_{tj})}{\sum_{l\in S_t} \exp(\mathbf{u}_{tl}\boldsymbol{\gamma}_l/\upsilon_t + \frac{\tau_l}{\upsilon_t} I_{tl})}$$

$$\Pr(C_1 = t) = \frac{\exp(\mathbf{z}_t\boldsymbol{\alpha}_t + \upsilon_t J_t)}{\sum_{l\in T} \exp(\mathbf{z}_l\boldsymbol{\alpha}_l + \upsilon_l J_l)}$$

We add a fourth index, i, for case and define the indicator variable y_{lijk}, $l = 1, \ldots, N$, to indicate the choice made by case i, $t \in T$, $j \in S_t$, and $k \in R_j$. The log likelihood for the `nlogit` RUM-consistent model is

$$\begin{aligned} \ell = \sum_{i=1}^{N} \sum_{t\in T} \sum_{j\in S_t} \sum_{k\in R_j} y_{itjk} \Bigg\{ & \mathbf{z}_{it}\boldsymbol{\alpha}_t + \upsilon_t J_{it} - \log\left(\sum_{m\in T} \mathbf{z}_{im}\boldsymbol{\alpha}_m + \upsilon_m J_{im} \right) + \\ & \mathbf{u}_{itj}\boldsymbol{\gamma}_j/\upsilon_t + \frac{\tau_j}{\upsilon_t} I_{itj} - \log\left(\sum_{m\in S_t} \mathbf{u}_{itm}\boldsymbol{\gamma}_m/\upsilon_t + \frac{\tau_m}{\upsilon_t} I_{itm} \right) + \\ & \mathbf{x}_{itjk}\boldsymbol{\beta}_k/\tau_k - \sum_{m\in R_t} \exp(\mathbf{x}_{itjm}\boldsymbol{\beta}_m/\tau_k) \Bigg\} \end{aligned}$$

and for the nonnormalized `nlogit` model the log likelihood is

$$\begin{aligned} \ell = \sum_{i=1}^{N} \sum_{t\in T} \sum_{j\in S_t} \sum_{k\in R_j} y_{itjk} \Bigg\{ & \mathbf{z}_{it}\boldsymbol{\alpha}_t + \upsilon_t J_{it} - \log\left(\sum_{m\in T} \mathbf{z}_{im}\boldsymbol{\alpha}_m + \upsilon_m J_{im} \right) + \\ & \mathbf{u}_{itj}\boldsymbol{\gamma}_j + \tau_j I_{itj} - \log\left(\sum_{m\in S_t} \mathbf{u}_{itm}\boldsymbol{\gamma}_m + \tau_m I_{itm} \right) + \\ & \mathbf{x}_{itjk}\boldsymbol{\beta}_k - \sum_{m\in R_t} \exp(\mathbf{x}_{itjm}\boldsymbol{\beta}_m) \Bigg\} \end{aligned}$$

Extending the model to more than three levels is straightforward, albeit notationally cumbersome.

This command supports the Huber/White/sandwich estimator of the variance and its clustered version using `vce(robust)` and `vce(cluster` *clustvar*`)`, respectively. See [P] **_robust**, particularly *Maximum likelihood estimators* and *Methods and formulas*.

References

Amemiya, T. 1985. *Advanced Econometrics*. Cambridge, MA: Harvard University Press.

Greene, W. H. 2012. *Econometric Analysis*. 7th ed. Upper Saddle River, NJ: Prentice Hall.

Hausman, J. A. 1978. Specification tests in econometrics. *Econometrica* 46: 1251–1271.

Hausman, J. A., and D. L. McFadden. 1984. Specification tests for the multinomial logit model. *Econometrica* 52: 1219–1240.

Heiss, F. 2002. Structural choice analysis with nested logit models. *Stata Journal* 2: 227–252.

Hensher, D. A., J. M. Rose, and W. H. Greene. 2005. *Applied Choice Analysis: A Primer*. New York: Cambridge University Press.

Kotz, S., and S. Nadarajah. 2000. *Extreme Value Distributions: Theory and Applications*. London: Imperial College Press.

Maddala, G. S. 1983. *Limited-Dependent and Qualitative Variables in Econometrics*. Cambridge: Cambridge University Press.

McFadden, D. L. 1977. Quantitative methods for analyzing travel behaviour of individuals: Some recent developments. Working paper 474, Cowles Foundation. http://cowles.econ.yale.edu/P/cd/d04b/d0474.pdf.

——. 1981. Econometric models of probabilistic choice. In *Structural Analysis of Discrete Data with Econometric Applications*, ed. C. F. Manski and D. McFadden, 198–272. Cambridge, MA: MIT Press.

Also see

[R] **nlogit postestimation** — Postestimation tools for nlogit

[R] **asclogit** — Alternative-specific conditional logit (McFadden's choice) model

[R] **clogit** — Conditional (fixed-effects) logistic regression

[R] **mlogit** — Multinomial (polytomous) logistic regression

[R] **ologit** — Ordered logistic regression

[R] **rologit** — Rank-ordered logistic regression

[R] **slogit** — Stereotype logistic regression

[U] **20 Estimation and postestimation commands**

Title

nlogit postestimation — Postestimation tools for nlogit

Description

The following postestimation command is of special interest after `nlogit`:

Command	Description
`estat alternatives`	alternative summary statistics

For information about this command, see [R] **asmprobit postestimation**.

The following standard postestimation commands are also available:

Command	Description
`estat`	AIC, BIC, VCE, and estimation sample summary
`estimates`	cataloging estimation results
`hausman`	Hausman's specification test
`lincom`	point estimates, standard errors, testing, and inference for linear combinations of coefficients
`lrtest`	likelihood-ratio test
`nlcom`	point estimates, standard errors, testing, and inference for nonlinear combinations of coefficients
`predict`	predictions, residuals, influence statistics, and other diagnostic measures
`predictnl`	point estimates, standard errors, testing, and inference for generalized predictions
`test`	Wald tests of simple and composite linear hypotheses
`testnl`	Wald tests of nonlinear hypotheses

See the corresponding entries in the *Base Reference Manual* for details.

Syntax for predict

predict [*type*] *newvar* [*if*] [*in*] [, *statistic* hlevel(#) altwise]

predict [*type*] { *stub** | *newvarlist* } [*if*] [*in*], scores

statistic	Description
Main	
pr	predicted probabilities of choosing the alternatives at all levels of the hierarchy or at level #, where # is specified by hlevel(#); the default
xb	linear predictors for all levels of the hierarchy or at level #, where # is specified by hlevel(#)
condp	predicted conditional probabilities at all levels of the hierarchy or at level #, where # is specified by hlevel(#)
iv	inclusive values for levels 2, ..., e(levels) or for hlevel(#)

The inclusive value for the first-level alternatives is not used in estimation; therefore, it is not calculated.

These statistics are available both in and out of sample; type predict ... if e(sample) ... if wanted only for the estimation sample.

Menu

Statistics > Postestimation > Predictions, residuals, etc.

Options for predict

Main

pr calculates the probability of choosing each alternative at each level of the hierarchy. Use the hlevel(#) option to compute the alternative probabilities at level #. When hlevel(#) is not specified, j new variables must be given, where j is the number of levels, or use the *stub** option to have predict generate j variables with the prefix *stub* and numbered from 1 to j. The pr option is the default and if one new variable is given, the probability of the bottom-level alternatives are computed. Otherwise, probabilities for all levels are computed and the *stub** option is still valid.

xb calculates the linear prediction for each alternative at each level. Use the hlevel(#) option to compute the linear predictor at level #. When hlevel(#) is not specified, j new variables must be given, where j is the number of levels, or use the *stub** option to have predict generate j variables with the prefix *stub* and numbered from 1 to j.

condp calculates the conditional probabilities for each alternative at each level. Use the hlevel(#) option to compute the conditional probabilities of the alternatives at level #. When hlevel(#) is not specified, j new variables must be given, where j is the number of levels, or use the *stub** option to have predict generate j variables with the prefix *stub* and numbered from 1 to j.

iv calculates the inclusive value for each alternative at each level. Use the hlevel(#) option to compute the inclusive value at level #. There is no inclusive value at level 1. If hlevel(#) is not used, $j-1$ new variables are required, where j is the number of levels, or use *stub** to have predict generate $j-1$ variables with the prefix *stub* and numbered from 2 to j. See *Methods and formulas* in [R] **nlogit** for a definition of the inclusive values.

hlevel(#) calculates the prediction only for hierarchy level #.

`altwise` specifies that alternativewise deletion be used when marking out observations due to missing values in your variables. The default is to use casewise deletion. The `xb` option always uses alternativewise deletion.

`scores` calculates the scores for each coefficient in `e(b)`. This option requires a new-variable list of length equal to the number of columns in `e(b)`. Otherwise, use the *stub*`*` option to have `predict` generate enumerated variables with prefix *stub*.

Remarks

`predict` may be used after `nlogit` to obtain the predicted values of the probabilities, the conditional probabilities, the linear predictions, and the inclusive values for each level of the nested logit model. Predicted probabilities for `nlogit` must be interpreted carefully. Probabilities are estimated for each case as a whole and not for individual observations.

▷ Example 1

Continuing with our model in example 3 of [R] **nlogit**, we refit the model and then examine a summary of the alternatives and their frequencies in the estimation sample.

```
. use http://www.stata-press.com/data/r12/restaurant
. nlogitgen type = restaurant(fast: Freebirds | MamasPizza,
> family: CafeEccell | LosNortenos | WingsNmore, fancy: Christophers | MadCows)
  (output omitted)
. nlogit chosen cost rating distance || type: income kids, base(family) ||
> restaurant:, noconst case(family_id)
  (output omitted)
. estat alternatives
  Alternatives summary for type
```

index	Alternative value	label	Cases present	Frequency selected	Percent selected
1	1	fast	600	27	9.00
2	2	family	900	222	74.00
3	3	fancy	600	51	17.00

```
  Alternatives summary for restaurant
```

index	Alternative value	label	Cases present	Frequency selected	Percent selected
1	1	Freebirds	300	12	4.00
2	2	MamasPizza	300	15	5.00
3	3	CafeEccell	300	78	26.00
4	4	LosNortenos	300	75	25.00
5	5	WingsNmore	300	69	23.00
6	6	Christophers	300	27	9.00
7	7	MadCows	300	24	8.00

Next we predict p2 = Pr(restaurant); p1 = Pr(type); condp = Pr(restaurant | type); xb2, the linear prediction for the bottom-level alternatives; xb1, the linear prediction for the first-level alternatives; and iv, the inclusive values for the bottom-level alternatives.

```
. predict p*
(option pr assumed)
. predict condp, condp hlevel(2)
. sort family_id type restaurant
. list restaurant type chosen p2 p1 condp in 1/14, sepby(family_id) divider
```

	restaurant	type	chosen	p2	p1	condp
1.	Freebirds	fast	1	.0642332	.1189609	.5399519
2.	MamasPizza	fast	0	.0547278	.1189609	.4600481
3.	CafeEccell	family	0	.284409	.7738761	.3675124
4.	LosNortenos	family	0	.3045242	.7738761	.3935051
5.	WingsNmore	family	0	.1849429	.7738761	.2389825
6.	Christophers	fancy	0	.0429508	.107163	.4007991
7.	MadCows	fancy	0	.0642122	.107163	.5992009
8.	Freebirds	fast	0	.0183578	.0488948	.3754559
9.	MamasPizza	fast	0	.030537	.0488948	.6245441
10.	CafeEccell	family	0	.2832149	.756065	.3745907
11.	LosNortenos	family	1	.3038883	.756065	.4019341
12.	WingsNmore	family	0	.1689618	.756065	.2234752
13.	Christophers	fancy	0	.1041277	.1950402	.533878
14.	MadCows	fancy	0	.0909125	.1950402	.466122

```
. predict xb*, xb
. predict iv, iv
. list restaurant type chosen xb* iv in 1/14, sepby(family_id) divider
```

	restaurant	type	chosen	xb1	xb2	iv
1.	Freebirds	fast	1	-1.124805	-1.476914	-.2459659
2.	MamasPizza	fast	0	-1.124805	-1.751229	-.2459659
3.	CafeEccell	family	0	0	-2.181112	.1303341
4.	LosNortenos	family	0	0	-2.00992	.1303341
5.	WingsNmore	family	0	0	-3.259229	.1303341
6.	Christophers	fancy	0	1.405185	-6.804211	-.745332
7.	MadCows	fancy	0	1.405185	-5.155514	-.745332
8.	Freebirds	fast	0	-1.804794	-2.552233	-.5104123
9.	MamasPizza	fast	0	-1.804794	-1.680583	-.5104123
10.	CafeEccell	family	0	0	-2.400434	.0237072
11.	LosNortenos	family	1	0	-2.223939	.0237072
12.	WingsNmore	family	0	0	-3.694409	.0237072
13.	Christophers	fancy	0	1.490775	-5.35932	-.6796131
14.	MadCows	fancy	0	1.490775	-5.915751	-.6796131

◁

Methods and formulas

All postestimation commands listed above are implemented as ado-files.

Also see

[R] **nlogit** — Nested logit regression

[U] **20 Estimation and postestimation commands**

Title

nlsur — Estimation of nonlinear systems of equations

Syntax

Interactive version

`nlsur` (*depvar_1* = <*sexp_1*>) (*depvar_2* = <*sexp_2*>) ... [*if*] [*in*] [*weight*] [, *options*]

Programmed substitutable expression version

`nlsur` *sexp_prog* : *depvar_1 depvar_2* ... [*varlist*] [*if*] [*in*] [*weight*] [, *options*]

Function evaluator program version

`nlsur` *func_prog* @ *depvar_1 depvar_2* ... [*varlist*] [*if*] [*in*] [*weight*] ,

`nequations(#)` { `parameters(`*namelist*`)` | `nparameters(#)`} [*options*]

where

depvar_j is the dependent variable for equation j;

<*sexp*>*_j* is the substitutable expression for equation j;

sexp_prog is a substitutable expression program; and

func_prog is a function evaluator program.

options	Description
Model	
fgnls	use two-step FGNLS estimator; the default
ifgnls	use iterative FGNLS estimator
nls	use NLS estimator
variables(*varlist*)	variables in model
initial(*initial_values*)	initial values for parameters
nequations(#)	number of equations in model (function evaluator program version only)
* parameters(*namelist*)	parameters in model (function evaluator program version only)
* nparameters(#)	number of parameters in model (function evaluator program version only)
sexp_options	options for substitutable expression program
func_options	options for function evaluator program
SE/Robust	
vce(*vcetype*)	*vcetype* may be gnr, robust, cluster *clustvar*, bootstrap, or jackknife
Reporting	
level(#)	set confidence level; default is level(95)
title(*string*)	display *string* as title above the table of parameter estimates
title2(*string*)	display *string* as subtitle
display_options	control column formats and line width
Optimization	
optimization_options	control the optimization process; seldom used
eps(#)	specify # for convergence criteria; default is eps(1e-5)
ifgnlsiterate(#)	set maximum number of FGNLS iterations
ifgnlseps(#)	specify # for FGNLS convergence criterion; default is ifgnlseps(1e-10)
delta(#)	specify stepsize # for computing derivatives; default is delta(4e-7)
noconstants	no equations have constant terms
hasconstants(*namelist*)	use *namelist* as constant terms
coeflegend	display legend instead of statistics

* You must specify parameters(*namelist*), nparameters(#), or both.
bootstrap, by, jackknife, rolling, and statsby are allowed; see [U] **11.1.10 Prefix commands**.
Weights are not allowed with the bootstrap prefix; see [R] **bootstrap**.
aweights are not allowed with the jackknife prefix; see [R] **jackknife**.
aweights, fweights, iweights, and pweights are allowed; see [U] **11.1.6 weight**.
coeflegend does not appear in the dialog box.
See [U] **20 Estimation and postestimation commands** for more capabilities of estimation commands.

Menu

Statistics > Linear models and related > Multiple-equation models > Nonlinear seemingly unrelated regression

Description

`nlsur` fits a system of nonlinear equations by feasible generalized nonlinear least squares (FGNLS). With the interactive version of the command, you enter the system of equations on the command line or in the dialog box by using *substitutable expressions*. If you have a system that you use regularly, you can write a *substitutable expression program* and use the second syntax to avoid having to reenter the system every time. The function evaluator program version gives you the most flexibility in exchange for increased complexity; with this version, your program is given a vector of parameters and a variable list, and your program computes the system of equations.

When you write a substitutable expression program or a function evaluator program, the first five letters of the name must be `nlsur`. *sexp_prog* and *func_prog* refer to the name of the program without the first five letters. For example, if you wrote a function evaluator program named `nlsurregss`, you would type `nlsur regss @` ... to estimate the parameters.

Options

Model

`fgnls` requests the two-step FGNLS estimator; this is the default.

`ifgnls` requests the iterative FGNLS estimator. For the nonlinear systems estimator, this is equivalent to maximum likelihood estimation.

`nls` requests the nonlinear least-squares (NLS) estimator.

`variables(`*varlist*`)` specifies the variables in the system. `nlsur` ignores observations for which any of these variables has missing values. If you do not specify `variables()`, `nlsur` issues an error message if the estimation sample contains any missing values.

`initial(`*initial_values*`)` specifies the initial values to begin the estimation. You can specify a $1 \times k$ matrix, where k is the total number of parameters in the system, or you can specify a parameter name, its initial value, another parameter name, its initial value, and so on. For example, to initialize `alpha` to 1.23 and `delta` to 4.57, you would type

```
. nlsur ..., initial(alpha 1.23 delta 4.57) ...
```

Initial values declared using this option override any that are declared within substitutable expressions. If you specify a matrix, the values must be in the same order in which the parameters are declared in your model. `nlsur` ignores the row and column names of the matrix.

`nequations(`*#*`)` specifies the number of equations in the system.

`parameters(`*namelist*`)` specifies the names of the parameters in the system. The names of the parameters must adhere to the naming conventions of Stata's variables; see [U] **11.3 Naming conventions**. If you specify both `parameters()` and `nparameters()`, the number of names in the former must match the number specified in the latter.

`nparameters(`*#*`)` specifies the number of parameters in the system. If you do not specify names with the `parameters()` options, `nlsur` names them `b1`, `b2`, ..., `b`*#*. If you specify both `parameters()` and `nparameters()`, the number of names in the former must match the number specified in the latter.

sexp_options refer to any options allowed by your *sexp_prog*.

func_options refer to any options allowed by your *func_prog*.

SE/Robust

vce(*vcetype*) specifies the type of standard error reported, which includes types that are derived from asymptotic theory, that are robust to some kinds of misspecification, that allow for intragroup correlation, and that use bootstrap or jackknife methods; see [R] ***vce_option***.

vce(gnr), the default, uses the conventionally derived variance estimator for nonlinear models fit using Gauss–Newton regression.

Reporting

level(*#*); see [R] **estimation options**.

title(*string*) specifies an optional title that will be displayed just above the table of parameter estimates.

title2(*string*) specifies an optional subtitle that will be displayed between the title specified in title() and the table of parameter estimates. If title2() is specified but title() is not, title2() has the same effect as title().

display_options: cformat(%*fmt*), pformat(%*fmt*), sformat(%*fmt*), and nolstretch; see [R] **estimation options**.

Optimization

optimization_options: iterate(*#*), [no]log, trace. iterate() specifies the maximum number of iterations to use for NLS at each round of FGNLS estimation. This option is different from ifgnlsiterate(), which controls the maximum rounds of FGNLS estimation to use when the ifgnls option is specified. log/nolog specifies whether to show the iteration log, and trace specifies that the iteration log should include the current parameter vector.

eps(*#*) specifies the convergence criterion for successive parameter estimates and for the residual sum of squares (RSS). The default is eps(1e-5) (0.00001). eps() also specifies the convergence criterion for successive parameter estimates between rounds of iterative FGNLS estimation when ifgnls is specified.

ifgnlsiterate(*#*) specifies the maximum number of FGNLS iterations to perform. The default is the number set using set maxiter (see [R] **maximize**), which is 16,000 by default. To use this option, you must also specify the ifgnls option.

ifgnlseps(*#*) specifies the convergence criterion for successive estimates of the error covariance matrix during iterative FGNLS estimation. The default is ifgnlseps(1e-10). To use this option, you must also specify the ifgnls option.

delta(*#*) specifies the relative change in a parameter, δ, to be used in computing the numeric derivatives. The derivative for parameter β_i is computed as

$$\left\{f_i\left(\mathbf{x}_i, \beta_1, \beta_2, \ldots, \beta_i + d, \beta_{i+1}, \ldots\right) - f_i\left(\mathbf{x}_i, \beta_1, \beta_2, \ldots, \beta_i, \beta_{i+1}, \ldots\right)\right\}/d$$

where $d = \delta(|\beta_i| + \delta)$. The default is delta(4e-7).

noconstants indicates that none of the equations in the system includes constant terms. This option is generally not needed, even if there are no constant terms in the system; though in rare cases without this option, nlsur may claim that there is one or more constant terms even if there are none.

hasconstants(*namelist*) indicates the parameters that are to be treated as constant terms in the system of equations. The number of elements of *namelist* must equal the number of equations in the system. The ith entry of *namelist* specifies the constant term in the ith equation. If an equation does not include a constant term, specify a period (.) instead of a parameter name. This option is seldom needed with the interactive and programmed substitutable expression versions, because in those cases `nlsur` can almost always find the constant terms automatically.

The following options are available with `nlsur` but are not shown in the dialog box:

`coeflegend`; see [R] **estimation options**.

Remarks

Remarks are presented under the following headings:

Introduction
Substitutable expression programs
Function evaluator programs

Introduction

`nlsur` fits a system of nonlinear equations by FGNLS. It can be viewed as a nonlinear variant of Zellner's seemingly unrelated regression model (Zellner 1962; Zellner and Huang 1962; Zellner 1963) and is therefore commonly called nonlinear SUR or nonlinear SURE. The model is also discussed in textbooks such as Davidson and MacKinnon (1993, 2004) and Greene (2012, 305–306). Formally, the model fit by `nlsur` is

$$\begin{aligned} y_{i1} &= f_1(\mathbf{x}_i, \beta) + u_{i1} \\ y_{i2} &= f_2(\mathbf{x}_i, \beta) + u_{i2} \\ \vdots &= \vdots \\ y_{iM} &= f_M(\mathbf{x}_i, \beta) + u_{iM} \end{aligned}$$

for $i = 1, \ldots, N$ observations and $m = 1, \ldots, M$ equations. The errors for the ith observation, $u_{i1}, u_{i2}, \ldots, u_{iM}$, may be correlated, so fitting the m equations jointly may lead to more efficient estimates. Moreover, fitting the equations jointly allows us to impose cross-equation restrictions on the parameters. Not all elements of the parameter vector β and data vector $\mathbf{x}_i$ must appear in all the equations, though each element of β must appear in at least one equation for β to be identified. For this model, iterative FGNLS estimation is equivalent to maximum likelihood estimation with multivariate normal disturbances.

The syntax you use with `nlsur` closely mirrors that used with `nl`. In particular, you use substitutable expressions with the interactive and programmed substitutable expression versions to define the functions in your system. See [R] **nl** for more information on substitutable expressions. Here we reiterate the three rules that you must follow:

1. Parameters of the model are bound in braces: `{b0}`, `{param}`, etc.
2. Initial values for parameters are given by including an equal sign and the initial value inside the braces: `{b0=1}`, `{param=3.571}`, etc. If you do not specify an initial value, that parameter is initialized to zero. The `initial()` option overrides initial values in substitutable expressions.

3. Linear combinations of variables can be included using the notation {*eqname*:*varlist*}, for example, `{xb: mpg price weight}`, `{score: w x z}`, etc. Parameters of linear combinations are initialized to zero.

▷ Example 1: Interactive version using two-step FGNLS estimator

We have data from an experiment in which two closely related types of bacteria were placed in a Petri dish, and the number of each type of bacteria were recorded every hour. We suspect a two-parameter exponential growth model can be used to model each type of bacteria, but because they shared the same dish, we want to allow for correlation in the error terms. We want to fit the system of equations

$$p_1 = \beta_1 {\beta_2}^t + u_1$$
$$p_2 = \gamma_1 {\gamma_2}^t + u_2$$

where p_1 and p_2 are the two populations and t is time, and we want to allow for nonzero correlation between u_1 and u_2. We type

```
. use http://www.stata-press.com/data/r12/petridish

. nlsur (p1 = {b1}*{b2}^t) (p2 = {g1}*{g2}^t)
(obs = 25)

Calculating NLS estimates...
Iteration 0:  Residual SS =  335.5286
Iteration 1:  Residual SS =  333.8583
Iteration 2:  Residual SS =  219.9233
Iteration 3:  Residual SS =  127.9355
Iteration 4:  Residual SS =  14.86765
Iteration 5:  Residual SS =  8.628459
Iteration 6:  Residual SS =  8.281268
Iteration 7:  Residual SS =   8.28098
Iteration 8:  Residual SS =  8.280979
Iteration 9:  Residual SS =  8.280979
Calculating FGNLS estimates...
Iteration 0:  Scaled RSS =  49.99892
Iteration 1:  Scaled RSS =  49.99892
Iteration 2:  Scaled RSS =  49.99892

FGNLS regression
```

	Equation	Obs	Parms	RMSE	R-sq	Constant
1	p1	25	2	.4337019	0.9734*	(none)
2	p2	25	2	.3783479	0.9776*	(none)

```
* Uncentered R-sq
```

	Coef.	Std. Err.	z	P>|z|	[95% Conf.	Interval]
/b1	.3926631	.064203	6.12	0.000	.2668275	.5184987
/b2	1.119593	.0088999	125.80	0.000	1.102149	1.137036
/g1	.5090441	.0669495	7.60	0.000	.3778256	.6402626
/g2	1.102315	.0072183	152.71	0.000	1.088167	1.116463

The header of the output contains a summary of each equation, including the number of observations and parameters and the root mean squared error of the residuals. `nlsur` checks to see whether each equation contains a constant term, and if an equation does contain a constant term, an R^2 statistic is presented. If an equation does not have a constant term, an uncentered R^2 is instead reported. The R^2 statistic for each equation measures the percentage of variance explained by the nonlinear function and may be useful for descriptive purposes, though it does not have the same formal interpretation in the context of FGNLS as it does with NLS estimation. As we would expect, β_2 and γ_2 are both greater than one, indicating the two bacterial populations increased in size over time.

◁

The model we fit in the next three examples is in fact linear in the parameters, so it could be fit using the `sureg` command. However, we will fit the model using `nlsur` so that we can focus on the mechanics of using the command. Moreover, using `nlsur` will obviate the need to generate several variables as well as the need to use the `constraint` command to impose parameter restrictions.

▷ Example 2: Interactive version using iterative FGNLS estimator—the translog production function

Greene (1997, sec. 15.6) discusses the transcendental logarithmic (translog) cost function and provides cost and input price data for capital, labor, energy, and materials for the U.S. economy. One way to fit the translog production function to these data is to fit the system of three equations

$$
\begin{aligned}
s_k &= \beta_k + \delta_{kk}\ln\left(\frac{p_k}{p_m}\right) + \delta_{kl}\ln\left(\frac{p_l}{p_m}\right) + \delta_{ke}\ln\left(\frac{p_e}{p_m}\right) + u_1 \\
s_l &= \beta_l + \delta_{kl}\ln\left(\frac{p_k}{p_m}\right) + \delta_{ll}\ln\left(\frac{p_l}{p_m}\right) + \delta_{le}\ln\left(\frac{p_e}{p_m}\right) + u_2 \\
s_e &= \beta_e + \delta_{ke}\ln\left(\frac{p_k}{p_m}\right) + \delta_{le}\ln\left(\frac{p_l}{p_m}\right) + \delta_{ee}\ln\left(\frac{p_e}{p_m}\right) + u_3
\end{aligned}
$$

where s_k is capital's cost share, s_l is labor's cost share, and s_e is energy's cost share; p_k, p_l, p_e, and p_m are the prices of capital, labor, energy, and materials, respectively; the u's are regression error terms; and the βs and δs are parameters to be estimated. There are three cross-equation restrictions on the parameters: δ_{kl}, δ_{ke}, and δ_{le} each appear in two equations. To fit this model by using the iterative FGNLS estimator, we type

```
. use http://www.stata-press.com/data/r12/mfgcost
. nlsur (s_k = {bk} + {dkk}*ln(pk/pm) + {dkl}*ln(pl/pm) + {dke}*ln(pe/pm))
>       (s_l = {bl} + {dkl}*ln(pk/pm) + {dll}*ln(pl/pm) + {dle}*ln(pe/pm))
>       (s_e = {be} + {dke}*ln(pk/pm) + {dle}*ln(pl/pm) + {dee}*ln(pe/pm)),
>       ifgnls
(obs = 25)
Calculating NLS estimates...
Iteration 0:  Residual SS =  .0009989
Iteration 1:  Residual SS =  .0009989
Calculating FGNLS estimates...
Iteration 0:  Scaled RSS =  65.45197
Iteration 1:  Scaled RSS =  65.45197
  (output omitted )
FGNLS iteration 10...
Iteration 0:  Scaled RSS =        75
Iteration 1:  Scaled RSS =        75
Iteration 2:  Scaled RSS =        75
Parameter change          =  4.076e-06
Covariance matrix change  =  6.264e-10
FGNLS regression
```

	Equation	Obs	Parms	RMSE	R-sq	Constant
1	s_k	25	4	.0031722	0.4776	bk
2	s_l	25	4	.0053963	0.8171	bl
3	s_e	25	4	.00177	0.6615	be

	Coef.	Std. Err.	z	P>|z|	[95% Conf.	Interval]
/bk	.0568925	.0013454	42.29	0.000	.0542556	.0595294
/dkk	.0294833	.0057956	5.09	0.000	.0181241	.0408425
/dkl	-.0000471	.0038478	-0.01	0.990	-.0075887	.0074945
/dke	-.0106749	.0033882	-3.15	0.002	-.0173157	-.0040341
/bl	.253438	.0020945	121.00	0.000	.2493329	.2575432
/dll	.0754327	.0067572	11.16	0.000	.0621889	.0886766
/dle	-.004756	.002344	-2.03	0.042	-.0093502	-.0001619
/be	.0444099	.0008533	52.04	0.000	.0427374	.0460823
/dee	.0183415	.0049858	3.68	0.000	.0085694	.0281135

We draw your attention to the iteration log at the top of the output. When iterative FGNLS estimation is used, the final scaled RSS will equal the product of the number of observations in the estimation sample and the number of equations; see *Methods and formulas* for details. Because the RSS is scaled by the error covariance matrix during each round of FGNLS estimation, the scaled RSS is not comparable from one FGNLS iteration to the next.

◁

❑ Technical note

You may have noticed that we mentioned having data for four factors of production, yet we fit only three share equations. Because the four shares sum to one, we must drop one of the equations to avoid having a singular error covariance matrix. The iterative FGNLS estimator is equivalent to maximum likelihood estimation, and thus it is invariant to which one of the four equations we choose to drop. The (linearly restricted) parameters of the fourth equation can be obtained using the `lincom` command. Nonlinear functions of the parameters, such as the elasticities of substitution, can be computed using `nlcom`.

❑

Substitutable expression programs

If you fit the same model repeatedly or you want to share code with colleagues, you can write a *substitutable expression program* to define your system of equations and avoid having to retype the system every time. The first five letters of the program's name must be `nlsur`, and the program must set the r-class macro `r(n_eq)` to the number of equations in your system. The first equation's substitutable expression must be returned in `r(eq_1)`, the second equation's in `r(eq_2)`, and so on. You may optionally set `r(title)` to label your output; that has the same effect as specifying the `title()` option.

▷ Example 3: Programmed substitutable expression version

We return to our translog cost function, for which a substitutable expression program is

```
program nlsurtranslog, rclass
        version 12
        syntax varlist(min=7 max=7) [if]
        tokenize `varlist'
        args sk sl se pk pl pe pm
        local pkpm ln(`pk'/`pm')
        local plpm ln(`pl'/`pm')
        local pepm ln(`pe'/`pm')
        return scalar n_eq = 3
        return local eq_1 "`sk'= {bk} + {dkk}*`pkpm' + {dkl}*`plpm' + {dke}*`pepm'"
        return local eq_2 "`sl'= {bl} + {dkl}*`pkpm' + {dll}*`plpm' + {dle}*`pepm'"
        return local eq_3 "`se'= {be} + {dke}*`pkpm' + {dle}*`plpm' + {dee}*`pepm'"
        return local title "4-factor translog cost function"
end
```

We made our program accept seven variables, for the three dependent variables s_k, s_l, and s_e, and the four factor prices p_k, p_l, p_m, and p_e. The `tokenize` command assigns to macros `` `1' ``, `` `2' ``, ..., `` `7' `` the seven variables stored in `` `varlist' ``, and the `args` command transfers those numbered macros to macros `` `sk' ``, `` `sl' ``, ..., `` `pm' ``. Because we knew our substitutable expressions were going to be somewhat long, we created local macros to hold the log price ratios. These are simply macros that hold strings such as `ln(pk/pm)`, not variables, and they will save us some repetitious typing when we define our substitutable expressions. Our program returns the number of equations in `r(n_eq)`, and we defined our substitutable expressions in `eq_1`, `eq_2`, and `eq_3`. We do not bind the expressions in parentheses as we do with the interactive version of `nlsur`. Finally, we put a title in `r(title)` to label our output.

Our `syntax` command also accepts an `if` clause, and that is how `nlsur` indicates the estimation sample to our program. In this application, we can safely ignore it, because our program does not compute initial values. However, had we used commands such as `summarize` or `regress` to obtain initial values, then we would need to restrict those commands to analyze only the estimation sample. In those cases, typically, you simply need to include `` `if' `` with the commands you are using. For example, instead of the command

```
summarize `depvar', meanonly
```

you would use

```
summarize `depvar' `if', meanonly
```

We can check our program by typing

```
. nlsurtranslog s_k s_l s_e pk pl pe pm
. return list
scalars:
              r(n_eq) =  3
macros:
             r(title) : "4-factor translog cost function"
              r(eq_3) : "s_e= {be} + {dke}*ln(pk/pm) + {dle}*ln(pl/pm) + {dee.."
              r(eq_2) : "s_l= {bl} + {dkl}*ln(pk/pm) + {dll}*ln(pl/pm) + {dle.."
              r(eq_1) : "s_k= {bk} + {dkk}*ln(pk/pm) + {dkl}*ln(pl/pm) + {dke.."
```

Now that we know that our program works, we fit our model by typing

```
. nlsur translog: s_k s_l s_e pk pl pe pm, ifgnls
(obs = 25)
Calculating NLS estimates...
Iteration 0:  Residual SS =  .0009989
Iteration 1:  Residual SS =  .0009989
Calculating FGNLS estimates...
Iteration 0:  Scaled RSS =  65.45197
Iteration 1:  Scaled RSS =  65.45197
FGNLS iteration 2...
Iteration 0:  Scaled RSS =  73.28311
Iteration 1:  Scaled RSS =  73.28311
Parameter change         =  6.537e-03
Covariance matrix change =  1.002e-06
  (output omitted)
FGNLS iteration 10...
Iteration 0:  Scaled RSS =        75
Iteration 1:  Scaled RSS =        75
Iteration 2:  Scaled RSS =        75
Parameter change         =  4.076e-06
Covariance matrix change =  6.264e-10
FGNLS regression
```

Equation		Obs	Parms	RMSE	R-sq	Constant
1	s_k	25	4	.0031722	0.4776	bk
2	s_l	25	4	.0053963	0.8171	bl
3	s_e	25	4	.00177	0.6615	be

4-factor translog cost function

	Coef.	Std. Err.	z	P>|z|	[95% Conf.	Interval]
/bk	.0568925	.0013454	42.29	0.000	.0542556	.0595294
/dkk	.0294833	.0057956	5.09	0.000	.0181241	.0408425
/dkl	-.0000471	.0038478	-0.01	0.990	-.0075887	.0074945
/dke	-.0106749	.0033882	-3.15	0.002	-.0173157	-.0040341
/bl	.253438	.0020945	121.00	0.000	.2493329	.2575432
/dll	.0754327	.0067572	11.16	0.000	.0621889	.0886766
/dle	-.004756	.002344	-2.03	0.042	-.0093502	-.0001619
/be	.0444099	.0008533	52.04	0.000	.0427374	.0460823
/dee	.0183415	.0049858	3.68	0.000	.0085694	.0281135

Because we set `r(title)` in our substitutable expression program, the coefficient table has a title attached to it. The estimates are identical to those we obtained in example 2.

◁

❑ Technical note

nlsur accepts frequency and analytic weights as well as pweights (sampling weights) and iweights (importance weights). You do not need to modify your substitutable expressions in any way to perform weighted estimation, though you must make two changes to your substitutable expression program. The general outline of a *sexp_prog* program is

```
program nlsur name, rclass
   version 12
   syntax varlist [fw aw pw iw] [if]
   // Obtain initial values incorporating weights.  For example,
   summarize varname [`weight'`exp'] `if'
   ...
   // Return n_eqn and substitutable expressions
   return scalar n_eq = #
   return local eq_1 = ...
   ...
end
```

First, we wrote the syntax statement to accept a weight expression. Here we allow all four types of weights, but if you know that your estimator is valid, say, for only frequency weights, then you should modify the syntax line to accept only fweights. Second, if your program computes starting values, then any commands you use must incorporate the weights passed to the program; you do that by including [`weight'`exp'] when calling those commands.

❑

Function evaluator programs

Although substitutable expressions are extremely flexible, there are some problems for which the nonlinear system cannot be defined using them. You can use the function evaluator program version of nlsur in these cases. We present two examples, a simple one to illustrate the mechanics of function evaluator programs and a more complicated one to illustrate the power of nlsur.

▷ Example 4: Function evaluator program version

Here we write a function evaluator program to fit the translog cost function used in examples 2 and 3. The function evaluator program is

```
program nlsurtranslog2
        version 12
        syntax varlist(min=7 max=7) [if], at(name)
        tokenize `varlist'
        args sk sl se pk pl pe pm
        tempname bk dkk dkl dke bl dll dle be dee
        scalar `bk'  = `at'[1,1]
        scalar `dkk' = `at'[1,2]
        scalar `dkl' = `at'[1,3]
        scalar `dke' = `at'[1,4]
        scalar `bl'  = `at'[1,5]
        scalar `dll' = `at'[1,6]
        scalar `dle' = `at'[1,7]
        scalar `be'  = `at'[1,8]
        scalar `dee' = `at'[1,9]
        local pkpm ln(`pk'/`pm')
        local plpm ln(`pl'/`pm')
        local pepm ln(`pe'/`pm')
        quietly {
                replace `sk' = `bk' + `dkk'*`pkpm' + `dkl'*`plpm' +     ///
                                      `dke'*`pepm' `if'
                replace `sl' = `bl' + `dkl'*`pkpm' + `dll'*`plpm' +     ///
                                      `dle'*`pepm' `if'
                replace `se' = `be' + `dke'*`pkpm' + `dle'*`plpm' +     ///
                                      `dee'*`pepm' `if'
        }
end
```

Unlike the substitutable expression program we wrote in example 3, `nlsurtranslog2` is not declared as r-class because we will not be returning any saved results. We are again expecting seven variables: three shares and four factor prices, and `nlsur` will again mark the estimation sample with an `if` expression.

Our function evaluator program also accepts an option named `at()`, which will receive a parameter vector at which we are to evaluate the system of equations. All function evaluator programs must accept this option. Our model has nine parameters to estimate, and we created nine temporary scalars to hold the elements of the `` `at' `` matrix.

Because our model has three equations, the first three variables passed to our program are the dependent variables that we are to fill in with the function values. We replaced only the observations in our estimation sample by including the `` `if' `` qualifier in the `replace` statements. Here we could have ignored the `` `if' `` qualifier because `nlsur` will skip over observations not in the estimation sample and we did not perform any computations requiring knowledge of the estimation sample. However, including the `` `if' `` is good practice and may result in a slight speed improvement if the functions of your model are complicated and the estimation sample is much smaller than the dataset in memory.

We could have avoided creating temporary scalars to hold our individual parameters by writing the `replace` statements as, for example,

```
replace `sk' = `at'[1,1] + `at'[1,2]*`pkpm' + `at'[1,3]*`plpm' + `at'[1,4]*`pepm' `if'
```

You can use whichever method you find more appealing, though giving the parameters descriptive names reduces the chance for mistakes and makes debugging easier.

To fit our model by using the function evaluator program version of nlsur, we type

```
. nlsur translog2 @ s_k s_l s_e pk pl pe pm, ifgnls nequations(3)
>       parameters(bk dkk dkl dke bl dll dle be dee)
>       hasconstants(bk bl be)
(obs = 25)

Calculating NLS estimates...
Iteration 0:  Residual SS =  .0009989
Iteration 1:  Residual SS =  .0009989
Calculating FGNLS estimates...
Iteration 0:  Scaled RSS =  65.45197
Iteration 1:  Scaled RSS =  65.45197
FGNLS iteration 2...
Iteration 0:  Scaled RSS =  73.28311
Iteration 1:  Scaled RSS =  73.28311
Parameter change         =  6.537e-03
Covariance matrix change =  1.002e-06
FGNLS iteration 3...
Iteration 0:  Scaled RSS =   74.7113
Iteration 1:  Scaled RSS =   74.7113
Parameter change         =  2.577e-03
Covariance matrix change =  3.956e-07
FGNLS iteration 4...
Iteration 0:  Scaled RSS =  74.95356
Iteration 1:  Scaled RSS =  74.95356
Parameter change         =  1.023e-03
Covariance matrix change =  1.571e-07
FGNLS iteration 5...
Iteration 0:  Scaled RSS =  74.99261
Iteration 1:  Scaled RSS =  74.99261
Iteration 2:  Scaled RSS =  74.99261
Iteration 3:  Scaled RSS =  74.99261
Parameter change         =  4.067e-04
Covariance matrix change =  6.250e-08
FGNLS iteration 6...
Iteration 0:  Scaled RSS =  74.99883
Iteration 1:  Scaled RSS =  74.99883
Parameter change         =  1.619e-04
Covariance matrix change =  2.489e-08
FGNLS iteration 7...
Iteration 0:  Scaled RSS =  74.99981
Iteration 1:  Scaled RSS =  74.99981
Parameter change         =  6.449e-05
Covariance matrix change =  9.912e-09
FGNLS iteration 8...
Iteration 0:  Scaled RSS =  74.99997
Iteration 1:  Scaled RSS =  74.99997
Iteration 2:  Scaled RSS =  74.99997
Parameter change         =  2.569e-05
Covariance matrix change =  3.948e-09
FGNLS iteration 9...
Iteration 0:  Scaled RSS =        75
Iteration 1:  Scaled RSS =        75
Iteration 2:  Scaled RSS =        75
Parameter change         =  1.023e-05
Covariance matrix change =  1.573e-09
FGNLS iteration 10...
Iteration 0:  Scaled RSS =        75
Iteration 1:  Scaled RSS =        75
Iteration 2:  Scaled RSS =        75
Parameter change         =  4.076e-06
Covariance matrix change =  6.264e-10
```

```
FGNLS regression

            Equation |       Obs  Parms        RMSE      R-sq     Constant
---------------------+----------------------------------------------------
1                s_k |        25      .    .0031722    0.4776           bk
2                s_l |        25      .    .0053963    0.8171           bl
3                s_e |        25      .      .00177    0.6615           be
--------------------------------------------------------------------------

------------------------------------------------------------------------------
             |      Coef.   Std. Err.      z    P>|z|     [95% Conf. Interval]
-------------+----------------------------------------------------------------
         /bk |   .0568925   .0013454    42.29   0.000     .0542556    .0595294
        /dkk |   .0294833   .0057956     5.09   0.000     .0181241    .0408425
        /dkl |  -.0000471   .0038478    -0.01   0.990    -.0075887    .0074945
        /dke |  -.0106749   .0033882    -3.15   0.002    -.0173157   -.0040341
         /bl |    .253438   .0020945   121.00   0.000     .2493329    .2575432
        /dll |   .0754327   .0067572    11.16   0.000     .0621889    .0886766
        /dle |   -.004756    .002344    -2.03   0.042    -.0093502   -.0001619
         /be |   .0444099   .0008533    52.04   0.000     .0427374    .0460823
        /dee |   .0183415   .0049858     3.68   0.000     .0085694    .0281135
------------------------------------------------------------------------------
```

When we use the function evaluator program version, `nlsur` requires us to specify the number of equations in `nequations()`, and it requires us to either specify names for each of our parameters or the number of parameters in the model. Here we used the `parameters()` option to name our parameters; the order in which we specified them in this option is the same as the order in which we extracted them from the `'at'` matrix in our program. Had we instead specified `nparameters(9)`, our parameters would have been labeled `/b1`, `/b2`, ..., `/b9` in the output.

`nlsur` has no way of telling how many parameters appear in each equation, so the `Parms` column in the header contains missing values. Moreover, the function evaluator program version of `nlsur` does not attempt to identify constant terms, so we used the `hasconstant` option to tell `nlsur` which parameter in each equation is a constant term.

The estimates are identical to those we obtained in examples 2 and 3.

◁

❑ Technical note

As with substitutable expression programs, if you intend to do weighted estimation with a function evaluator program, you must modify your *func_prog* program's `syntax` statement to accept weights. Moreover, if you use any statistical commands when computing your nonlinear functions, then you must include the weight expression with those commands.

❑

▷ Example 5: Fitting the basic AIDS model using nlsur

Poi (2002) showed how to fit a quadratic almost ideal demand system (AIDS) by using the `ml` command. Here we show how to fit the basic AIDS model by using `nlsur`. Poi (2008) shows how to fit the quadratic AIDS model using `nlsur`. The dataset `food.dta` contains household expenditures, expenditure shares, and log prices for four broad food groups. For a four-good demand system, we need to fit the following system of three equations:

$$w_1 = \alpha_1 + \gamma_{11}\,\mathrm{ln}p_1 + \gamma_{12}\,\mathrm{ln}p_2 + \gamma_{13}\,\mathrm{ln}p_3 + \beta_1 \ln\left\{\frac{m}{P(\mathbf{p})}\right\} + u_1$$

$$w_2 = \alpha_2 + \gamma_{12}\,\mathrm{ln}p_1 + \gamma_{22}\,\mathrm{ln}p_2 + \gamma_{23}\,\mathrm{ln}p_3 + \beta_2 \ln\left\{\frac{m}{P(\mathbf{p})}\right\} + u_2$$

$$w_3 = \alpha_3 + \gamma_{13}\,\mathrm{ln}p_1 + \gamma_{23}\,\mathrm{ln}p_2 + \gamma_{33}\,\mathrm{ln}p_3 + \beta_3 \ln\left\{\frac{m}{P(\mathbf{p})}\right\} + u_3$$

where w_k denotes a household's fraction of expenditures on good k, $\mathrm{ln}p_k$ denotes the logarithm of the price paid for good k, m denotes a household's total expenditure on all four goods, the u's are regression error terms, and

$$\mathrm{ln}P(\mathbf{p}) = \alpha_0 + \sum_{i=1}^{4} \alpha_i\,\mathrm{ln}p_i + \frac{1}{2}\sum_{i=1}^{4}\sum_{j=1}^{4} \gamma_{ij}\,\mathrm{ln}p_i\,\mathrm{ln}p_j$$

The parameters for the fourth good's share equation can be recovered from the following constraints that are imposed by economic theory:

$$\sum_{i=1}^{4} \alpha_i = 1 \qquad \sum_{i=1}^{4} \beta_i = 0 \qquad \gamma_{ij} = \gamma_{ji} \qquad \text{and} \qquad \sum_{i=1}^{4} \gamma_{ij} = 0 \ \text{ for all } j$$

Our model has a total of 12 unrestricted parameters. We will not estimate α_0 directly. Instead, we will set it equal to 5 as was done in Poi (2002); see Deaton and Muellbauer (1980) for a discussion of why treating α_0 as fixed is acceptable.

Our function evaluator program is

```
program nlsuraids
        version 12
        syntax varlist(min=8 max=8) if, at(name)
        tokenize `varlist'
        args w1 w2 w3 lnp1 lnp2 lnp3 lnp4 lnm
        tempname a1 a2 a3 a4
        scalar `a1' = `at'[1,1]
        scalar `a2' = `at'[1,2]
        scalar `a3' = `at'[1,3]
        scalar `a4' = 1 - `a1' - `a2' - `a3'
        tempname b1 b2 b3
        scalar `b1' = `at'[1,4]
        scalar `b2' = `at'[1,5]
        scalar `b3' = `at'[1,6]
        tempname g11 g12 g13 g14
        tempname g21 g22 g23 g24
        tempname g31 g32 g33 g34
        tempname g41 g42 g43 g44
        scalar `g11' = `at'[1,7]
        scalar `g12' = `at'[1,8]
        scalar `g13' = `at'[1,9]
        scalar `g14' = -`g11'-`g12'-`g13'
        scalar `g21' = `g12'
        scalar `g22' = `at'[1,10]
        scalar `g23' = `at'[1,11]
        scalar `g24' = -`g21'-`g22'-`g23'
        scalar `g31' = `g13'
        scalar `g32' = `g23'
        scalar `g33' = `at'[1,12]
        scalar `g34' = -`g31'-`g32'-`g33'
        scalar `g41' = `g14'
        scalar `g42' = `g24'
        scalar `g43' = `g34'
        scalar `g44' = -`g41'-`g42'-`g43'
        quietly {
                tempvar lnpindex
                gen double `lnpindex' = 5 + `a1'*`lnp1' + `a2'*`lnp2' + ///
                                            `a3'*`lnp3' + `a4'*`lnp4'
                forvalues i = 1/4 {
                        forvalues j = 1/4 {
                                replace `lnpindex' = `lnpindex' + ///
                                        0.5*`g`i'`j''*`lnp`i''*`lnp`j''
                        }
                }
                replace `w1' = `a1' + `g11'*`lnp1' + `g12'*`lnp2' + ///
                                      `g13'*`lnp3' + `g14'*`lnp4' + ///
                                      `b1'*(`lnm' - `lnpindex')
                replace `w2' = `a2' + `g21'*`lnp1' + `g22'*`lnp2' + ///
                                      `g23'*`lnp3' + `g24'*`lnp4' + ///
                                      `b2'*(`lnm' - `lnpindex')
                replace `w3' = `a3' + `g31'*`lnp1' + `g32'*`lnp2' + ///
                                      `g33'*`lnp3' + `g34'*`lnp4' + ///
                                      `b3'*(`lnm' - `lnpindex')
        }
end
```

The `syntax` statement accepts eight variables: three expenditure share variables, all four log-price variables, and a variable for log expenditures ($\ln m$). Most of the code simply extracts the parameters from the `at` matrix. Although we are estimating only 12 parameters, to calculate the price index term and the expenditure share equations, we need the restricted parameters as well. Notice how we impose the constraints on the parameters. We then created a temporary variable to hold $\ln P(\mathbf{p})$, and we filled the three dependent variables with the predicted expenditure shares.

To fit our model, we type

```
. use http://www.stata-press.com/data/r12/food

. nlsur aids @ w1 w2 w3 lnp1 lnp2 lnp3 lnp4 lnexp,
>             parameters(a1 a2 a3 b1 b2 b3
>                        g11 g12 g13 g22 g32 g33)
>             neq(3) ifgnls
(obs = 4048)

Calculating NLS estimates...
Iteration 0:  Residual SS =  126.9713
Iteration 1:  Residual SS =   125.669
Iteration 2:  Residual SS =   125.669
Iteration 3:  Residual SS =   125.669
Iteration 4:  Residual SS =   125.669
Calculating FGNLS estimates...
Iteration 0:  Scaled RSS =  12080.14
Iteration 1:  Scaled RSS =  12080.14
Iteration 2:  Scaled RSS =  12080.14
Iteration 3:  Scaled RSS =  12080.14
FGNLS iteration 2...
Iteration 0:  Scaled RSS =  12143.99
Iteration 1:  Scaled RSS =  12143.99
Iteration 2:  Scaled RSS =  12143.99
Parameter change         =  1.972e-04
Covariance matrix change =  2.936e-06
FGNLS iteration 3...
Iteration 0:  Scaled RSS =     12144
Iteration 1:  Scaled RSS =     12144
Parameter change         =  2.178e-06
Covariance matrix change =  3.467e-08

FGNLS regression
```

	Equation	Obs	Parms	RMSE	R-sq	Constant
1	w1	4048	.	.1333175	0.9017*	(none)
2	w2	4048	.	.1024166	0.8480*	(none)
3	w3	4048	.	.053777	0.7906*	(none)

* Uncentered R-sq

	Coef.	Std. Err.	z	P>|z|	[95% Conf.	Interval]
/a1	.3163958	.0073871	42.83	0.000	.3019175	.3308742
/a2	.2712501	.0056938	47.64	0.000	.2600904	.2824097
/a3	.1039898	.0029004	35.85	0.000	.0983051	.1096746
/b1	.0161044	.0034153	4.72	0.000	.0094105	.0227983
/b2	-.0260771	.002623	-9.94	0.000	-.0312181	-.0209361
/b3	.0014538	.0013776	1.06	0.291	-.0012463	.004154
/g11	.1215838	.0057186	21.26	0.000	.1103756	.1327921
/g12	-.0522943	.0039305	-13.30	0.000	-.0599979	-.0445908
/g13	-.0351292	.0021788	-16.12	0.000	-.0393996	-.0308588
/g22	.0644298	.0044587	14.45	0.000	.0556909	.0731687
/g32	-.0011786	.0019767	-0.60	0.551	-.0050528	.0026957
/g33	.0424381	.0017589	24.13	0.000	.0389909	.0458854

To get the restricted parameters for the fourth share equation, we can use lincom. For example, to obtain α_4, we type

```
. lincom 1 - [a1]_cons - [a2]_cons - [a3]_cons
 ( 1) - [a1]_cons - [a2]_cons - [a3]_cons = -1
```

	Coef.	Std. Err.	z	P>|z|	[95% Conf.	Interval]
(1)	.3083643	.0052611	58.61	0.000	.2980528	.3186758

For more information on lincom, see [R] **lincom**.

◁

Saved results

nlsur saves the following in e():

Scalars

e(N)	number of observations
e(k)	number of parameters
e(k_#)	number of parameters for equation #
e(k_eq)	number of equation names in e(b)
e(k_eq_model)	number of equations in overall model test
e(n_eq)	number of equations
e(mss_#)	model sum of squares for equation #
e(rss_#)	RSS for equation #
e(rmse_#)	root mean squared error for equation #
e(r2_#)	R^2 for equation #
e(ll)	Gaussian log likelihood (iflgs version only)
e(N_clust)	number of clusters
e(rank)	rank of e(V)
e(converge)	1 if converged, 0 otherwise

Macros

e(cmd)	nlsur
e(cmdline)	command as typed
e(method)	fgnls, ifgnls, or nls
e(depvar)	names of dependent variables
e(depvar_#)	dependent variable for equation #
e(wtype)	weight type
e(wexp)	weight expression
e(title)	title in estimation output
e(title_2)	secondary title in estimation output
e(clustvar)	name of cluster variable
e(vce)	*vcetype* specified in vce()
e(vcetype)	title used in label Std. Err.
e(type)	1 = interactively entered expression 2 = substitutable expression program 3 = function evaluator program
e(sexpprog)	substitutable expression program
e(sexp_#)	substitutable expression for equation #
e(params)	names of all parameters
e(params_#)	parameters in equation #
e(funcprog)	function evaluator program
e(rhs)	contents of variables()
e(constants)	identifies constant terms
e(properties)	b V
e(predict)	program used to implement predict

Matrices

`e(b)`	coefficient vector
`e(init)`	initial values vector
`e(Sigma)`	error covariance matrix ($\widehat{\boldsymbol{\Sigma}}$)
`e(V)`	variance–covariance matrix of the estimators

Functions

`e(sample)`	marks estimation sample

Methods and formulas

`nlsur` is implemented as an ado-file.

Write the system of equations for the ith observation as

$$\mathbf{y}_i = \mathbf{f}(\mathbf{x}_i, \beta) + \mathbf{u}_i \tag{1}$$

where $\mathbf{y}_i$ and $\mathbf{u}_i$ are $1 \times M$ vectors, for $i = 1, \ldots, N$; $\mathbf{f}$ is a function that returns a $1 \times M$ vector; $\mathbf{x}_i$ represents all the exogenous variables in the system; and β is a $1 \times k$ vector of parameters. The generalized nonlinear least-squares system estimator is defined as

$$\widehat{\beta} \equiv \text{argmin}_\beta \sum_{i=1}^{N} \{\mathbf{y}_i - \mathbf{f}(\mathbf{x}_i, \beta)\} \boldsymbol{\Sigma}^{-1} \{\mathbf{y}_i - \mathbf{f}(\mathbf{x}_i, \beta)\}'$$

where $\boldsymbol{\Sigma} = E(\mathbf{u}_i'\mathbf{u}_i)$ is an $M \times M$ positive-definite weight matrix. Let $\mathbf{T}$ be the Cholesky decomposition of $\boldsymbol{\Sigma}^{-1}$; that is, $\mathbf{T}\mathbf{T}' = \boldsymbol{\Sigma}^{-1}$. Postmultiply (1) by $\mathbf{T}$:

$$\mathbf{y}_i\mathbf{T} = \mathbf{f}(\mathbf{x}_i, \beta)\mathbf{T} + \mathbf{u}_i\mathbf{T} \tag{2}$$

Because $E(\mathbf{T}'\mathbf{u}_i'\mathbf{u}_i\mathbf{T}) = \mathbf{I}$, we can "stack" the columns of (2) and write

$$\begin{aligned}
\mathbf{y}_1\mathbf{T}_1 &= \mathbf{f}(\mathbf{x}_1, \beta)\mathbf{T}_1 + \widetilde{u}_{11} \\
\mathbf{y}_1\mathbf{T}_2 &= \mathbf{f}(\mathbf{x}_1, \beta)\mathbf{T}_2 + \widetilde{u}_{12} \\
\vdots &= \vdots \\
\mathbf{y}_1\mathbf{T}_M &= \mathbf{f}(\mathbf{x}_1, \beta)\mathbf{T}_M + \widetilde{u}_{1M} \\
\vdots &= \vdots \\
\mathbf{y}_N\mathbf{T}_1 &= \mathbf{f}(\mathbf{x}_N, \beta)\mathbf{T}_1 + \widetilde{u}_{N1} \\
\mathbf{y}_N\mathbf{T}_2 &= \mathbf{f}(\mathbf{x}_N, \beta)\mathbf{T}_2 + \widetilde{u}_{N2} \\
\vdots &= \vdots \\
\mathbf{y}_N\mathbf{T}_M &= \mathbf{f}(\mathbf{x}_N, \beta)\mathbf{T}_M + \widetilde{u}_{NM}
\end{aligned} \tag{3}$$

where $\mathbf{T}_j$ denotes the jth column of $\mathbf{T}$. By construction, all $\widetilde{u}_{ij}$ are independently distributed with unit variance. As a result, by transforming the model in (1) to that shown in (3), we have reduced the multivariate generalized nonlinear least-squares system estimator to a univariate nonlinear least-squares problem; and the same parameter estimation technique used by `nl` can be used here. See [R] **nl** for the details. Moreover, because the $\widetilde{u}_{ij}$ all have variance 1, the final scaled RSS reported by `nlsur` is equal to NM.

To make the estimator feasible, we require an estimate $\widehat{\boldsymbol{\Sigma}}$ of $\boldsymbol{\Sigma}$. `nlsur` first sets $\widehat{\boldsymbol{\Sigma}} = \mathbf{I}$. Although not efficient, the resulting estimate, $\widehat{\beta}_{\text{NLS}}$, is consistent. If the `nls` option is specified, estimation is complete. Otherwise, the residuals

$$\widehat{\mathbf{u}}_i = \mathbf{y}_i - \mathbf{f}(\mathbf{x}_i, \widehat{\beta}_{\text{NLS}})$$

are calculated and used to compute

$$\widehat{\boldsymbol{\Sigma}} = \frac{1}{N}\sum_{i=1}^{N} \widehat{\mathbf{u}}_i' \widehat{\mathbf{u}}_i$$

With $\widehat{\boldsymbol{\Sigma}}$ in hand, a new estimate $\widehat{\beta}$ is then obtained.

If the `ifgnls` option is specified, the new $\widehat{\beta}$ is used to recompute the residuals and obtain a new estimate of $\widehat{\boldsymbol{\Sigma}}$, from which $\widehat{\beta}$ can then be reestimated. Iterations stop when the relative change in $\widehat{\beta}$ is less than `eps()`, the relative change in $\widehat{\boldsymbol{\Sigma}}$ is less than `ifgnlseps()`, or if `ifgnlsiterate()` iterations have been performed.

If the `vce(robust)` and `vce(cluster` *clustvar*`)` options were not specified, then

$$V(\widehat{\beta}) = \left(\sum_{i=1}^{N} \mathbf{X}_i' \widehat{\boldsymbol{\Sigma}}^{-1} \mathbf{X}_i \right)^{-1}$$

where the $M \times k$ matrix $\mathbf{X}_i$ has typical element X_{ist}, the derivative of the sth element of $\mathbf{f}$ with respect to the tth element of β, evaluated at $\mathbf{x}_i$ and $\widehat{\beta}$. As a practical matter, once the model is written in the form of (3), the variance–covariance matrix can be calculated via a Gauss–Newton regression; see Davidson and MacKinnon (1993, chap. 6).

If `robust` is specified, then

$$V_R(\widehat{\beta}) = \left(\sum_{i=1}^{N} \mathbf{X}_i' \widehat{\boldsymbol{\Sigma}}^{-1} \mathbf{X}_i \right)^{-1} \sum_{i=1}^{N} \mathbf{X}_i' \widehat{\boldsymbol{\Sigma}}^{-1} \widehat{\mathbf{u}}_i' \widehat{\mathbf{u}}_i \widehat{\boldsymbol{\Sigma}}^{-1} \mathbf{X}_i \left(\sum_{i=1}^{N} \mathbf{X}_i' \widehat{\boldsymbol{\Sigma}}^{-1} \mathbf{X}_i \right)^{-1}$$

The cluster–robust variance matrix is

$$V_C(\widehat{\beta}) = \left(\sum_{i=1}^{N} \mathbf{X}_i' \widehat{\boldsymbol{\Sigma}}^{-1} \mathbf{X}_i \right)^{-1} \sum_{c=1}^{N_C} \mathbf{w}_c' \mathbf{w}_c \left(\sum_{i=1}^{N} \mathbf{X}_i' \widehat{\boldsymbol{\Sigma}}^{-1} \mathbf{X}_i \right)^{-1}$$

where N_C is the number of clusters and

$$\mathbf{w}_c = \sum_{j \in C_k} \mathbf{X}_j' \widehat{\boldsymbol{\Sigma}}^{-1} \widehat{\mathbf{u}}_j'$$

with C_k denoting the set of observations in the kth cluster. In evaluating these formulas, we use the value of $\widehat{\boldsymbol{\Sigma}}$ used in calculating the final estimate of $\widehat{\beta}$. That is, we do not recalculate $\widehat{\boldsymbol{\Sigma}}$ after we obtain the final value of $\widehat{\beta}$.

The RSS for the jth equation, RSS_j, is

$$\text{RSS}_j = \sum_{i=1}^{N} (\widehat{y}_{ij} - y_{ij})^2$$

where $\widehat{y}_{ij}$ is the predicted value of the ith observation on the jth dependent variable; the total sum of squares (TSS) for the jth equation, TSS_j, is

$$\text{TSS}_j = \sum_{i=1}^{N} (y_{ij} - \bar{y}_j)^2$$

if there is a constant term in the jth equation, where $\bar{y}_j$ is the sample mean of the jth dependent variable, and

$$\text{TSS}_j = \sum_{i=1}^{N} y_{ij}^2$$

if there is no constant term in the jth equation; and the model sum of squares (MSS) for the jth equation, MSS_j, is $\text{TSS}_j - \text{RSS}_j$.

The R^2 for the jth equation is $\text{MSS}_j/\text{TSS}_j$. If an equation does not have a constant term, then the reported R^2 for that equation is "uncentered" and based on the latter definition of TSS_j.

Under the assumption that the $\mathbf{u}_i$ are independent and identically distributed $N(\mathbf{0}, \widehat{\boldsymbol{\Sigma}})$, the log likelihood for the model is

$$\ln L = -\frac{MN}{2}\{1 + \ln(2\pi)\} - \frac{N}{2}\ln\left|\widehat{\boldsymbol{\Sigma}}\right|$$

The log likelihood is reported only when the `ifgnls` option is specified.

References

Davidson, R., and J. G. MacKinnon. 1993. *Estimation and Inference in Econometrics.* New York: Oxford University Press.

——. 2004. *Econometric Theory and Methods.* New York: Oxford University Press.

Deaton, A., and J. Muellbauer. 1980. An almost ideal demand system. *American Economic Review* 70: 312–326.

Greene, W. H. 1997. *Econometric Analysis.* 3rd ed. Upper Saddle River, NJ: Prentice Hall.

——. 2012. *Econometric Analysis.* 7th ed. Upper Saddle River, NJ: Prentice Hall.

Poi, B. P. 2002. From the help desk: Demand system estimation. *Stata Journal* 2: 403–410.

——. 2008. Demand-system estimation: Update. *Stata Journal* 8: 554–556.

Zellner, A. 1962. An efficient method of estimating seemingly unrelated regressions and tests for aggregation bias. *Journal of the American Statistical Association* 57: 348–368.

——. 1963. Estimators for seemingly unrelated regression equations: Some exact finite sample results. *Journal of the American Statistical Association* 58: 977–992.

Zellner, A., and D. S. Huang. 1962. Further properties of efficient estimators for seemingly unrelated regression equations. *International Economic Review* 3: 300–313.

Also see

[R] **nlsur postestimation** — Postestimation tools for nlsur

[R] **nl** — Nonlinear least-squares estimation

[R] **gmm** — Generalized method of moments estimation

[R] **sureg** — Zellner's seemingly unrelated regression

[R] **reg3** — Three-stage estimation for systems of simultaneous equations

[R] **ml** — Maximum likelihood estimation

[U] **20 Estimation and postestimation commands**

Title

nlsur postestimation — Postestimation tools for nlsur

Description

The following postestimation commands are available after `nlsur`:

Command	Description
`estat`	AIC, BIC, VCE, and estimation sample summary
`estimates`	cataloging estimation results
`lincom`	point estimates, standard errors, testing, and inference for linear combinations of coefficients
`lrtest`	likelihood-ratio test
`margins`[1]	marginal means, predictive margins, marginal effects, and average marginal effects
`marginsplot`	graph the results from margins (profile plots, interaction plots, etc.)
`nlcom`	point estimates, standard errors, testing, and inference for nonlinear combinations of coefficients
`predict`	predictions, residuals, influence statistics, and other diagnostic measures
`predictnl`	point estimates, standard errors, testing, and inference for generalized predictions
`test`	Wald tests of simple and composite linear hypotheses
`testnl`	Wald tests of nonlinear hypotheses

[1] You must specify the `variables()` option with `nlsur`.

See the corresponding entries in the *Base Reference Manual* for details.

Syntax for predict

`predict` [*type*] *newvar* [*if*] [*in*] [, `equation(#`*eqno*`)` `yhat` `residuals`]

These statistics are available both in and out of sample; type `predict ... if e(sample) ...` if wanted only for the estimation sample.

Menu

Statistics > Postestimation > Predictions, residuals, etc.

Options for predict

Main

`equation(#`*eqno*`)` specifies to which equation you are referring. `equation(#1)` would mean that the calculation is to be made for the first equation, `equation(#2)` would mean the second, and so on. If you do not specify `equation()`, results are the same as if you had specified `equation(#1)`.

`yhat`, the default, calculates the fitted values for the specified equation.

`residuals` calculates the residuals for the specified equation.

Remarks

▷ Example 1

In example 2 of [R] **nlsur**, we fit a four-factor translog cost function to data for the U.S. economy. The own-price elasticity for a factor measures the percentage change in its usage as a result of a 1% increase in the factor's price, assuming that output is held constant. For the translog production function, the own-price factor elasticities are

$$\eta_i = \frac{\delta_{ii} + s_i(s_i - 1)}{s_i}$$

Here we compute the elasticity for capital at the sample mean of capital's factor share. First, we use `summarize` to get the mean of `s_k` and store that value in a scalar:

```
. summarize s_k
    Variable |       Obs        Mean    Std. Dev.       Min        Max
-------------+--------------------------------------------------------
         s_k |        25     .053488    .0044795     .04602     .06185
. scalar kmean = r(mean)
```

Now we can use `nlcom` to calculate the elasticity:

```
. nlcom (([dkk]_cons + kmean*(kmean-1)) / kmean)
       _nl_1:  ([dkk]_cons + kmean*(kmean-1)) / kmean

------------------------------------------------------------------------------
             |      Coef.   Std. Err.      z    P>|z|     [95% Conf. Interval]
-------------+----------------------------------------------------------------
       _nl_1 |  -.3952986   .1083535    -3.65   0.000    -.6076676   -.1829295
------------------------------------------------------------------------------
```

If the price of capital increases by 1%, its usage will decrease by about 0.4%. To maintain its current level of output, a firm would increase its usage of other inputs to compensate for the lower capital usage. The standard error reported by `nlcom` reflects the sampling variance of the estimated parameter $\widehat{\delta_{kk}}$, but `nlcom` treats the sample mean of `s_k` as a fixed parameter that does not contribute to the sampling variance of the estimated elasticity.

◁

Methods and formulas

All postestimation commands listed above are implemented as ado-files.

Also see

[R] **nlsur** — Estimation of nonlinear systems of equations

[U] **20 Estimation and postestimation commands**

Title

nptrend — Test for trend across ordered groups

Syntax

`nptrend` *varname* [*if*] [*in*] , `by(`*groupvar*`)` [`nodetail` `score(`*scorevar*`)`]

Menu

Statistics > Nonparametric analysis > Tests of hypotheses > Trend test across ordered groups

Description

`nptrend` performs a nonparametric test for trend across ordered groups.

Options

Main

`by(`*groupvar*`)` is required; it specifies the group on which the data are to be ordered.

`nodetail` suppresses the listing of group rank sums.

`score(`*scorevar*`)` defines scores for groups. When it is not specified, the values of *groupvar* are used for the scores.

Remarks

`nptrend` performs the nonparametric test for trend across ordered groups developed by Cuzick (1985), which is an extension of the Wilcoxon rank-sum test (see [R] **ranksum**). A correction for ties is incorporated into the test. `nptrend` is a useful adjunct to the Kruskal–Wallis test; see [R] **kwallis**.

If your data are not grouped, you can test for trend with the `signtest` and `spearman` commands; see [R] **signrank** and [R] **spearman**. With `signtest`, you can perform the Cox and Stuart test, a sign test applied to differences between equally spaced observations of *varname*. With `spearman`, you can perform the Daniels test, a test of zero Spearman correlation between *varname* and a time index. See Conover (1999, 169–175, 323) for a discussion of these tests and their asymptotic relative efficiency.

▷ Example 1

The following data (Altman 1991, 217) show ocular exposure to ultraviolet radiation for 32 pairs of sunglasses classified into three groups according to the amount of visible light transmitted.

Group	Transmission of visible light	Ocular exposure to ultraviolet radiation
1	$< 25\%$	1.4 1.4 1.4 1.6 2.3 2.3
2	25 to 35%	0.9 1.0 1.1 1.1 1.2 1.2 1.5 1.9 2.2 2.6 2.6 2.6 2.8 2.8 3.2 3.5 4.3 5.1
3	$> 35\%$	0.8 1.7 1.7 1.7 3.4 7.1 8.9 13.5

Entering these data into Stata, we have

```
. use http://www.stata-press.com/data/r12/sg
. list, sep(6)

     +------------------+
     | group   exposure |
     |------------------|
  1. |     1        1.4 |
  2. |     1        1.4 |
  3. |     1        1.4 |
  4. |     1        1.6 |
  5. |     1        2.3 |
  6. |     1        2.3 |
     |------------------|
  7. |     2         .9 |
  (output omitted )
     |------------------|
 31. |     3        8.9 |
 32. |     3       13.5 |
     +------------------+
```

We use nptrend to test for a trend of (increasing) exposure across the three groups by typing

```
. nptrend exposure, by(group)
     group      score       obs      sum of ranks
         1          1         6             76
         2          2        18            290
         3          3         8            162

           z  =  1.52
  Prob > |z|  = 0.129
```

When the groups are given any equally spaced scores (such as -1, 0, 1), we will obtain the same answer as above. To illustrate the effect of changing scores, an analysis of these data with scores 1, 2, and 5 (admittedly not sensible here) produces

```
. gen mysc = cond(group==3,5,group)
. nptrend exposure, by(group) score(mysc)
     group      score       obs      sum of ranks
         1          1         6             76
         2          2        18            290
         3          5         8            162

           z  =  1.46
  Prob > |z|  = 0.143
```

This example suggests that the analysis is not all that sensitive to the scores chosen.

◁

❑ Technical note

The grouping variable may be either a string variable or a numeric variable. If it is a string variable and no score variable is specified, the natural numbers 1, 2, 3, ... are assigned to the groups in the sort order of the string variable. This may not always be what you expect. For example, the sort order of the strings "one", "two", "three" is "one", "three", "two".

❑

Saved results

`nptrend` saves the following in `r()`:

Scalars

`r(N)`	number of observations	`r(z)`	z statistic
`r(p)`	two-sided p-value	`r(T)`	test statistic

Methods and formulas

`nptrend` is implemented as an ado-file.

`nptrend` is based on a method in Cuzick (1985). The following description of the statistic is from Altman (1991, 215–217). We have k groups of sample sizes n_i ($i = 1, \ldots, k$). The groups are given scores, l_i, which reflect their ordering, such as 1, 2, and 3. The scores do not have to be equally spaced, but they usually are. $N = \sum n_i$ observations are ranked from 1 to N, and the sums of the ranks in each group, R_i, are obtained. L, the weighted sum of all the group scores, is

$$L = \sum_{i=1}^{k} l_i n_i$$

The statistic T is calculated as

$$T = \sum_{i=1}^{k} l_i R_i$$

Under the null hypothesis, the expected value of T is $E(T) = 0.5(N+1)L$, and its standard error is

$$\text{se}(T) = \sqrt{\frac{n+1}{12}\left(N\sum_{i=1}^{k} l_i^2 n_i - L^2\right)}$$

so that the test statistic, z, is given by $z = \{\,T - E(T)\,\}/\text{se}(T)$, which has an approximately standard normal distribution when the null hypothesis of no trend is true.

The correction for ties affects the standard error of T. Let $\widetilde{N}$ be the number of unique values of the variable being tested ($\widetilde{N} \le N$), and let t_j be the number of times the jth unique value of the variable appears in the data. Define

$$a = \frac{\sum_{j=1}^{\widetilde{N}} t_j(t_j^2 - 1)}{N(N^2 - 1)}$$

The corrected standard error of T is $\widetilde{\text{se}}(T) = \sqrt{1-a}\ \text{se}(T)$.

Acknowledgments

`nptrend` was written by K. A. Stepniewska and D. G. Altman (1992) of the Cancer Research UK.

References

Altman, D. G. 1991. *Practical Statistics for Medical Research*. London: Chapman & Hall/CRC.

Conover, W. J. 1999. *Practical Nonparametric Statistics*. 3rd ed. New York: Wiley.

Cuzick, J. 1985. A Wilcoxon-type test for trend. *Statistics in Medicine* 4: 87–90.

Sasieni, P. 1996. snp12: Stratified test for trend across ordered groups. *Stata Technical Bulletin* 33: 24–27. Reprinted in *Stata Technical Bulletin Reprints*, vol. 6, pp. 196–200. College Station, TX: Stata Press.

Sasieni, P., K. A. Stepniewska, and D. G. Altman. 1996. snp11: Test for trend across ordered groups revisited. *Stata Technical Bulletin* 32: 27–29. Reprinted in *Stata Technical Bulletin Reprints*, vol. 6, pp. 193–196. College Station, TX: Stata Press.

Stepniewska, K. A., and D. G. Altman. 1992. snp4: Non-parametric test for trend across ordered groups. *Stata Technical Bulletin* 9: 21–22. Reprinted in *Stata Technical Bulletin Reprints*, vol. 2, p. 169. College Station, TX: Stata Press.

Also see

[R] **kwallis** — Kruskal–Wallis equality-of-populations rank test

[R] **signrank** — Equality tests on matched data

[R] **spearman** — Spearman's and Kendall's correlations

[R] **symmetry** — Symmetry and marginal homogeneity tests

[ST] **epitab** — Tables for epidemiologists

[ST] **strate** — Tabulate failure rates and rate ratios

Title

ologit — Ordered logistic regression

Syntax

ologit *depvar* [*indepvars*] [*if*] [*in*] [*weight*] [, *options*]

options	Description
Model	
offset(*varname*)	include *varname* in model with coefficient constrained to 1
constraints(*constraints*)	apply specified linear constraints
collinear	keep collinear variables
SE/Robust	
vce(*vcetype*)	*vcetype* may be oim, robust, cluster *clustvar*, bootstrap, or jackknife
Reporting	
level(#)	set confidence level; default is level(95)
or	report odds ratios
nocnsreport	do not display constraints
display_options	control column formats, row spacing, line width, and display of omitted variables and base and empty cells
Maximization	
maximize_options	control the maximization process; seldom used
coeflegend	display legend instead of statistics

indepvars may contain factor variables; see [U] **11.4.3 Factor variables**.

depvar and *indepvars* may contain time-series operators; see [U] **11.4.4 Time-series varlists**.

bootstrap, by, fracpoly, jackknife, mfp, mi estimate, nestreg, rolling, statsby, stepwise, and svy are allowed; see [U] **11.1.10 Prefix commands**.

vce(bootstrap) and vce(jackknife) are not allowed with the mi estimate prefix; see [MI] **mi estimate**.

Weights are not allowed with the bootstrap prefix; see [R] **bootstrap**.

vce() and weights are not allowed with the svy prefix; see [SVY] **svy**.

fweights, iweights, and pweights are allowed; see [U] **11.1.6 weight**.

coeflegend does not appear in the dialog box.

See [U] **20 Estimation and postestimation commands** for more capabilities of estimation commands.

Menu

Statistics > Ordinal outcomes > Ordered logistic regression

Description

ologit fits ordered logit models of ordinal variable *depvar* on the independent variables *indepvars*. The actual values taken on by the dependent variable are irrelevant, except that larger values are assumed to correspond to "higher" outcomes.

See [R] **logistic** for a list of related estimation commands.

Options

Model

`offset(`*varname*`)`, `constraints(`*constraints*`)`, `collinear`; see [R] **estimation options**.

SE/Robust

`vce(`*vcetype*`)` specifies the type of standard error reported, which includes types that are derived from asymptotic theory, that are robust to some kinds of misspecification, that allow for intragroup correlation, and that use bootstrap or jackknife methods; see [R] ***vce_option***.

Reporting

`level(`*#*`)`; see [R] **estimation options**.

`or` reports the estimated coefficients transformed to odds ratios, that is, e^b rather than b. Standard errors and confidence intervals are similarly transformed. This option affects how results are displayed, not how they are estimated. `or` may be specified at estimation or when replaying previously estimated results.

`nocnsreport`; see [R] **estimation options**.

display_options: `noomitted`, `vsquish`, `noemptycells`, `baselevels`, `allbaselevels`, `cformat(`*%fmt*`)`, `pformat(`*%fmt*`)`, `sformat(`*%fmt*`)`, and `nolstretch`; see [R] **estimation options**.

Maximization

maximize_options: `difficult`, `technique(`*algorithm_spec*`)`, `iterate(`*#*`)`, `[no]log`, `trace`, `gradient`, `showstep`, `hessian`, `showtolerance`, `tolerance(`*#*`)`, `ltolerance(`*#*`)`, `nrtolerance(`*#*`)`, `nonrtolerance`, and `from(`*init_specs*`)`; see [R] **maximize**. These options are seldom used.

The following option is available with `ologit` but is not shown in the dialog box:

`coeflegend`; see [R] **estimation options**.

Remarks

Ordered logit models are used to estimate relationships between an ordinal dependent variable and a set of independent variables. An *ordinal* variable is a variable that is categorical and ordered, for instance, “poor”, “good”, and “excellent”, which might indicate a person’s current health status or the repair record of a car. If there are only two outcomes, see [R] **logistic**, [R] **logit**, and [R] **probit**. This entry is concerned only with more than two outcomes. If the outcomes cannot be ordered (for example, residency in the north, east, south, or west), see [R] **mlogit**. This entry is concerned only with models in which the outcomes can be ordered.

In ordered logit, an underlying score is estimated as a linear function of the independent variables and a set of cutpoints. The probability of observing outcome i corresponds to the probability that the estimated linear function, plus random error, is within the range of the cutpoints estimated for the outcome:

$$\Pr(\text{outcome}_j = i) = \Pr(\kappa_{i-1} < \beta_1 x_{1j} + \beta_2 x_{2j} + \cdots + \beta_k x_{kj} + u_j \le \kappa_i)$$

u_j is assumed to be logistically distributed in ordered logit. In either case, we estimate the coefficients β_1, β_2, $\dots$, β_k together with the cutpoints κ_1, κ_2, $\dots$, κ_{k-1}, where k is the number of possible outcomes. κ_0 is taken as $-\infty$, and κ_k is taken as $+\infty$. All of this is a direct generalization of the ordinary two-outcome logit model.

▷ Example 1

We wish to analyze the 1977 repair records of 66 foreign and domestic cars. The data are a variation of the automobile dataset described in [U] **1.2.2 Example datasets**. The 1977 repair records, like those in 1978, take on values "Poor", "Fair", "Average", "Good", and "Excellent". Here is a cross-tabulation of the data:

```
. use http://www.stata-press.com/data/r12/fullauto
(Automobile Models)

. tabulate rep77 foreign, chi2

    Repair |
    Record |        Foreign
      1977 |  Domestic    Foreign |     Total
-----------+----------------------+----------
      Poor |         2          1 |         3
      Fair |        10          1 |        11
   Average |        20          7 |        27
      Good |        13          7 |        20
 Excellent |         0          5 |         5
-----------+----------------------+----------
     Total |        45         21 |        66

          Pearson chi2(4) =  13.8619   Pr = 0.008
```

Although it appears that `foreign` takes on the values "`Domestic`" and "`Foreign`", it is actually a numeric variable taking on the values 0 and 1. Similarly, `rep77` takes on the values 1, 2, 3, 4, and 5, corresponding to "`Poor`", "`Fair`", and so on. The more meaningful words appear because we have attached value labels to the data; see [U] **12.6.3 Value labels**.

Because the chi-squared value is significant, we could claim that there is a relationship between `foreign` and `rep77`. Literally, however, we can only claim that the distributions are different; the chi-squared test is not directional. One way to model these data is to model the categorization that took place when the data were created. Cars have a true frequency of repair, which we will assume is given by $S_j = \beta\,\texttt{foreign}_j + u_j$, and a car is categorized as "poor" if $S_j \leq \kappa_0$, as "fair" if $\kappa_0 < S_j \leq \kappa_1$, and so on:

```
. ologit rep77 foreign
Iteration 0:   log likelihood = -89.895098
Iteration 1:   log likelihood = -85.951765
Iteration 2:   log likelihood = -85.908227
Iteration 3:   log likelihood = -85.908161
Iteration 4:   log likelihood = -85.908161

Ordered logistic regression                       Number of obs   =         66
                                                  LR chi2(1)      =       7.97
                                                  Prob > chi2     =     0.0047
Log likelihood = -85.908161                       Pseudo R2       =     0.0444

------------------------------------------------------------------------------
       rep77 |      Coef.   Std. Err.      z    P>|z|     [95% Conf. Interval]
-------------+----------------------------------------------------------------
     foreign |   1.455878   .5308951     2.74   0.006     .4153425    2.496413
-------------+----------------------------------------------------------------
       /cut1 |  -2.765562   .5988208                     -3.939229   -1.591895
       /cut2 |  -.9963603   .3217706                     -1.627019   -.3657016
       /cut3 |   .9426153   .3136398                      .3278925    1.557338
       /cut4 |   3.123351   .5423257                      2.060412     4.18629
------------------------------------------------------------------------------
```

Our model is $S_j = 1.46\,\texttt{foreign}_j + u_j$; the expected value for foreign cars is 1.46 and, for domestic cars, 0; foreign cars have better repair records.

The estimated cutpoints tell us how to interpret the score. For a foreign car, the probability of a poor record is the probability that $1.46 + u_j \leq -2.77$, or equivalently, $u_j \leq -4.23$. Making this calculation requires familiarity with the logistic distribution: the probability is $1/(1 + e^{4.23}) = 0.014$. On the other hand, for domestic cars, the probability of a poor record is the probability $u_j \leq -2.77$, which is 0.059.

This, it seems to us, is a far more reasonable prediction than we would have made based on the table alone. The table showed that 2 of 45 domestic cars had poor records, whereas 1 of 21 foreign cars had poor records—corresponding to probabilities $2/45 = 0.044$ and $1/21 = 0.048$. The predictions from our model imposed a smoothness assumption—foreign cars should not, overall, have better repair records without the difference revealing itself in each category. In our data, the fractions of foreign and domestic cars in the poor category are virtually identical only because of the randomness associated with small samples.

Thus if we were asked to predict the true fractions of foreign and domestic cars that would be classified in the various categories, we would choose the numbers implied by the ordered logit model:

	tabulate		logit	
	Domestic	Foreign	Domestic	Foreign
Poor	0.044	0.048	0.059	0.014
Fair	0.222	0.048	0.210	0.065
Average	0.444	0.333	0.450	0.295
Good	0.289	0.333	0.238	0.467
Excellent	0.000	0.238	0.043	0.159

See [R] **ologit postestimation** for a more complete explanation of how to generate predictions from an ordered logit model.

◁

❑ Technical note

Here ordered logit provides an alternative to ordinary two-outcome logistic models with an arbitrary dichotomization, which might otherwise have been tempting. We could, for instance, have summarized these data by converting the five-outcome `rep77` variable to a two-outcome variable, combining cars in the average, fair, and poor categories to make one outcome and combining cars in the good and excellent categories to make the second.

Another even less appealing alternative would have been to use ordinary regression, arbitrarily labeling "excellent" as 5, "good" as 4, and so on. The problem is that with different but equally valid labelings (say, 10 for "excellent"), we would obtain different estimates. We would have no way of choosing one metric over another. That assertion is not, however, true of `ologit`. The actual values used to label the categories make no difference other than through the order they imply.

In fact, our labeling was 5 for "excellent", 4 for "good", and so on. The words "excellent" and "good" appear in our output because we attached a value label to the variables; see [U] **12.6.3 Value labels**. If we were to now go back and type `replace rep77=10 if rep77==5`, changing all the 5s to 10s, we would still obtain the same results when we refit our model.

❑

▷ Example 2

In the example above, we used ordered logit as a way to model a table. We are not, however, limited to including only one explanatory variable or to including only categorical variables. We can explore the relationship of rep77 with any of the variables in our data. We might, for instance, model rep77 not only in terms of the origin of manufacture, but also including length (a proxy for size) and mpg:

```
. ologit rep77 foreign length mpg
Iteration 0:   log likelihood = -89.895098
Iteration 1:   log likelihood = -78.775147
Iteration 2:   log likelihood = -78.254294
Iteration 3:   log likelihood = -78.250719
Iteration 4:   log likelihood = -78.250719
Ordered logistic regression                       Number of obs   =         66
                                                  LR chi2(3)      =      23.29
                                                  Prob > chi2     =     0.0000
Log likelihood = -78.250719                       Pseudo R2       =     0.1295
```

rep77	Coef.	Std. Err.	z	P>|z|	[95% Conf.	Interval]
foreign	2.896807	.7906411	3.66	0.000	1.347179	4.446435
length	.0828275	.02272	3.65	0.000	.0382972	.1273579
mpg	.2307677	.0704548	3.28	0.001	.0926788	.3688566
/cut1	17.92748	5.551191			7.047344	28.80761
/cut2	19.86506	5.59648			8.896161	30.83396
/cut3	22.10331	5.708936			10.914	33.29262
/cut4	24.69213	5.890754			13.14647	36.2378

foreign still plays a role—and an even larger role than previously. We find that larger cars tend to have better repair records, as do cars with better mileage ratings.

◁

Saved results

`ologit` saves the following in `e()`:

Scalars

e(N)	number of observations
e(N_cd)	number of completely determined observations
e(k_cat)	number of categories
e(k)	number of parameters
e(k_aux)	number of auxiliary parameters
e(k_eq)	number of equations in e(b)
e(k_eq_model)	number of equations in overall model test
e(k_dv)	number of dependent variables
e(df_m)	model degrees of freedom
e(r2_p)	pseudo-R-squared
e(ll)	log likelihood
e(ll_0)	log likelihood, constant-only model
e(N_clust)	number of clusters
e(chi2)	χ^2
e(p)	significance of model test
e(rank)	rank of e(V)
e(ic)	number of iterations
e(rc)	return code
e(converged)	1 if converged, 0 otherwise

Macros

e(cmd)	ologit
e(cmdline)	command as typed
e(depvar)	name of dependent variable
e(wtype)	weight type
e(wexp)	weight expression
e(title)	title in estimation output
e(clustvar)	name of cluster variable
e(offset)	linear offset variable
e(chi2type)	Wald or LR; type of model χ^2 test
e(vce)	*vcetype* specified in vce()
e(vcetype)	title used to label Std. Err.
e(opt)	type of optimization
e(which)	max or min; whether optimizer is to perform maximization or minimization
e(ml_method)	type of ml method
e(user)	name of likelihood-evaluator program
e(technique)	maximization technique
e(properties)	b V
e(predict)	program used to implement predict
e(asbalanced)	factor variables fvset as asbalanced
e(asobserved)	factor variables fvset as asobserved

Matrices

e(b)	coefficient vector
e(Cns)	constraints matrix
e(ilog)	iteration log (up to 20 iterations)
e(gradient)	gradient vector
e(cat)	category values
e(V)	variance–covariance matrix of the estimators
e(V_modelbased)	model-based variance

Functions

e(sample)	marks estimation sample

Methods and formulas

ologit is implemented as an ado-file.

See Long and Freese (2006, chap. 5) for a discussion of models for ordinal outcomes and examples that use Stata. Cameron and Trivedi (2005, chap. 15) describe multinomial models, including the model fit by ologit. When you have a qualitative dependent variable, several estimation procedures are available. A popular choice is multinomial logistic regression (see [R] **mlogit**), but if you use this procedure when the response variable is ordinal, you are discarding information because multinomial logit ignores the ordered aspect of the outcome. Ordered logit and probit models provide a means to exploit the ordering information.

There is more than one "ordered logit" model. The model fit by ologit, which we will call the ordered logit model, is also known as the proportional odds model. Another popular choice, not fit by ologit, is known as the stereotype model; see [R] **slogit**. All ordered logit models have been derived by starting with a binary logit/probit model and generalizing it to allow for more than two outcomes.

The proportional-odds ordered logit model is so called because, if we consider the odds $\text{odds}(k) = P(Y \leq k)/P(Y > k)$, then $\text{odds}(k_1)$ and $\text{odds}(k_2)$ have the same ratio for all independent variable combinations. The model is based on the principle that the only effect of combining adjoining categories in ordered categorical regression problems should be a loss of efficiency in estimating the regression parameters (McCullagh 1980). This model was also described by McKelvey and Zavoina (1975) and, previously, by Aitchison and Silvey (1957) in a different algebraic form. Brant (1990) offers a set of diagnostics for the model.

Peterson and Harrell (1990) suggest a model that allows nonproportional odds for a subset of the explanatory variables. ologit does not allow this, but a model similar to this was implemented by Fu (1998).

The stereotype model rejects the principle on which the ordered logit model is based. Anderson (1984) argues that there are two distinct types of ordered categorical variables: "grouped continuous", such as income, where the "type a" model applies; and "assessed", such as extent of pain relief, where the stereotype model applies. Greenland (1985) independently developed the same model. The stereotype model starts with a multinomial logistic regression model and imposes constraints on this model.

Goodness of fit for ologit can be evaluated by comparing the likelihood value with that obtained by fitting the model with mlogit. Let $\ln L_1$ be the log-likelihood value reported by ologit, and let $\ln L_0$ be the log-likelihood value reported by mlogit. If there are p independent variables (excluding the constant) and k categories, mlogit will estimate $p(k-1)$ additional parameters. We can then perform a "likelihood-ratio test", that is, calculate $-2(\ln L_1 - \ln L_0)$, and compare it with $\chi^2\{p(k-2)\}$. This test is suggestive only because the ordered logit model is not nested within the multinomial logit model. A large value of $-2(\ln L_1 - \ln L_0)$ should, however, be taken as evidence of poorness of fit. Marginally large values, on the other hand, should not be taken too seriously.

The coefficients and cutpoints are estimated using maximum likelihood as described in [R] **maximize**. In our parameterization, no constant appears, because the effect is absorbed into the cutpoints.

ologit and oprobit begin by tabulating the dependent variable. Category $i = 1$ is defined as the minimum value of the variable, $i = 2$ as the next ordered value, and so on, for the empirically determined k categories.

The probability of a given observation for ordered logit is

$$p_{ij} = \Pr(y_j = i) = \Pr\left(\kappa_{i-1} < \mathbf{x}_j\boldsymbol{\beta} + u \le \kappa_i\right)$$
$$= \frac{1}{1+\exp(-\kappa_i + \mathbf{x}_j\boldsymbol{\beta})} - \frac{1}{1+\exp(-\kappa_{i-1} + \mathbf{x}_j\boldsymbol{\beta})}$$

κ_0 is defined as $-\infty$ and κ_k as $+\infty$.

For ordered probit, the probability of a given observation is

$$p_{ij} = \Pr(y_j = i) = \Pr\left(\kappa_{i-1} < \mathbf{x}_j\boldsymbol{\beta} + u \le \kappa_i\right)$$
$$= \Phi\left(\kappa_i - \mathbf{x}_j\boldsymbol{\beta}\right) - \Phi\left(\kappa_{i-1} - \mathbf{x}_j\boldsymbol{\beta}\right)$$

where $\Phi(\cdot)$ is the standard normal cumulative distribution function.

The log likelihood is

$$\ln L = \sum_{j=1}^{N} w_j \sum_{i=1}^{k} I_i(y_j) \ln p_{ij}$$

where w_j is an optional weight and

$$I_i(y_j) = \begin{cases} 1, \text{ if } y_j = i \\ 0, \text{ otherwise} \end{cases}$$

`ologit` and `oprobit` support the Huber/White/sandwich estimator of the variance and its clustered version using `vce(robust)` and `vce(cluster` *clustvar*`)`, respectively. See [P] **_robust**, particularly *Maximum likelihood estimators* and *Methods and formulas*.

These commands also support estimation with survey data. For details on VCEs with survey data, see [SVY] **variance estimation**.

References

Aitchison, J., and S. D. Silvey. 1957. The generalization of probit analysis to the case of multiple responses. *Biometrika* 44: 131–140.

Anderson, J. A. 1984. Regression and ordered categorical variables (with discussion). *Journal of the Royal Statistical Society, Series B* 46: 1–30.

Brant, R. 1990. Assessing proportionality in the proportional odds model for ordinal logistic regression. *Biometrics* 46: 1171–1178.

Cameron, A. C., and P. K. Trivedi. 2005. *Microeconometrics: Methods and Applications*. New York: Cambridge University Press.

Fu, V. K. 1998. sg88: Estimating generalized ordered logit models. *Stata Technical Bulletin* 44: 27–30. Reprinted in *Stata Technical Bulletin Reprints*, vol. 8, pp. 160–164. College Station, TX: Stata Press.

Goldstein, R. 1997. sg59: Index of ordinal variation and Neyman–Barton GOF. *Stata Technical Bulletin* 33: 10–12. Reprinted in *Stata Technical Bulletin Reprints*, vol. 6, pp. 145–147. College Station, TX: Stata Press.

Greenland, S. 1985. An application of logistic models to the analysis of ordinal responses. *Biometrical Journal* 27: 189–197.

Kleinbaum, D. G., and M. Klein. 2010. *Logistic Regression: A Self-Learning Text.* 3rd ed. New York: Springer.

Long, J. S. 1997. *Regression Models for Categorical and Limited Dependent Variables.* Thousand Oaks, CA: Sage.

Long, J. S., and J. Freese. 2006. *Regression Models for Categorical Dependent Variables Using Stata.* 2nd ed. College Station, TX: Stata Press.

Lunt, M. 2001. sg163: Stereotype ordinal regression. *Stata Technical Bulletin* 61: 12–18. Reprinted in *Stata Technical Bulletin Reprints*, vol. 10, pp. 298–307. College Station, TX: Stata Press.

McCullagh, P. 1977. A logistic model for paired comparisons with ordered categorical data. *Biometrika* 64: 449–453.

——. 1980. Regression models for ordinal data (with discussion). *Journal of the Royal Statistical Society, Series B* 42: 109–142.

McCullagh, P., and J. A. Nelder. 1989. *Generalized Linear Models.* 2nd ed. London: Chapman & Hall/CRC.

McKelvey, R. D., and W. Zavoina. 1975. A statistical model for the analysis of ordinal level dependent variables. *Journal of Mathematical Sociology* 4: 103–120.

Miranda, A., and S. Rabe-Hesketh. 2006. Maximum likelihood estimation of endogenous switching and sample selection models for binary, ordinal, and count variables. *Stata Journal* 6: 285–308.

Peterson, B., and F. E. Harrell, Jr. 1990. Partial proportional odds models for ordinal response variables. *Applied Statistics* 39: 205–217.

Williams, R. 2006. Generalized ordered logit/partial proportional odds models for ordinal dependent variables. *Stata Journal* 6: 58–82.

——. 2010. Fitting heterogeneous choice models with oglm. *Stata Journal* 10: 540–567.

Wolfe, R. 1998. sg86: Continuation-ratio models for ordinal response data. *Stata Technical Bulletin* 44: 18–21. Reprinted in *Stata Technical Bulletin Reprints*, vol. 8, pp. 149–153. College Station, TX: Stata Press.

Wolfe, R., and W. W. Gould. 1998. sg76: An approximate likelihood-ratio test for ordinal response models. *Stata Technical Bulletin* 42: 24–27. Reprinted in *Stata Technical Bulletin Reprints*, vol. 7, pp. 199–204. College Station, TX: Stata Press.

Xu, J., and J. S. Long. 2005. Confidence intervals for predicted outcomes in regression models for categorical outcomes. *Stata Journal* 5: 537–559.

Also see

[R] **ologit postestimation** — Postestimation tools for ologit

[R] **clogit** — Conditional (fixed-effects) logistic regression

[R] **logistic** — Logistic regression, reporting odds ratios

[R] **logit** — Logistic regression, reporting coefficients

[R] **mlogit** — Multinomial (polytomous) logistic regression

[R] **oprobit** — Ordered probit regression

[R] **rologit** — Rank-ordered logistic regression

[R] **slogit** — Stereotype logistic regression

[MI] **estimation** — Estimation commands for use with mi estimate

[SVY] **svy estimation** — Estimation commands for survey data

[U] **20 Estimation and postestimation commands**

Title

ologit postestimation — Postestimation tools for ologit

Description

The following postestimation commands are available after `ologit`:

Command	Description
`contrast`	contrasts and ANOVA-style joint tests of estimates
`estat`	AIC, BIC, VCE, and estimation sample summary
`estat` (svy)	postestimation statistics for survey data
`estimates`	cataloging estimation results
`lincom`	point estimates, standard errors, testing, and inference for linear combinations of coefficients
`linktest`	link test for model specification
`lrtest`[1]	likelihood-ratio test
`margins`	marginal means, predictive margins, marginal effects, and average marginal effects
`marginsplot`	graph the results from margins (profile plots, interaction plots, etc.)
`nlcom`	point estimates, standard errors, testing, and inference for nonlinear combinations of coefficients
`predict`	predictions, residuals, influence statistics, and other diagnostic measures
`predictnl`	point estimates, standard errors, testing, and inference for generalized predictions
`pwcompare`	pairwise comparisons of estimates
`suest`	seemingly unrelated estimation
`test`	Wald tests of simple and composite linear hypotheses
`testnl`	Wald tests of nonlinear hypotheses

[1] `lrtest` is not appropriate with `svy` estimation results.

See the corresponding entries in the *Base Reference Manual* for details, but see [SVY] **estat** for details about `estat` (svy).

Syntax for predict

`predict` [*type*] { *stub** | *newvar* | *newvarlist* } [*if*] [*in*] [, *statistic* `outcome(`*outcome*`)` `nooffset`]

`predict` [*type*] { *stub** | *newvarlist* } [*if*] [*in*], `scores`

statistic	Description
Main	
`pr`	predicted probabilities; the default
`xb`	linear prediction
`stdp`	standard error of the linear prediction

If you do not specify outcome(), pr (with one new variable specified) assumes outcome(#1).

You specify one or k new variables with pr, where k is the number of outcomes.

You specify one new variable with xb and stdp.

These statistics are available both in and out of sample; type predict ... if e(sample) ... if wanted only for the estimation sample.

Menu

Statistics > Postestimation > Predictions, residuals, etc.

Options for predict

Main

pr, the default, calculates the predicted probabilities. If you do not also specify the outcome() option, you specify k new variables, where k is the number of categories of the dependent variable. Say that you fit a model by typing ologit result x1 x2, and result takes on three values. Then you could type predict p1 p2 p3 to obtain all three predicted probabilities. If you specify the outcome() option, you must specify one new variable. Say that result takes on the values 1, 2, and 3. Typing predict p1, outcome(1) would produce the same p1.

xb calculates the linear prediction. You specify one new variable, for example, predict linear, xb. The linear prediction is defined, ignoring the contribution of the estimated cutpoints.

stdp calculates the standard error of the linear prediction. You specify one new variable, for example, predict se, stdp.

outcome(*outcome*) specifies for which outcome the predicted probabilities are to be calculated. outcome() should contain either one value of the dependent variable or one of #1, #2, ..., with #1 meaning the first category of the dependent variable, #2 meaning the second category, etc.

nooffset is relevant only if you specified offset(*varname*) for ologit. It modifies the calculations made by predict so that they ignore the offset variable; the linear prediction is treated as $\mathbf{x}_j\mathbf{b}$ rather than as $\mathbf{x}_j\mathbf{b} + \text{offset}_j$.

scores calculates equation-level score variables. The number of score variables created will equal the number of outcomes in the model. If the number of outcomes in the model was k, then

the first new variable will contain $\partial \ln L/\partial(\mathbf{x}_j\mathbf{b})$;

the second new variable will contain $\partial \ln L/\partial\kappa_1$;

the third new variable will contain $\partial \ln L/\partial\kappa_2$;

...

and the kth new variable will contain $\partial \ln L/\partial\kappa_{k-1}$, where κ_i refers to the ith cutpoint.

Remarks

See [U] **20 Estimation and postestimation commands** for instructions on obtaining the variance–covariance matrix of the estimators, predicted values, and hypothesis tests. Also see [R] **lrtest** for performing likelihood-ratio tests.

▷ Example 1

In example 2 of [R] **ologit**, we fit the model `ologit rep77 foreign length mpg`. The `predict` command can be used to obtain the predicted probabilities.

We type `predict` followed by the names of the new variables to hold the predicted probabilities, ordering the names from low to high. In our data, the lowest outcome is "poor", and the highest is "excellent". We have five categories, so we must type five names following `predict`; the choice of names is up to us:

```
. predict poor fair avg good exc
(option pr assumed; predicted probabilities)
. list exc good make model rep78 if rep77>=., sep(4) divider
```

	exc	good	make	model	rep78
3.	.0033341	.0393056	AMC	Spirit	.
10.	.0098392	.1070041	Buick	Opel	.
32.	.0023406	.0279497	Ford	Fiesta	Good
44.	.015697	.1594413	Merc.	Monarch	Average
53.	.065272	.4165188	Peugeot	604	.
56.	.005187	.059727	Plym.	Horizon	Average
57.	.0261461	.2371826	Plym.	Sapporo	.
63.	.0294961	.2585825	Pont.	Phoenix	.

The eight cars listed were introduced after 1977, so they do not have 1977 repair records in our data. We predicted what their 1977 repair records might have been using the fitted model. We see that, based on its characteristics, the Peugeot 604 had about a $41.65 + 6.53 \approx 48.2\%$ chance of a good or excellent repair record. The Ford Fiesta, which had only a 3% chance of a good or excellent repair record, in fact, had a good record when it was introduced in the following year.

◁

❑ Technical note

For ordered logit, `predict, xb` produces $S_j = x_{1j}\beta_1 + x_{2j}\beta_2 + \cdots + x_{kj}\beta_k$. The ordered-logit predictions are then the probability that $S_j + u_j$ lies between a pair of cutpoints, κ_{i-1} and κ_i. Some handy formulas are

$$\Pr(S_j + u_j < \kappa) = 1/(1 + e^{S_j - \kappa})$$
$$\Pr(S_j + u_j > \kappa) = 1 - 1/(1 + e^{S_j - \kappa})$$
$$\Pr(\kappa_1 < S_j + u_j < \kappa_2) = 1/(1 + e^{S_j - \kappa_2}) - 1/(1 + e^{S_j - \kappa_1})$$

Rather than using `predict` directly, we could calculate the predicted probabilities by hand. If we wished to obtain the predicted probability that the repair record is excellent and the probability that it is good, we look back at `ologit`'s output to obtain the cutpoints. We find that "good" corresponds to the interval /cut3 $< S_j + u <$ /cut4 and "excellent" to the interval $S_j + u >$ /cut4:

```
. predict score, xb
. generate probgood = 1/(1+exp(score-_b[/cut4])) - 1/(1+exp(score-_b[/cut3]))
. generate probexc = 1 - 1/(1+exp(score-_b[/cut4]))
```

The results of our calculation will be the same as those produced in the previous example. We refer to the estimated cutpoints just as we would any coefficient, so `_b[/cut3]` refers to the value of the `/cut3` coefficient; see [U] **13.5 Accessing coefficients and standard errors**.

❑

Methods and formulas

All postestimation commands listed above are implemented as ado-files.

Also see

[R] **ologit** — Ordered logistic regression

[U] **20 Estimation and postestimation commands**

Title

oneway — One-way analysis of variance

Syntax

oneway *response_var factor_var* [*if*] [*in*] [*weight*] [, *options*]

options	Description
Main	
bonferroni	Bonferroni multiple-comparison test
scheffe	Scheffé multiple-comparison test
sidak	Šidák multiple-comparison test
tabulate	produce summary table
[no]means	include or suppress means; default is means
[no]standard	include or suppress standard deviations; default is standard
[no]freq	include or suppress frequencies; default is freq
[no]obs	include or suppress number of obs; default is obs if data are weighted
noanova	suppress the ANOVA table
nolabel	show numeric codes, not labels
wrap	do not break wide tables
missing	treat missing values as categories

by is allowed; see [D] **by**.

aweights and fweights are allowed; see [U] **11.1.6 weight**.

Menu

Statistics > Linear models and related > ANOVA/MANOVA > One-way ANOVA

Description

The oneway command reports one-way analysis-of-variance (ANOVA) models and performs multiple-comparison tests.

If you wish to fit more complicated ANOVA layouts or wish to fit analysis-of-covariance (ANCOVA) models, see [R] **anova**.

See [D] **encode** for examples of fitting ANOVA models on string variables.

See [R] **loneway** for an alternative oneway command with slightly different features.

Options

Main

bonferroni reports the results of a Bonferroni multiple-comparison test.

scheffe reports the results of a Scheffé multiple-comparison test.

sidak reports the results of a Šidák multiple-comparison test.

`tabulate` produces a table of summary statistics of the *response_var* by levels of the *factor_var*. The table includes the mean, standard deviation, frequency, and, if the data are weighted, the number of observations. Individual elements of the table may be included or suppressed by using the [no]`means`, [no]`standard`, [no]`freq`, and [no]`obs` options. For example, typing

```
oneway response factor, tabulate means standard
```

produces a summary table that contains only the means and standard deviations. You could achieve the same result by typing

```
oneway response factor, tabulate nofreq
```

[no]`means` includes or suppresses only the means from the table produced by the `tabulate` option. See `tabulate` above.

[no]`standard` includes or suppresses only the standard deviations from the table produced by the `tabulate` option. See `tabulate` above.

[no]`freq` includes or suppresses only the frequencies from the table produced by the `tabulate` option. See `tabulate` above.

[no]`obs` includes or suppresses only the reported number of observations from the table produced by the `tabulate` option. If the data are not weighted, only the frequency is reported. If the data are weighted, the frequency refers to the sum of the weights. See `tabulate` above.

`noanova` suppresses the display of the ANOVA table.

`nolabel` causes the numeric codes to be displayed rather than the value labels in the ANOVA and multiple-comparison test tables.

`wrap` requests that Stata not break up wide tables to make them more readable.

`missing` requests that missing values of *factor_var* be treated as a category rather than as observations to be omitted from the analysis.

Remarks

Remarks are presented under the following headings:

Introduction
Obtaining observed means
Multiple-comparison tests
Weighted data

Introduction

The `oneway` command reports one-way ANOVA models. To perform a one-way layout of a variable called `endog` on `exog`, type `oneway endog exog`.

▷ Example 1

We run an experiment varying the amount of fertilizer used in growing apple trees. We test four concentrations, using each concentration in three groves of 12 trees each. Later in the year, we measure the average weight of the fruit.

If all had gone well, we would have had 3 observations on the average weight for each of the four concentrations. Instead, two of the groves were mistakenly leveled by a confused man on a large bulldozer. We are left with the following dataset:

```
. use http://www.stata-press.com/data/r12/apple
(Apple trees)

. describe
Contains data from http://www.stata-press.com/data/r12/apple.dta
  obs:            10                          Apple trees
 vars:             2                          16 Jan 2011 11:23
 size:           100
-------------------------------------------------------------------------------
              storage  display     value
variable name   type   format      label      variable label
-------------------------------------------------------------------------------
treatment       int    %8.0g                  Fertilizer
weight          double %10.0g                 Average weight in grams
-------------------------------------------------------------------------------
Sorted by:

. list, abbreviate(10)

     +--------------------+
     | treatment   weight |
     |--------------------|
  1. |         1    117.5 |
  2. |         1    113.8 |
  3. |         1    104.4 |
  4. |         2     48.9 |
  5. |         2     50.4 |
     |--------------------|
  6. |         2     58.9 |
  7. |         3     70.4 |
  8. |         3     86.9 |
  9. |         4     87.7 |
 10. |         4     67.3 |
     +--------------------+
```

To obtain the one-way ANOVA results, we type

```
. oneway weight treatment
                        Analysis of Variance
    Source              SS         df      MS            F     Prob > F
------------------------------------------------------------------------
Between groups      5295.54433      3   1765.18144     21.46     0.0013
 Within groups      493.591667      6   82.2652778
------------------------------------------------------------------------
    Total             5789.136      9   643.237333

Bartlett's test for equal variances:  chi2(3) =   1.3900  Prob>chi2 = 0.708
```

We find significant (at better than the 1% level) differences among the four concentrations. ◁

❑ Technical note

Rather than using the oneway command, we could have performed this analysis by using anova. Example 1 in [R] **anova** repeats this same analysis. You may wish to compare the output.

You will find the oneway command quicker than the anova command, and, as you will learn, oneway allows you to perform multiple-comparison tests. On the other hand, anova will let you generate predictions, examine the covariance matrix of the estimators, and perform more general hypothesis tests.

❑

❑ Technical note

Although the output is a usual ANOVA table, let's run through it anyway. The between-group sum of squares for the model is 5295.5 with 3 degrees of freedom, resulting in a mean square of $5295.5/3 \approx 1765.2$. The corresponding F statistic is 21.46 and has a significance level of 0.0013. Thus the model appears to be significant at the 0.13% level.

The second line summarizes the within-group (residual) variation. The within-group sum of squares is 493.59 with 6 degrees of freedom, resulting in a mean squared error of 82.27.

The between- and residual-group variations sum to the total sum of squares (TSS), which is reported as 5789.1 in the last line of the table. This is the TSS of `weight` after removal of the mean. Similarly, the between plus residual degrees of freedom sum to the total degrees of freedom, 9. Remember that there are 10 observations. Subtracting 1 for the mean, we are left with 9 total degrees of freedom.

At the bottom of the table, Bartlett's test for equal variances is reported. The value of the statistic is 1.39. The corresponding significance level (χ^2 with 3 degrees of freedom) is 0.708, so we cannot reject the assumption that the variances are homogeneous.

❑

Obtaining observed means

▷ Example 2

We typed `oneway weight treatment` to obtain an ANOVA table of weight of fruit by fertilizer concentration. Although we obtained the table, we obtained no information on which fertilizer seems to work the best. If we add the `tabulate` option, we obtain that additional information:

```
. oneway weight treatment, tabulate

            |  Summary of Average weight in grams
 Fertilizer |        Mean   Std. Dev.       Freq.
------------+------------------------------------
          1 |       111.9   6.7535176           3
          2 |   52.733333   5.3928966           3
          3 |       78.65   11.667262           2
          4 |        77.5   14.424978           2
------------+------------------------------------
      Total |       80.62   25.362124          10

                        Analysis of Variance
    Source              SS         df      MS            F     Prob > F
------------------------------------------------------------------------
Between groups      5295.54433      3   1765.18144     21.46     0.0013
 Within groups      493.591667      6   82.2652778
------------------------------------------------------------------------
    Total             5789.136      9   643.237333

Bartlett's test for equal variances:  chi2(3) =   1.3900  Prob>chi2 = 0.708
```

We find that the average weight was largest when we used fertilizer concentration 1.

◁

Multiple-comparison tests

▷ Example 3

oneway can also perform multiple-comparison tests using either Bonferroni, Scheffé, or Šidák normalizations. For instance, to obtain the Bonferroni multiple-comparison test, we specify the bonferroni option:

```
. oneway weight treatment, bonferroni
                        Analysis of Variance
    Source              SS         df      MS            F     Prob > F
------------------------------------------------------------------------
Between groups      5295.54433      3   1765.18144     21.46     0.0013
 Within groups      493.591667      6   82.2652778
------------------------------------------------------------------------
    Total             5789.136      9   643.237333
Bartlett's test for equal variances:  chi2(3) =   1.3900  Prob>chi2 = 0.708
            Comparison of Average weight in grams by Fertilizer
                               (Bonferroni)
Row Mean-|
Col Mean |          1          2          3
---------+---------------------------------
       2 |   -59.1667
         |      0.001
         |
       3 |     -33.25    25.9167
         |      0.042      0.122
         |
       4 |      -34.4    24.7667      -1.15
         |      0.036      0.146      1.000
```

The results of the Bonferroni test are presented as a matrix. The first entry, −59.17, represents the difference between fertilizer concentrations 2 and 1 (labeled "Row Mean - Col Mean" in the upper stub of the table). Remember that in the previous example we requested the tabulate option. Looking back, we find that the means of concentrations 1 and 2 are 111.90 and 52.73, respectively. Thus $52.73 - 111.90 = -59.17$.

Underneath that number is reported "0.001". This is the Bonferroni-adjusted significance of the difference. The difference is significant at the 0.1% level. Looking down the column, we see that concentration 3 is also worse than concentration 1 (4.2% level), as is concentration 4 (3.6% level).

On the basis of this evidence, we would use concentration 1 if we grew apple trees.

◁

▷ Example 4

We can just as easily obtain the Scheffé-adjusted significance levels. Rather than specifying the bonferroni option, we specify the scheffe option.

We will also add the noanova option to prevent Stata from redisplaying the ANOVA table:

```
. oneway weight treatment, noanova scheffe
                  Comparison of Average weight in grams by Fertilizer
                                      (Scheffe)
Row Mean-|
Col Mean |          1           2           3
---------+------------------------------------
       2 |   -59.1667
         |      0.001
         |
       3 |     -33.25     25.9167
         |      0.039       0.101
         |
       4 |      -34.4     24.7667       -1.15
         |      0.034       0.118       0.999
```

The differences are the same as those we obtained in the Bonferroni output, but the significance levels are not. According to the Bonferroni-adjusted numbers, the significance of the difference between fertilizer concentrations 1 and 3 is 4.2%. The Scheffé-adjusted significance level is 3.9%.

We will leave it to you to decide which results are more accurate.

◁

▷ Example 5

Let's conclude this example by obtaining the Šidák-adjusted multiple-comparison tests. We do this to illustrate Stata's capabilities to calculate these results, because searching across adjustment methods until you find the results you want is not a valid technique for obtaining significance levels.

```
. oneway weight treatment, noanova sidak
                  Comparison of Average weight in grams by Fertilizer
                                       (Sidak)
Row Mean-|
Col Mean |          1           2           3
---------+------------------------------------
       2 |   -59.1667
         |      0.001
         |
       3 |     -33.25     25.9167
         |      0.041       0.116
         |
       4 |      -34.4     24.7667       -1.15
         |      0.035       0.137       1.000
```

We find results that are similar to the Bonferroni-adjusted numbers.

◁

Henry Scheffé (1907–1977) was born in New York. He studied mathematics at the University of Wisconsin, gaining a doctorate with a dissertation on differential equations. He taught mathematics at Wisconsin, Oregon State University, and Reed College, but his interests changed to statistics and he joined Wilks at Princeton. After periods at Syracuse, UCLA, and Columbia, Scheffé settled in Berkeley from 1953. His research increasingly focused on linear models and particularly ANOVA, on which he produced a celebrated monograph. His death was the result of a bicycle accident.

Weighted data

▷ Example 6

oneway can work with both weighted and unweighted data. Let's assume that we wish to perform a one-way layout of the death rate on the four census regions of the United States using state data. Our data contain three variables, drate (the death rate), region (the region), and pop (the population of the state).

To fit the model, we type oneway drate region [weight=pop], although we typically abbreviate weight as w. We will also add the tabulate option to demonstrate how the table of summary statistics differs for weighted data:

```
. use http://www.stata-press.com/data/r12/census8
(1980 Census data by state)

. oneway drate region [w=pop], tabulate
(analytic weights assumed)

     Census  |             Summary of Death Rate
     region  |        Mean   Std. Dev.        Freq.         Obs.
-------------+-----------------------------------------------------
         NE  |       97.15        5.82     49135283            9
    N Cntrl  |       88.10        5.58     58865670           12
      South  |       87.05       10.40     74734029           16
       West  |       75.65        8.23     43172490           13
-------------+-----------------------------------------------------
      Total  |       87.34       10.43    2.259e+08           50

                        Analysis of Variance
    Source              SS         df      MS            F     Prob > F
------------------------------------------------------------------------
Between groups      2360.92281      3   786.974272     12.17     0.0000
 Within groups      2974.09635     46   64.6542685
------------------------------------------------------------------------
    Total           5335.01916     49   108.877942

Bartlett's test for equal variances:  chi2(3) =   5.4971  Prob>chi2 = 0.139
```

When the data are weighted, the summary table has four columns rather than three. The column labeled "Freq." reports the sum of the weights. The overall frequency is 2.259×10^8, meaning that there are approximately 226 million people in the United States.

The ANOVA table is appropriately weighted. Also see [U] **11.1.6 weight**.

◁

Saved results

oneway saves the following in r():

Scalars

r(N)	number of observations	r(df_m)	between-group degrees of freedom
r(F)	F statistic	r(rss)	within-group sum of squares
r(df_r)	within-group degrees of freedom	r(chi2bart)	Bartlett's χ^2
r(mss)	between-group sum of squares	r(df_bart)	Bartlett's degrees of freedom

Methods and formulas

Methods and formulas are presented under the following headings:

One-way analysis of variance
Bartlett's test
Multiple-comparison tests

One-way analysis of variance

The model of one-way ANOVA is

$$y_{ij} = \mu + \alpha_i + \epsilon_{ij}$$

for levels $i = 1, \ldots, k$ and observations $j = 1, \ldots, n_i$. Define $\overline{y}_i$ as the (weighted) mean of y_{ij} over j and $\overline{y}$ as the overall (weighted) mean of y_{ij}. Define w_{ij} as the weight associated with y_{ij}, which is 1 if the data are unweighted. w_{ij} is normalized to sum to $n = \sum_i n_i$ if aweights are used and is otherwise not normalized. w_i refers to $\sum_j w_{ij}$, and w refers to $\sum_i w_i$.

The between-group sum of squares is then

$$S_1 = \sum_i w_i (\overline{y}_i - \overline{y})^2$$

The TSS is

$$S = \sum_i \sum_j w_{ij} (y_{ij} - \overline{y})^2$$

The within-group sum of squares is given by $S_e = S - S_1$.

The between-group mean square is $s_1^2 = S_1/(k-1)$, and the within-group mean square is $s_e^2 = S_e/(w-k)$. The test statistic is $F = s_1^2/s_e^2$. See, for instance, Snedecor and Cochran (1989).

Bartlett's test

Bartlett's test assumes that you have m independent, normal, random samples and tests the hypothesis $\sigma_1^2 = \sigma_2^2 = \cdots = \sigma_m^2$. The test statistic, M, is defined as

$$M = \frac{(T-m)\ln\widehat{\sigma}^2 - \sum (T_i - 1)\ln\widehat{\sigma}_i^2}{1 + \frac{1}{3(m-1)}\left\{\left(\sum \frac{1}{T_i - 1}\right) - \frac{1}{T-m}\right\}}$$

where there are T overall observations, T_i observations in the ith group, and

$$(T_i - 1)\widehat{\sigma}_i^2 = \sum_{j=1}^{T_i} (y_{ij} - \overline{y}_i)^2$$

$$(T-m)\widehat{\sigma}^2 = \sum_{i=1}^{m} (T_i - 1)\widehat{\sigma}_i^2$$

An approximate test of the homogeneity of variance is based on the statistic M with critical values obtained from the χ^2 distribution of $m-1$ degrees of freedom. See Bartlett (1937) or Draper and Smith (1998, 56–57).

Multiple-comparison tests

Let's begin by reviewing the logic behind these adjustments. The "standard" t statistic for the comparison of two means is

$$t = \frac{\overline{y}_i - \overline{y}_j}{s\sqrt{\frac{1}{n_i} + \frac{1}{n_j}}}$$

where s is the overall standard deviation, $\overline{y}_i$ is the measured average of y in group i, and n_i is the number of observations in the group. We perform hypothesis tests by calculating this t statistic. We simultaneously choose a critical level, α, and look up the t statistic corresponding to that level in a table. We reject the hypothesis if our calculated t exceeds the value we looked up. Alternatively, because we have a computer at our disposal, we calculate the significance level e corresponding to our calculated t statistic, and if $e < \alpha$, we reject the hypothesis.

This logic works well when we are performing one test. Now consider what happens when we perform several separate tests, say, n of them. Let's assume, just for discussion, that we set α equal to 0.05 and that we will perform six tests. For each test, we have a 0.05 probability of falsely rejecting the equality-of-means hypothesis. Overall, then, our chances of falsely rejecting *at least one* of the hypotheses is $1 - (1 - 0.05)^6 \approx 0.26$ if the tests are independent.

The idea behind multiple-comparison tests is to control for the fact that we will perform multiple tests and to reduce our overall chances of falsely rejecting each hypothesis to α rather than letting our chances increase with each additional test. (See Miller [1981] and Hochberg and Tamhane [1987] for rather advanced texts on multiple-comparison procedures.)

The Bonferroni adjustment (see Miller [1981]; also see van Belle et al. [2004, 534–537]) does this by (falsely but approximately) asserting that the critical level we should use, a, is the true critical level, α, divided by the number of tests, n; that is, $a = \alpha/n$. For instance, if we are going to perform six tests, each at the 0.05 significance level, we want to adopt a critical level of $0.05/6 \approx 0.00833$.

We can just as easily apply this logic to e, the significance level associated with our t statistic, as to our critical level α. If a comparison has a calculated significance of e, then its "real" significance, adjusted for the fact of n comparisons, is $n \times e$. If a comparison has a significance level of, say, 0.012, and we perform six tests, then its "real" significance is 0.072. If we adopt a critical level of 0.05, we cannot reject the hypothesis. If we adopt a critical level of 0.10, we can reject it.

Of course, this calculation can go above 1, but that just means that there is no $\alpha < 1$ for which we could reject the hypothesis. (This situation arises because of the crude nature of the Bonferroni adjustment.) Stata handles this case by simply calling the significance level 1. Thus the formula for the Bonferroni significance level is

$$e_b = \min(1, en)$$

where $n = k(k-1)/2$ is the number of comparisons.

The Šidák adjustment (Šidák [1967]; also see Winer, Brown, and Michels [1991, 165–166]) is slightly different and provides a tighter bound. It starts with the assertion that

$$a = 1 - (1 - \alpha)^{1/n}$$

Turning this formula around and substituting calculated significance levels, we obtain

$$e_s = \min\left\{1, 1 - (1 - e)^n\right\}$$

For example, if the calculated significance is 0.012 and we perform six tests, the "real" significance is approximately 0.07.

The Scheffé test (Scheffé [1953, 1959]; also see Kuehl [2000, 97–98]) differs in derivation, but it attacks the same problem. Let there be k means for which we want to make all the pairwise tests. Two means are declared significantly different if

$$t \geq \sqrt{(k-1)F(\alpha; k-1, \nu)}$$

where $F(\alpha; k-1, \nu)$ is the α-critical value of the F distribution with $k-1$ numerator and ν denominator degrees of freedom. Scheffé's test has the nicety that it never declares a contrast significant if the overall F test is not significant.

Turning the test around, Stata calculates a significance level

$$\widehat{e} = F\left(\frac{t^2}{k-1}, k-1, \nu\right)$$

For instance, you have a calculated t statistic of 4.0 with 50 degrees of freedom. The simple t test says that the significance level is 0.00021. The F test equivalent, 16 with 1 and 50 degrees of freedom, says the same. If you are comparing three means, however, you calculate an F test of 8.0 with 2 and 50 degrees of freedom, which says that the significance level is 0.0010.

References

Acock, A. C. 2010. *A Gentle Introduction to Stata*. 3rd ed. College Station, TX: Stata Press.

Altman, D. G. 1991. *Practical Statistics for Medical Research*. London: Chapman & Hall/CRC.

Bartlett, M. S. 1937. Properties of sufficiency and statistical tests. *Proceedings of the Royal Society, Series A* 160: 268–282.

Daniel, C., and E. L. Lehmann. 1979. Henry Scheffé 1907–1977. *Annals of Statistics* 7: 1149–1161.

Draper, N., and H. Smith. 1998. *Applied Regression Analysis*. 3rd ed. New York: Wiley.

Hochberg, Y., and A. C. Tamhane. 1987. *Multiple Comparison Procedures*. New York: Wiley.

Kuehl, R. O. 2000. *Design of Experiments: Statistical Principles of Research Design and Analysis*. 2nd ed. Belmont, CA: Duxbury.

Marchenko, Y. V. 2006. Estimating variance components in Stata. *Stata Journal* 6: 1–21.

Miller, R. G., Jr. 1981. *Simultaneous Statistical Inference*. 2nd ed. New York: Springer.

Scheffé, H. 1953. A method for judging all contrasts in the analysis of variance. *Biometrika* 40: 87–104.

——. 1959. *The Analysis of Variance*. New York: Wiley.

Šidák, Z. 1967. Rectangular confidence regions for the means of multivariate normal distributions. *Journal of the American Statistical Association* 62: 626–633.

Snedecor, G. W., and W. G. Cochran. 1989. *Statistical Methods*. 8th ed. Ames, IA: Iowa State University Press.

van Belle, G., L. D. Fisher, P. J. Heagerty, and T. S. Lumley. 2004. *Biostatistics: A Methodology for the Health Sciences*. 2nd ed. New York: Wiley.

Winer, B. J., D. R. Brown, and K. M. Michels. 1991. *Statistical Principles in Experimental Design*. 3rd ed. New York: McGraw–Hill.

Also see

[R] **anova** — Analysis of variance and covariance

[R] **loneway** — Large one-way ANOVA, random effects, and reliability

Title

oprobit — Ordered probit regression

Syntax

`oprobit` *depvar* [*indepvars*] [*if*] [*in*] [*weight*] [, *options*]

options	Description
Model	
`offset(`*varname*`)`	include *varname* in model with coefficient constrained to 1
`constraints(`*constraints*`)`	apply specified linear constraints
`collinear`	keep collinear variables
SE/Robust	
`vce(`*vcetype*`)`	*vcetype* may be `oim`, `robust`, `cluster` *clustvar*, `bootstrap`, or `jackknife`
Reporting	
`level(`*#*`)`	set confidence level; default is `level(95)`
`nocnsreport`	do not display constraints
display_options	control column formats, row spacing, line width, and display of omitted variables and base and empty cells
Maximization	
maximize_options	control the maximization process; seldom used
`coeflegend`	display legend instead of statistics

indepvars may contain factor variables; see [U] **11.4.3 Factor variables**.

depvar and *indepvars* may contain time-series operators; see [U] **11.4.4 Time-series varlists**.

`bootstrap`, `by`, `fracpoly`, `jackknife`, `mfp`, `mi estimate`, `nestreg`, `rolling`, `statsby`, `stepwise`, and `svy` are allowed; see [U] **11.1.10 Prefix commands**.

`vce(bootstrap)` and `vce(jackknife)` are not allowed with the `mi estimate` prefix; see [MI] **mi estimate**.

Weights are not allowed with the `bootstrap` prefix; see [R] **bootstrap**.

`vce()` and weights are not allowed with the `svy` prefix; see [SVY] **svy**.

`fweight`s, `iweight`s, and `pweight`s are allowed; see [U] **11.1.6 weight**.

`coeflegend` does not appear in the dialog box.

See [U] **20 Estimation and postestimation commands** for more capabilities of estimation commands.

Menu

Statistics > Ordinal outcomes > Ordered probit regression

Description

`oprobit` fits ordered probit models of ordinal variable *depvar* on the independent variables *indepvars*. The actual values taken on by the dependent variable are irrelevant, except that larger values are assumed to correspond to "higher" outcomes.

See [R] **logistic** for a list of related estimation commands.

Options

Model

`offset(`*varname*`)`, `constraints(`*constraints*`)`, `collinear`; see [R] **estimation options**.

SE/Robust

`vce(`*vcetype*`)` specifies the type of standard error reported, which includes types that are derived from asymptotic theory, that are robust to some kinds of misspecification, that allow for intragroup correlation, and that use bootstrap or jackknife methods; see [R] ***vce_option***.

Reporting

`level(`*#*`)`; see [R] **estimation options**.

`nocnsreport`; see [R] **estimation options**.

display_options: `noomitted`, `vsquish`, `noemptycells`, `baselevels`, `allbaselevels`, `cformat(%`*fmt*`)`, `pformat(%`*fmt*`)`, `sformat(%`*fmt*`)`, and `nolstretch`; see [R] **estimation options**.

Maximization

maximize_options: `difficult`, `technique(`*algorithm_spec*`)`, `iterate(`*#*`)`, [`no`]`log`, `trace`, `gradient`, `showstep`, `hessian`, `showtolerance`, `tolerance(`*#*`)`, `ltolerance(`*#*`)`, `nrtolerance(`*#*`)`, `nonrtolerance`, and `from(`*init_specs*`)`; see [R] **maximize**. These options are seldom used.

The following option is available with `oprobit` but is not shown in the dialog box:

`coeflegend`; see [R] **estimation options**.

Remarks

An ordered probit model is used to estimate relationships between an ordinal dependent variable and a set of independent variables. An *ordinal* variable is a variable that is categorical and ordered, for instance, "poor", "good", and "excellent", which might indicate a person's current health status or the repair record of a car. If there are only two outcomes, see [R] **logistic**, [R] **logit**, and [R] **probit**. This entry is concerned only with more than two outcomes. If the outcomes cannot be ordered (for example, residency in the north, east, south, or west), see [R] **mlogit**. This entry is concerned only with models in which the outcomes can be ordered.

In ordered probit, an underlying score is estimated as a linear function of the independent variables and a set of cutpoints. The probability of observing outcome i corresponds to the probability that the estimated linear function, plus random error, is within the range of the cutpoints estimated for the outcome:

$$\Pr(\text{outcome}_j = i) = \Pr(\kappa_{i-1} < \beta_1 x_{1j} + \beta_2 x_{2j} + \cdots + \beta_k x_{kj} + u_j \leq \kappa_i)$$

u_j is assumed to be normally distributed. In either case, we estimate the coefficients β_1, β_2, $\dots$, β_k together with the cutpoints κ_1, κ_2, $\dots$, κ_{I-1}, where I is the number of possible outcomes. κ_0 is taken as $-\infty$, and κ_I is taken as $+\infty$. All of this is a direct generalization of the ordinary two-outcome probit model.

▷ Example 1

In example 2 of [R] **ologit**, we use a variation of the automobile dataset (see [U] **1.2.2 Example datasets**) to analyze the 1977 repair records of 66 foreign and domestic cars. We use ordered logit to explore the relationship of `rep77` in terms of `foreign` (origin of manufacture), `length` (a proxy for size), and `mpg`. Here we fit the same model using ordered probit rather than ordered logit:

```
. use http://www.stata-press.com/data/r12/fullauto
(Automobile Models)
. oprobit rep77 foreign length mpg

Iteration 0:   log likelihood = -89.895098
Iteration 1:   log likelihood = -78.106316
Iteration 2:   log likelihood = -78.020086
Iteration 3:   log likelihood = -78.020025
Iteration 4:   log likelihood = -78.020025
Ordered probit regression                         Number of obs   =         66
                                                  LR chi2(3)      =      23.75
                                                  Prob > chi2     =     0.0000
Log likelihood = -78.020025                       Pseudo R2       =     0.1321

------------------------------------------------------------------------------
       rep77 |      Coef.   Std. Err.      z    P>|z|     [95% Conf. Interval]
-------------+----------------------------------------------------------------
     foreign |   1.704861   .4246796     4.01   0.000     .8725037    2.537217
      length |   .0468675    .012648     3.71   0.000      .022078    .0716571
         mpg |   .1304559   .0378628     3.45   0.001     .0562463    .2046656
-------------+----------------------------------------------------------------
       /cut1 |    10.1589   3.076754                      4.128577    16.18923
       /cut2 |   11.21003   3.107527                      5.119389    17.30067
       /cut3 |   12.54561   3.155233                      6.361467    18.72975
       /cut4 |   13.98059   3.218793                      7.671874    20.28931
------------------------------------------------------------------------------
```

We find that foreign cars have better repair records, as do larger cars and cars with better mileage ratings.

◁

Saved results

oprobit saves the following in e():

Scalars	
e(N)	number of observations
e(N_cd)	number of completely determined observations
e(k_cat)	number of categories
e(k)	number of parameters
e(k_aux)	number of auxiliary parameters
e(k_eq)	number of equations in e(b)
e(k_eq_model)	number of equations in overall model test
e(k_dv)	number of dependent variables
e(df_m)	model degrees of freedom
e(r2_p)	pseudo-R-squared
e(ll)	log likelihood
e(ll_0)	log likelihood, constant-only model
e(N_clust)	number of clusters
e(chi2)	χ^2
e(p)	significance of model test
e(rank)	rank of e(V)
e(ic)	number of iterations
e(rc)	return code
e(converged)	1 if converged, 0 otherwise

Macros	
e(cmd)	oprobit
e(cmdline)	command as typed
e(depvar)	name of dependent variable
e(wtype)	weight type
e(wexp)	weight expression
e(title)	title in estimation output
e(clustvar)	name of cluster variable
e(offset)	linear offset variable
e(chi2type)	Wald or LR; type of model χ^2 test
e(vce)	*vcetype* specified in vce()
e(vcetype)	title used to label Std. Err.
e(opt)	type of optimization
e(which)	max or min; whether optimizer is to perform maximization or minimization
e(ml_method)	type of ml method
e(user)	name of likelihood-evaluator program
e(technique)	maximization technique
e(properties)	b V
e(predict)	program used to implement predict
e(asbalanced)	factor variables fvset as asbalanced
e(asobserved)	factor variables fvset as asobserved

Matrices

e(b)	coefficient vector
e(Cns)	constraints matrix
e(ilog)	iteration log (up to 20 iterations)
e(gradient)	gradient vector
e(cat)	category values
e(V)	variance–covariance matrix of the estimators
e(V_modelbased)	model-based variance

Functions

e(sample)	marks estimation sample

Methods and formulas

oprobit is implemented as an ado-file.

See *Methods and formulas* of [R] **ologit**.

References

Aitchison, J., and S. D. Silvey. 1957. The generalization of probit analysis to the case of multiple responses. *Biometrika* 44: 131–140.

Cameron, A. C., and P. K. Trivedi. 2005. *Microeconometrics: Methods and Applications*. New York: Cambridge University Press.

Chiburis, R., and M. Lokshin. 2007. Maximum likelihood and two-step estimation of an ordered-probit selection model. *Stata Journal* 7: 167–182.

Goldstein, R. 1997. sg59: Index of ordinal variation and Neyman–Barton GOF. *Stata Technical Bulletin* 33: 10–12. Reprinted in *Stata Technical Bulletin Reprints*, vol. 6, pp. 145–147. College Station, TX: Stata Press.

Long, J. S. 1997. *Regression Models for Categorical and Limited Dependent Variables*. Thousand Oaks, CA: Sage.

Long, J. S., and J. Freese. 2006. *Regression Models for Categorical Dependent Variables Using Stata*. 2nd ed. College Station, TX: Stata Press.

Miranda, A., and S. Rabe-Hesketh. 2006. Maximum likelihood estimation of endogenous switching and sample selection models for binary, ordinal, and count variables. *Stata Journal* 6: 285–308.

Stewart, M. B. 2004. Semi-nonparametric estimation of extended ordered probit models. *Stata Journal* 4: 27–39.

Williams, R. 2010. Fitting heterogeneous choice models with oglm. *Stata Journal* 10: 540–567.

Wolfe, R. 1998. sg86: Continuation-ratio models for ordinal response data. *Stata Technical Bulletin* 44: 18–21. Reprinted in *Stata Technical Bulletin Reprints*, vol. 8, pp. 149–153. College Station, TX: Stata Press.

Wolfe, R., and W. W. Gould. 1998. sg76: An approximate likelihood-ratio test for ordinal response models. *Stata Technical Bulletin* 42: 24–27. Reprinted in *Stata Technical Bulletin Reprints*, vol. 7, pp. 199–204. College Station, TX: Stata Press.

Xu, J., and J. S. Long. 2005. Confidence intervals for predicted outcomes in regression models for categorical outcomes. *Stata Journal* 5: 537–559.

Also see

[R] **oprobit postestimation** — Postestimation tools for oprobit

[R] **logistic** — Logistic regression, reporting odds ratios

[R] **mlogit** — Multinomial (polytomous) logistic regression

[R] **mprobit** — Multinomial probit regression

[R] **ologit** — Ordered logistic regression

[R] **probit** — Probit regression

[MI] **estimation** — Estimation commands for use with mi estimate

[SVY] **svy estimation** — Estimation commands for survey data

[U] **20 Estimation and postestimation commands**

Title

oprobit postestimation — Postestimation tools for oprobit

Description

The following postestimation commands are available after `oprobit`:

Command	Description
`contrast`	contrasts and ANOVA-style joint tests of estimates
`estat`	AIC, BIC, VCE, and estimation sample summary
`estat` (svy)	postestimation statistics for survey data
`estimates`	cataloging estimation results
`lincom`	point estimates, standard errors, testing, and inference for linear combinations of coefficients
`linktest`	link test for model specification
`lrtest`[1]	likelihood-ratio test
`margins`	marginal means, predictive margins, marginal effects, and average marginal effects
`marginsplot`	graph the results from margins (profile plots, interaction plots, etc.)
`nlcom`	point estimates, standard errors, testing, and inference for nonlinear combinations of coefficients
`predict`	predictions, residuals, influence statistics, and other diagnostic measures
`predictnl`	point estimates, standard errors, testing, and inference for generalized predictions
`pwcompare`	pairwise comparisons of estimates
`suest`	seemingly unrelated estimation
`test`	Wald tests of simple and composite linear hypotheses
`testnl`	Wald tests of nonlinear hypotheses

[1] `lrtest` is not appropriate with `svy` estimation results.

See the corresponding entries in the *Base Reference Manual* for details, but see [SVY] **estat** for details about `estat` (svy).

Syntax for predict

`predict` [*type*] { *stub** | *newvar* | *newvarlist* } [*if*] [*in*] [, *statistic* `outcome(`*outcome*`)` `nooffset`]

`predict` [*type*] { *stub** | *newvarlist* } [*if*] [*in*], `scores`

statistic	Description
Main	
`pr`	predicted probabilities; the default
`xb`	linear prediction
`stdp`	standard error of the linear prediction

If you do not specify outcome(), pr (with one new variable specified) assumes outcome(#1).

You specify one or k new variables with pr, where k is the number of outcomes.

You specify one new variable with xb and stdp.

These statistics are available both in and out of sample; type predict ... if e(sample) ... if wanted only for the estimation sample.

Menu

Statistics > Postestimation > Predictions, residuals, etc.

Options for predict

Main

pr, the default, calculates the predicted probabilities. If you do not also specify the outcome() option, you specify k new variables, where k is the number of categories of the dependent variable. Say that you fit a model by typing oprobit result x1 x2, and result takes on three values. Then you could type predict p1 p2 p3 to obtain all three predicted probabilities. If you specify the outcome() option, you must specify one new variable. Say that result takes on values 1, 2, and 3. Typing predict p1, outcome(1) would produce the same p1.

xb calculates the linear prediction. You specify one new variable, for example, predict linear, xb. The linear prediction is defined ignoring the contribution of the estimated cutpoints.

stdp calculates the standard error of the linear prediction. You specify one new variable, for example, predict se, stdp.

outcome(*outcome*) specifies for which outcome the predicted probabilities are to be calculated. outcome() should contain either one value of the dependent variable or one of #1, #2, ..., with #1 meaning the first category of the dependent variable, #2 meaning the second category, etc.

nooffset is relevant only if you specified offset(*varname*) for oprobit. It modifies the calculations made by predict so that they ignore the offset variable; the linear prediction is treated as $\mathbf{x}_j\mathbf{b}$ rather than as $\mathbf{x}_j\mathbf{b} + \text{offset}_j$.

scores calculates equation-level score variables. The number of score variables created will equal the number of outcomes in the model. If the number of outcomes in the model was k, then

the first new variable will contain $\partial \ln L / \partial(\mathbf{x}_j\mathbf{b})$;

the second new variable will contain $\partial \ln L / \partial\kappa_1$;

the third new variable will contain $\partial \ln L / \partial\kappa_2$;

...

and the kth new variable will contain $\partial \ln L / \partial\kappa_{k-1}$, where κ_i refers to the ith cutpoint.

Remarks

See [U] **20 Estimation and postestimation commands** for instructions on obtaining the variance–covariance matrix of the estimators, predicted values, and hypothesis tests. Also see [R] **lrtest** for performing likelihood-ratio tests.

▷ Example 1

In example 1 of [R] **oprobit**, we fit the model `oprobit rep77 foreign length mpg`. The `predict` command can be used to obtain the predicted probabilities. We type `predict` followed by the names of the new variables to hold the predicted probabilities, ordering the names from low to high. In our data, the lowest outcome is "poor" and the highest is "excellent". We have five categories, so we must type five names following `predict`; the choice of names is up to us:

```
. predict poor fair avg good exc
(option pr assumed; predicted probabilities)
. list make model exc good if rep77>=., sep(4) divider
```

	make	model	exc	good
3.	AMC	Spirit	.0006044	.0351813
10.	Buick	Opel	.0043803	.1133763
32.	Ford	Fiesta	.0002927	.0222789
44.	Merc.	Monarch	.0093209	.1700846
53.	Peugeot	604	.0734199	.4202766
56.	Plym.	Horizon	.001413	.0590294
57.	Plym.	Sapporo	.0197543	.2466034
63.	Pont.	Phoenix	.0234156	.266771

◁

❑ Technical note

For ordered probit, `predict, xb` produces $S_j = x_{1j}\beta_1 + x_{2j}\beta_2 + \cdots + x_{kj}\beta_k$. Ordered probit is identical to ordered logit, except that we use different distribution functions for calculating probabilities. The ordered-probit predictions are then the probability that $S_j + u_j$ lies between a pair of cutpoints κ_{i-1} and κ_i. The formulas for ordered probit are

$$\Pr(S_j + u < \kappa) = \Phi(\kappa - S_j)$$
$$\Pr(S_j + u > \kappa) = 1 - \Phi(\kappa - S_j) = \Phi(S_j - \kappa)$$
$$\Pr(\kappa_1 < S_j + u < \kappa_2) = \Phi(\kappa_2 - S_j) - \Phi(\kappa_1 - S_j)$$

Rather than using `predict` directly, we could calculate the predicted probabilities by hand.

```
. predict pscore, xb
. generate probexc = normal(pscore-_b[/cut4])
. generate probgood = normal(_b[/cut4]-pscore) - normal(_b[/cut3]-pscore)
```

❑

Methods and formulas

All postestimation tools listed above are implemented as ado-files.

Also see

[R] **oprobit** — Ordered probit regression

[U] **20 Estimation and postestimation commands**

Title

orthog — Orthogonalize variables and compute orthogonal polynomials

Syntax

Orthogonalize variables

orthog [*varlist*] [*if*] [*in*] [*weight*] , generate(*newvarlist*) [matrix(*matname*)]

Compute orthogonal polynomial

orthpoly *varname* [*if*] [*in*] [*weight*] ,

{ generate(*newvarlist*) | poly(*matname*) } [degree(#)]

orthpoly requires that generate(*newvarlist*) or poly(*matname*), or both, be specified.

varlist may contain time-series operators; see [U] **11.4.4 Time-series varlists**.

iweights, fweights, pweights, and aweights are allowed, see [U] **11.1.6 weight**.

Menu

orthog

Data > Create or change data > Other variable-creation commands > Orthogonalize variables

orthpoly

Data > Create or change data > Other variable-creation commands > Orthogonal polynomials

Description

orthog orthogonalizes a set of variables, creating a new set of orthogonal variables (all of type double), using a modified Gram–Schmidt procedure (Golub and Van Loan 1996). The order of the variables determines the orthogonalization; hence, the "most important" variables should be listed first.

Execution time is proportional to the square of the number of variables. With many (>10) variables, orthog will be fairly slow.

orthpoly computes orthogonal polynomials for one variable.

Options for orthog

Main

generate(*newvarlist*) is required. generate() creates new orthogonal variables of type double. For orthog, *newvarlist* will contain the orthogonalized *varlist*. If *varlist* contains d variables, then so will *newvarlist*. *newvarlist* can be specified by giving a list of exactly d new variable names, or it can be abbreviated using the styles *newvar*1-*newvard* or *newvar**. For these two styles of abbreviation, new variables *newvar*1, *newvar*2, ..., *newvard* are generated.

`matrix(`*matname*`)` creates a $(d+1) \times (d+1)$ matrix containing the matrix R defined by $X = QR$, where X is the $N \times (d+1)$ matrix representation of *varlist* plus a column of ones and Q is the $N \times (d+1)$ matrix representation of *newvarlist* plus a column of ones (d = number of variables in *varlist*, and N = number of observations).

Options for orthpoly

Main

`generate(`*newvarlist*`)` or `poly()`, or both, must be specified. `generate()` creates new orthogonal variables of type `double`. *newvarlist* will contain orthogonal polynomials of degree 1, 2, ..., d evaluated at *varname*, where d is as specified by `degree(`d`)`. *newvarlist* can be specified by giving a list of exactly d new variable names, or it can be abbreviated using the styles *newvar*`1`-*newvard* or *newvar*`*`. For these two styles of abbreviation, new variables *newvar*`1`, *newvar*`2`, ..., *newvard* are generated.

`poly(`*matname*`)` creates a $(d+1) \times (d+1)$ matrix called *matname* containing the coefficients of the orthogonal polynomials. The orthogonal polynomial of degree $i \leq d$ is

matname`[`i`, `$d+1$`]` + *matname*`[`i`, 1]*`*varname* + *matname*`[`i`, 2]*`*varname*^{2}
+ ⋯ + *matname*`[`i`, `i`]*`*varname*^{i}

The coefficients corresponding to the constant term are placed in the last column of the matrix. The last row of the matrix is all zeros, except for the last column, which corresponds to the constant term.

`degree(#)` specifies the highest-degree polynomial to include. Orthogonal polynomials of degree 1, 2, ..., $d = \#$ are computed. The default is $d = 1$.

Remarks

Orthogonal variables are useful for two reasons. The first is numerical accuracy for highly collinear variables. Stata's `regress` and other estimation commands can face much collinearity and still produce accurate results. But, at some point, these commands will drop variables because of collinearity. If you know with certainty that the variables are not perfectly collinear, you may want to retain all their effects in the model. If you use `orthog` or `orthpoly` to produce a set of orthogonal variables, all variables will be present in the estimation results.

Users are more likely to find orthogonal variables useful for the second reason: ease of interpreting results. `orthog` and `orthpoly` create a set of variables such that the "effects" of all the preceding variables have been removed from each variable. For example, if we issue the command

```
. orthog x1 x2 x3, generate(q1 q2 q3)
```

the effect of the constant is removed from `x1` to produce `q1`; the constant and `x1` are removed from `x2` to produce `q2`; and finally the constant, `x1`, and `x2` are removed from `x3` to produce `q3`. Hence,

$$q1 = r_{01} + r_{11}\,x1$$
$$q2 = r_{02} + r_{12}\,x1 + r_{22}\,x2$$
$$q3 = r_{03} + r_{13}\,x1 + r_{23}\,x2 + r_{33}\,x3$$

This effect can be generalized and written in matrix notation as

$$X = QR$$

where X is the $N \times (d+1)$ matrix representation of *varlist* plus a column of ones, and Q is the $N \times (d+1)$ matrix representation of *newvarlist* plus a column of ones ($d =$ number of variables in *varlist* and $N =$ number of observations). The $(d+1) \times (d+1)$ matrix R is a permuted upper-triangular matrix, that is, R would be upper triangular if the constant were first, but the constant is last, so the first row/column has been permuted with the last row/column. Because Stata's estimation commands list the constant term last, this allows R, obtained via the `matrix()` option, to be used to transform estimation results.

▷ Example 1

Consider Stata's `auto.dta` dataset. Suppose that we postulate a model in which `price` depends on the car's `length`, `weight`, `headroom`, and trunk size (`trunk`). These predictors are collinear, but not extremely so—the correlations are not that close to 1:

```
. use http://www.stata-press.com/data/r12/auto
(1978 Automobile Data)

. correlate length weight headroom trunk
(obs=74)

             |   length   weight headroom    trunk
-------------+------------------------------------
      length |   1.0000
      weight |   0.9460   1.0000
    headroom |   0.5163   0.4835   1.0000
       trunk |   0.7266   0.6722   0.6620   1.0000
```

`regress` certainly has no trouble fitting this model:

```
. regress price length weight headroom trunk

      Source |       SS       df       MS              Number of obs =      74
-------------+------------------------------           F(  4,    69) =   10.20
       Model |   236016580     4    59004145           Prob > F      =  0.0000
    Residual |   399048816    69  5783316.17           R-squared     =  0.3716
-------------+------------------------------           Adj R-squared =  0.3352
       Total |   635065396    73  8699525.97           Root MSE      =  2404.9

------------------------------------------------------------------------------
       price |      Coef.   Std. Err.      t    P>|t|     [95% Conf. Interval]
-------------+----------------------------------------------------------------
      length |  -101.7092   42.12534    -2.41   0.018     -185.747   -17.67147
      weight |   4.753066   1.120054     4.24   0.000     2.518619    6.987512
    headroom |  -711.5679   445.0204    -1.60   0.114    -1599.359    176.2236
       trunk |   114.0859   109.9488     1.04   0.303    -105.2559    333.4277
       _cons |   11488.47   4543.902     2.53   0.014     2423.638    20553.31
------------------------------------------------------------------------------
```

However, we may believe a priori that `length` is the most important predictor, followed by `weight`, `headroom`, and `trunk`. We would like to remove the "effect" of `length` from all the other predictors, remove `weight` from `headroom` and `trunk`, and remove `headroom` from `trunk`. We can do this by running `orthog`, and then we fit the model again using the orthogonal variables:

```
. orthog length weight headroom trunk, gen(olength oweight oheadroom otrunk)
> matrix(R)

. regress price olength oweight oheadroom otrunk

      Source |       SS       df       MS              Number of obs =      74
-------------+------------------------------           F(  4,    69) =   10.20
       Model |   236016580     4    59004145           Prob > F      =  0.0000
    Residual |   399048816    69  5783316.17           R-squared     =  0.3716
-------------+------------------------------           Adj R-squared =  0.3352
       Total |   635065396    73  8699525.97           Root MSE      =  2404.9

------------------------------------------------------------------------------
       price |      Coef.   Std. Err.      t    P>|t|     [95% Conf. Interval]
-------------+----------------------------------------------------------------
     olength |   1265.049   279.5584     4.53   0.000     707.3454    1822.753
     oweight |   1175.765   279.5584     4.21   0.000     618.0617    1733.469
   oheadroom |  -349.9916   279.5584    -1.25   0.215    -907.6955    207.7122
      otrunk |   290.0776   279.5584     1.04   0.303    -267.6262    847.7815
       _cons |   6165.257   279.5584    22.05   0.000     5607.553    6722.961
------------------------------------------------------------------------------
```

Using the matrix R, we can transform the results obtained using the orthogonal predictors back to the metric of original predictors:

```
. matrix b = e(b)*inv(R)'

. matrix list b

b[1,5]
        length       weight     headroom        trunk        _cons
y1  -101.70924    4.7530659   -711.56789    114.08591    11488.475
```

◁

❑ Technical note

The matrix R obtained using the `matrix()` option with `orthog` can also be used to recover X (the original *varlist*) from Q (the orthogonalized *newvarlist*), one variable at a time. Continuing with the previous example, we illustrate how to recover the `trunk` variable:

```
. matrix C = R[1...,"trunk"]'

. matrix score double rtrunk = C

. compare rtrunk trunk

                                        ---------- difference ----------
                            count       minimum      average     maximum
------------------------------------------------------------------------
rtrunk>trunk                   74      1.42e-14     2.27e-14    3.55e-14
                       ----------
jointly defined                74      1.42e-14     2.27e-14    3.55e-14
                       ----------
total                          74
```

Here the recovered variable `rtrunk` is almost exactly the same as the original `trunk` variable. When you are orthogonalizing many variables, this procedure can be performed to check the numerical soundness of the orthogonalization. Because of the ordering of the orthogonalization procedure, the last variable and the variables near the end of the *varlist* are the most important ones to check.

❑

The `orthpoly` command effectively does for polynomial terms what the `orthog` command does for an arbitrary set of variables.

▷ Example 2

Again consider the auto.dta dataset. Suppose that we wish to fit the model

$$\texttt{mpg} = \beta_0 + \beta_1\,\texttt{weight} + \beta_2\,\texttt{weight}^2 + \beta_3\,\texttt{weight}^3 + \beta_4\,\texttt{weight}^4 + \epsilon$$

We will first compute the regression with natural polynomials:

```
. gen double w1 = weight
. gen double w2 = w1*w1
. gen double w3 = w2*w1
. gen double w4 = w3*w1
. correlate w1-w4
(obs=74)

             |       w1       w2       w3       w4
-------------+------------------------------------
          w1 |   1.0000
          w2 |   0.9915   1.0000
          w3 |   0.9665   0.9916   1.0000
          w4 |   0.9279   0.9679   0.9922   1.0000
. regress mpg w1-w4
      Source |       SS       df       MS              Number of obs =      74
-------------+------------------------------           F(  4,    69) =   36.06
       Model |  1652.73666     4  413.184164           Prob > F      =  0.0000
    Residual |  790.722803    69  11.4597508           R-squared     =  0.6764
-------------+------------------------------           Adj R-squared =  0.6576
       Total |  2443.45946    73  33.4720474           Root MSE      =  3.3852

------------------------------------------------------------------------------
         mpg |      Coef.   Std. Err.      t    P>|t|     [95% Conf. Interval]
-------------+----------------------------------------------------------------
          w1 |   .0289302   .1161939     0.25   0.804    -.2028704    .2607307
          w2 |  -.0000229   .0000566    -0.40   0.687    -.0001359    .0000901
          w3 |   5.74e-09   1.19e-08     0.48   0.631    -1.80e-08    2.95e-08
          w4 |  -4.86e-13   9.14e-13    -0.53   0.596    -2.31e-12    1.34e-12
       _cons |   23.94421   86.60667     0.28   0.783    -148.8314    196.7198
------------------------------------------------------------------------------
```

Some of the correlations among the powers of weight are very large, but this does not create any problems for regress. However, we may wish to look at the quadratic trend with the constant removed, the cubic trend with the quadratic and constant removed, etc. orthpoly will generate polynomial terms with this property:

```
. orthpoly weight, generate(pw*) deg(4) poly(P)
. regress mpg pw1-pw4
      Source |       SS       df       MS              Number of obs =      74
-------------+------------------------------           F(  4,    69) =   36.06
       Model |  1652.73666     4  413.184164           Prob > F      =  0.0000
    Residual |  790.722803    69  11.4597508           R-squared     =  0.6764
-------------+------------------------------           Adj R-squared =  0.6576
       Total |  2443.45946    73  33.4720474           Root MSE      =  3.3852

------------------------------------------------------------------------------
         mpg |      Coef.   Std. Err.      t    P>|t|     [95% Conf. Interval]
-------------+----------------------------------------------------------------
         pw1 |  -4.638252   .3935245   -11.79   0.000    -5.423312   -3.853192
         pw2 |   .8263545   .3935245     2.10   0.039     .0412947    1.611414
         pw3 |  -.3068616   .3935245    -0.78   0.438    -1.091921    .4781982
         pw4 |   -.209457   .3935245    -0.53   0.596    -.9945168    .5756028
       _cons |    21.2973   .3935245    54.12   0.000     20.51224    22.08236
------------------------------------------------------------------------------
```

Compare the p-values of the terms in the natural polynomial regression with those in the orthogonal polynomial regression. With orthogonal polynomials, it is easy to see that the pure cubic and quartic trends are not significant and that the constant, linear, and quadratic terms each have $p < 0.05$.

The matrix P obtained with the `poly()` option can be used to transform coefficients for orthogonal polynomials to coefficients for natural polynomials:

```
. orthpoly weight, poly(P) deg(4)
. matrix b = e(b)*P
. matrix list b
b[1,5]
          deg1        deg2        deg3        deg4       _cons
y1   .02893016  -.00002291   5.745e-09  -4.862e-13   23.944212
```

◁

Methods and formulas

`orthog` and `orthpoly` are implemented as ado-files.

`orthog`'s orthogonalization can be written in matrix notation as

$$X = QR$$

where X is the $N \times (d+1)$ matrix representation of *varlist* plus a column of ones and Q is the $N \times (d+1)$ matrix representation of *newvarlist* plus a column of ones (d = number of variables in *varlist*, and N = number of observations). The $(d+1) \times (d+1)$ matrix R is a permuted upper-triangular matrix; that is, R would be upper triangular if the constant were first, but the constant is last, so the first row/column has been permuted with the last row/column.

Q and R are obtained using a modified Gram–Schmidt procedure; see Golub and Van Loan (1996, 218–219) for details. The traditional Gram–Schmidt procedure is notoriously unsound, but the modified procedure is good. `orthog` performs two passes of this procedure.

`orthpoly` uses the Christoffel–Darboux recurrence formula (Abramowitz and Stegun 1972).

Both `orthog` and `orthpoly` normalize the orthogonal variables such that

$$Q'WQ = MI$$

where $W = \text{diag}(w_1, w_2, \dots, w_N)$ with weights $w_1, w_2, \dots, w_N$ (all 1 if weights are not specified), and M is the sum of the weights (the number of observations if weights are not specified).

References

Abramowitz, M., and I. A. Stegun, ed. 1972. *Handbook of Mathematical Functions with Formulas, Graphs, and Mathematical Tables*. 10th ed. Washington, DC: National Bureau of Standards.

Golub, G. H., and C. F. Van Loan. 1996. *Matrix Computations*. 3rd ed. Baltimore: Johns Hopkins University Press.

Sribney, W. M. 1995. sg37: Orthogonal polynomials. *Stata Technical Bulletin* 25: 17–18. Reprinted in *Stata Technical Bulletin Reprints*, vol. 5, pp. 96–98. College Station, TX: Stata Press.

Also see

[R] **regress** — Linear regression

Title

pcorr — Partial and semipartial correlation coefficients

Syntax

`pcorr` *varname*$_1$ *varlist* [*if*] [*in*] [*weight*]

varname$_1$ and *varlist* may contain time-series operators; see [U] **11.4.4 Time-series varlists**.
`by` is allowed; see [D] **by**.
`aweights` and `fweights` are allowed; see [U] **11.1.6 weight**.

Menu

Statistics > Summaries, tables, and tests > Summary and descriptive statistics > Partial correlations

Description

`pcorr` displays the partial and semipartial correlation coefficients of *varname*$_1$ with each variable in *varlist* after removing the effects of all other variables in *varlist*. The squared correlations and corresponding significance are also reported.

Remarks

Assume that y is determined by x_1, x_2, ..., x_k. The partial correlation between y and x_1 is an attempt to estimate the correlation that would be observed between y and x_1 if the other x's did not vary. The semipartial correlation, also called part correlation, between y and x_1 is an attempt to estimate the correlation that would be observed between y and x_1 after the effects of all other x's are removed from x_1 but not from y.

Both squared correlations estimate the proportion of the variance of y that is explained by each predictor. The squared semipartial correlation between y and x_1 represents the proportion of variance in y that is explained by x_1 only. This squared correlation can also be interpreted as the decrease in the model's R^2 value that results from removing x_1 from the full model. Thus one could use the squared semipartial correlations as criteria for model selection. The squared partial correlation between y and x_1 represents the proportion of variance in y not associated with any other x's that is explained by x_1. Thus the squared partial correlation gives an estimate of how much of the variance of y not explained by the other x's is explained by x_1.

▷ Example 1

Using our automobile dataset (described in [U] **1.2.2 Example datasets**), we can obtain the simple correlations between `price`, `mpg`, `weight`, and `foreign` from `correlate` (see [R] **correlate**):

```
. use http://www.stata-press.com/data/r12/auto
(1978 Automobile Data)

. correlate price mpg weight foreign
(obs=74)

             |    price      mpg   weight  foreign
-------------+------------------------------------
       price |   1.0000
         mpg |  -0.4686   1.0000
      weight |   0.5386  -0.8072   1.0000
     foreign |   0.0487   0.3934  -0.5928   1.0000
```

Although `correlate` gave us the full correlation matrix, our interest is in just the first column. We find, for instance, that the higher the `mpg`, the lower the `price`. We obtain the partial and semipartial correlation coefficients by using `pcorr`:

```
. pcorr price mpg weight foreign
(obs=74)

Partial and semipartial correlations of price with

               Partial   Semipartial      Partial   Semipartial   Significance
   Variable |    Corr.         Corr.      Corr.^2       Corr.^2          Value
------------+-----------------------------------------------------------------
        mpg |   0.0352        0.0249       0.0012        0.0006         0.7693
     weight |   0.5488        0.4644       0.3012        0.2157         0.0000
    foreign |   0.5402        0.4541       0.2918        0.2062         0.0000
```

We now find that the partial and semipartial correlations of `price` with `mpg` are near 0. In the simple correlations, we found that `price` and `foreign` were virtually uncorrelated. In the partial and semipartial correlations, we find that `price` and `foreign` are positively correlated. The nonsignificance of `mpg` tells us that the amount in which R^2 decreases by removing `mpg` from the model is not significant. We find that removing either `weight` or `foreign` results in a significant drop in the R^2 of the model.

◁

❑ Technical note

Use caution when interpreting the above results. As we said at the outset, the partial and semipartial correlation coefficients are an *attempt* to estimate the correlation that would be observed if the effects of all other variables were taken out of both y and x or only x. `pcorr` makes it too easy to ignore the fact that we are fitting a model. In the example above, the model is

$$\texttt{price} = \beta_0 + \beta_1\texttt{mpg} + \beta_2\texttt{weight} + \beta_3\texttt{foreign} + \epsilon$$

which is, in all honesty, a rather silly model. Even if we accept the implied economic assumptions of the model—that consumers value `mpg`, `weight`, and `foreign`—do we really believe that consumers place equal value on every extra 1,000 pounds of weight? That is, have we correctly parameterized the model? If we have not, then the estimated partial and semipartial correlation coefficients may not represent what they claim to represent. Partial and semipartial correlation coefficients are a reasonable way to summarize data if we are convinced that the underlying model is reasonable. We should not, however, pretend that there is no underlying model and that these correlation coefficients are unaffected by the assumptions and parameterization.

❑

Saved results

pcorr saves the following in r():

Scalars

r(N)	number of observations
r(df)	degrees of freedom

Matrices

r(p_corr)	partial correlation coefficient vector
r(sp_corr)	semipartial correlation coefficient vector

Methods and formulas

pcorr is implemented as an ado-file.

Results are obtained by fitting a linear regression of *varname*$_1$ on *varlist*; see [R] **regress**. The partial correlation coefficient between *varname*$_1$ and each variable in *varlist* is then calculated as

$$\frac{t}{\sqrt{t^2 + n - k}}$$

(Greene 2012, 37), where t is the t statistic, n is the number of observations, and k is the number of independent variables, including the constant but excluding any dropped variables.

The semipartial correlation coefficient between *varname*$_1$ and each variable in *varlist* is calculated as

$$\text{sign}(t)\sqrt{\frac{t^2(1 - R^2)}{n - k}}$$

(Cohen et al. 2003, 89), where R^2 is the model R^2 value, and t, n, and k are as described above.

The significance is given by $2\text{Pr}(t_{n-k} > |t|)$, where t_{n-k} follows a Student's t distribution with $n - k$ degrees of freedom.

Acknowledgment

The addition of semipartial correlation coefficients to pcorr is based on the pcorr2 command by Richard Williams, University of Notre Dame.

References

Cohen, J., P. Cohen, S. G. West, and L. S. Aiken. 2003. *Applied Multiple Regression/Correlation Analysis for the Behavioral Sciences*. 3rd ed. Hillsdale, NJ: Erlbaum.

Greene, W. H. 2012. *Econometric Analysis*. 7th ed. Upper Saddle River, NJ: Prentice Hall.

Also see

[R] **correlate** — Correlations (covariances) of variables or coefficients

[R] **spearman** — Spearman's and Kendall's correlations

Title

permute — Monte Carlo permutation tests

Syntax

Compute permutation test

`permute` *permvar exp_list* [, *options*] : *command*

Report saved results

`permute` [*varlist*] [`using` *filename*] [, *display_options*]

options	Description
Main	
`reps(`*#*`)`	perform # random permutations; default is `reps(100)`
`left` | `right`	compute one-sided *p*-values; default is two-sided
Options	
`strata(`*varlist*`)`	permute within strata
`saving(`*filename*`, ...)`	save results to *filename*; save statistics in double precision; save results to *filename* every # replications
Reporting	
`level(`*#*`)`	set confidence level; default is `level(95)`
`noheader`	suppress table header
`nolegend`	suppress table legend
`verbose`	display full table legend
`nodrop`	do not drop observations
`nodots`	suppress replication dots
`noisily`	display any output from *command*
`trace`	trace *command*
`title(`*text*`)`	use *text* as title for permutation results
Advanced	
`eps(`*#*`)`	numerical tolerance; seldom used
`nowarn`	do not warn when `e(sample)` is not set
`force`	do not check for *weights* or `svy` commands; seldom used
`reject(`*exp*`)`	identify invalid results
`seed(`*#*`)`	set random-number seed to #

weights are not allowed in *command*.

display_options	Description
`left` | `right`	compute one-sided p-values; default is two-sided
`level(#)`	set confidence level; default is `level(95)`
`noheader`	suppress table header
`nolegend`	suppress table legend
`verbose`	display full table legend
`title(`*text*`)`	use *text* as title for results
`eps(#)`	numerical tolerance; seldom used

exp_list contains	(*name*: *elist*)
	elist
	eexp
elist contains	*newvar* = (*exp*)
	(*exp*)
eexp is	*specname*
	[*eqno*]*specname*
specname is	`_b`
	`_b[]`
	`_se`
	`_se[]`
eqno is	`##`
	name

exp is a standard Stata expression; see [U] **13 Functions and expressions**.

Distinguish between `[]`, which are to be typed, and [ ], which indicate optional arguments.

Menu

Statistics > Resampling > Permutation tests

Description

`permute` estimates p-values for permutation tests on the basis of Monte Carlo simulations. Typing

```
. permute permvar exp_list, reps(#): command
```

randomly permutes the values in *permvar* # times, each time executing *command* and collecting the associated values from the expression in *exp_list*.

These p-value estimates can be one-sided: $\Pr(T^* \leq T)$ or $\Pr(T^* \geq T)$. The default is two-sided: $\Pr(|T^*| \geq |T|)$. Here T^* denotes the value of the statistic from a randomly permuted dataset, and T denotes the statistic as computed on the original data.

permvar identifies the variable whose observed values will be randomly permuted.

command defines the statistical command to be executed. Most Stata commands and user-written programs can be used with `permute`, as long as they follow standard Stata syntax; see [U] **11 Language syntax**. The `by` prefix may not be part of *command*.

exp_list specifies the statistics to be collected from the execution of *command*.

`permute` may be used for replaying results, but this feature is appropriate only when a dataset generated by `permute` is currently in memory or is identified by the `using` option. The variables specified in *varlist* in this context must be present in the respective dataset.

Options

Main

`reps(#)` specifies the number of random permutations to perform. The default is 100.

`left` or `right` requests that one-sided p-values be computed. If `left` is specified, an estimate of $\Pr(T^* \leq T)$ is produced, where T^* is the test statistic and T is its observed value. If `right` is specified, an estimate of $\Pr(T^* \geq T)$ is produced. By default, two-sided p-values are computed; that is, $\Pr(|T^*| \geq |T|)$ is estimated.

Options

`strata(`*varlist*`)` specifies that the permutations be performed within each stratum defined by the values of *varlist*.

`saving(`*filename* [, *suboptions*]`)` creates a Stata data file (`.dta` file) consisting of (for each statistic in *exp_list*) a variable containing the replicates.

`double` specifies that the results for each replication be stored as `double`s, meaning 8-byte reals. By default, they are stored as `float`s, meaning 4-byte reals.

`every(#)` specifies that results are to be written to disk every #th replication. `every()` should be specified only in conjunction with `saving()` when *command* takes a long time for each replication. This will allow recovery of partial results should some other software crash your computer. See [P] **postfile**.

`replace` specifies that *filename* be overwritten if it exists. This option does not appear in the dialog box.

Reporting

`level(#)` specifies the confidence level, as a percentage, for confidence intervals. The default is `level(95)` or as set by `set level`; see [R] **level**.

`noheader` suppresses display of the table header. This option implies the `nolegend` option.

`nolegend` suppresses display of the table legend. The table legend identifies the rows of the table with the expressions they represent.

`verbose` requests that the full table legend be displayed. By default, coefficients and standard errors are not displayed.

`nodrop` prevents `permute` from dropping observations outside the `if` and `in` qualifiers. `nodrop` will also cause `permute` to ignore the contents of `e(sample)` if it exists as a result of running *command*. By default, `permute` temporarily drops out-of-sample observations.

`nodots` suppresses display of the replication dots. By default, one dot character is displayed for each successful replication. A red 'x' is displayed if *command* returns an error or if one of the values in *exp_list* is missing.

`noisily` requests that any output from *command* be displayed. This option implies the `nodots` option.

`trace` causes a trace of the execution of *command* to be displayed. This option implies the `noisily` option.

`title(`*text*`)` specifies a title to be displayed above the table of permutation results; the default title is `Monte Carlo permutation results`.

Advanced

`eps(`*#*`)` specifies the numerical tolerance for testing $|T^*| \geq |T|$, $T^* \leq T$, or $T^* \geq T$. These are considered true if, respectively, $|T^*| \geq |T| - \#$, $T^* \leq T + \#$, or $T^* \geq T - \#$. The default is `1e-7`. You will not have to specify `eps()` under normal circumstances.

`nowarn` suppresses the printing of a warning message when *command* does not set `e(sample)`.

`force` suppresses the restriction that *command* may not specify weights or be a `svy` command. `permute` is not suited for weighted estimation, thus `permute` should not be used with weights or `svy`. `permute` reports an error when it encounters weights or `svy` in *command* if the `force` option is not specified. This is a seldom used option, so use it only if you know what you are doing!

`reject(`*exp*`)` identifies an expression that indicates when results should be rejected. When *exp* is true, the resulting values are reset to missing values.

`seed(`*#*`)` sets the random-number seed. Specifying this option is equivalent to typing the following command prior to calling `permute`:

```
. set seed #
```

Remarks

Permutation tests determine the significance of the observed value of a test statistic in light of rearranging the order (permuting) of the observed values of a variable.

▷ Example 1

Suppose that we conducted an experiment to determine the effect of a treatment on the development of cells. Further suppose that we are restricted to six experimental units because of the extreme cost of the experiment. Thus three units are to be given a placebo, and three units are given the treatment. The measurement is the number of newly developed healthy cells. The following listing gives the hypothetical data, along with some summary statistics.

```
. input y treatment
              y  treatment
  1. 7 0
  2. 9 0
  3. 11 0
  4. 10 1
  5. 12 1
  6. 14 1
  7. end
. sort treatment
```

```
. summarize y
    Variable |       Obs        Mean    Std. Dev.       Min        Max
-------------+--------------------------------------------------------
           y |         6        10.5    2.428992          7         14
. by treatment: summarize y

----------------------------------------------------------------------------
-> treatment = 0
    Variable |       Obs        Mean    Std. Dev.       Min        Max
-------------+--------------------------------------------------------
           y |         3           9           2          7         11

----------------------------------------------------------------------------
-> treatment = 1
    Variable |       Obs        Mean    Std. Dev.       Min        Max
-------------+--------------------------------------------------------
           y |         3          12           2         10         14
```

Clearly, there are more cells in the treatment group than in the placebo group, but a statistical test is needed to conclude that the treatment does affect the development of cells. If the sum of the treatment measures is our test statistic, we can use `permute` to determine the probability of observing 36 or more cells, given the observed data and assuming that there is no effect due to the treatment.

```
. set seed 1234
. permute y sum=r(sum), saving(permdish) right nodrop nowarn: sum y if treatment
(running summarize on estimation sample)
Permutation replications (100)
----+--- 1 ---+--- 2 ---+--- 3 ---+--- 4 ---+--- 5
..................................................    50
..................................................   100
Monte Carlo permutation results                 Number of obs    =          6
      command:  summarize y if treatment
          sum:  r(sum)
  permute var:  y

------------------------------------------------------------------------------
T            |     T(obs)       c       n   p=c/n   SE(p) [95% Conf. Interval]
-------------+----------------------------------------------------------------
         sum |         36      10     100  0.1000  0.0300  .0490047   .1762226
------------------------------------------------------------------------------
Note:  confidence interval is with respect to p=c/n.
Note:  c = #{T >= T(obs)}
```

We see that 10 of the 100 randomly permuted datasets yielded sums from the treatment group larger than or equal to the observed sum of 36. Thus the evidence is not strong enough, at the 5% level, to reject the null hypothesis that there is no effect of the treatment.

Because of the small size of this experiment, we could have calculated the exact permutation p-value from all possible permutations. There are six units, but we want the sum of the treatment units. Thus there are $\binom{6}{3} = 20$ permutation sums from the possible unique permutations.

$7+9+10=26$	$7+10+12=29$	$9+10+11=30$	$9+12+14=35$
$7+9+11=27$	$7+10+14=31$	$9+10+12=31$	$10+11+12=33$
$7+9+12=28$	$7+11+12=30$	$9+10+14=33$	$10+11+14=35$
$7+9+14=30$	$7+11+14=32$	$9+11+12=32$	$10+12+14=36$
$7+10+11=28$	$7+12+14=33$	$9+11+14=34$	$11+12+14=37$

Two of the 20 permutation sums are greater than or equal to 36. Thus the exact p-value for this permutation test is 0.1. Tied values will decrease the number of unique permutations.

When the `saving()` option is supplied, `permute` saves the values of the permutation statistic to the indicated file, in our case, `permdish.dta`. This file can be used to replay the result of `permute`. The `level()` option controls the confidence level of the confidence interval for the permutation p-value. This confidence interval is calculated using `cii` with the reported `n` (number of nonmissing replications) and `c` (the counter for events of significance).

```
. permute using permdish, level(80)
Monte Carlo permutation results                   Number of obs    =          6

      command:  summarize y if treatment
          sum:  r(sum)
  permute var:  y

--------------------------------------------------------------------------------
T            |     T(obs)       c       n   p=c/n   SE(p) [80% Conf. Interval]
-------------+------------------------------------------------------------------
         sum |         36      10     100  0.1000  0.0300  .0631113   .1498826
--------------------------------------------------------------------------------
Note:  confidence interval is with respect to p=c/n.
Note:  c = #{|T| >= |T(obs)|}
```

◁

▷ Example 2

Consider some fictional data from a randomized complete-block design in which we wish to determine the significance of five treatments.

```
. use http://www.stata-press.com/data/r12/permute1, clear
. list y treatment in 1/10, abbrev(10)

     +----------------------+
     |        y   treatment |
     |----------------------|
  1. | 4.407557           1 |
  2. | 5.693386           1 |
  3. | 7.099699           1 |
  4. |  3.12132           1 |
  5. | 5.242648           1 |
     |----------------------|
  6. | 4.280349           2 |
  7. | 4.508785           2 |
  8. | 4.079967           2 |
  9. | 5.904368           2 |
 10. | 3.010556           2 |
     +----------------------+
```

These data may be analyzed using anova.

```
. anova y treatment subject

                           Number of obs =      50     R-squared     =  0.3544
                           Root MSE      = .914159     Adj R-squared =  0.1213

                  Source | Partial SS    df       MS           F     Prob > F
              -----------+----------------------------------------------------
                   Model | 16.5182188    13  1.27063221       1.52     0.1574
                         |
               treatment | 13.0226706     9  1.44696341       1.73     0.1174
                 subject | 3.49554813     4  .873887032       1.05     0.3973
                         |
                Residual | 30.0847503    36  .835687509
              -----------+----------------------------------------------------
                   Total | 46.6029691    49  .951081002
```

Suppose that we want to compute the significance of the F statistic for `treatment` by using `permute`. All we need to do is write a short program that will save the result of this statistic for `permute` to use. For example,

```
program panova, rclass
        version 12
        args response fac_intrst fac_other
        anova `response' `fac_intrst' `fac_other'
        return scalar Fmodel = e(F)
        test `fac_intrst'
        return scalar F = r(F)
end
```

Now in panova, `test` saves the F statistic for the factor of interest in `r(F)`. This is different from `e(F)`, which is the overall model F statistic for the model fit by anova that panova saves in `r(Fmodel)`. In the following example, we use the `strata()` option so that the treatments are randomly rearranged within each `subject`. It should not be too surprising that the estimated p-values are equal for this example, because the two F statistics are equivalent when controlling for differences between subjects. However, we would not expect to always get the same p-values every time we reran `permute`.

```
. set seed 1234

. permute treatment treatmentF=r(F) modelF=e(F), reps(1000) strata(subject)
> saving(permanova) nodots: panova y treatment subject

Monte Carlo permutation results

Number of strata =          5                  Number of obs    =         50

      command:  panova y treatment subject
   treatmentF:  r(F)
       modelF:  e(F)
  permute var:  treatment

------------------------------------------------------------------------------
T            |     T(obs)       c       n   p=c/n   SE(p) [95% Conf. Interval]
-------------+----------------------------------------------------------------
  treatmentF |   1.731465     118    1000  0.1180  0.0102  .0986525   .1396277
      modelF |   1.520463     118    1000  0.1180  0.0102  .0986525   .1396277
------------------------------------------------------------------------------
Note:  confidence intervals are with respect to p=c/n.
Note:  c = #{|T| >= |T(obs)|}
```

◁

▷ Example 3

As a final example, let's consider estimating the p-value of the Z statistic returned by `ranksum`. Suppose that we collected data from some experiment: `y` is some measure we took on 17 individuals, and `group` identifies the group that an individual belongs to.

```
. use http://www.stata-press.com/data/r12/permute2
. list
```

	group	y
1.	1	6
2.	1	11
3.	1	20
4.	1	2
5.	1	9
6.	1	5
7.	0	2
8.	0	1
9.	0	6
10.	0	0
11.	0	2
12.	0	3
13.	0	3
14.	0	12
15.	0	4
16.	0	1
17.	0	5

Next we analyze the data using `ranksum` and notice that the observed value of the test statistic (saved as `r(z)`) is −2.02 with an approximate p-value of 0.0434.

```
. ranksum y, by(group)
Two-sample Wilcoxon rank-sum (Mann-Whitney) test
       group |      obs    rank sum    expected
-------------+---------------------------------
           0 |       11          79          99
           1 |        6          74          54
-------------+---------------------------------
    combined |       17         153         153
unadjusted variance        99.00
adjustment for ties        -0.97
                        ----------
adjusted variance          98.03
Ho: y(group==0) = y(group==1)
             z =  -2.020
    Prob > |z| =   0.0434
```

The observed value of the rank-sum statistic is 79, with an expected value (under the null hypothesis of no group effect) of 99. There are 17 observations, so the permutation distribution contains $\binom{17}{6} = 12{,}376$ possible values of the rank-sum statistic if we ignore ties. With ties, we have fewer possible values but still too many to want to count them. Thus we use `permute` with 10,000 replications and see that the Monte Carlo permutation test agrees with the result of the test based on the normal approximation.

```
. set seed 18385766
. permute y z=r(z), reps(10000) nowarn nodots: ranksum y, by(group)
Monte Carlo permutation results                 Number of obs    =         17

      command:  ranksum y, by(group)
            z:  r(z)
  permute var:  y

T            |    T(obs)       c       n   p=c/n   SE(p) [95% Conf. Interval]
-------------+----------------------------------------------------------------
           z | -2.020002     468   10000  0.0468  0.0021  .0427429   .0511236
------------------------------------------------------------------------------
Note:  confidence interval is with respect to p=c/n.
Note:  c = #{|T| >= |T(obs)|}
```

◁

For an application of a permutation test to a problem in epidemiology, see Hayes and Moulton (2009, 190–193).

❑ Technical note

permute reports confidence intervals for p to emphasize that it is based on the binomial estimator for proportions. When the variability implied by the confidence interval makes conclusions difficult, you may increase the number of replications to determine more precisely the significance of the test statistic of interest. In other words, the value of p from permute will converge to the true permutation p-value as the number of replications gets arbitrarily large. ❑

Saved results

permute saves the following in r():

Scalars

r(N)	sample size	r(k_exp)	number of standard expressions
r(N_reps)	number of requested replications	r(k_eexp)	number of _b/_se expressions
r(level)	confidence level		

Macros

r(cmd)	permute	r(left)	left or empty
r(command)	*command* following colon	r(right)	right or empty
r(permvar)	permutation variable	r(seed)	initial random-number seed
r(title)	title in output	r(event)	T <= T(obs), T >= T(obs), or |T| <= |T(obs)|
r(exp#)	#th expression		

Matrices

r(b)	observed statistics	r(p)	observed proportions
r(c)	count when r(event) is true	r(se)	standard errors of observed proportions
r(reps)	number of nonmissing results	r(ci)	confidence intervals of observed proportions

Methods and formulas

permute is implemented as an ado-file.

References

Ängquist, L. 2010. Stata tip 92: Manual implementation of permutations and bootstraps. *Stata Journal* 10: 686–688.

Good, P. I. 2006. *Resampling Methods: A Practical Guide to Data Analysis*. 3rd ed. Boston: Birkhäuser.

Hayes, R. J., and L. H. Moulton. 2009. *Cluster Randomised Trials*. Boca Raton, FL: Chapman & Hall/CRC.

Kaiser, J. 2007. An exact and a Monte Carlo proposal to the Fisher–Pitman permutation tests for paired replicates and for independent samples. *Stata Journal* 7: 402–412.

Kaiser, J., and M. G. Lacy. 2009. A general-purpose method for two-group randomization tests. *Stata Journal* 9: 70–85.

Also see

[R] **bootstrap** — Bootstrap sampling and estimation

[R] **jackknife** — Jackknife estimation

[R] **simulate** — Monte Carlo simulations

Title

pk — Pharmacokinetic (biopharmaceutical) data

Description

The term pk refers to pharmacokinetic data and the Stata commands, all of which begin with the letters pk, designed to do some of the analyses commonly performed in the pharmaceutical industry. The system is intended for the analysis of pharmacokinetic data, although some of the commands are for general use.

The pk commands are

`pkexamine`	[R] **pkexamine**	Calculate pharmacokinetic measures
`pksumm`	[R] **pksumm**	Summarize pharmacokinetic data
`pkshape`	[R] **pkshape**	Reshape (pharmacokinetic) Latin-square data
`pkcross`	[R] **pkcross**	Analyze crossover experiments
`pkequiv`	[R] **pkequiv**	Perform bioequivalence tests
`pkcollapse`	[R] **pkcollapse**	Generate pharmacokinetic measurement dataset

Remarks

Several types of clinical trials are commonly performed in the pharmaceutical industry. Examples include combination trials, multicenter trials, equivalence trials, and active control trials. For each type of trial, there is an optimal study design for estimating the effects of interest. Currently, the pk system can be used to analyze equivalence trials, which are usually conducted using a crossover design; however, it is possible to use a parallel design and still draw conclusions about equivalence.

Equivalence trials assess bioequivalence between two drugs. Although proving that two drugs behave the same is impossible, the United States Food and Drug Administration believes that if the absorption properties of two drugs are similar, the two drugs will produce similar effects and have similar safety profiles. Generally, the goal of an equivalence trial is to assess the equivalence of a generic drug to an existing drug. This goal is commonly accomplished by comparing a confidence interval about the difference between a pharmacokinetic measurement of two drugs with a confidence limit constructed from U.S. federal regulations. If the confidence interval is entirely within the confidence limit, the drugs are declared bioequivalent. Another approach to assessing bioequivalence is to use the method of interval hypotheses testing. `pkequiv` is used to conduct these tests of bioequivalence.

Several pharmacokinetic measures can be used to ascertain how available a drug is for cellular absorption. The most common measure is the area under the time-versus-concentration curve (AUC). Another common measure of drug availability is the maximum concentration ($C_{\max}$) achieved by the drug during the follow-up period. Stata reports these and other less common measures of drug availability, including the time at which the maximum drug concentration was observed and the duration of the period during which the subject was being measured. Stata also reports the elimination rate, that is, the rate at which the drug is metabolized, and the drug's half-life, that is, the time it takes for the drug concentration to fall to one-half of its maximum concentration.

pkexamine computes and reports all the pharmacokinetic measures that Stata produces, including four calculations of the area under the time-versus-concentration curve. The standard area under the curve from 0 to the maximum observed time ($\mathrm{AUC}_{0,t_{\max}}$) is computed using cubic splines or the trapezoidal rule. Additionally, pkexamine also computes the area under the curve from 0 to infinity by extending the standard time-versus-concentration curve from the maximum observed time by using three different methods. The first method simply extends the standard curve by using a least-squares linear fit through the last few data points. The second method extends the standard curve by fitting a decreasing exponential curve through the last few data points. Finally, the third method extends the curve by fitting a least-squares linear regression line on the log concentration. The mathematical details of these extensions are described in *Methods and formulas* of [R] **pkexamine**.

Data from an equivalence trial may also be analyzed using methods appropriate to the particular study design. When you have a crossover design, pkcross can be used to fit an appropriate ANOVA model. As an aside, a crossover design is simply a restricted Latin square; therefore, pkcross can also be used to analyze any Latin-square design.

There are some practical concerns when dealing with data from equivalence trials. Primarily, the data must be organized in a manner that Stata can use. The pk commands include pkcollapse and pkshape, which are designed to help transform data from a common format to one that is suitable for analysis with Stata.

In the following example, we illustrate several different data formats that are often encountered in pharmaceutical research and describe how these formats can be transformed to formats that can be analyzed with Stata.

▷ Example 1

Assume that we have one subject and are interested in determining the drug profile for that subject. A reasonable experiment would be to give the subject the drug and then measure the concentration of the drug in the subject's blood over a given period. For example, here is a part of a dataset from Chow and Liu (2009, 13):

```
. use http://www.stata-press.com/data/r12/auc
. list, abbrev(14)
```

	id	time	concentration
1.	1	0	0
2.	1	.5	0
3.	1	1	2.8
4.	1	1.5	4.4
5.	1	2	4.4
6.	1	3	4.7
7.	1	4	4.1
8.	1	6	4
9.	1	8	3.6
10.	1	12	3
11.	1	16	2.5
12.	1	24	2
13.	1	32	1.6

Examining these data, we notice that the concentration quickly increases, plateaus for a short period, and then slowly decreases over time. pkexamine is used to calculate the pharmacokinetic measures of interest. pkexamine is explained in detail in [R] **pkexamine**. The output is

```
. pkexamine time conc
                         Maximum concentration =          4.7
                 Time of maximum concentration =            3
             Time of last observation (Tmax) =           32
                              Elimination rate =       0.0279
                                     Half life =      24.8503

Area under the curve
```

AUC [0, Tmax]	AUC [0, inf.) Linear of log conc.	AUC [0, inf.) Linear fit	AUC [0, inf.) Exponential fit
85.24	142.603	107.759	142.603

```
Fit based on last 3 points.
```

Clinical trials, however, require that data be collected on more than one subject. There are several ways to enter raw measured data collected on several subjects. It would be reasonable to enter for each subject the drug concentration value at specific points in time. Such data could be

id	conc1	conc2	conc3	conc4	conc5	conc6	conc7
1	0	1	4	7	5	3	1
2	0	2	6	5	4	3	2
3	0	1	2	3	5	4	1

where conc1 is the concentration at the first measured time, conc2 is the concentration at the second measured time, etc. This format requires that each drug concentration measurement be made at the same time on each subject. Another more flexible way to enter the data is to have an observation with three variables for each time measurement on a subject. Each observation would have a subject ID, the time at which the measurement was made, and the corresponding drug concentration at that time. The data would be

```
. use http://www.stata-press.com/data/r12/pkdata
. list id concA time, sepby(id)
```

```
      +----------------------+
      | id      concA   time |
      |----------------------|
   1. |  1          0      0 |
   2. |  1   3.073403     .5 |
   3. |  1   5.188444      1 |
   4. |  1   5.898577    1.5 |
   5. |  1   5.096378      2 |
   6. |  1   6.094085      3 |
   7. |  1   5.158772      4 |
   8. |  1     5.7065      6 |
   9. |  1   5.272467      8 |
  10. |  1     4.4576     12 |
  11. |  1   5.146423     16 |
  12. |  1   4.947427     24 |
  13. |  1   1.920421     32 |
      |----------------------|
  14. |  2          0      0 |
  15. |  2    2.48462     .5 |
  16. |  2   4.883569      1 |
  17. |  2   7.253442    1.5 |
  18. |  2   5.849345      2 |
  19. |  2   6.761085      3 |
  20. |  2    4.33839      4 |
  21. |  2    5.04199      6 |
  22. |  2    4.25128      8 |
  23. |  2   6.205004     12 |
  24. |  2   5.566165     16 |
  25. |  2   3.689007     24 |
  26. |  2   3.644063     32 |
      |----------------------|
  27. |  3          0      0 |
           (output omitted)
 207. | 20   4.673281     24 |
 208. | 20   3.487347     32 |
      +----------------------+
```

Stata expects the data to be organized in the second form. If your data are organized as described in the first dataset, you will need to use `reshape` to change the data to the second form; see [D] **reshape**. Because the data in the second (or long) format contain information for one drug on several subjects, `pksumm` can be used to produce summary statistics of the pharmacokinetic measurements. The output is

```
. pksumm id time concA
................

Summary statistics for the pharmacokinetic measures

                                        Number of observations =      16

    Measure |    Mean     Median      Variance  Skewness   Kurtosis   p-value
------------+------------------------------------------------------------------
        auc |  151.63     152.18        127.58     -0.34       2.07      0.55
    aucline |  397.09     219.83     178276.59      2.69       9.61      0.00
     aucexp |  668.60     302.96     720356.98      2.67       9.54      0.00
     auclog |  665.95     298.03     752573.34      2.71       9.70      0.00
       half |   90.68      29.12      17750.70      2.36       7.92      0.00
         ke |    0.02       0.02          0.00      0.88       3.87      0.08
       cmax |    7.37       7.42          0.40     -0.64       2.75      0.36
       tomc |    3.38       3.00          7.25      2.27       7.70      0.00
       tmax |   32.00      32.00          0.00         .          .         .
```

Until now, we have been concerned with the profile of only one drug. We have characterized the profile of that drug by individual subjects by using pkexamine and by a group of subjects by using pksumm. The goal of an equivalence trial, however, is to compare two drugs, which we will do in the rest of this example.

For equivalence trials, the study design most often used is the crossover design. For a complete discussion of crossover designs, see Ratkowsky, Evans, and Alldredge (1993).

In brief, crossover designs require that each subject be given both treatments at two different times. The order in which the treatments are applied changes between groups. For example, if we had 20 subjects numbered 1–20, the first 10 would receive treatment A during the first period of the study, and then they would be given treatment B. The second 10 subjects would be given treatment B during the first period of the study, and then they would be given treatment A. Each subject in the study will have four variables that describe the observation: a subject identifier, a sequence identifier that indicates the order of treatment, and two outcome variables, one for each treatment. The outcome variables for each subject are the pharmacokinetic measures. The data must be transformed from a series of measurements on individual subjects to data containing the pharmacokinetic measures for each subject. In Stata parlance, this is referred to as a collapse, which can be done with pkcollapse; see [R] **pkcollapse**.

Here is a part of our data:

```
. list, sepby(id)
```

	id	seq	time	concA	concB
1.	1	1	0	0	0
2.	1	1	.5	3.073403	3.712592
3.	1	1	1	5.188444	6.230602
4.	1	1	1.5	5.898577	7.885944
5.	1	1	2	5.096378	9.241735
6.	1	1	3	6.094085	13.10507
7.	1	1	4	5.158772	.169429
8.	1	1	6	5.7065	8.759894
9.	1	1	8	5.272467	7.985409
10.	1	1	12	4.4576	7.740126
11.	1	1	16	5.146423	7.607208
12.	1	1	24	4.947427	7.588428
13.	1	1	32	1.920421	2.791115
14.	2	1	0	0	0
15.	2	1	.5	2.48462	.9209593
16.	2	1	1	4.883569	5.925818
17.	2	1	1.5	7.253442	8.710549
18.	2	1	2	5.849345	10.90552
19.	2	1	3	6.761085	8.429898
20.	2	1	4	4.33839	5.573152
21.	2	1	6	5.04199	6.32341
22.	2	1	8	4.25128	.5251224
23.	2	1	12	6.205004	7.415988
24.	2	1	16	5.566165	6.323938
25.	2	1	24	3.689007	1.133553
26.	2	1	32	3.644063	5.759489
27.	3	1	0	0	0
			(output omitted)		
207.	20	2	24	4.673281	6.059818
208.	20	2	32	3.487347	5.213639

This format is similar to the second format described above, except that now we have measurements for two drugs at each time for each subject. We transform these data with pkcollapse:

```
. pkcollapse time concA concB, id(id) keep(seq) stat(auc)
................................
. list, sep(8) abbrev(10)
```

	id	seq	auc_concA	auc_concB
1.	1	1	150.9643	218.5551
2.	2	1	146.7606	133.3201
3.	3	1	160.6548	126.0635
4.	4	1	157.8622	96.17461
5.	5	1	133.6957	188.9038
6.	7	1	160.639	223.6922
7.	8	1	131.2604	104.0139
8.	9	1	168.5186	237.8962
9.	10	2	137.0627	139.7382
10.	12	2	153.4038	202.3942
11.	13	2	163.4593	136.7848
12.	14	2	146.0462	104.5191
13.	15	2	158.1457	165.8654
14.	18	2	147.1977	139.235
15.	19	2	164.9988	166.2391
16.	20	2	145.3823	158.5146

For this example, we chose to use the AUC for two drugs as our pharmacokinetic measure. We could have used any of the measures computed by pkexamine. In addition to the AUCs, the dataset also contains a sequence variable for each subject indicating when each treatment was administered.

The data produced by pkcollapse are in what Stata calls wide format; that is, there is one observation per subject containing two or more outcomes. To use pkcross and pkequiv, we need to transform these data to long format. This goal can be accomplished using pkshape; see [R] **pkshape**.

Consider the first subject in the dataset. This subject is in sequence one, which means that treatment A was applied during the first period of the study and treatment B was applied in the second period of the study. We need to split the first observation into two observations so that the outcome measure is only in one variable. Also we need two new variables, one indicating the treatment the subject received and another recording the period of the study when the subject received that treatment. We might expect the expansion of the first subject to be

id	sequence	auc	treat	period
1	1	150.9643	A	1
1	1	218.5551	B	2

We see that subject number 1 was in sequence 1, had an AUC of 150.9643 when treatment A was applied in the first period of the study, and had an AUC of 218.5551 when treatment B was applied.

Similarly, the expansion of subject 10 (the first subject in sequence 2) would be

id	sequence	auc	treat	period
10	2	137.0627	B	1
10	2	139.7382	A	2

Here treatment B was applied to the subject during the first period of the study, and treatment A was applied to the subject during the second period of the study.

An additional complication is common in crossover study designs. The treatment applied in the first period of the study might still have some effect on the outcome in the second period. In this example, each subject was given one treatment followed by another treatment. To get accurate estimates of treatment effects, it is necessary to account for the effect that the first treatment has in the second period of the study. This is called the carryover effect. We must, therefore, have a variable that indicates which treatment was applied in the first treatment period. `pkshape` creates a variable that indicates the carryover effect. For treatments applied during the first treatment period, there will never be a carryover effect. Thus the expanded data created by `pkshape` for subject 1 will be

```
 id   sequence    outcome      treat     period      carry
  1          1   150.9643          A          1          0
  1          1   218.5551          B          2          A
```

and the data for subject 10 will be

```
 id   sequence    outcome      treat     period      carry
 10          2   137.0627          B          1          0
 10          2   139.7382          A          2          B
```

We `pkshape` the data:

```
. pkshape id seq auc*, order(ab ba)
. sort id sequence period
. list, sep(16)
```

	id	sequence	outcome	treat	carry	period
1.	1	1	150.9643	1	0	1
2.	1	1	218.5551	2	1	2
3.	2	1	146.7606	1	0	1
4.	2	1	133.3201	2	1	2
5.	3	1	160.6548	1	0	1
6.	3	1	126.0635	2	1	2
7.	4	1	157.8622	1	0	1
8.	4	1	96.17461	2	1	2
9.	5	1	133.6957	1	0	1
10.	5	1	188.9038	2	1	2
11.	7	1	160.639	1	0	1
12.	7	1	223.6922	2	1	2
13.	8	1	131.2604	1	0	1
14.	8	1	104.0139	2	1	2
15.	9	1	168.5186	1	0	1
16.	9	1	237.8962	2	1	2
17.	10	2	137.0627	2	0	1
18.	10	2	139.7382	1	2	2
19.	12	2	153.4038	2	0	1
20.	12	2	202.3942	1	2	2
21.	13	2	163.4593	2	0	1
22.	13	2	136.7848	1	2	2
23.	14	2	146.0462	2	0	1
24.	14	2	104.5191	1	2	2
25.	15	2	158.1457	2	0	1
26.	15	2	165.8654	1	2	2
27.	18	2	147.1977	2	0	1
28.	18	2	139.235	1	2	2
29.	19	2	164.9988	2	0	1
30.	19	2	166.2391	1	2	2
31.	20	2	145.3823	2	0	1
32.	20	2	158.5146	1	2	2

As an aside, crossover designs do not require that each subject receive each treatment, but if they do, the crossover design is referred to as a complete crossover design.

The last dataset is organized in a manner that can be analyzed with Stata. To fit an ANOVA model to these data, we can use `anova` or `pkcross`. To conduct equivalence tests, we can use `pkequiv`. This example is further analyzed in [R] **pkcross** and [R] **pkequiv**.

◁

References

Chow, S.-C., and J.-P. Liu. 2009. *Design and Analysis of Bioavailability and Bioequivalence Studies*. 3rd ed. Boca Raton, FL: Chapman & Hall/CRC.

Ratkowsky, D. A., M. A. Evans, and J. R. Alldredge. 1993. *Cross-over Experiments: Design, Analysis, and Application*. New York: Dekker.

Title

pkcollapse — Generate pharmacokinetic measurement dataset

Syntax

`pkcollapse` *time concentration* [*if*] , `id(`*id_var*`)` [*options*]

options	Description
Main	
* `id(`*id_var*`)`	subject ID variable
`stat(`*measures*`)`	create specified *measures*; default is all
`trapezoid`	use trapezoidal rule; default is cubic splines
`fit(`*#*`)`	use # points to estimate $AUC_{0,\infty}$; default is `fit(3)`
`keep(`*varlist*`)`	keep variables in *varlist*
`force`	force collapse
`nodots`	suppress dots during calculation

* `id(`*id_var*`)` is required.

measures	Description
`auc`	area under the concentration-time curve ($AUC_{0,\infty}$)
`aucline`	area under the concentration-time curve from 0 to ∞ using a linear extension
`aucexp`	area under the concentration-time curve from 0 to ∞ using an exponential extension
`auclog`	area under the log-concentration-time curve extended with a linear fit
`half`	half-life of the drug
`ke`	elimination rate
`cmax`	maximum concentration
`tmax`	time at last concentration
`tomc`	time of maximum concentration

Menu

Statistics > Epidemiology and related > Other > Generate pharmacokinetic measurement dataset

Description

`pkcollapse` generates new variables with the pharmacokinetic summary measures of interest.

`pkcollapse` is one of the pk commands. Please read [R] **pk** before reading this entry.

Options

Main

`id(`*id_var*`)` is required and specifies the variable that contains the subject ID over which `pkcollapse` is to operate.

`stat(`*measures*`)` specifies the measures to be generated. The default is to generate all the measures.

`trapezoid` tells Stata to use the trapezoidal rule when calculating the AUC. The default is to use cubic splines, which give better results for most functions. When the curve is irregular, `trapezoid` may give better results.

`fit(`*#*`)` specifies the number of points to use in estimating the $\text{AUC}_{0,\infty}$. The default is `fit(3)`, the last three points. This number should be viewed as a minimum; the appropriate number of points will depend on your data.

`keep(`*varlist*`)` specifies the variables to be kept during the collapse. Variables not specified with the `keep()` option will be dropped. When `keep()` is specified, the keep variables are checked to ensure that all values of the variables are the same within *id_var*.

`force` forces the collapse, even when the values of the `keep()` variables are different within the *id_var*.

`nodots` suppresses the display of dots during calculation.

Remarks

`pkcollapse` generates all the summary pharmacokinetic measures.

▷ Example 1

We demonstrate the use of `pkcollapse` with the data described in [R] **pk**. We have drug concentration data on 15 subjects. Each subject is measured at 13 time points over a 32-hour period. Some of the records are

```
. use http://www.stata-press.com/data/r12/pkdata
. list, sep(0)
```

	id	seq	time	concA	concB
1.	1	1	0	0	0
2.	1	1	.5	3.073403	3.712592
3.	1	1	1	5.188444	6.230602
4.	1	1	1.5	5.898577	7.885944
5.	1	1	2	5.096378	9.241735
6.	1	1	3	6.094085	13.10507
			(output omitted)		
14.	2	1	0	0	0
15.	2	1	.5	2.48462	.9209593
16.	2	1	1	4.883569	5.925818
17.	2	1	1.5	7.253442	8.710549
18.	2	1	2	5.849345	10.90552
19.	2	1	3	6.761085	8.429898
			(output omitted)		
207.	20	2	24	4.673281	6.059818
208.	20	2	32	3.487347	5.213639

Although pksumm allows us to view all the pharmacokinetic measures, we can create a dataset with the measures by using pkcollapse.

```
. pkcollapse time concA concB, id(id) stat(auc) keep(seq)
................................
. list, sep(8) abbrev(10)
```

	id	seq	auc_concA	auc_concB
1.	1	1	150.9643	218.5551
2.	2	1	146.7606	133.3201
3.	3	1	160.6548	126.0635
4.	4	1	157.8622	96.17461
5.	5	1	133.6957	188.9038
6.	7	1	160.639	223.6922
7.	8	1	131.2604	104.0139
8.	9	1	168.5186	237.8962
9.	10	2	137.0627	139.7382
10.	12	2	153.4038	202.3942
11.	13	2	163.4593	136.7848
12.	14	2	146.0462	104.5191
13.	15	2	158.1457	165.8654
14.	18	2	147.1977	139.235
15.	19	2	164.9988	166.2391
16.	20	2	145.3823	158.5146

The resulting dataset, which we will call pkdata2, contains 1 observation per subject. This dataset is in wide format. If we want to use pkcross or pkequiv, we must transform these data to long format, which we do in the last example of [R] **pkshape**.

◁

Methods and formulas

pkcollapse is implemented as an ado-file.

The statistics generated by pkcollapse are described in [R] **pkexamine**.

Also see

[R] **pk** — Pharmacokinetic (biopharmaceutical) data

Title

pkcross — Analyze crossover experiments

Syntax

`pkcross` *outcome* [*if*] [*in*] [, *options*]

options	Description
Model	
`sequence(`*varname*`)`	sequence variable; default is `sequence(sequence)`
`treatment(`*varname*`)`	treatment variable; default is `treatment(treat)`
`period(`*varname*`)`	period variable; default is `period(period)`
`id(`*varname*`)`	ID variable
`carryover(`*varname*`)`	name of carryover variable; default is `carryover(carry)`
`carryover(none)`	omit carryover effects from model; default is `carryover(carry)`
`model(`*string*`)`	specify the model to fit
`sequential`	estimate sequential instead of partial sums of squares
Parameterization	
`param(3)`	estimate mean and the period, treatment, and sequence effects; assume no carryover effects exist; the default
`param(1)`	estimate mean and the period, treatment, and carryover effects; assume no sequence effects exist
`param(2)`	estimate mean, period and treatment effects, and period-by-treatment interaction; assume no sequence or carryover effects exist
`param(4)`	estimate mean, period and treatment effects, and period-by-treatment interaction; assume no period or crossover effects exist

Menu

Statistics > Epidemiology and related > Other > Analyze crossover experiments

Description

`pkcross` analyzes data from a crossover design experiment. When analyzing pharmaceutical trial data, if the treatment, carryover, and sequence variables are known, the omnibus test for separability of the treatment and carryover effects is calculated.

`pkcross` is one of the pk commands. Please read [R] **pk** before reading this entry.

Options

Model

`sequence(`*varname*`)` specifies the variable that contains the sequence in which the treatment was administered. If this option is not specified, `sequence(sequence)` is assumed.

treatment(*varname*) specifies the variable that contains the treatment information. If this option is not specified, treatment(treat) is assumed.

period(*varname*) specifies the variable that contains the period information. If this option is not specified, period(period) is assumed.

id(*varname*) specifies the variable that contains the subject identifiers. If this option is not specified, id(id) is assumed.

carryover(*varname* | none) specifies the variable that contains the carryover information. If carry(none) is specified, the carryover effects are omitted from the model. If this option is not specified, carryover(carry) is assumed.

model(*string*) specifies the model to be fit. For higher-order crossover designs, this option can be useful if you want to fit a model other than the default. However, anova (see [R] **anova**) can also be used to fit a crossover model. The default model for higher-order crossover designs is outcome predicted by sequence, period, treatment, and carryover effects. By default, the model statement is model(sequence period treat carry).

sequential specifies that sequential sums of squares be estimated.

Parameterization

param(#) specifies which of the four parameterizations to use for the analysis of a 2×2 crossover experiment. This option is ignored with higher-order crossover designs. The default is param(3). See the technical note for 2×2 crossover designs for more details.

param(3) estimates the overall mean, the period effects, the treatment effects, and the sequence effects, assuming that no carryover effects exist. This is the default parameterization.

param(1) estimates the overall mean, the period effects, the treatment effects, and the carryover effects, assuming that no sequence effects exist.

param(2) estimates the overall mean, the period effects, the treatment effects, and the period-by-treatment interaction, assuming that no sequence or carryover effects exist.

param(4) estimates the overall mean, the sequence effects, the treatment effects, and the sequence-by-treatment interaction, assuming that no period or crossover effects exist. When the sequence by treatment is equivalent to the period effect, this reduces to the third parameterization.

Remarks

pkcross is designed to analyze crossover experiments. Use pkshape first to reshape your data; see [R] **pkshape**. pkcross assumes that the data were reshaped by pkshape or are organized in the same manner as produced with pkshape. Washout periods are indicated by the number 0. See the technical note in this entry for more information on analyzing 2×2 crossover experiments.

❑ Technical note

The 2×2 crossover design cannot be used to estimate more than four parameters because there are only four pieces of information (the four cell means) collected. pkcross uses ANOVA models to analyze the data, so one of the four parameters must be the overall mean of the model, leaving just 3 degrees of freedom to estimate the remaining effects (period, sequence, treatment, and carryover). Thus the model is overparameterized. Estimation of treatment and carryover effects requires the assumption of either no period effects or no sequence effects. Some researchers maintain that it estimating carryover effects at the expense of other effects is a bad idea. This is a limitation of this design. pkcross implements four parameterizations for this model. They are numbered sequentially from one to four and are described in *Options*.

❑

▷ Example 1

Consider the example data published in Chow and Liu (2009, 71) and described in [R] **pkshape**. We have entered and reshaped the data with pkshape and have variables that identify the subjects, periods, treatments, sequence, and carryover treatment. To compute the ANOVA table, use pkcross:

```
. use http://www.stata-press.com/data/r12/chowliu
. pkshape id seq period1 period2, order(ab ba)
. pkcross outcome
                                    sequence variable = sequence
                                      period variable = period
                                   treatment variable = treat
                                   carryover variable = carry
                                          id variable = id
          Analysis of variance (ANOVA) for a 2x2 crossover study
Source of Variation        SS        df        MS            F      Prob > F

Intersubjects
     Sequence effect     276.00       1      276.00        0.37      0.5468
           Residuals   16211.49      22      736.89        4.41      0.0005

Intrasubjects
    Treatment effect      62.79       1       62.79        0.38      0.5463
       Period effect      35.97       1       35.97        0.22      0.6474
           Residuals    3679.43      22      167.25

               Total   20265.68      47
Omnibus measure of separability of treatment and carryover =   29.2893%
```

There is evidence of intersubject variability, but there are no other significant effects. The omnibus test for separability is a measure reflecting the degree to which the study design allows the treatment effects to be estimated independently of the carryover effects. The measure of separability of the treatment and carryover effects indicates approximately 29% separability, which can be interpreted as the degree to which the treatment and carryover effects are orthogonal. This is a characteristic of the design of the study. For a complete discussion, see Ratkowsky, Evans, and Alldredge (1993). Compared to the output in Chow and Liu (2009), the sequence effect is mislabeled as a carryover effect. See Ratkowsky, Evans, and Alldredge (1993, sec. 3.2) for a complete discussion of the mislabeling.

By specifying param(1), we obtain parameterization 1 for this model.

```
. pkcross outcome, param(1)
                                   sequence variable = sequence
                                     period variable = period
                                  treatment variable = treat
                                  carryover variable = carry
                                         id variable = id

           Analysis of variance (ANOVA) for a 2x2 crossover study
 Source of Variation  | Partial SS   df        MS           F     Prob > F
                      |
      Treatment effect|    301.04     1      301.04       0.67     0.4189
         Period effect|    255.62     1      255.62       0.57     0.4561
      Carryover effect|    276.00     1      276.00       0.61     0.4388
             Residuals|  19890.92    44      452.07
                      |
                 Total|  20265.68    47

Omnibus measure of separability of treatment and carryover =   29.2893%
```

◁

▷ Example 2

Consider the case of a two-treatment, four-sequence, two-period crossover design. This design is commonly referred to as Balaam's design (Balaam 1968). Ratkowsky, Evans, and Alldredge (1993, 140) published the following data from an amantadine trial, originally published by Taka and Armitage (1983):

```
. use http://www.stata-press.com/data/r12/balaam, clear
. list, sep(0)
```

	id	seq	period1	period2	period3
1.	1	-ab	9	8.75	8.75
2.	2	-ab	12	10.5	9.75
3.	3	-ab	17	15	18.5
4.	4	-ab	21	21	21.5
5.	1	-ba	23	22	18
6.	2	-ba	15	15	13
7.	3	-ba	13	14	13.75
8.	4	-ba	24	22.75	21.5
9.	5	-ba	18	17.75	16.75
10.	1	-aa	14	12.5	14
11.	2	-aa	27	24.25	22.5
12.	3	-aa	19	17.25	16.25
13.	4	-aa	30	28.25	29.75
14.	1	-bb	21	20	19.51
15.	2	-bb	11	10.5	10
16.	3	-bb	20	19.5	20.75
17.	4	-bb	25	22.5	23.5

The sequence identifier must be a string with zeros to indicate washout or baseline periods, or a number. If the sequence identifier is numeric, the order option must be specified with pkshape. If the sequence identifier is a string, pkshape will create sequence, period, and treatment identifiers without the order option. In this example, the dash is used to indicate a baseline period, which is an invalid code for this purpose. As a result, the data must be encoded; see [D] **encode**.

```
. encode seq, gen(num_seq)
. pkshape id num_seq period1 period2 period3, order(0aa 0ab 0ba 0bb)
. pkcross outcome, se
                                    sequence variable = sequence
                                      period variable = period
                                   treatment variable = treat
                                   carryover variable = carry
                                          id variable = id

            Analysis of variance (ANOVA) for a crossover study
  Source of Variation |    SS        df        MS          F      Prob > F
  --------------------+-----------------------------------------------------
  Intersubjects       |
      Sequence effect |   285.82      3       95.27       1.01      0.4180
            Residuals |  1221.49     13       93.96      59.96      0.0000
  --------------------+-----------------------------------------------------
  Intrasubjects       |
        Period effect |    15.13      2        7.56       6.34      0.0048
     Treatment effect |     8.48      1        8.48       8.86      0.0056
     Carryover effect |     0.11      1        0.11       0.12      0.7366
            Residuals |    29.56     30        0.99
  --------------------+-----------------------------------------------------
                Total |  1560.59     50
Omnibus measure of separability of treatment and carryover =   64.6447%
```

In this example, the sequence specifier used dashes instead of zeros to indicate a baseline period during which no treatment was given. For pkcross to work, we need to encode the string sequence variable and then use the order option with pkshape. A word of caution: encode does not necessarily choose the first sequence to be sequence 1, as in this example. Always double-check the sequence numbering when using encode.

◁

▷ Example 3

Continuing with the example from [R] **pkshape**, we fit an ANOVA model.

```
. use http://www.stata-press.com/data/r12/pkdata3, clear
. list, sep(8)
```

	id	sequence	outcome	treat	carry	period
1.	1	1	150.9643	A	0	1
2.	2	1	146.7606	A	0	1
3.	3	1	160.6548	A	0	1
4.	4	1	157.8622	A	0	1
5.	5	1	133.6957	A	0	1
6.	7	1	160.639	A	0	1
7.	8	1	131.2604	A	0	1
8.	9	1	168.5186	A	0	1
9.	10	2	137.0627	B	0	1
10.	12	2	153.4038	B	0	1
11.	13	2	163.4593	B	0	1
12.	14	2	146.0462	B	0	1
13.	15	2	158.1457	B	0	1
14.	18	2	147.1977	B	0	1
15.	19	2	164.9988	B	0	1
16.	20	2	145.3823	B	0	1
17.	1	1	218.5551	B	A	2
18.	2	1	133.3201	B	A	2
19.	3	1	126.0635	B	A	2
20.	4	1	96.17461	B	A	2
21.	5	1	188.9038	B	A	2
22.	7	1	223.6922	B	A	2
23.	8	1	104.0139	B	A	2
24.	9	1	237.8962	B	A	2
25.	10	2	139.7382	A	B	2
26.	12	2	202.3942	A	B	2
27.	13	2	136.7848	A	B	2
28.	14	2	104.5191	A	B	2
29.	15	2	165.8654	A	B	2
30.	18	2	139.235	A	B	2
31.	19	2	166.2391	A	B	2
32.	20	2	158.5146	A	B	2

The ANOVA model is fit using pkcross:

```
. pkcross outcome
                                   sequence variable = sequence
                                     period variable = period
                                  treatment variable = treat
                                  carryover variable = carry
                                         id variable = id

          Analysis of variance (ANOVA) for a 2x2 crossover study
 Source of Variation |      SS        df        MS          F      Prob > F
---------------------+--------------------------------------------------------
 Intersubjects       |
      Sequence effect|    378.04       1     378.04      0.29      0.5961
            Residuals|  17991.26      14    1285.09      1.40      0.2691
---------------------+--------------------------------------------------------
 Intrasubjects       |
     Treatment effect|    455.04       1     455.04      0.50      0.4931
        Period effect|    419.47       1     419.47      0.46      0.5102
            Residuals|  12860.78      14     918.63
---------------------+--------------------------------------------------------
                Total|  32104.59      31
Omnibus measure of separability of treatment and carryover =   29.2893%
```

◁

▷ Example 4

Consider the case of a six-treatment crossover trial in which the squares are not variance balanced. The following dataset is from a partially balanced crossover trial published by Patterson and Lucas (1962) and reproduced in Ratkowsky, Evans, and Alldredge (1993, 231):

```
. use http://www.stata-press.com/data/r12/nobalance
. list, sep(4)
```

	cow	seq	period1	period2	period3	period4	block
1.	1	adbe	38.7	37.4	34.3	31.3	1
2.	2	baed	48.9	46.9	42	39.6	1
3.	3	ebda	34.6	32.3	28.5	27.1	1
4.	4	deab	35.2	33.5	28.4	25.1	1
5.	1	dafc	32.9	33.1	27.5	25.1	2
6.	2	fdca	30.4	29.5	26.7	23.1	2
7.	3	cfad	30.8	29.3	26.4	23.2	2
8.	4	acdf	25.7	26.1	23.4	18.7	2
9.	1	efbc	25.4	26	23.9	19.9	3
10.	2	becf	21.8	23.9	21.7	17.6	3
11.	3	fceb	21.4	22	19.4	16.6	3
12.	4	cbfe	22.8	21	18.6	16.1	3

When there is no variance balance in the design, a square or blocking variable is needed to indicate in which treatment cell a sequence was observed, but the mechanical steps are the same.

```
. pkshape cow seq period1 period2 period3 period4
. pkcross outcome, model(block cow|block period|block treat carry) se
                         Number of obs =      48     R-squared     =  0.9965
                         Root MSE      = .740408     Adj R-squared =  0.9903

                Source |  Seq. SS      df       MS            F     Prob > F
          -------------+-----------------------------------------------------
                 Model |  2650.1331    30  88.3377701      161.14     0.0000
                       |
                 block |  1607.01128    2  803.505642    1465.71     0.0000
             cow|block |  628.706274    9  69.8562527     127.43     0.0000
          period|block |  408.031253    9  45.3368059      82.70     0.0000
                 treat |  2.50000057    5  .500000114       0.91     0.4964
                 carry |  3.88428906    5  .776857812       1.42     0.2680
                       |
              Residual |  9.31945887   17  .548203463
          -------------+-----------------------------------------------------
                 Total |  2659.45256   47   56.584097
```

When the model statement is used and the omnibus measure of separability is desired, specify the variables in the `treatment()`, `carryover()`, and `sequence()` options to `pkcross`. ◁

Methods and formulas

`pkcross` is implemented as an ado-file.

`pkcross` uses ANOVA to fit models for crossover experiments; see [R] **anova**.

The omnibus measure of separability is

$$S = 100(1 - V)\%$$

where V is Cramér's V and is defined as

$$V = \left\{ \frac{\frac{\chi^2}{N}}{\min(r-1, c-1)} \right\}^{\frac{1}{2}}$$

The χ^2 is calculated as

$$\chi^2 = \sum_i \sum_j \left\{ \frac{(O_{ij} - E_{ij})^2}{E_{ij}} \right\}$$

where O and E are the observed and expected counts in a table of the number of times each treatment is followed by the other treatments.

References

Balaam, L. N. 1968. A two-period design with t^2 experimental units. *Biometrics* 24: 61–73.

Chow, S.-C., and J.-P. Liu. 2009. *Design and Analysis of Bioavailability and Bioequivalence Studies*. 3rd ed. Boca Raton, FL: Chapman & Hall/CRC.

Kutner, M. H., C. J. Nachtsheim, J. Neter, and W. Li. 2005. *Applied Linear Statistical Models*. 5th ed. New York: McGraw–Hill/Irwin.

Patterson, H. D., and H. L. Lucas. 1962. Change-over designs. Technical Bulletin 147, North Carolina Agricultural Experiment Station and the USDA.

Ratkowsky, D. A., M. A. Evans, and J. R. Alldredge. 1993. *Cross-over Experiments: Design, Analysis, and Application*. New York: Dekker.

Taka, M. T., and P. Armitage. 1983. Autoregressive models in clinical trials. *Communications in Statistics, Theory and Methods* 12: 865–876.

Also see

[R] **pk** — Pharmacokinetic (biopharmaceutical) data

Title

pkequiv — Perform bioequivalence tests

Syntax

`pkequiv` *outcome treatment period sequence id* [*if*] [*in*] [, *options*]

options	Description
Options	
`compare(`*string*`)`	compare the two specified values of the treatment variable
`limit(`*#*`)`	equivalence limit (between 0.10 and 0.99); default is 0.2
`level(`*#*`)`	set confidence level; default is `level(90)`
`fieller`	calculate confidence interval by Fieller's theorem
`symmetric`	calculate symmetric equivalence interval
`anderson`	Anderson and Hauck hypothesis test for bioequivalence
`tost`	two one-sided hypothesis tests for bioequivalence
`noboot`	do not estimate probability that CI lies within confidence limits

Menu

Statistics > Epidemiology and related > Other > Bioequivalence tests

Description

`pkequiv` performs bioequivalence testing for two treatments. By default, `pkequiv` calculates a standard confidence interval symmetric about the difference between the two treatment means. `pkequiv` also calculates confidence intervals symmetric about zero and intervals based on Fieller's theorem. Also, `pkequiv` can perform interval hypothesis tests for bioequivalence.

`pkequiv` is one of the pk commands. Please read [R] **pk** before reading this entry.

Options

Options

`compare(`*string*`)` specifies the two treatments to be tested for equivalence. Sometimes there may be more than two treatments, but the equivalence can be determined only between any two treatments.

`limit(`*#*`)` specifies the equivalence limit. The default is 0.2. The equivalence limit can be changed only symmetrically; that is, it is not possible to have a 0.15 lower limit and a 0.2 upper limit in the same test.

`level(`*#*`)` specifies the confidence level, as a percentage, for confidence intervals. The default is `level(90)`. This setting is not controlled by the `set level` command.

`fieller` specifies that an equivalence interval based on Fieller's theorem be calculated.

`symmetric` specifies that a symmetric equivalence interval be calculated.

`anderson` specifies that the Anderson and Hauck (1983) hypothesis test for bioequivalence be computed. This option is ignored when calculating equivalence intervals based on Fieller's theorem or when calculating a confidence interval that is symmetric about zero.

`tost` specifies that the two one-sided hypothesis tests for bioequivalence be computed. This option is ignored when calculating equivalence intervals based on Fieller's theorem or when calculating a confidence interval that is symmetric about zero.

`noboot` prevents the estimation of the probability that the confidence interval lies within the confidence limits. If this option is not specified, this probability is estimated by resampling the data.

Remarks

`pkequiv` is designed to conduct tests for bioequivalence based on data from a crossover experiment. `pkequiv` requires that the user specify the *outcome*, *treatment*, *period*, *sequence*, and *id* variables. The data must be in the same format as that produced by `pkshape`; see [R] **pkshape**.

▷ Example 1

Continuing with example 4 from [R] **pkshape**, we will conduct equivalence testing.

```
. use http://www.stata-press.com/data/r12/pkdata3
. list, sep(4)
```

	id	sequence	outcome	treat	carry	period
1.	1	1	150.9643	A	0	1
2.	2	1	146.7606	A	0	1
3.	3	1	160.6548	A	0	1
4.	4	1	157.8622	A	0	1
5.	5	1	133.6957	A	0	1
6.	7	1	160.639	A	0	1
7.	8	1	131.2604	A	0	1
8.	9	1	168.5186	A	0	1
9.	10	2	137.0627	B	0	1
10.	12	2	153.4038	B	0	1
11.	13	2	163.4593	B	0	1
12.	14	2	146.0462	B	0	1
13.	15	2	158.1457	B	0	1
14.	18	2	147.1977	B	0	1
15.	19	2	164.9988	B	0	1
16.	20	2	145.3823	B	0	1
17.	1	1	218.5551	B	A	2
18.	2	1	133.3201	B	A	2
19.	3	1	126.0635	B	A	2
20.	4	1	96.17461	B	A	2
21.	5	1	188.9038	B	A	2
22.	7	1	223.6922	B	A	2
23.	8	1	104.0139	B	A	2
24.	9	1	237.8962	B	A	2
25.	10	2	139.7382	A	B	2
26.	12	2	202.3942	A	B	2
27.	13	2	136.7848	A	B	2
28.	14	2	104.5191	A	B	2
29.	15	2	165.8654	A	B	2
30.	18	2	139.235	A	B	2
31.	19	2	166.2391	A	B	2
32.	20	2	158.5146	A	B	2

Now we can conduct a bioequivalence test between treat = A and treat = B.

```
. set seed 1
. pkequiv outcome treat period seq id
     Classic confidence interval for bioequivalence

                      [equivalence limits]         [     test limits    ]

          difference:    -30.296       30.296        -11.332        26.416
               ratio:        80%         120%        92.519%      117.439%

     probability test limits are within equivalence limits =      0.6410
     note: reference treatment = 1
```

The default output for pkequiv shows a confidence interval for the difference of the means (test limits), the ratio of the means, and the federal equivalence limits. The classic confidence interval can be constructed around the difference between the average measure of effect for the two drugs or around the ratio of the average measure of effect for the two drugs. pkequiv reports both the difference measure and the ratio measure. For these data, U.S. federal government regulations state that the confidence interval for the difference must be entirely contained within the range $[-30.296, 30.296]$ and between 80% and 120% for the ratio. Here the test limits are within the equivalence limits. Although the test limits are inside the equivalence limits, there is only a 64% assurance that the observed confidence interval will be within the equivalence limits in the long run. This is an interesting case because, although this sample shows bioequivalence, the evaluation of the long-run performance indicates possible problems. These fictitious data were generated with high intersubject variability, which causes poor long-run performance.

If we conduct a bioequivalence test with the data published in Chow and Liu (2009, 71), which we introduced in [R] **pk** and fully described in [R] **pkshape**, we observe that the probability that the test limits are within the equivalence limits is high.

```
. use http://www.stata-press.com/data/r12/chowliu2
. set seed 1
. pkequiv outcome treat period seq id
     Classic confidence interval for bioequivalence

                      [equivalence limits]         [     test limits    ]

          difference:    -16.512       16.512         -8.698         4.123
               ratio:        80%         120%        89.464%      104.994%

     probability test limits are within equivalence limits =      0.9980
     note: reference treatment = 1
```

For these data, the test limits are well within the equivalence limits, and the probability that the test limits are within the equivalence limits is 99.8%.

◁

▷ Example 2

We compute a confidence interval that is symmetric about zero:

```
. pkequiv outcome treat period seq id, symmetric
        Westlake's symmetric confidence interval for bioequivalence
        ------------------------------------------------------------------
                             [Equivalence limits]      [    Test mean    ]
        ------------------------------------------------------------------
        Test formulation:      75.145     89.974              80.272
        ------------------------------------------------------------------
        note: reference treatment = 1
```

The reported equivalence limit is constructed symmetrically about the reference mean, which is equivalent to constructing a confidence interval symmetric about zero for the difference in the two drugs. In the output above, we see that the test formulation mean of 80.272 is within the equivalence limits, indicating that the test drug is bioequivalent to the reference drug.

pkequiv displays interval hypothesis tests of bioequivalence if you specify the tost or the anderson option, or both. For example,

```
. set seed 1
. pkequiv outcome treat period seq id, tost anderson
        Classic confidence interval for bioequivalence
        ------------------------------------------------------------------
                             [equivalence limits]       [    test limits    ]
        ------------------------------------------------------------------
             difference:   -16.512        16.512         -8.698        4.123
                  ratio:       80%          120%        89.464%     104.994%
        ------------------------------------------------------------------
        probability test limits are within equivalence limits =     0.9980
        Schuirmann's two one-sided tests
        ------------------------------------------------------------------
        upper test statistic =    -5.036                 p-value =     0.000
        lower test statistic =     3.810                 p-value =     0.001
        Anderson and Hauck's test
        ------------------------------------------------------------------
        noncentrality parameter =     4.423
                 test statistic =    -0.613     empirical p-value =    0.0005
        note: reference treatment = 1
```

Both of Schuirmann's one-sided tests are highly significant, suggesting that the two drugs are bioequivalent. A similar conclusion is drawn from the Anderson and Hauck test of bioequivalence.

◁

Saved results

pkequiv saves the following in r():

Scalars

r(stddev)	pooled-sample standard deviation of period differences from both sequences
r(uci)	upper confidence interval for a classic interval
r(lci)	lower confidence interval for a classic interval
r(delta)	delta value used in calculating a symmetric confidence interval
r(u3)	upper confidence interval for Fieller's confidence interval
r(l3)	lower confidence interval for Fieller's confidence interval

Methods and formulas

pkequiv is implemented as an ado-file.

The lower confidence interval for the difference in the two treatments for the classic shortest confidence interval is

$$L_1 = \left(\overline{Y}_T - \overline{Y}_R\right) - t_{(\alpha, n_1+n_2-2)}\widehat{\sigma}_d\sqrt{\frac{1}{n_1} + \frac{1}{n_2}}$$

The upper limit is

$$U_1 = \left(\overline{Y}_T - \overline{Y}_R\right) + t_{(\alpha, n_1+n_2-2)}\widehat{\sigma}_d\sqrt{\frac{1}{n_1} + \frac{1}{n_2}}$$

The limits for the ratio measure are

$$L_2 = \left(\frac{L_1}{\overline{Y}_R} + 1\right) 100\%$$

and

$$U_2 = \left(\frac{U_1}{\overline{Y}_R} + 1\right) 100\%$$

where $\overline{Y}_T$ is the mean of the test formulation of the drug, $\overline{Y}_R$ is the mean of the reference formulation of the drug, and $t_{(\alpha, n_1+n_2-2)}$ is the t distribution with $n_1 + n_2 - 2$ degrees of freedom. $\widehat{\sigma}_d$ is the pooled sample variance of the period differences from both sequences, defined as

$$\widehat{\sigma}_d = \frac{1}{n_1 + n_2 - 2}\sum_{k=1}^{2}\sum_{i=1}^{n_k}\left(d_{ik} - \overline{d}_{.k}\right)^2$$

The upper and lower limits for the symmetric confidence interval are $\overline{Y}_R + \Delta$ and $\overline{Y}_R - \Delta$, where

$$\Delta = k_1\widehat{\sigma}_d\sqrt{\frac{1}{n_1} + \frac{1}{n_2}} - \left(\overline{Y}_T - \overline{Y}_R\right)$$

and (simultaneously)

$$\Delta = -k_2\widehat{\sigma}_d\sqrt{\frac{1}{n_1} + \frac{1}{n_2}} + 2\left(\overline{Y}_T - \overline{Y}_R\right)$$

and k_1 and k_2 are computed iteratively to satisfy the above equalities and the condition

$$\int_{k_1}^{k_2} f(t)dt = 1 - 2\alpha$$

where $f(t)$ is the probability density function of the t distribution with $n_1 + n_2 - 2$ degrees of freedom.

See Chow and Liu (2009, 88–92) for details about calculating the confidence interval based on Fieller's theorem.

The two test statistics for the two one-sided tests of equivalence are

$$T_L = \frac{\left(\overline{Y}_T - \overline{Y}_R\right) - \theta_L}{\widehat{\sigma}_d \sqrt{\frac{1}{n_1} + \frac{1}{n_2}}}$$

and

$$T_U = \frac{\left(\overline{Y}_T - \overline{Y}_R\right) - \theta_U}{\widehat{\sigma}_d \sqrt{\frac{1}{n_1} + \frac{1}{n_2}}}$$

where $-\theta_L = \theta_U$ and are the regulated confidence limits.

The logic of the Anderson and Hauck test is tricky; see Chow and Liu (2009) for a complete explanation. However, the test statistic is

$$T_{AH} = \frac{\left(\overline{Y}_T - \overline{Y}_R\right) - \left(\frac{\theta_L + \theta_U}{2}\right)}{\widehat{\sigma}_d \sqrt{\frac{1}{n_1} + \frac{1}{n_2}}}$$

and the noncentrality parameter is estimated by

$$\widehat{\delta} = \frac{\theta_U - \theta_L}{2\widehat{\sigma}_d \sqrt{\frac{1}{n_1} + \frac{1}{n_2}}}$$

The empirical p-value is calculated as

$$p = F_t\left(|T_{AH}| - \widehat{\delta}\right) - F_t\left(-|T_{AH}| - \widehat{\delta}\right)$$

where F_t is the cumulative distribution function of the t distribution with $n_1 + n_2 - 2$ degrees of freedom.

References

Anderson, S., and W. W. Hauck. 1983. A new procedure for testing equivalence in comparative bioavailability and other clinical trials. *Communications in Statistics, Theory and Methods* 12: 2663–2692.

Chow, S.-C., and J.-P. Liu. 2009. *Design and Analysis of Bioavailability and Bioequivalence Studies*. 3rd ed. Boca Raton, FL: Chapman & Hall/CRC.

Fieller, E. C. 1954. Some problems in interval estimation. *Journal of the Royal Statistical Society, Series B* 16: 175–185.

Kutner, M. H., C. J. Nachtsheim, J. Neter, and W. Li. 2005. *Applied Linear Statistical Models*. 5th ed. New York: McGraw–Hill/Irwin.

Locke, C. S. 1984. An exact confidence interval from untransformed data for the ratio of two formulation means. *Journal of Pharmacokinetics and Biopharmaceutics* 12: 649–655.

Schuirmann, D. J. 1989. Confidence intervals for the ratio of two means from a cross-over study. In *Proceedings of the Biopharmaceutical Section*, 121–126. Washington, DC: American Statistical Association.

Westlake, W. J. 1976. Symmetrical confidence intervals for bioequivalence trials. *Biometrics* 32: 741–744.

Also see

[R] **pk** — Pharmacokinetic (biopharmaceutical) data

Title

pkexamine — Calculate pharmacokinetic measures

Syntax

`pkexamine` *time concentration* [*if*] [*in*] [, *options*]

options	Description
Main	
`fit(#)`	use # points to estimate $\text{AUC}_{0,\infty}$; default is `fit(3)`
`trapezoid`	use trapezoidal rule; default is cubic splines
`graph`	graph the AUC
`line`	graph the linear extension
`log`	graph the log extension
`exp(#)`	plot the exponential fit for the $\text{AUC}_{0,\infty}$
AUC plot	
cline_options	affect rendition of plotted points connected by lines
marker_options	change look of markers (color, size, etc.)
marker_label_options	add marker labels; change look or position
Add plots	
`addplot(`*plot*`)`	add other plots to the generated graph
Y axis, X axis, Titles, Legend, Overall	
twoway_options	any options other than `by()` documented in [G-3] ***twoway_options***

`by` is allowed; see [D] **by**.

Menu

Statistics > Epidemiology and related > Other > Pharmacokinetic measures

Description

`pkexamine` calculates pharmacokinetic measures from time-and-concentration subject-level data. `pkexamine` computes and displays the maximum measured concentration, the time at the maximum measured concentration, the time of the last measurement, the elimination time, the half-life, and the area under the concentration-time curve (AUC). Three estimates of the area under the concentration-time curve from 0 to infinity ($\text{AUC}_{0,\infty}$) are also calculated.

`pkexamine` is one of the pk commands. Please read [R] **pk** before reading this entry.

Options

Main

`fit(#)` specifies the number of points, counting back from the last measurement, to use in fitting the extension to estimate the $\text{AUC}_{0,\infty}$. The default is `fit(3)`, or the last three points. This value should be viewed as a minimum; the appropriate number of points will depend on your data.

`trapezoid` specifies that the trapezoidal rule be used to calculate the AUC. The default is cubic splines, which give better results for most functions. When the curve is irregular, `trapezoid` may give better results.

`graph` tells `pkexamine` to graph the concentration-time curve.

`line` and `log` specify the estimates of the $\text{AUC}_{0,\infty}$ to display when graphing the $\text{AUC}_{0,\infty}$. These options are ignored, unless they are specified with the `graph` option.

`exp(#)` specifies that the exponential fit for the $\text{AUC}_{0,\infty}$ be plotted. You must specify the maximum time value to which you want to plot the curve, and this time value must be greater than the maximum time measurement in the data. If you specify 0, the curve will be plotted to the point at which the linear extension would cross the x axis. This option is not valid with the `line` or `log` option and is ignored, unless the `graph` option is also specified.

AUC plot

cline_options affect the rendition of the plotted points connected by lines; see [G-3] ***cline_options***.

marker_options specify the look of markers. This look includes the marker symbol, the marker size, and its color and outline; see [G-3] ***marker_options***.

marker_label_options specify if and how the markers are to be labeled; see [G-3] ***marker_label_options***.

Add plots

`addplot(`*plot*`)` provides a way to add other plots to the generated graph. See [G-3] ***addplot_option***.

Y axis, X axis, Titles, Legend, Overall

twoway_options are any of the options documented in [G-3] ***twoway_options***, excluding `by()`. These include options for titling the graph (see [G-3] ***title_options***) and for saving the graph to disk (see [G-3] ***saving_option***).

Remarks

`pkexamine` computes summary statistics for a given patient in a pharmacokinetic trial. If `by` *idvar*`:` is specified, statistics will be displayed for each subject in the data.

▷ Example 1

Chow and Liu (2009, 13) present data on a study examining primidone concentrations versus time for a subject over a 32-hour period after dosing.

```
. use http://www.stata-press.com/data/r12/auc
. list, abbrev(14)
```

	id	time	concentration
1.	1	0	0
2.	1	.5	0
3.	1	1	2.8
4.	1	1.5	4.4
5.	1	2	4.4
6.	1	3	4.7
7.	1	4	4.1
8.	1	6	4
9.	1	8	3.6
10.	1	12	3
11.	1	16	2.5
12.	1	24	2
13.	1	32	1.6

We use `pkexamine` to produce the summary statistics:

```
. pkexamine time conc, graph
                        Maximum concentration =        4.7
                Time of maximum concentration =          3
             Time of last observation (Tmax) =         32
                             Elimination rate =     0.0279
                                    Half life =    24.8503
```

Area under the curve

AUC [0, Tmax]	AUC [0, inf.) Linear of log conc.	AUC [0, inf.) Linear fit	AUC [0, inf.) Exponential fit
85.24	142.603	107.759	142.603

Fit based on last 3 points.

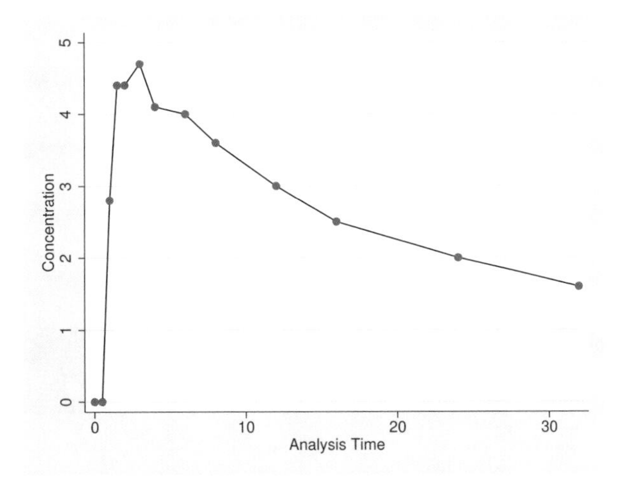

The maximum concentration of 4.7 occurs at time 3, and the time of the last observation (Tmax) is 32. In addition to the AUC, which is calculated from 0 to the maximum value of `time`, `pkexamine` also reports the area under the curve, computed by extending the curve with each of three methods: a linear fit to the log of the concentration, a linear regression line, and a decreasing exponential regression line. See *Methods and formulas* for details on these three methods.

By default, all extensions to the AUC are based on the last three points. Looking at the graph for these data, it seems more appropriate to use the last seven points to estimate the $AUC_{0,\infty}$:

```
. pkexamine time conc, fit(7)
                                     Maximum concentration =        4.7
                             Time of maximum concentration =          3
                           Time of last observation (Tmax) =         32
                                          Elimination rate =     0.0349
                                                 Half life =    19.8354

         Area under the curve
```

AUC [0, Tmax]	AUC [0, inf.) Linear of log conc.	AUC [0, inf.) Linear fit	AUC [0, inf.) Exponential fit
85.24	131.027	96.805	129.181

```
Fit based on last 7 points.
```

This approach decreased the estimate of the $AUC_{0,\infty}$ for all extensions. To see a graph of the $AUC_{0,\infty}$ using a linear extension, specify the `graph` and `line` options.

```
. pkexamine time conc, fit(7) graph line
                                     Maximum concentration =        4.7
                             Time of maximum concentration =          3
                           Time of last observation (Tmax) =         32
                                          Elimination rate =     0.0349
                                                 Half life =    19.8354

         Area under the curve
```

AUC [0, Tmax]	AUC [0, inf.) Linear of log conc.	AUC [0, inf.) Linear fit	AUC [0, inf.) Exponential fit
85.24	131.027	96.805	129.181

```
Fit based on last 7 points.
```

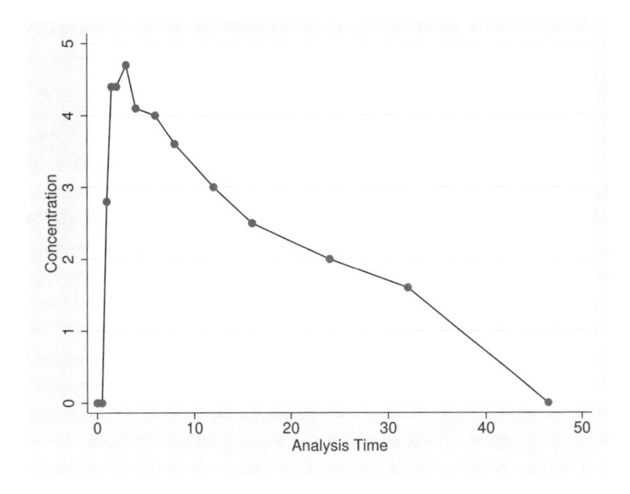

◁

Saved results

pkexamine saves the following in r():

Scalars

r(auc)	area under the concentration curve
r(half)	half-life of the drug
r(ke)	elimination rate
r(tmax)	time at last concentration measurement
r(cmax)	maximum concentration
r(tomc)	time of maximum concentration
r(auc_line)	$\text{AUC}_{0,\infty}$ estimated with a linear fit
r(auc_exp)	$\text{AUC}_{0,\infty}$ estimated with an exponential fit
r(auc_ln)	$\text{AUC}_{0,\infty}$ estimated with a linear fit of the natural log

Methods and formulas

pkexamine is implemented as an ado-file.

Let i index the observations sorted by time, let k be the number of observations, and let f be the number of points specified in the fit(#) option.

The $\text{AUC}_{0,t_{\max}}$ is defined as

$$\text{AUC}_{0,t_{\max}} = \int_0^{t_{\max}} C_t dt$$

where C_t is the concentration at time t. By default, the integral is calculated numerically using cubic splines. However, if the trapezoidal rule is used, the $\text{AUC}_{0,t_{\max}}$ is given as

$$\text{AUC}_{0,t_{\max}} = \sum_{i=2}^{k} \frac{C_{i-1} + C_i}{2} (t_i - t_{i-1})$$

The $\text{AUC}_{0,\infty}$ is the $\text{AUC}_{0,t_{\max}}$ + $\text{AUC}_{t_{\max},\infty}$, or

$$\text{AUC}_{0,\infty} = \int_0^{t_{\max}} C_t dt + \int_{t_{\max}}^{\infty} C_t dt$$

When using the linear extension to the $\text{AUC}_{0,t_{\max}}$, the integration is cut off when the line crosses the x axis. The log extension is a linear extension on the log concentration scale. The area for the exponential extension is

$$\text{AUC}_{0,\infty} = \int_{t_{\max}}^{\infty} e^{-(\beta_0+t\beta_1)} dt = -\frac{e^{-(\beta_0+t_{\max}\beta_1)}}{\beta_1}$$

The elimination rate K_{eq} is the negative of the slope from a linear regression of log concentration on time fit to the number of points specified in the `fit(#)` option:

$$K_{\text{eq}} = -\frac{\sum_{i=k-f+1}^{k} \left(t_i - \overline{t}\right)\left(\ln C_i - \overline{\ln C}\right)}{\sum_{i=k-f+1}^{k} \left(t_i - \overline{t}\right)^2}$$

The half-life is

$$t_{\text{half}} = \frac{\ln 2}{K_{\text{eq}}}$$

Reference

Chow, S.-C., and J.-P. Liu. 2009. *Design and Analysis of Bioavailability and Bioequivalence Studies*. 3rd ed. Boca Raton, FL: Chapman & Hall/CRC.

Also see

[R] **pk** — Pharmacokinetic (biopharmaceutical) data

Title

pkshape — Reshape (pharmacokinetic) Latin-square data

Syntax

pkshape *id sequence period1 period2* [*period list*] [, *options*]

options	Description
order(*string*)	apply treatments in specified order
outcome(*newvar*)	name for outcome variable; default is outcome(outcome)
treatment(*newvar*)	name for treatment variable; default is treatment(treat)
carryover(*newvar*)	name for carryover variable; default is carryover(carry)
sequence(*newvar*)	name for sequence variable; default is sequence(sequence)
period(*newvar*)	name for period variable; default is period(period)

Menu

Statistics > Epidemiology and related > Other > Reshape pharmacokinetic latin-square data

Description

pkshape reshapes the data for use with anova, pkcross, and pkequiv; see [R] **anova**, [R] **pkcross**, and [R] **pkequiv**. Latin-square and crossover data are often organized in a manner that cannot be analyzed easily with Stata. pkshape reorganizes the data in memory for use in Stata.

pkshape is one of the pk commands. Please read [R] **pk** before reading this entry.

Options

order(*string*) specifies the order in which treatments were applied. If the sequence() specifier is a string variable that specifies the order, this option is not necessary. Otherwise, order() specifies how to generate the treatment and carryover variables. Any string variable can be used to specify the order. For crossover designs, any washout periods can be indicated with the number 0.

outcome(*newvar*) specifies the name for the outcome variable in the reorganized data. By default, outcome(outcome) is used.

treatment(*newvar*) specifies the name for the treatment variable in the reorganized data. By default, treatment(treat) is used.

carryover(*newvar*) specifies the name for the carryover variable in the reorganized data. By default, carryover(carry) is used.

sequence(*newvar*) specifies the name for the sequence variable in the reorganized data. By default, sequence(sequence) is used.

period(*newvar*) specifies the name for the period variable in the reorganized data. By default, period(period) is used.

Remarks

Often data from a Latin-square experiment are naturally organized in a manner that Stata cannot manage easily. pkshape reorganizes Latin-square data so that they can be used with anova (see [R] **anova**) or any pk command. This includes the classic 2×2 crossover design commonly used in pharmaceutical research, as well as many other Latin-square designs.

▷ Example 1

Consider the example data published in Chow and Liu (2009, 71). There are 24 patients, 12 in each sequence. Sequence 1 consists of the reference formulation followed by the test formulation; sequence 2 is the test formulation followed by the reference formulation. The measurements reported are the $\text{AUC}_{0-t_{\max}}$ for each patient and for each period.

```
. use http://www.stata-press.com/data/r12/chowliu
. list, sep(4)
```

	id	seq	period1	period2
1.	1	1	74.675	73.675
2.	4	1	96.4	93.25
3.	5	1	101.95	102.125
4.	6	1	79.05	69.45
5.	11	1	79.05	69.025
6.	12	1	85.95	68.7
7.	15	1	69.725	59.425
8.	16	1	86.275	76.125
9.	19	1	112.675	114.875
10.	20	1	99.525	116.25
11.	23	1	89.425	64.175
12.	24	1	55.175	74.575
13.	2	2	74.825	37.35
14.	3	2	86.875	51.925
15.	7	2	81.675	72.175
16.	8	2	92.7	77.5
17.	9	2	50.45	71.875
18.	10	2	66.125	94.025
19.	13	2	122.45	124.975
20.	14	2	99.075	85.225
21.	17	2	86.35	95.925
22.	18	2	49.925	67.1
23.	21	2	42.7	59.425
24.	22	2	91.725	114.05

Because the outcome for one person is in two different variables, the treatment that was applied to an individual is a function of the period and the sequence. To analyze this treatment using anova, all the outcomes must be in one variable, and each covariate must be in its own variable. To reorganize these data, use pkshape:

```
. pkshape id seq period1 period2, order(ab ba)
. sort seq id treat
```

```
. list, sep(8)

     +-------------------------------------------------------+
     | id   sequence   outcome   treat   carry   period |
     |-------------------------------------------------------|
  1. |  1          1    74.675       1       0        1 |
  2. |  1          1    73.675       2       1        2 |
  3. |  4          1      96.4       1       0        1 |
  4. |  4          1     93.25       2       1        2 |
  5. |  5          1    101.95       1       0        1 |
  6. |  5          1   102.125       2       1        2 |
  7. |  6          1     79.05       1       0        1 |
  8. |  6          1     69.45       2       1        2 |
     |-------------------------------------------------------|
  9. | 11          1     79.05       1       0        1 |
 10. | 11          1    69.025       2       1        2 |
 11. | 12          1     85.95       1       0        1 |
 12. | 12          1      68.7       2       1        2 |
 13. | 15          1    69.725       1       0        1 |
 14. | 15          1    59.425       2       1        2 |
 15. | 16          1    86.275       1       0        1 |
 16. | 16          1    76.125       2       1        2 |
     |-------------------------------------------------------|
 17. | 19          1   112.675       1       0        1 |
 18. | 19          1   114.875       2       1        2 |
 19. | 20          1    99.525       1       0        1 |
 20. | 20          1    116.25       2       1        2 |
 21. | 23          1    89.425       1       0        1 |
 22. | 23          1    64.175       2       1        2 |
 23. | 24          1    55.175       1       0        1 |
 24. | 24          1    74.575       2       1        2 |
     |-------------------------------------------------------|
 25. |  2          2     37.35       1       2        2 |
 26. |  2          2    74.825       2       0        1 |
 27. |  3          2    51.925       1       2        2 |
 28. |  3          2    86.875       2       0        1 |
 29. |  7          2    72.175       1       2        2 |
 30. |  7          2    81.675       2       0        1 |
 31. |  8          2      77.5       1       2        2 |
 32. |  8          2      92.7       2       0        1 |
     |-------------------------------------------------------|
 33. |  9          2    71.875       1       2        2 |
 34. |  9          2     50.45       2       0        1 |
 35. | 10          2    94.025       1       2        2 |
 36. | 10          2    66.125       2       0        1 |
 37. | 13          2   124.975       1       2        2 |
 38. | 13          2    122.45       2       0        1 |
 39. | 14          2    85.225       1       2        2 |
 40. | 14          2    99.075       2       0        1 |
     |-------------------------------------------------------|
 41. | 17          2    95.925       1       2        2 |
 42. | 17          2     86.35       2       0        1 |
 43. | 18          2      67.1       1       2        2 |
 44. | 18          2    49.925       2       0        1 |
 45. | 21          2    59.425       1       2        2 |
 46. | 21          2      42.7       2       0        1 |
 47. | 22          2    114.05       1       2        2 |
 48. | 22          2    91.725       2       0        1 |
     +-------------------------------------------------------+
```

Now the data are organized into separate variables that indicate each factor level for each of the covariates, so the data may be used with `anova` or `pkcross`; see [R] **anova** and [R] **pkcross**.

◁

▷ Example 2

Consider the study of background music on bank teller productivity published in Kutner et al. (2005). The data are

Week	Monday	Tuesday	Wednesday	Thursday	Friday
1	18(D)	17(C)	14(A)	21(B)	17(E)
2	13(C)	34(B)	21(E)	16(A)	15(D)
3	7(A)	29(D)	32(B)	27(E)	13(C)
4	17(E)	13(A)	24(C)	31(D)	25(B)
5	21(B)	26(E)	26(D)	31(C)	7(A)

The numbers are the productivity scores, and the letters represent the treatment. We entered the data into Stata:

```
. use http://www.stata-press.com/data/r12/music, clear
. list
```

	id	seq	day1	day2	day3	day4	day5
1.	1	dcabe	18	17	14	21	17
2.	2	cbead	13	34	21	16	15
3.	3	adbec	7	29	32	27	13
4.	4	eacdb	17	13	24	31	25
5.	5	bedca	21	26	26	31	7

We reshape these data with pkshape:

```
. pkshape id seq day1 day2 day3 day4 day5
. list, sep(0)
```

	id	sequence	outcome	treat	carry	period
1.	3	1	7	1	0	1
2.	5	2	21	3	0	1
3.	2	3	13	5	0	1
4.	1	4	18	2	0	1
5.	4	5	17	4	0	1
6.	3	1	29	2	1	2
7.	5	2	26	4	3	2
8.	2	3	34	3	5	2
9.	1	4	17	5	2	2
10.	4	5	13	1	4	2
11.	3	1	32	3	2	3
12.	5	2	26	2	4	3
13.	2	3	21	4	3	3
14.	1	4	14	1	5	3
15.	4	5	24	5	1	3
16.	3	1	27	4	3	4
17.	5	2	31	5	2	4
18.	2	3	16	1	4	4
19.	1	4	21	3	1	4
20.	4	5	31	2	5	4
21.	3	1	13	5	4	5
22.	5	2	7	1	5	5
23.	2	3	15	2	1	5
24.	1	4	17	4	3	5
25.	4	5	25	3	2	5

Here the `sequence` variable is a string variable that specifies how the treatments were applied, so the `order` option is not used. When the sequence variable is a string and the `order` is specified, the arguments from the `order` option are used. We could now produce an ANOVA table:

```
. anova outcome seq period treat

                           Number of obs =      25     R-squared     =  0.8666
                           Root MSE      = 3.96232     Adj R-squared =  0.7331

                  Source |  Partial SS    df       MS           F     Prob > F
              -----------+----------------------------------------------------
                   Model |      1223.6    12  101.966667       6.49     0.0014
                         |
                sequence |          82     4        20.5       1.31     0.3226
                  period |       477.2     4       119.3       7.60     0.0027
                   treat |       664.4     4       166.1      10.58     0.0007
                         |
                Residual |       188.4    12        15.7
              -----------+----------------------------------------------------
                   Total |        1412    24  58.8333333
```

◁

▷ Example 3

Consider the Latin-square crossover example published in Kutner et al. (2005). The example is about apple sales given different methods for displaying apples.

Pattern	Store	Week 1	Week 2	Week 3
1	1	9(B)	12(C)	15(A)
	2	4(B)	12(C)	9(A)
2	1	12(A)	14(B)	3(C)
	2	13(A)	14(B)	3(C)
3	1	7(C)	18(A)	6(B)
	2	5(C)	20(A)	4(B)

We entered the data into Stata:

```
. use http://www.stata-press.com/data/r12/applesales, clear

. list, sep(2)

     +-------------------------------------+
     | id   seq   p1   p2   p3      square |
     |-------------------------------------|
  1. |  1     1    9   12   15           1 |
  2. |  2     1    4   12    9           2 |
     |-------------------------------------|
  3. |  3     2   12   14    3           1 |
  4. |  4     2   13   14    3           2 |
     |-------------------------------------|
  5. |  5     3    7   18    6           1 |
  6. |  6     3    5   20    4           2 |
     +-------------------------------------+
```

Now the data can be reorganized using descriptive names for the outcome variables.

```
. pkshape id seq p1 p2 p3, order(bca abc cab) seq(pattern) period(order)
> treat(displays)
```

```
. anova outcome pattern order display id|pattern
                         Number of obs =         18     R-squared     =  0.9562
                         Root MSE      =    1.59426     Adj R-squared =  0.9069

                  Source   Partial SS    df       MS           F     Prob > F

                   Model   443.666667     9   49.2962963     19.40     0.0002

                 pattern   .333333333     2   .166666667      0.07     0.9370
                   order   233.333333     2   116.666667     45.90     0.0000
                displays          189     2         94.5     37.18     0.0001
              id|pattern           21     3            7      2.75     0.1120

                Residual   20.3333333     8   2.54166667

                   Total          464    17   27.2941176
```

These are the same results reported by Kutner et al. (2005).

◁

▷ Example 4

We continue with example 1 from [R] **pkcollapse**; the data are

```
. use http://www.stata-press.com/data/r12/pkdata2, clear
. list, sep(4) abbrev(10)
```

	id	seq	auc_concA	auc_concB
1.	1	1	150.9643	218.5551
2.	2	1	146.7606	133.3201
3.	3	1	160.6548	126.0635
4.	4	1	157.8622	96.17461
5.	5	1	133.6957	188.9038
6.	7	1	160.639	223.6922
7.	8	1	131.2604	104.0139
8.	9	1	168.5186	237.8962
9.	10	2	137.0627	139.7382
10.	12	2	153.4038	202.3942
11.	13	2	163.4593	136.7848
12.	14	2	146.0462	104.5191
13.	15	2	158.1457	165.8654
14.	18	2	147.1977	139.235
15.	19	2	164.9988	166.2391
16.	20	2	145.3823	158.5146

```
. pkshape id seq auc_concA auc_concB, order(ab ba)
. sort period id
```

```
. list, sep(4)
```

	id	sequence	outcome	treat	carry	period
1.	1	1	150.9643	1	0	1
2.	2	1	146.7606	1	0	1
3.	3	1	160.6548	1	0	1
4.	4	1	157.8622	1	0	1
5.	5	1	133.6957	1	0	1
6.	7	1	160.639	1	0	1
7.	8	1	131.2604	1	0	1
8.	9	1	168.5186	1	0	1
9.	10	2	137.0627	2	0	1
10.	12	2	153.4038	2	0	1
11.	13	2	163.4593	2	0	1
12.	14	2	146.0462	2	0	1
13.	15	2	158.1457	2	0	1
14.	18	2	147.1977	2	0	1
15.	19	2	164.9988	2	0	1
16.	20	2	145.3823	2	0	1
17.	1	1	218.5551	2	1	2
18.	2	1	133.3201	2	1	2
19.	3	1	126.0635	2	1	2
20.	4	1	96.17461	2	1	2
21.	5	1	188.9038	2	1	2
22.	7	1	223.6922	2	1	2
23.	8	1	104.0139	2	1	2
24.	9	1	237.8962	2	1	2
25.	10	2	139.7382	1	2	2
26.	12	2	202.3942	1	2	2
27.	13	2	136.7848	1	2	2
28.	14	2	104.5191	1	2	2
29.	15	2	165.8654	1	2	2
30.	18	2	139.235	1	2	2
31.	19	2	166.2391	1	2	2
32.	20	2	158.5146	1	2	2

◁

We call the resulting dataset `pkdata3`. We conduct equivalence testing on the data in [R] **pkequiv**, and we fit an ANOVA model to these data in the third example of [R] **pkcross**.

Methods and formulas

`pkshape` is implemented as an ado-file.

References

Chow, S.-C., and J.-P. Liu. 2009. *Design and Analysis of Bioavailability and Bioequivalence Studies*. 3rd ed. Boca Raton, FL: Chapman & Hall/CRC.

Kutner, M. H., C. J. Nachtsheim, J. Neter, and W. Li. 2005. *Applied Linear Statistical Models*. 5th ed. New York: McGraw–Hill/Irwin.

Also see

[R] **pk** — Pharmacokinetic (biopharmaceutical) data

Title

pksumm — Summarize pharmacokinetic data

Syntax

pksumm *id time concentration* [*if*] [*in*] [, *options*]

options	Description
Main	
`trapezoid`	use trapezoidal rule to calculate AUC; default is cubic splines
`fit(#)`	use # points to estimate AUC; default is `fit(3)`
`notimechk`	do not check whether follow-up time for all subjects is the same
`nodots`	suppress the dots during calculation
`graph`	graph the distribution of *statistic*
`stat(`*statistic*`)`	graph the specified statistic; default is `stat(auc)`
Histogram, Density plots, Y axis, X axis, Titles, Legend, Overall	
histogram_options	any option other than `by()` documented in [R] **histogram**

statistic	Description
`auc`	area under the concentration-time curve ($AUC_{0,\infty}$); the default
`aucline`	area under the concentration-time curve from 0 to ∞ using a linear extension
`aucexp`	area under the concentration-time curve from 0 to ∞ using an exponential extension
`auclog`	area under the log-concentration-time curve extended with a linear fit
`half`	half-life of the drug
`ke`	elimination rate
`cmax`	maximum concentration
`tmax`	time at last concentration
`tomc`	time of maximum concentration

Menu

Statistics > Epidemiology and related > Other > Summarize pharmacokinetic data

Description

`pksumm` obtains summary measures based on the first four moments from the empirical distribution of each pharmacokinetic measurement and tests the null hypothesis that the distribution of that measurement is normally distributed.

`pksumm` is one of the pk commands. Please read [R] **pk** before reading this entry.

Options

Main

trapezoid specifies that the trapezoidal rule be used to calculate the AUC. The default is cubic splines, which give better results for most situations. When the curve is irregular, the trapezoidal rule may give better results.

fit(*#*) specifies the number of points, counting back from the last time measurement, to use in fitting the extension to estimate the $\text{AUC}_{0,\infty}$. The default is fit(3), the last three points. This default should be viewed as a minimum; the appropriate number of points will depend on the data.

notimechk suppresses the check that the follow-up time for all subjects is the same. By default, pksumm expects the maximum follow-up time to be equal for all subjects.

nodots suppresses the progress dots during calculation. By default, a period is displayed for every call to calculate the pharmacokinetic measures.

graph requests a graph of the distribution of the statistic specified with stat().

stat(*statistic*) specifies the statistic that pksumm should graph. The default is stat(auc). If the graph option is not specified, this option is ignored.

Histogram, Density plots, Y axis, X axis, Titles, Legend, Overall

histogram_options are any of the options documented in [R] **histogram**, excluding by(). For pksumm, fraction is the default, not density.

Remarks

pksumm produces summary statistics for the distribution of nine common pharmacokinetic measurements. If there are more than eight subjects, pksumm also computes a test for normality on each measurement. The nine measurements summarized by pksumm are listed above and are described in *Methods and formulas* of [R] **pkexamine**.

▷ Example 1

We demonstrate the use of pksumm on a variation of the data described in [R] **pk**. We have drug concentration data on 15 subjects, each measured at 13 time points over a 32-hour period. A few of the records are

```
. use http://www.stata-press.com/data/r12/pksumm
. list, sep(0)
```

	id	time	conc
1.	1	0	0
2.	1	.5	3.073403
3.	1	1	5.188444
4.	1	1.5	5.898577
5.	1	2	5.096378
6.	1	3	6.094085
	(output omitted)		
183.	15	0	0
184.	15	.5	3.86493
185.	15	1	6.432444
186.	15	1.5	6.969195
187.	15	2	6.307024
188.	15	3	6.509584
189.	15	4	6.555091
190.	15	6	7.318319
191.	15	8	5.329813
192.	15	12	5.411624
193.	15	16	3.891397
194.	15	24	5.167516
195.	15	32	2.649686

We can use pksumm to view the summary statistics for all the pharmacokinetic parameters.

```
. pksumm id time conc
...............
Summary statistics for the pharmacokinetic measures
                                                Number of observations =    15
```

Measure	Mean	Median	Variance	Skewness	Kurtosis	p-value
auc	150.74	150.96	123.07	-0.26	2.10	0.69
aucline	408.30	214.17	188856.87	2.57	8.93	0.00
aucexp	691.68	297.08	762679.94	2.56	8.87	0.00
auclog	688.98	297.67	797237.24	2.59	9.02	0.00
half	94.84	29.39	18722.13	2.26	7.37	0.00
ke	0.02	0.02	0.00	0.89	3.70	0.09
cmax	7.36	7.42	0.42	-0.60	2.56	0.44
tomc	3.47	3.00	7.62	2.17	7.18	0.00
tmax	32.00	32.00	0.00	.	.	.

For the 15 subjects, the mean $\text{AUC}_{0,t_{\max}}$ is 150.74, and $\sigma^2 = 123.07$. The skewness of -0.26 indicates that the distribution is slightly skewed left. The p-value of 0.69 for the χ^2 test of normality indicates that we cannot reject the null hypothesis that the distribution is normal.

If we were to consider any of the three variants of the $\text{AUC}_{0,\infty}$, we would see that there is huge variability and that the distribution is heavily skewed. A skewness different from 0 and a kurtosis different from 3 are expected because the distribution of the $\text{AUC}_{0,\infty}$ is not normal.

We now graph the distribution of $\text{AUC}_{0,t_{\max}}$ by specifying the graph option.

```
. pksumm id time conc, graph bin(20)
...............
 Summary statistics for the pharmacokinetic measures
                                          Number of observations =     15
     Measure |    Mean     Median       Variance  Skewness   Kurtosis   p-value
-------------+-------------------------------------------------------------------
         auc |  150.74     150.96         123.07     -0.26       2.10      0.69
     aucline |  408.30     214.17      188856.87      2.57       8.93      0.00
      aucexp |  691.68     297.08      762679.94      2.56       8.87      0.00
      auclog |  688.98     297.67      797237.24      2.59       9.02      0.00
        half |   94.84      29.39       18722.13      2.26       7.37      0.00
          ke |    0.02       0.02           0.00      0.89       3.70      0.09
        cmax |    7.36       7.42           0.42     -0.60       2.56      0.44
        tomc |    3.47       3.00           7.62      2.17       7.18      0.00
        tmax |   32.00      32.00           0.00         .          .         .
```

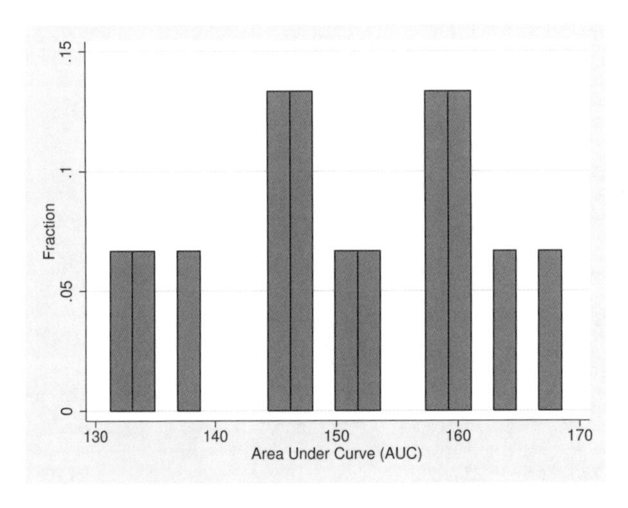

graph, by default, plots $\text{AUC}_{0,t_{\max}}$. To plot a graph of one of the other pharmacokinetic measurements, we need to specify the stat() option. For example, we can ask Stata to produce a plot of the $\text{AUC}_{0,\infty}$ using the log extension:

```
. pksumm id time conc, stat(auclog) graph bin(20)
...............
 Summary statistics for the pharmacokinetic measures
                                          Number of observations =     15
     Measure |    Mean     Median       Variance  Skewness   Kurtosis   p-value
-------------+-------------------------------------------------------------------
         auc |  150.74     150.96         123.07     -0.26       2.10      0.69
     aucline |  408.30     214.17      188856.87      2.57       8.93      0.00
      aucexp |  691.68     297.08      762679.94      2.56       8.87      0.00
      auclog |  688.98     297.67      797237.24      2.59       9.02      0.00
        half |   94.84      29.39       18722.13      2.26       7.37      0.00
          ke |    0.02       0.02           0.00      0.89       3.70      0.09
        cmax |    7.36       7.42           0.42     -0.60       2.56      0.44
        tomc |    3.47       3.00           7.62      2.17       7.18      0.00
        tmax |   32.00      32.00           0.00         .          .         .
```

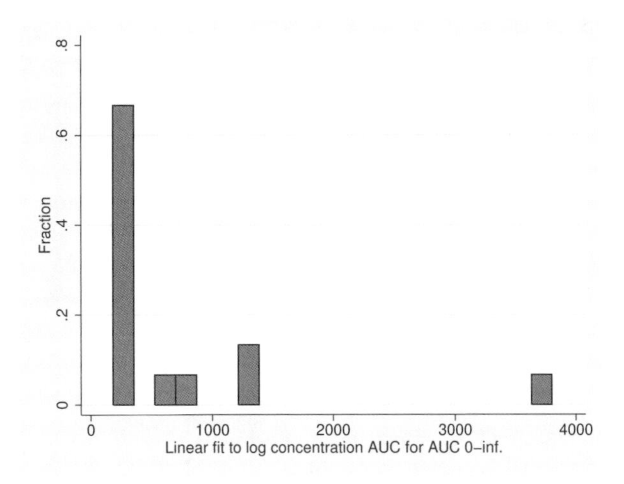

◁

Methods and formulas

pksumm is implemented as an ado-file.

The χ^2 test for normality is conducted with sktest; see [R] **sktest** for more information on the test of normality.

The statistics reported by pksumm are identical to those reported by summarize and sktest; see [R] **summarize** and [R] **sktest**.

Also see

[R] **pk** — Pharmacokinetic (biopharmaceutical) data

Title

poisson — Poisson regression

Syntax

`poisson` *depvar* [*indepvars*] [*if*] [*in*] [*weight*] [, *options*]

options	Description
Model	
`noconstant`	suppress constant term
`exposure(`$varname_e$`)`	include ln($varname_e$) in model with coefficient constrained to 1
`offset(`$varname_o$`)`	include $varname_o$ in model with coefficient constrained to 1
`constraints(`*constraints*`)`	apply specified linear constraints
`collinear`	keep collinear variables
SE/Robust	
`vce(`*vcetype*`)`	*vcetype* may be `oim`, `robust`, `cluster` *clustvar*, `opg`, `bootstrap`, or `jackknife`
Reporting	
`level(`*#*`)`	set confidence level; default is `level(95)`
`irr`	report incidence-rate ratios
`nocnsreport`	do not display constraints
display_options	control column formats, row spacing, line width, and display of omitted variables and base and empty cells
Maximization	
maximize_options	control the maximization process; seldom used
`coeflegend`	display legend instead of statistics

indepvars may contain factor variables; see [U] **11.4.3 Factor variables**.

depvar, *indepvars*, $varname_e$, and $varname_o$ may contain time-series operators; see [U] **11.4.4 Time-series varlists**.

`bootstrap`, `by`, `fracpoly`, `jackknife`, `mfp`, `mi estimate`, `nestreg`, `rolling`, `statsby`, `stepwise`, and `svy` are allowed; see [U] **11.1.10 Prefix commands**.

`vce(bootstrap)` and `vce(jackknife)` are not allowed with the `mi estimate` prefix; see [MI] **mi estimate**.

Weights are not allowed with the `bootstrap` prefix; see [R] **bootstrap**.

`vce()` and weights are not allowed with the `svy` prefix; see [SVY] **svy**.

`fweight`s, `iweight`s, and `pweight`s are allowed; see [U] **11.1.6 weight**.

`coeflegend` does not appear in the dialog box.

See [U] **20 Estimation and postestimation commands** for more capabilities of estimation commands.

Menu

Statistics > Count outcomes > Poisson regression

Description

poisson fits a Poisson regression of *depvar* on *indepvars*, where *depvar* is a nonnegative count variable.

If you have panel data, see [XT] **xtpoisson**.

Options

Model

noconstant, exposure(*varname*$_e$), offset(*varname*$_o$), constraints(*constraints*), collinear; see [R] **estimation options**.

SE/Robust

vce(*vcetype*) specifies the type of standard error reported, which includes types that are derived from asymptotic theory, that are robust to some kinds of misspecification, that allow for intragroup correlation, and that use bootstrap or jackknife methods; see [R] ***vce_option***.

Reporting

level(*#*); see [R] **estimation options**.

irr reports estimated coefficients transformed to incidence-rate ratios, that is, e^{β_i} rather than β_i. Standard errors and confidence intervals are similarly transformed. This option affects how results are displayed, not how they are estimated or stored. irr may be specified at estimation or when replaying previously estimated results.

nocnsreport; see [R] **estimation options**.

display_options: noomitted, vsquish, noemptycells, baselevels, allbaselevels, cformat(*%fmt*), pformat(*%fmt*), sformat(*%fmt*), and nolstretch; see [R] **estimation options**.

Maximization

maximize_options: difficult, technique(*algorithm_spec*), iterate(*#*), [no]log, trace, gradient, showstep, hessian, showtolerance, tolerance(*#*), ltolerance(*#*), nrtolerance(*#*), nonrtolerance, and from(*init_specs*); see [R] **maximize**. These options are seldom used.

Setting the optimization type to technique(bhhh) resets the default *vcetype* to vce(opg).

The following option is available with poisson but is not shown in the dialog box:

coeflegend; see [R] **estimation options**.

Remarks

The basic idea of Poisson regression was outlined by Coleman (1964, 378–379). See Cameron and Trivedi (1998; 2010, chap. 17) and Johnson, Kemp, and Kotz (2005, chap. 4) for information about the Poisson distribution. See Cameron and Trivedi (1998), Long (1997, chap. 8), Long and Freese (2006, chap. 8), McNeil (1996, chap. 6), and Selvin (2004, chap. 9) for an introduction to Poisson regression. Also see Selvin (2004, chap. 5) for a discussion of the analysis of spatial distributions, which includes a discussion of the Poisson distribution. An early example of Poisson regression was Cochran (1940).

Poisson regression fits models of the number of occurrences (counts) of an event. The Poisson distribution has been applied to diverse events, such as the number of soldiers kicked to death by horses in the Prussian army (von Bortkewitsch 1898); the pattern of hits by buzz bombs launched against London during World War II (Clarke 1946); telephone connections to a wrong number (Thorndike 1926); and disease incidence, typically with respect to time, but occasionally with respect to space. The basic assumptions are as follows:

1. There is a quantity called the *incidence rate* that is the rate at which events occur. Examples are 5 per second, 20 per 1,000 person-years, 17 per square meter, and 38 per cubic centimeter.
2. The incidence rate can be multiplied by *exposure* to obtain the expected number of observed events. For example, a rate of 5 per second multiplied by 30 seconds means that 150 events are expected; a rate of 20 per 1,000 person-years multiplied by 2,000 person-years means that 40 events are expected; and so on.
3. Over very small exposures ϵ, the probability of finding more than one event is small compared with ϵ.
4. Nonoverlapping exposures are mutually independent.

With these assumptions, to find the probability of k events in an exposure of size E, you divide E into n subintervals E_1, E_2, $\ldots$, E_n, and approximate the answer as the binomial probability of observing k successes in n trials. If you let $n \to \infty$, you obtain the Poisson distribution.

In the Poisson regression model, the incidence rate for the jth observation is assumed to be given by

$$r_j = e^{\beta_0+\beta_1 x_{1,j}+\cdots+\beta_k x_{k,j}}$$

If E_j is the exposure, the expected number of events, C_j, will be

$$\begin{aligned} C_j &= E_j e^{\beta_0+\beta_1 x_{1,j}+\cdots+\beta_k x_{k,j}} \\ &= e^{\ln(E_j)+\beta_0+\beta_1 x_{1,j}+\cdots+\beta_k x_{k,j}} \end{aligned}$$

This model is fit by `poisson`. Without the `exposure()` or `offset()` options, E_j is assumed to be 1 (equivalent to assuming that exposure is unknown), and controlling for exposure, if necessary, is your responsibility.

Comparing rates is most easily done by calculating *incidence-rate ratios* (IRRs). For instance, what is the relative incidence rate of chromosome interchanges in cells as the intensity of radiation increases; the relative incidence rate of telephone connections to a wrong number as load increases; or the relative incidence rate of deaths due to cancer for females relative to males? That is, you want to hold all the x's in the model constant except one, say, the ith. The IRR for a one-unit change in x_i is

$$\frac{e^{\ln(E)+\beta_1 x_1+\cdots+\beta_i(x_i+1)+\cdots+\beta_k x_k}}{e^{\ln(E)+\beta_1 x_1+\cdots+\beta_i x_i+\cdots+\beta_k x_k}} = e^{\beta_i}$$

More generally, the IRR for a Δx_i change in x_i is $e^{\beta_i \Delta x_i}$. The `lincom` command can be used after `poisson` to display incidence-rate ratios for any group relative to another; see [R] **lincom**.

▷ Example 1

Chatterjee and Hadi (2006, 162) give the number of injury incidents and the proportion of flights for each airline out of the total number of flights from New York for nine major U.S. airlines in one year:

```
. use http://www.stata-press.com/data/r12/airline
. list

     +----------------------------------------+
     | airline   injuries        n   XYZowned |
     |----------------------------------------|
  1. |       1         11   0.0950          1 |
  2. |       2          7   0.1920          0 |
  3. |       3          7   0.0750          0 |
  4. |       4         19   0.2078          0 |
  5. |       5          9   0.1382          0 |
     |----------------------------------------|
  6. |       6          4   0.0540          1 |
  7. |       7          3   0.1292          0 |
  8. |       8          1   0.0503          0 |
  9. |       9          3   0.0629          1 |
     +----------------------------------------+
```

To their data, we have added a fictional variable, XYZowned. We will imagine that an accusation is made that the airlines owned by XYZ Company have a higher injury rate.

```
. poisson injuries XYZowned, exposure(n) irr
Iteration 0:   log likelihood = -23.027197
Iteration 1:   log likelihood = -23.027177
Iteration 2:   log likelihood = -23.027177
Poisson regression                                Number of obs   =          9
                                                  LR chi2(1)      =       1.77
                                                  Prob > chi2     =     0.1836
Log likelihood = -23.027177                       Pseudo R2       =     0.0370

------------------------------------------------------------------------------
    injuries |        IRR   Std. Err.      z    P>|z|     [95% Conf. Interval]
-------------+----------------------------------------------------------------
    XYZowned |   1.463467    .406872     1.37   0.171     .8486578    2.523675
       _cons |   58.04416   8.558145    27.54   0.000     43.47662    77.49281
       ln(n) |          1  (exposure)
------------------------------------------------------------------------------
```

We specified irr to see the IRRs rather than the underlying coefficients. We estimate that XYZ Airlines' injury rate is 1.46 times larger than that for other airlines, but the 95% confidence interval is 0.85 to 2.52; we cannot even reject the hypothesis that XYZ Airlines has a lower injury rate.

◁

❑ Technical note

In example 1, we assumed that each airline's exposure was proportional to its fraction of flights out of New York. What if "large" airlines, however, also used larger planes, and so had even more passengers than would be expected, given this measure of exposure? A better measure would be each airline's fraction of passengers on flights out of New York, a number that we do not have. Even so, we suppose that n represents this number to some extent, so a better estimate of the effect might be

```
. gen lnN=ln(n)
. poisson injuries XYZowned lnN
Iteration 0:   log likelihood = -22.333875
Iteration 1:   log likelihood = -22.332276
Iteration 2:   log likelihood = -22.332276
Poisson regression                                Number of obs   =          9
                                                  LR chi2(2)      =      19.15
                                                  Prob > chi2     =     0.0001
Log likelihood = -22.332276                       Pseudo R2       =     0.3001

------------------------------------------------------------------------------
    injuries |      Coef.   Std. Err.      z    P>|z|     [95% Conf. Interval]
-------------+----------------------------------------------------------------
    XYZowned |   .6840667   .3895877     1.76   0.079    -.0795111    1.447645
         lnN |   1.424169   .3725155     3.82   0.000     .6940517    2.154285
       _cons |   4.863891   .7090501     6.86   0.000     3.474178    6.253603
------------------------------------------------------------------------------
```

Here rather than specifying the exposure() option, we explicitly included the variable that would normalize for exposure in the model. We did not specify the irr option, so we see coefficients rather than IRRs. We started with the model

$$\text{rate} = e^{\beta_0+\beta_1 \text{XYZowned}}$$

The observed counts are therefore

$$\text{count} = ne^{\beta_0+\beta_1 \text{XYZowned}} = e^{\ln(n)+\beta_0+\beta_1 \text{XYZowned}}$$

which amounts to constraining the coefficient on ln(n) to 1. This is what was estimated when we specified the exposure(n) option. In the above model, we included the normalizing exposure ourselves and, rather than constraining the coefficient to be 1, estimated the coefficient.

The estimated coefficient is 1.42, a respectable distance away from 1, and is consistent with our speculation that larger airlines also use larger airplanes. With this small amount of data, however, we also have a wide confidence interval that includes 1.

Our estimated *coefficient* on XYZowned is now 0.684, and the implied IRR is $e^{0.684} \approx 1.98$ (which we could also see by typing poisson, irr). The 95% confidence interval for the coefficient still includes 0 (the interval for the IRR includes 1), so although the point estimate is now larger, we still cannot be certain of our results.

Our expert opinion would be that, although there is not enough evidence to support the charge, there is enough evidence to justify collecting more data.

❑

▷ Example 2

In a famous age-specific study of coronary disease deaths among male British doctors, Doll and Hill (1966) reported the following data (reprinted in Rothman, Greenland, and Lash [2008, 264]):

	Smokers		Nonsmokers	
Age	Deaths	Person-years	Deaths	Person-years
35–44	32	52,407	2	18,790
45–54	104	43,248	12	10,673
55–64	206	28,612	28	5,710
65–74	186	12,663	28	2,585
75–84	102	5,317	31	1,462

The first step is to enter these data into Stata, which we have done:

```
. use http://www.stata-press.com/data/r12/dollhill3, clear
. list

     +-----------------------------------+
     | agecat   smokes   deaths   pyears |
     |-----------------------------------|
  1. |      1        1       32   52,407 |
  2. |      2        1      104   43,248 |
  3. |      3        1      206   28,612 |
  4. |      4        1      186   12,663 |
  5. |      5        1      102    5,317 |
     |-----------------------------------|
  6. |      1        0        2   18,790 |
  7. |      2        0       12   10,673 |
  8. |      3        0       28    5,710 |
  9. |      4        0       28    2,585 |
 10. |      5        0       31    1,462 |
     +-----------------------------------+
```

agecat 1 corresponds to 35–44, agecat 2 to 45–54, and so on. The most "natural" analysis of these data would begin by introducing indicator variables for each age category and one indicator for smoking:

```
. poisson deaths smokes i.agecat, exposure(pyears) irr
Iteration 0:   log likelihood = -33.823284
Iteration 1:   log likelihood = -33.600471
Iteration 2:   log likelihood = -33.600153
Iteration 3:   log likelihood = -33.600153
Poisson regression                                Number of obs   =         10
                                                  LR chi2(5)      =     922.93
                                                  Prob > chi2     =     0.0000
Log likelihood = -33.600153                       Pseudo R2       =     0.9321

------------------------------------------------------------------------------
      deaths |        IRR   Std. Err.      z    P>|z|     [95% Conf. Interval]
-------------+----------------------------------------------------------------
      smokes |   1.425519   .1530638     3.30   0.001     1.154984    1.759421
             |
      agecat |
          2  |   4.410584   .8605197     7.61   0.000     3.009011    6.464997
          3  |    13.8392   2.542638    14.30   0.000     9.654328    19.83809
          4  |   28.51678   5.269878    18.13   0.000     19.85177    40.96395
          5  |   40.45121   7.775511    19.25   0.000     27.75326    58.95885
             |
       _cons |   .0003636   .0000697   -41.30   0.000     .0002497    .0005296
  ln(pyears) |          1  (exposure)
------------------------------------------------------------------------------
```

In the above, we specified irr to obtain IRRs. We estimate that smokers have 1.43 times the mortality rate of nonsmokers. See, however, example 1 in [R] **poisson postestimation**.

◁

Saved results

poisson saves the following in e():

Scalars	
e(N)	number of observations
e(k)	number of parameters
e(k_eq)	number of equations in e(b)
e(k_eq_model)	number of equations in overall model test
e(k_dv)	number of dependent variables
e(df_m)	model degrees of freedom
e(r2_p)	pseudo-R-squared
e(ll)	log likelihood
e(ll_0)	log likelihood, constant-only model
e(N_clust)	number of clusters
e(chi2)	χ^2
e(p)	significance
e(rank)	rank of e(V)
e(ic)	number of iterations
e(rc)	return code
e(converged)	1 if converged, 0 otherwise
Macros	
e(cmd)	poisson
e(cmdline)	command as typed
e(depvar)	name of dependent variable
e(wtype)	weight type
e(wexp)	weight expression
e(title)	title in estimation output
e(clustvar)	name of cluster variable
e(offset)	linear offset variable
e(chi2type)	Wald or LR; type of model χ^2 test
e(vce)	*vcetype* specified in vce()
e(vcetype)	title used to label Std. Err.
e(opt)	type of optimization
e(which)	max or min; whether optimizer is to perform maximization or minimization
e(ml_method)	type of ml method
e(user)	name of likelihood-evaluator program
e(technique)	maximization technique
e(properties)	b V
e(estat_cmd)	program used to implement estat
e(predict)	program used to implement predict
e(asbalanced)	factor variables fvset as asbalanced
e(asobserved)	factor variables fvset as asobserved
Matrices	
e(b)	coefficient vector
e(Cns)	constraints matrix
e(ilog)	iteration log (up to 20 iterations)
e(gradient)	gradient vector
e(V)	variance–covariance matrix of the estimators
e(V_modelbased)	model-based variance
Functions	
e(sample)	marks estimation sample

Siméon-Denis Poisson (1781–1840) was a French mathematician and physicist who contributed to several fields: his name is perpetuated in Poisson brackets, Poisson's constant, Poisson's differential equation, Poisson's integral, and Poisson's ratio. Among many other results, he produced a version of the law of large numbers. His rather misleadingly titled *Recherches sur la probabilité des jugements* embraces a complete treatise on probability, as the subtitle indicates, including what is now known as the Poisson distribution. That, however, was discovered earlier by the Huguenot–British mathematician Abraham de Moivre (1667–1754).

Methods and formulas

`poisson` is implemented as an ado-file.

The log likelihood (with weights w_j and offsets) is given by

$$\Pr(Y = y) = \frac{e^{-\lambda}\lambda^y}{y!}$$

$$\xi_j = \mathbf{x}_j\boldsymbol{\beta} + \text{offset}_j$$

$$f(y_j) = \frac{e^{-\exp(\xi_j)}e^{\xi_j y_j}}{y_j!}$$

$$\ln L = \sum_{j=1}^{n} w_j \left\{-e^{\xi_j} + \xi_j y_j - \ln(y_j!)\right\}$$

This command supports the Huber/White/sandwich estimator of the variance and its clustered version using `vce(robust)` and `vce(cluster` *clustvar*`)`, respectively. See [P] **_robust**, particularly *Maximum likelihood estimators* and *Methods and formulas*.

`poisson` also supports estimation with survey data. For details on VCEs with survey data, see [SVY] **variance estimation**.

References

Bru, B. 2001. Siméon-Denis Poisson. In *Statisticians of the Centuries*, ed. C. C. Heyde and E. Seneta, 123–126. New York: Springer.

Cameron, A. C., and P. K. Trivedi. 1998. *Regression Analysis of Count Data*. Cambridge: Cambridge University Press.

——. 2010. *Microeconometrics Using Stata*. Rev. ed. College Station, TX: Stata Press.

Chatterjee, S., and A. S. Hadi. 2006. *Regression Analysis by Example*. 4th ed. New York: Wiley.

Clarke, R. D. 1946. An application of the Poisson distribution. *Journal of the Institute of Actuaries* 72: 481.

Cochran, W. G. 1940. The analysis of variance when experimental errors follow the Poisson or binomial laws. *Annals of Mathematical Statistics* 11: 335–347.

——. 1982. *Contributions to Statistics*. New York: Wiley.

Coleman, J. S. 1964. *Introduction to Mathematical Sociology*. New York: Free Press.

Doll, R., and A. B. Hill. 1966. Mortality of British doctors in relation to smoking: Observations on coronary thrombosis. *Journal of the National Cancer Institute, Monographs* 19: 205–268.

Hilbe, J. M. 1998. sg91: Robust variance estimators for MLE Poisson and negative binomial regression. *Stata Technical Bulletin* 45: 26–28. Reprinted in *Stata Technical Bulletin Reprints*, vol. 8, pp. 177–180. College Station, TX: Stata Press.

——. 1999. sg102: Zero-truncated Poisson and negative binomial regression. *Stata Technical Bulletin* 47: 37–40. Reprinted in *Stata Technical Bulletin Reprints*, vol. 8, pp. 233–236. College Station, TX: Stata Press.

Hilbe, J. M., and D. H. Judson. 1998. sg94: Right, left, and uncensored Poisson regression. *Stata Technical Bulletin* 46: 18–20. Reprinted in *Stata Technical Bulletin Reprints*, vol. 8, pp. 186–189. College Station, TX: Stata Press.

Johnson, N. L., A. W. Kemp, and S. Kotz. 2005. *Univariate Discrete Distributions*. 3rd ed. New York: Wiley.

Long, J. S. 1997. *Regression Models for Categorical and Limited Dependent Variables*. Thousand Oaks, CA: Sage.

Long, J. S., and J. Freese. 2001. Predicted probabilities for count models. *Stata Journal* 1: 51–57.

——. 2006. *Regression Models for Categorical Dependent Variables Using Stata*. 2nd ed. College Station, TX: Stata Press.

McNeil, D. 1996. *Epidemiological Research Methods*. Chichester, UK: Wiley.

Miranda, A., and S. Rabe-Hesketh. 2006. Maximum likelihood estimation of endogenous switching and sample selection models for binary, ordinal, and count variables. *Stata Journal* 6: 285–308.

Newman, S. C. 2001. *Biostatistical Methods in Epidemiology*. New York: Wiley.

Poisson, S. D. 1837. *Recherches sur la probabilité des jugements en matière criminelle et en matière civile: précédées des règles générales du calcul des probabilités*. Paris: Bachelier.

Raciborski, R. 2011. Right-censored Poisson regression model. *Stata Journal* 11: 95–105.

Rodríguez, G. 1993. sbe10: An improvement to poisson. *Stata Technical Bulletin* 11: 11–14. Reprinted in *Stata Technical Bulletin Reprints*, vol. 2, pp. 94–98. College Station, TX: Stata Press.

Rogers, W. H. 1991. sbe1: Poisson regression with rates. *Stata Technical Bulletin* 1: 11–12. Reprinted in *Stata Technical Bulletin Reprints*, vol. 1, pp. 62–64. College Station, TX: Stata Press.

Rothman, K. J., S. Greenland, and T. L. Lash. 2008. *Modern Epidemiology*. 3rd ed. Philadelphia: Lippincott Williams & Wilkins.

Rutherford, E., J. Chadwick, and C. D. Ellis. 1930. *Radiations from Radioactive Substances*. Cambridge: Cambridge University Press.

Rutherford, M. J., P. C. Lambert, and J. R. Thompson. 2010. Age–period–cohort modeling. *Stata Journal* 10: 606–627.

Schonlau, M. 2005. Boosted regression (boosting): An introductory tutorial and a Stata plugin. *Stata Journal* 5: 330–354.

Selvin, S. 2004. *Statistical Analysis of Epidemiologic Data*. 3rd ed. New York: Oxford University Press.

Thorndike, F. 1926. Applications of Poisson's probability summation. *Bell System Technical Journal* 5: 604–624.

Tobías, A., and M. J. Campbell. 1998. sg90: Akaike's information criterion and Schwarz's criterion. *Stata Technical Bulletin* 45: 23–25. Reprinted in *Stata Technical Bulletin Reprints*, vol. 8, pp. 174–177. College Station, TX: Stata Press.

von Bortkewitsch, L. 1898. *Das Gesetz der Kleinen Zahlen*. Leipzig: Teubner.

Also see

[R] **poisson postestimation** — Postestimation tools for poisson

[R] **glm** — Generalized linear models

[R] **nbreg** — Negative binomial regression

[R] **tpoisson** — Truncated Poisson regression

[R] **zip** — Zero-inflated Poisson regression

[MI] **estimation** — Estimation commands for use with mi estimate

[SVY] **svy estimation** — Estimation commands for survey data

[XT] **xtpoisson** — Fixed-effects, random-effects, and population-averaged Poisson models

[U] **20 Estimation and postestimation commands**

Title

poisson postestimation — Postestimation tools for poisson

Description

The following postestimation command is of special interest after `poisson`:

Command	Description
`estat gof`	goodness-of-fit test

`estat gof` is not appropriate after the `svy` prefix. For information about `estat gof`, see below.

The following standard postestimation commands are also available:

Command	Description
`contrast`	contrasts and ANOVA-style joint tests of estimates
`estat`	AIC, BIC, VCE, and estimation sample summary
`estat` (svy)	postestimation statistics for survey data
`estimates`	cataloging estimation results
`lincom`	point estimates, standard errors, testing, and inference for linear combinations of coefficients
`linktest`	link test for model specification
`lrtest`[1]	likelihood-ratio test
`margins`	marginal means, predictive margins, marginal effects, and average marginal effects
`marginsplot`	graph the results from margins (profile plots, interaction plots, etc.)
`nlcom`	point estimates, standard errors, testing, and inference for nonlinear combinations of coefficients
`predict`	predictions, residuals, influence statistics, and other diagnostic measures
`predictnl`	point estimates, standard errors, testing, and inference for generalized predictions
`pwcompare`	pairwise comparisons of estimates
`suest`	seemingly unrelated estimation
`test`	Wald tests of simple and composite linear hypotheses
`testnl`	Wald tests of nonlinear hypotheses

[1] `lrtest` is not appropriate with `svy` estimation results.

See the corresponding entries in the *Base Reference Manual* for details, but see [SVY] **estat** for details about `estat` (svy).

Special-interest postestimation command

`estat gof` performs a goodness-of-fit test of the model. Both the deviance statistic and the Pearson statistic are reported. If the tests are significant, the Poisson regression model is inappropriate. Then you could try a negative binomial model; see [R] **nbreg**.

Syntax for predict

predict [*type*] *newvar* [*if*] [*in*] [, *statistic* nooffset]

statistic	Description
Main	
n	number of events; the default
ir	incidence rate
pr(*n*)	probability $\Pr(y_j = n)$
pr(*a*,*b*)	probability $\Pr(a \leq y_j \leq b)$
xb	linear prediction
stdp	standard error of the linear prediction
score	first derivative of the log likelihood with respect to $\mathbf{x}_j\boldsymbol{\beta}$

These statistics are available both in and out of sample; type predict ... if e(sample) ... if wanted only for the estimation sample.

Menu

Statistics > Postestimation > Predictions, residuals, etc.

Options for predict

Main

n, the default, calculates the predicted number of events, which is $\exp(\mathbf{x}_j\boldsymbol{\beta})$ if neither offset() nor exposure() was specified when the model was fit; $\exp(\mathbf{x}_j\boldsymbol{\beta} + \text{offset}_j)$ if offset() was specified; or $\exp(\mathbf{x}_j\boldsymbol{\beta}) \times \text{exposure}_j$ if exposure() was specified.

ir calculates the incidence rate $\exp(\mathbf{x}_j\boldsymbol{\beta})$, which is the predicted number of events when exposure is 1. Specifying ir is equivalent to specifying n when neither offset() nor exposure() was specified when the model was fit.

pr(*n*) calculates the probability $\Pr(y_j = n)$, where *n* is a nonnegative integer that may be specified as a number or a variable.

pr(*a*,*b*) calculates the probability $\Pr(a \leq y_j \leq b)$, where *a* and *b* are nonnegative integers that may be specified as numbers or variables;

b missing ($b \geq .$) means $+\infty$;
pr(20,.) calculates $\Pr(y_j \geq 20)$;
pr(20,*b*) calculates $\Pr(y_j \geq 20)$ in observations for which $b \geq .$ and calculates $\Pr(20 \leq y_j \leq b)$ elsewhere.

pr(.,*b*) produces a syntax error. A missing value in an observation of the variable *a* causes a missing value in that observation for pr(*a*,*b*).

xb calculates the linear prediction, which is $\mathbf{x}_j\boldsymbol{\beta}$ if neither offset() nor exposure() was specified; $\mathbf{x}_j\boldsymbol{\beta} + \text{offset}_j$ if offset() was specified; or $\mathbf{x}_j\boldsymbol{\beta} + \ln(\text{exposure}_j)$ if exposure() was specified; see nooffset below.

stdp calculates the standard error of the linear prediction.

score calculates the equation-level score, $\partial \ln L/\partial(\mathbf{x}_j\boldsymbol{\beta})$.

`nooffset` is relevant only if you specified `offset()` or `exposure()` when you fit the model. It modifies the calculations made by `predict` so that they ignore the offset or exposure variable; the linear prediction is treated as $\mathbf{x}_j\boldsymbol{\beta}$ rather than as $\mathbf{x}_j\boldsymbol{\beta}+\text{offset}_j$ or $\mathbf{x}_j\boldsymbol{\beta}+\ln(\text{exposure}_j)$. Specifying `predict ..., nooffset` is equivalent to specifying `predict ..., ir`.

Syntax for estat gof

```
estat gof
```

Menu

Statistics > Postestimation > Reports and statistics

Remarks

▷ Example 1

Continuing with example 2 of [R] **poisson**, we use `estat gof` to determine whether the model fits the data well.

```
. use http://www.stata-press.com/data/r12/dollhill3
. poisson deaths smokes i.agecat, exp(pyears) irr
  (output omitted)
. estat gof
         Deviance goodness-of-fit =  12.13244
         Prob > chi2(4)           =    0.0164
         Pearson goodness-of-fit  =  11.15533
         Prob > chi2(4)           =    0.0249
```

The deviance goodness-of-fit test tells us that, given the model, we can reject the hypothesis that these data are Poisson distributed at the 1.64% significance level. The Pearson goodness-of-fit test tells us that we can reject the hypothesis at the 2.49% significance level.

So let us now back up and be more careful. We can most easily obtain the incidence-rate ratios within age categories by using `ir`; see [ST] **epitab**:

```
. ir deaths smokes pyears, by(agecat) nohet
          agecat |        IRR       [95% Conf. Interval]   M-H Weight
-----------------+-------------------------------------------------
               1 |   5.736638       1.463557   49.40468     1.472169 (exact)
               2 |   2.138812       1.173714   4.272545     9.624747 (exact)
               3 |    1.46824       .9863624   2.264107     23.34176 (exact)
               4 |    1.35606       .9081925   2.096412     23.25315 (exact)
               5 |   .9047304       .6000757   1.399687     24.31435 (exact)
-----------------+-------------------------------------------------
           Crude |   1.719823       1.391992    2.14353              (exact)
     M-H combined |   1.424682       1.154703   1.757784
```

We find that the mortality incidence ratios are greatly different within age category, being highest for the youngest categories and actually dropping below 1 for the oldest. (In the last case, we might argue that those who smoke and who have not died by age 75 are self-selected to be particularly robust.)

Seeing this, we will now parameterize the smoking effects separately for each age category, although we will begin by constraining the smoking effects on age categories 3 and 4 to be equivalent:

```
. constraint 1 smokes#3.agecat = smokes#4.agecat
. poisson deaths c.smokes#agecat i.agecat, exposure(pyears) irr constraints(1)
Iteration 0:   log likelihood =  -31.95424
Iteration 1:   log likelihood = -27.796801
Iteration 2:   log likelihood = -27.574177
Iteration 3:   log likelihood = -27.572645
Iteration 4:   log likelihood = -27.572645
Poisson regression                                Number of obs   =         10
                                                  Wald chi2(8)    =     632.14
Log likelihood = -27.572645                       Prob > chi2     =     0.0000
 ( 1)  [deaths]3.agecat#c.smokes - [deaths]4.agecat#c.smokes = 0

------------------------------------------------------------------------------
      deaths |        IRR   Std. Err.      z    P>|z|     [95% Conf. Interval]
-------------+----------------------------------------------------------------
agecat#c.smokes |
           1 |   5.736637   4.181256     2.40   0.017     1.374811    23.93711
           2 |   2.138812   .6520701     2.49   0.013     1.176691    3.887609
           3 |   1.412229   .2017485     2.42   0.016     1.067343    1.868557
           4 |   1.412229   .2017485     2.42   0.016     1.067343    1.868557
           5 |   .9047304   .1855513    -0.49   0.625     .6052658     1.35236
             |
      agecat |
           2 |    10.5631   8.067701     3.09   0.002     2.364153    47.19623
           3 |     47.671   34.37409     5.36   0.000     11.60056    195.8978
           4 |   98.22765   70.85012     6.36   0.000     23.89324    403.8244
           5 |   199.2099   145.3356     7.26   0.000     47.67693    832.3648
             |
       _cons |   .0001064   .0000753   -12.94   0.000     .0000266    .0004256
  ln(pyears) |          1  (exposure)
------------------------------------------------------------------------------

. estat gof
         Deviance goodness-of-fit =  .0774185
         Prob > chi2(1)           =    0.7808
         Pearson goodness-of-fit  =  .0773882
         Prob > chi2(1)           =    0.7809
```

The goodness-of-fit is now small; we are no longer running roughshod over the data. Let us now consider simplifying the model. The point estimate of the incidence-rate ratio for smoking in age category 1 is much larger than that for smoking in age category 2, but the confidence interval for `smokes#1.agecat` is similarly wide. Is the difference real?

```
. test smokes#1.agecat = smokes#2.agecat
 ( 1)  [deaths]1b.agecat#c.smokes - [deaths]2.agecat#c.smokes = 0
           chi2(  1) =    1.56
         Prob > chi2 =    0.2117
```

The point estimates may be far apart, but there is insufficient data, and we may be observing random differences. With that success, might we also combine the smokers in age categories 3 and 4 with those in 1 and 2?

```
. test smokes#2.agecat = smokes#3.agecat, accum
 ( 1)  [deaths]1b.agecat#c.smokes - [deaths]2.agecat#c.smokes = 0
 ( 2)  [deaths]2.agecat#c.smokes - [deaths]3.agecat#c.smokes = 0
           chi2(  2) =    4.73
         Prob > chi2 =    0.0938
```

Combining age categories 1–4 may be overdoing it—the 9.38% significance level is enough to stop us, although others may disagree.

Thus we now fit our final model:

```
. constraint 2 smokes#1.agecat = smokes#2.agecat
. poisson deaths c.smokes#agecat i.agecat, exposure(pyears) irr constraints(1/2)
Iteration 0:   log likelihood = -31.550722
Iteration 1:   log likelihood = -28.525057
Iteration 2:   log likelihood = -28.514535
Iteration 3:   log likelihood = -28.514535
Poisson regression                                Number of obs   =         10
                                                  Wald chi2(7)    =     642.25
Log likelihood = -28.514535                       Prob > chi2     =     0.0000
 ( 1)  [deaths]3.agecat#c.smokes - [deaths]4.agecat#c.smokes = 0
 ( 2)  [deaths]1b.agecat#c.smokes - [deaths]2.agecat#c.smokes = 0
```

deaths	IRR	Std. Err.	z	P>|z|	[95% Conf.	Interval]
agecat#c.smokes						
1	2.636259	.7408403	3.45	0.001	1.519791	4.572907
2	2.636259	.7408403	3.45	0.001	1.519791	4.572907
3	1.412229	.2017485	2.42	0.016	1.067343	1.868557
4	1.412229	.2017485	2.42	0.016	1.067343	1.868557
5	.9047304	.1855513	-0.49	0.625	.6052658	1.35236
agecat						
2	4.294559	.8385329	7.46	0.000	2.928987	6.296797
3	23.42263	7.787716	9.49	0.000	12.20738	44.94164
4	48.26309	16.06939	11.64	0.000	25.13068	92.68856
5	97.87965	34.30881	13.08	0.000	49.24123	194.561
_cons	.0002166	.0000652	-28.03	0.000	.0001201	.0003908
ln(pyears)	1	(exposure)				

The above strikes us as a fair representation of the data. The probabilities of observing the deaths seen in these data are estimated using the following `predict` command:

```
. predict p, pr(0, deaths)
. list deaths p
```

	deaths	p
1.	32	.6891766
2.	104	.4456625
3.	206	.5455328
4.	186	.4910622
5.	102	.5263011
6.	2	.227953
7.	12	.7981917
8.	28	.4772961
9.	28	.6227565
10.	31	.5475718

The probability $\Pr(y \leq \texttt{deaths})$ ranges from 0.23 to 0.80.

◁

Methods and formulas

All postestimation commands listed above are implemented as ado-files.

In the following, we use the same notation as in [R] **poisson**.

The equation-level scores are given by

$$\text{score}(\mathbf{x}\boldsymbol{\beta})_j = y_j - e^{\xi_j}$$

The deviance (D) and Pearson (P) goodness-of-fit statistics are given by

$$\begin{aligned}
\ln L_{\max} &= \sum_{j=1}^{n} w_j \left[-y_j \{ \ln(y_j) - 1 \} - \ln(y_j!) \right] \\
\chi_D^2 &= -2\{ \ln L - \ln L_{\max} \} \\
\chi_P^2 &= \sum_{j=1}^{n} \frac{w_j (y_j - e^{\xi_j})^2}{e^{\xi_j}}
\end{aligned}$$

Also see

[R] **poisson** — Poisson regression

[U] **20 Estimation and postestimation commands**

Title

predict — Obtain predictions, residuals, etc., after estimation

Syntax

After single-equation (SE) *models*

predict [*type*] *newvar* [*if*] [*in*] [, *single_options*]

After multiple-equation (ME) *models*

predict [*type*] *newvar* [*if*] [*in*] [, *multiple_options*]

predict [*type*] { *stub** | $newvar_1$ ... $newvar_q$ } [*if*] [*in*] , scores

single_options	Description
Main	
xb	calculate linear prediction
stdp	calculate standard error of the prediction
score	calculate first derivative of the log likelihood with respect to $\mathbf{x}_j\beta$
Options	
nooffset	ignore any offset() or exposure() variable
other_options	command-specific options

multiple_options	Description
Main	
equation(*eqno*[, *eqno*])	specify equations
xb	calculate linear prediction
stdp	calculate standard error of the prediction
stddp	calculate the difference in linear predictions
Options	
nooffset	ignore any offset() or exposure() variable
other_options	command-specific options

Menu

Statistics > Postestimation > Predictions, residuals, etc.

Description

predict calculates predictions, residuals, influence statistics, and the like after estimation. Exactly what predict can do is determined by the previous estimation command; command-specific options are documented with each estimation command. Regardless of command-specific options, the actions of predict share certain similarities across estimation commands:

1. `predict` *newvar* creates *newvar* containing "predicted values"—numbers related to the $E(y_j|\mathbf{x}_j)$. For instance, after linear regression, `predict` *newvar* creates $\mathbf{x}_j\mathbf{b}$ and, after probit, creates the probability $\Phi(\mathbf{x}_j\mathbf{b})$.
2. `predict` *newvar*`, xb` creates *newvar* containing $\mathbf{x}_j\mathbf{b}$. This may be the same result as option 1 (for example, linear regression) or different (for example, probit), but regardless, option `xb` is allowed.
3. `predict` *newvar*`, stdp` creates *newvar* containing the standard error of the linear prediction $\mathbf{x}_j\mathbf{b}$.
4. `predict` *newvar*`,` *other_options* may create *newvar* containing other useful quantities; see `help` or the reference manual entry for the particular estimation command to find out about other available options.
5. `nooffset` added to any of the above commands requests that the calculation ignore any offset or exposure variable specified by including the `offset(`*varname*$_o$`)` or `exposure(`*varname*$_e$`)` option when you fit the model.

`predict` can be used to make in-sample or out-of-sample predictions:

6. `predict` calculates the requested statistic for all possible observations, whether they were used in fitting the model or not. `predict` does this for standard options 1–3 and generally does this for estimator-specific options 4.
7. `predict` *newvar* `if e(sample), ...` restricts the prediction to the estimation subsample.
8. Some statistics make sense only with respect to the estimation subsample. In such cases, the calculation is automatically restricted to the estimation subsample, and the documentation for the specific option states this. Even so, you can still specify `if e(sample)` if you are uncertain.
9. `predict` can make out-of-sample predictions even using other datasets. In particular, you can

```
. use ds1
. (fit a model)
. use two                          /* another dataset          */
. predict yhat, ...                /* fill in the predictions */
```

Options

Main

`xb` calculates the linear prediction from the fitted model. That is, all models can be thought of as estimating a set of parameters $b_1, b_2, \ldots, b_k$, and the linear prediction is $\widehat{y}_j = b_1x_{1j} + b_2x_{2j} + \cdots + b_kx_{kj}$, often written in matrix notation as $\widehat{\mathbf{y}}_j = \mathbf{x}_j\mathbf{b}$. For linear regression, the values $\widehat{y}_j$ are called the predicted values or, for out-of-sample predictions, the forecast. For logit and probit, for example, $\widehat{y}_j$ is called the logit or probit index.

$x_{1j}, x_{2j}, \ldots, x_{kj}$ are obtained from the data currently in memory and do not necessarily correspond to the data on the independent variables used to fit the model (obtaining $b_1, b_2, \ldots, b_k$).

`stdp` calculates the standard error of the linear prediction. Here the prediction means the same thing as the "index", namely, $\mathbf{x}_j\mathbf{b}$. The statistic produced by `stdp` can be thought of as the standard error of the predicted expected value, or mean index, for the observation's covariate pattern. The standard error of the prediction is also commonly referred to as the standard error of the fitted value. The calculation can be made in or out of sample.

`stddp` is allowed only after you have previously fit a multiple-equation model. The standard error of the difference in linear predictions $(\mathbf{x}_{1j}\mathbf{b} - \mathbf{x}_{2j}\mathbf{b})$ between equations 1 and 2 is calculated. This option requires that `equation(`*eqno*$_1$`,`*eqno*$_2$`)` be specified.

`score` calculates the equation-level score, $\partial\ln L/\partial(\mathbf{x}_j\boldsymbol{\beta})$. Here $\ln L$ refers to the log-likelihood function.

`scores` is the ME model equivalent of the `score` option, resulting in multiple equation-level score variables. An equation-level score variable is created for each equation in the model; ancillary parameters—such as $\ln\sigma$ and $\text{atanh}\rho$—make up separate equations.

`equation(`*eqno*[`,`*eqno*]`)`—synonym `outcome()`—is relevant only when you have previously fit a multiple-equation model. It specifies the equation to which you are referring.

`equation()` is typically filled in with one *eqno*—it would be filled in that way with options `xb` and `stdp`, for instance. `equation(#1)` would mean the calculation is to be made for the first equation, `equation(#2)` would mean the second, and so on. You could also refer to the equations by their names. `equation(income)` would refer to the equation named income and `equation(hours)` to the equation named hours.

If you do not specify `equation()`, results are the same as if you specified `equation(#1)`.

Other statistics, such as `stddp`, refer to between-equation concepts. In those cases, you might specify `equation(#1,#2)` or `equation(income,hours)`. When two equations must be specified, `equation()` is required.

Options

`nooffset` may be combined with most statistics and specifies that the calculation should be made, ignoring any offset or exposure variable specified when the model was fit.

This option is available, even if it is not documented for `predict` after a specific command. If neither the `offset(`*varname*$_o$`)` option nor the `exposure(`*varname*$_e$`)` option was specified when the model was fit, specifying `nooffset` does nothing.

other_options refers to command-specific options that are documented with each command.

Remarks

Remarks are presented under the following headings:

Estimation-sample predictions
Out-of-sample predictions
Residuals
Single-equation (SE) models
SE model scores
Multiple-equation (ME) models
ME model scores

Most of the examples are presented using linear regression, but the general syntax is applicable to all estimators.

You can think of any estimation command as estimating a set of coefficients b_1, b_2, ..., b_k corresponding to the variables x_1, x_2, ..., x_k, along with a (possibly empty) set of ancillary statistics γ_1, γ_2, ..., γ_m. All estimation commands save the b_is and γ_is. `predict` accesses that saved information and combines it with the data currently in memory to make various calculations. For instance, `predict` can calculate the linear prediction, $\widehat{y}_j = b_1x_{1j} + b_2x_{2j} + \cdots + b_kx_{kj}$. The data on which `predict` makes the calculation can be the same data used to fit the model or a different dataset—it does not matter. `predict` uses the saved parameter estimates from the model, obtains the corresponding values of x for each observation in the data, and then combines them to produce the desired result.

Estimation-sample predictions

▷ Example 1

We have a 74-observation dataset on automobiles, including the mileage rating (`mpg`), the car's weight (`weight`), and whether the car is foreign (`foreign`). We fit the model

```
. use http://www.stata-press.com/data/r12/auto
(1978 Automobile Data)

. regress mpg weight if foreign

      Source |       SS       df       MS              Number of obs =      22
-------------+------------------------------           F(  1,    20) =   17.47
       Model |  427.990298     1  427.990298           Prob > F      =  0.0005
    Residual |  489.873338    20  24.4936669           R-squared     =  0.4663
-------------+------------------------------           Adj R-squared =  0.4396
       Total |  917.863636    21  43.7077922           Root MSE      =  4.9491

------------------------------------------------------------------------------
         mpg |      Coef.   Std. Err.      t    P>|t|     [95% Conf. Interval]
-------------+----------------------------------------------------------------
      weight |   -.010426   .0024942    -4.18   0.000    -.0156287   -.0052232
       _cons |    48.9183   5.871851     8.33   0.000     36.66983    61.16676
------------------------------------------------------------------------------
```

If we were to type `predict pmpg` now, we would obtain the linear predictions for all 74 observations. To obtain the predictions just for the sample on which we fit the model, we could type

```
. predict pmpg if e(sample)
(option xb assumed; fitted values)
(52 missing values generated)
```

Here `e(sample)` is true only for foreign cars because we typed `if foreign` when we fit the model and because there are no missing values among the relevant variables. If there had been missing values, `e(sample)` would also account for those.

By the way, the `if e(sample)` restriction can be used with any Stata command, so we could obtain summary statistics on the estimation sample by typing

```
. summarize if e(sample)
  (output omitted)
```

◁

Out-of-sample predictions

By out-of-sample predictions, we mean predictions extending beyond the estimation sample. In the example above, typing `predict pmpg` would generate linear predictions using all 74 observations.

`predict` will work on other datasets, too. You can use a new dataset and type `predict` to obtain results for that sample.

▷ Example 2

Using the same auto dataset, assume that we wish to fit the model

$$\texttt{mpg} = \beta_1\texttt{weight} + \beta_2\ln(\texttt{weight}) + \beta_3\texttt{foreign} + \beta_4$$

We first create the ln(weight) variable, and then type the regress command:

```
. use http://www.stata-press.com/data/r12/auto, clear
(1978 Automobile Data)

. generate lnweight = ln(weight)

. regress mpg weight lnweight foreign

      Source |       SS       df       MS              Number of obs =      74
-------------+------------------------------           F(  3,    70) =   52.36
       Model |  1690.27997     3  563.426657           Prob > F      =  0.0000
    Residual |  753.179489    70   10.759707           R-squared     =  0.6918
-------------+------------------------------           Adj R-squared =  0.6785
       Total |  2443.45946    73  33.4720474           Root MSE      =  3.2802

------------------------------------------------------------------------------
         mpg |      Coef.   Std. Err.      t    P>|t|     [95% Conf. Interval]
-------------+----------------------------------------------------------------
      weight |    .003304   .0038995     0.85   0.400    -.0044734    .0110813
    lnweight |  -29.59133   11.52018    -2.57   0.012     -52.5676   -6.615061
     foreign |  -2.125299   1.052324    -2.02   0.047    -4.224093   -.0265044
       _cons |   248.0548   80.37079     3.09   0.003     87.76035    408.3493
------------------------------------------------------------------------------
```

If we typed predict pmpg now, we would obtain predictions for all 74 cars in the current data. Instead, we are going to use a new dataset.

The dataset newautos.dta contains the make, weight, and place of manufacture of two cars, the Pontiac Sunbird and the Volvo 260. Let's use the dataset and create the predictions:

```
. use http://www.stata-press.com/data/r12/newautos, clear
(New Automobile Models)

. list

     +-----------------------------------+
     |          make   weight    foreign |
     |-----------------------------------|
  1. | Pont. Sunbird     2690   Domestic |
  2. |     Volvo 260     3170    Foreign |
     +-----------------------------------+

. predict mpg
(option xb assumed; fitted values)
variable lnweight not found
r(111);
```

Things did not work. We typed predict mpg, and Stata responded with the message "variable lnweight not found". predict can calculate predicted values on a different dataset only if that dataset contains the variables that went into the model. Here our dataset does not contain a variable called lnweight. lnweight is just the log of weight, so we can create it and try again:

```
. generate lnweight = ln(weight)

. predict mpg
(option xb assumed; fitted values)
```

```
. list

     +---------------------------------------------------------------+
     |          make   weight    foreign   lnweight            mpg  |
     |---------------------------------------------------------------|
  1. | Pont. Sunbird     2690   Domestic   7.897296       23.25097  |
  2. |     Volvo 260     3170    Foreign   8.061487       17.85295  |
     +---------------------------------------------------------------+
```

We obtained our predicted values. The Pontiac Sunbird has a predicted mileage rating of 23.3 mpg, whereas the Volvo 260 has a predicted rating of 17.9 mpg.

◁

Residuals

▷ Example 3

With many estimators, `predict` can calculate more than predicted values. With most regression-type estimators, we can, for instance, obtain residuals. Using our regression example, we return to our original data and obtain residuals by typing

```
. use http://www.stata-press.com/data/r12/auto, clear
(1978 Automobile Data)
. generate lnweight = ln(weight)
. regress mpg weight lnweight foreign
  (output omitted)
. predict double resid, residuals
. summarize resid

    Variable |       Obs        Mean    Std. Dev.       Min        Max
-------------+--------------------------------------------------------
       resid |        74   -1.51e-15    3.212091  -5.453078   13.83719
```

We could do this without refitting the model. Stata always remembers the last set of estimates, even as we use new datasets.

It was not necessary to type the `double` in `predict double resid, residuals`, but we wanted to remind you that you can specify the type of a variable in front of the variable's name; see [U] **11.4.2 Lists of new variables**. We made the new variable `resid` a `double` rather than the default `float`.

If you want your residuals to have a mean as close to zero as possible, remember to request the extra precision of `double`. If we had not specified `double`, the mean of `resid` would have been roughly 10^{-9} rather than 10^{-14}. Although 10^{-14} sounds more precise than 10^{-9}, the difference really does not matter.

◁

For linear regression, `predict` can also calculate standardized residuals and Studentized residuals with the options `rstandard` and `rstudent`; for examples, see [R] **regress postestimation**.

Single-equation (SE) models

If you have not read the discussion above on using predict after linear regression, please do so. And predict's default calculation almost always produces a statistic in the same metric as the dependent variable of the fitted model—for example, predicted counts for Poisson regression. In any case, xb can always be specified to obtain the linear prediction.

predict can calculate the standard error of the prediction, which is obtained by using the covariance matrix of the estimators.

▷ Example 4

After most binary outcome models (for example, logistic, logit, probit, cloglog, scobit), predict calculates the probability of a positive outcome if we do not tell it otherwise. We can specify the xb option if we want the linear prediction (also known as the logit or probit index). The odd abbreviation xb is meant to suggest $\mathbf{x}\boldsymbol{\beta}$. In logit and probit models, for example, the predicted probability is $p = F(\mathbf{x}\boldsymbol{\beta})$, where $F()$ is the logistic or normal cumulative distribution function, respectively.

```
. logistic foreign mpg weight
  (output omitted)
. predict phat
(option pr assumed; Pr(foreign))
. predict idxhat, xb
. summarize foreign phat idxhat
```

Variable	Obs	Mean	Std. Dev.	Min	Max
foreign	74	.2972973	.4601885	0	1
phat	74	.2972973	.3052979	.000729	.8980594
idxhat	74	-1.678202	2.321509	-7.223107	2.175845

Because this is a logit model, we could obtain the predicted probabilities ourselves from the predicted index

```
. generate phat2 = exp(idxhat)/(1+exp(idxhat))
```

but using predict without options is easier.

◁

▷ Example 5

For all models, predict attempts to produce a predicted value in the same metric as the dependent variable of the model. We have seen that for dichotomous outcome models, the default statistic produced by predict is the probability of a success. Similarly, for Poisson regression, the default statistic produced by predict is the predicted count for the dependent variable. You can always specify the xb option to obtain the linear combination of the coefficients with an observation's x values (the inner product of the coefficients and x values). For poisson (without an explicit exposure), this is the natural log of the count.

```
. use http://www.stata-press.com/data/r12/airline, clear
. poisson injuries XYZowned
  (output omitted)
. predict injhat
(option n assumed; predicted number of events)
```

```
. predict idx, xb
. generate exp_idx = exp(idx)
. summarize injuries injhat exp_idx idx

    Variable |       Obs        Mean    Std. Dev.       Min        Max
-------------+--------------------------------------------------------
    injuries |         9    7.111111    5.487359          1         19
      injhat |         9    7.111111    .8333333          6   7.666667
     exp_idx |         9    7.111111    .8333333          6   7.666667
         idx |         9    1.955174    .1225612   1.791759   2.036882
```

We note that our "hand-computed" prediction of the count (`exp_idx`) matches what was produced by the default operation of `predict`.

If our model has an exposure-time variable, we can use `predict` to obtain the linear prediction with or without the exposure. Let's verify what we are getting by obtaining the linear prediction with and without exposure, transforming these predictions to count predictions and comparing them with the default count prediction from `predict`. We must remember to multiply by the exposure time when using `predict ..., nooffset`.

```
. use http://www.stata-press.com/data/r12/airline, clear
. poisson injuries XYZowned, exposure(n)
  (output omitted)
. predict double injhat
(option n assumed; predicted number of events)
. predict double idx, xb
. generate double exp_idx = exp(idx)
. predict double idxn, xb nooffset
. generate double exp_idxn = exp(idxn)*n
. summarize injuries injhat exp_idx exp_idxn idx idxn

    Variable |       Obs        Mean    Std. Dev.       Min        Max
-------------+--------------------------------------------------------
    injuries |         9    7.111111    5.487359          1         19
      injhat |         9    7.111111     3.10936   2.919621   12.06158
     exp_idx |         9    7.111111     3.10936   2.919621   12.06158
    exp_idxn |         9    7.111111     3.10936   2.919621   12.06158
         idx |         9    1.869722    .4671044   1.071454   2.490025
-------------+--------------------------------------------------------
        idxn |         9     4.18814    .1904042   4.061204   4.442013
```

Looking at the identical means and standard deviations for `injhat`, `exp_idx`, and `exp_idxn`, we see that we can reproduce the default computations of `predict` for `poisson` estimations. We have also demonstrated the relationship between the count predictions and the linear predictions with and without exposure.

◁

SE model scores

▷ Example 6

With most maximum likelihood estimators, `predict` can calculate equation-level scores. The first derivative of the log likelihood with respect to $\mathbf{x}_j\boldsymbol{\beta}$ is the equation-level score.

```
. use http://www.stata-press.com/data/r12/auto, clear
(1978 Automobile Data)
. logistic foreign mpg weight
  (output omitted)
```

```
. predict double sc, score
. summarize sc
    Variable |       Obs        Mean    Std. Dev.       Min        Max
-------------+--------------------------------------------------------
          sc |        74   -1.37e-12    .3533133  -.8760856   .8821309
```

See [P] **_robust** and [SVY] **variance estimation** for details regarding the role equation-level scores play in linearization-based variance estimators.

◁

❑ Technical note

`predict` after some estimation commands, such as `regress` and `cnsreg`, allows the `score` option as a synonym for the `residuals` option.

❑

Multiple-equation (ME) models

If you have not read the above discussion on using `predict` after SE models, please do so. With the exception of the ability to select specific equations to predict from, the use of `predict` after ME models follows almost the same form that it does for SE models.

▷ Example 7

The details of prediction statistics that are specific to particular ME models are documented with the estimation command. If you are using ME commands that do not have separate discussions on obtaining predictions, read *Obtaining predicted values* in [R] **mlogit postestimation**, even if your interest is not in multinomial logistic regression. As a general introduction to the ME models, we will demonstrate `predict` after `sureg`:

```
. use http://www.stata-press.com/data/r12/auto, clear
(1978 Automobile Data)
. sureg (price foreign displ) (weight foreign length)
Seemingly unrelated regression
----------------------------------------------------------------------
Equation          Obs  Parms        RMSE    "R-sq"       chi2        P
----------------------------------------------------------------------
price              74      2    2202.447    0.4348      45.21   0.0000
weight             74      2    245.5238    0.8988     658.85   0.0000
----------------------------------------------------------------------

------------------------------------------------------------------------------
             |      Coef.   Std. Err.      z    P>|z|     [95% Conf. Interval]
-------------+----------------------------------------------------------------
price        |
     foreign |   3137.894   697.3805     4.50   0.000     1771.054    4504.735
displacement |   23.06938   3.443212     6.70   0.000     16.32081    29.81795
       _cons |   680.8438   859.8142     0.79   0.428    -1004.361    2366.049
-------------+----------------------------------------------------------------
weight       |
     foreign |   -154.883    75.3204    -2.06   0.040    -302.5082   -7.257674
      length |   30.67594   1.531981    20.02   0.000     27.67331    33.67856
       _cons |  -2699.498   302.3912    -8.93   0.000    -3292.173   -2106.822
------------------------------------------------------------------------------
```

sureg estimated two equations, one called price and the other weight; see [R] **sureg**.

```
. predict pred_p, equation(price)
(option xb assumed; fitted values)
. predict pred_w, equation(weight)
(option xb assumed; fitted values)
. summarize price pred_p weight pred_w
```

Variable	Obs	Mean	Std. Dev.	Min	Max
price	74	6165.257	2949.496	3291	15906
pred_p	74	6165.257	1678.805	2664.81	10485.33
weight	74	3019.459	777.1936	1760	4840
pred_w	74	3019.459	726.0468	1501.602	4447.996

You may specify the equation by name, as we did above, or by number: equation(#1) means the same thing as equation(price) in this case.

◁

ME model scores

▷ Example 8

For ME models, predict allows you to specify a stub when generating equation-level score variables. predict generates new variables using this stub by appending an equation index. Depending upon the command, the index will start with 0 or 1. Here is an example where predict starts indexing the score variables with 0.

```
. ologit rep78 mpg weight
  (output omitted)
. predict double sc*, scores
. summarize sc*
```

Variable	Obs	Mean	Std. Dev.	Min	Max
sc0	69	-1.33e-11	.5337363	-.9854088	.921433
sc1	69	-7.69e-13	.186919	-.2738537	.9854088
sc2	69	-2.87e-11	.4061637	-.5188487	1.130178
sc3	69	-1.04e-10	.5315368	-1.067351	.8194842
sc4	69	1.47e-10	.360525	-.921433	.6140182

Although it involves much more typing, we could also specify the new variable names individually.

```
. predict double (sc_xb sc_1 sc_2 sc_3 sc_4), scores
. summarize sc_*
```

Variable	Obs	Mean	Std. Dev.	Min	Max
sc_xb	69	-1.33e-11	.5337363	-.9854088	.921433
sc_1	69	-7.69e-13	.186919	-.2738537	.9854088
sc_2	69	-2.87e-11	.4061637	-.5188487	1.130178
sc_3	69	-1.04e-10	.5315368	-1.067351	.8194842
sc_4	69	1.47e-10	.360525	-.921433	.6140182

◁

Methods and formulas

predict is implemented as an ado-file.

Denote the previously estimated coefficient vector as $\mathbf{b}$ and its estimated variance matrix as $\mathbf{V}$. predict works by recalling various aspects of the model, such as $\mathbf{b}$, and combining that information with the data currently in memory. Let's write $\mathbf{x}_j$ for the jth observation currently in memory.

The *predicted value* (xb option) is defined as $\widehat{y}_j = \mathbf{x}_j\mathbf{b} + \text{offset}_j$

The *standard error of the prediction* (the stdp option) is defined as $s_{p_j} = \sqrt{\mathbf{x}_j\mathbf{V}\mathbf{x}_j'}$

The *standard error of the difference in linear predictions* between equations 1 and 2 is defined as

$$s_{dp_j} = \{(\mathbf{x}_{1j}, -\mathbf{x}_{2j}, \mathbf{0}, \ldots, \mathbf{0})\ \mathbf{V}\ (\mathbf{x}_{1j}, -\mathbf{x}_{2j}, \mathbf{0}, \ldots, \mathbf{0})'\}^{\frac{1}{2}}$$

See the individual estimation commands for information about calculating command-specific predict statistics.

Also see

[R] **predictnl** — Obtain nonlinear predictions, standard errors, etc., after estimation

[P] **_predict** — Obtain predictions, residuals, etc., after estimation programming command

[U] **20 Estimation and postestimation commands**

Title

predictnl — Obtain nonlinear predictions, standard errors, etc., after estimation

Syntax

`predictnl` [*type*] *newvar* = *pnl_exp* [*if*] [*in*] [, *options*]

options	Description
Main	
`se(`*newvar*`)`	create *newvar* containing standard errors
`variance(`*newvar*`)`	create *newvar* containing variances
`wald(`*newvar*`)`	create *newvar* containing the Wald test statistic
`p(`*newvar*`)`	create *newvar* containing the significance level (p-value) of the Wald test
`ci(`*newvars*`)`	create *newvars* containing lower and upper confidence intervals
`level(`*#*`)`	set confidence level; default is `level(95)`
`g(`*stub*`)`	create *stub*`1`, *stub*`2`, ..., *stub*`k` variables containing observation-specific derivatives
Advanced	
`iterate(`*#*`)`	maximum iterations for finding optimal step size; default is 100
`force`	calculate standard errors, etc., even when possibly inappropriate

Menu

Statistics > Postestimation > Nonlinear predictions

Description

`predictnl` calculates (possibly) nonlinear predictions after any Stata estimation command and optionally calculates the variances, standard errors, Wald test statistics, significance levels, and confidence limits for these predictions. Unlike its companion nonlinear postestimation commands `testnl` and `nlcom`, `predictnl` generates functions of the data (that is, predictions), not scalars. The quantities generated by `predictnl` are thus vectorized over the observations in the data.

Consider some general prediction, $g(\boldsymbol{\theta}, \mathbf{x}_i)$, for $i = 1, \dots, n$, where $\boldsymbol{\theta}$ are the model parameters and $\mathbf{x}_i$ are some data for the ith observation; $\mathbf{x}_i$ is assumed fixed. Typically, $g(\boldsymbol{\theta}, \mathbf{x}_i)$ is estimated by $g(\widehat{\boldsymbol{\theta}}, \mathbf{x}_i)$, where $\widehat{\boldsymbol{\theta}}$ are the estimated model parameters, which are stored in `e(b)` following any Stata estimation command.

In its most common use, `predictnl` generates two variables: one containing the estimated prediction, $g(\widehat{\boldsymbol{\theta}}, \mathbf{x}_i)$, the other containing the estimated standard error of $g(\widehat{\boldsymbol{\theta}}, \mathbf{x}_i)$. The calculation of standard errors (and other obtainable quantities that are based on the standard errors, such as test statistics) is based on the delta method, an approximation appropriate in large samples; see *Methods and formulas*.

predictnl can be used with svy estimation results (assuming that predict is also allowed), see [SVY] **svy postestimation**.

The specification of $g(\widehat{\boldsymbol{\theta}}, \mathbf{x}_i)$ is handled by specifying *pnl_exp*, and the values of $g(\widehat{\boldsymbol{\theta}}, \mathbf{x}_i)$ are stored in the new variable *newvar* of storage type *type*. *pnl_exp* is any valid Stata expression and may also contain calls to two special functions unique to predictnl:

1. predict([*predict_options*]): When you are evaluating *pnl_exp*, predict() is a convenience function that replicates the calculation performed by the command

   predict ..., *predict_options*

   As such, the predict() function may be used either as a shorthand for the formula used to make this prediction or when the formula is not readily available. When used without arguments, predict() replicates the default prediction for that particular estimation command.

2. xb([*eqno*]): The xb() function replicates the calculation of the linear predictor $\mathbf{x}_i\mathbf{b}$ for equation *eqno*. If xb() is specified without *eqno*, the linear predictor for the first equation (or the only equation in single-equation estimation) is obtained.

   For example, xb(#1) (or equivalently, xb() with no arguments) translates to the linear predictor for the first equation, xb(#2) for the second, and so on. You could also refer to the equations by their names, such as xb(income).

   When specifying *pnl_exp*, both of these functions may be used repeatedly, in combination, and in combination with other Stata functions and expressions. See *Remarks* for examples that use both of these functions.

Options

Main

se(*newvar*) adds *newvar* of storage type *type*, where for each i in the prediction sample, *newvar*[i] contains the estimated standard error of $g(\widehat{\boldsymbol{\theta}}, \mathbf{x}_i)$.

variance(*newvar*) adds *newvar* of storage type *type*, where for each i in the prediction sample, *newvar*[i] contains the estimated variance of $g(\widehat{\boldsymbol{\theta}}, \mathbf{x}_i)$.

wald(*newvar*) adds *newvar* of storage type *type*, where for each i in the prediction sample, *newvar*[i] contains the Wald test statistic for the test of the hypothesis $H_0\colon g(\boldsymbol{\theta}, \mathbf{x}_i) = 0$.

p(*newvar*) adds *newvar* of storage type *type*, where *newvar*[i] contains the significance level (p-value) of the Wald test of $H_0\colon g(\boldsymbol{\theta}, \mathbf{x}_i) = 0$ versus the two-sided alternative.

ci(*newvars*) requires the specification of two *newvars*, such that the ith observation of each will contain the left and right endpoints (respectively) of a confidence interval for $g(\boldsymbol{\theta}, \mathbf{x}_i)$. The level of the confidence intervals is determined by level(*#*).

level(*#*) specifies the confidence level, as a percentage, for confidence intervals. The default is level(95) or as set by set level; see [U] **20.7 Specifying the width of confidence intervals**.

g(*stub*) specifies that new variables, *stub*1, *stub*2, ..., *stub*k be created, where k is the dimension of $\boldsymbol{\theta}$. *stub*1 will contain the observation-specific derivatives of $g(\boldsymbol{\theta}, \mathbf{x}_i)$ with respect to the first element, θ_1, of $\boldsymbol{\theta}$; *stub*2 will contain the derivatives of $g(\boldsymbol{\theta}, \mathbf{x}_i)$ with respect to θ_2, etc.; If the derivative of $g(\boldsymbol{\theta}, \mathbf{x}_i)$ with respect to a particular coefficient in $\boldsymbol{\theta}$ equals zero for all observations in the prediction sample, the *stub* variable for that coefficient is not created. The ordering of the parameters in $\boldsymbol{\theta}$ is precisely that of the stored vector of parameter estimates e(b).

Advanced

`iterate(#)` specifies the maximum number of iterations used to find the optimal step size in the calculation of numerical derivatives of $g(\boldsymbol{\theta}, \mathbf{x}_i)$ with respect to $\boldsymbol{\theta}$. By default, the maximum number of iterations is 100, but convergence is usually achieved after only a few iterations. You should rarely have to use this option.

`force` forces the calculation of standard errors and other inference-related quantities in situations where `predictnl` would otherwise refuse to do so. The calculation of standard errors takes place by evaluating (at $\widehat{\boldsymbol{\theta}}$) the numerical derivative of $g(\boldsymbol{\theta}, \mathbf{x}_i)$ with respect to $\boldsymbol{\theta}$. If `predictnl` detects that $g()$ is possibly a function of random quantities other than $\widehat{\boldsymbol{\theta}}$, it will refuse to calculate standard errors or any other quantity derived from them. The `force` option forces the calculation to take place anyway. If you use the `force` option, there is no guarantee that any inference quantities (for example, standard errors) will be correct or that the values obtained can be interpreted.

Remarks

Remarks are presented under the following headings:

Introduction
Nonlinear transformations and standard errors
Using xb() and predict()
Multiple-equation (ME) estimators
Test statistics and significance levels
Manipulability
Confidence intervals

Introduction

`predictnl` and `nlcom` both use the delta method. They take a nonlinear transformation of the estimated parameter vector from some fitted model and apply the delta method to calculate the variance, standard error, Wald test statistic, etc., of this transformation. `nlcom` is designed for scalar functions of the parameters, and `predictnl` is designed for functions of the parameters and of the data, that is, for predictions.

Nonlinear transformations and standard errors

We begin by fitting a probit model to the low-birthweight data of Hosmer and Lemeshow (2000, 25). The data are described in detail in example 1 of [R] **logistic**.

```
. use http://www.stata-press.com/data/r12/lbw
(Hosmer & Lemeshow data)

. probit low lwt smoke ptl ht

Iteration 0:   log likelihood =   -117.336
Iteration 1:   log likelihood = -106.75886
Iteration 2:   log likelihood = -106.67852
Iteration 3:   log likelihood = -106.67851

Probit regression                                 Number of obs   =        189
                                                  LR chi2(4)      =      21.31
                                                  Prob > chi2     =     0.0003
Log likelihood = -106.67851                       Pseudo R2       =     0.0908

------------------------------------------------------------------------------
         low |      Coef.   Std. Err.      z    P>|z|     [95% Conf. Interval]
-------------+----------------------------------------------------------------
         lwt |  -.0095164   .0036875    -2.58   0.010    -.0167438   -.0022891
       smoke |   .3487004   .2041772     1.71   0.088    -.0514794    .7488803
         ptl |    .365667   .1921201     1.90   0.057    -.0108815    .7422154
          ht |   1.082355    .410673     2.64   0.008     .2774503    1.887259
       _cons |   .4238985   .4823224     0.88   0.379    -.5214361    1.369233
------------------------------------------------------------------------------
```

After we fit such a model, we first would want to generate the predicted probabilities of a low birthweight, given the covariate values in the estimation sample. This is easily done using `predict` after `probit`, but it doesn't answer the question, "What are the standard errors of those predictions?"

For the time being, we will consider ourselves ignorant of any automated way to obtain the predicted probabilities after `probit`. The formula for the prediction is

$$\Pr(y \neq 0|\mathbf{x}_i) = \Phi(\mathbf{x}_i\boldsymbol{\beta})$$

where Φ is the standard cumulative normal. Thus for this example, $g(\boldsymbol{\theta}, \mathbf{x}_i) = \Phi(\mathbf{x}_i\boldsymbol{\beta})$. Armed with the formula, we can use `predictnl` to generate the predictions and their standard errors:

```
. predictnl phat = normal(_b[_cons] + _b[ht]*ht + _b[ptl]*ptl +
> _b[smoke]*smoke + _b[lwt]*lwt), se(phat_se)

. list phat phat_se lwt smoke ptl ht in -10/l

     +-----------------------------------------------------+
     |      phat    phat_se   lwt   smoke   ptl   ht |
     |-----------------------------------------------------|
180. | .2363556     .042707   120       0     0    0 |
181. | .6577712    .1580714   154       0     1    1 |
182. | .2793261    .0519958   106       0     0    0 |
183. | .1502118    .0676338   190       1     0    0 |
184. | .5702871    .0819911   101       1     1    0 |
     |-----------------------------------------------------|
185. | .4477045     .079889    95       1     0    0 |
186. | .2988379    .0576306   100       0     0    0 |
187. | .4514706     .080815    94       1     0    0 |
188. | .5615571    .1551051   142       0     0    1 |
189. | .7316517    .1361469   130       1     0    1 |
     +-----------------------------------------------------+
```

Thus subject 180 in our data has an estimated probability of low birthweight of 23.6% with standard error 4.3%.

Used without options, `predictnl` is not much different from `generate`. By specifying the `se(phat_se)` option, we were able to obtain a variable containing the standard errors of the predictions; therein lies the utility of `predictnl`.

Using xb() and predict()

As was the case above, a prediction is often not a function of a few isolated parameters and their corresponding variables but instead is some (possibly elaborate) function of the entire linear predictor. For models with many predictors, the brute-force expression for the linear predictor can be cumbersome to type. An alternative is to use the inline function `xb()`. `xb()` is a shortcut for having to type `_b[_cons] + _b[ht]*ht + _b[ptl]*ptl + ...`,

```
. drop phat phat_se
. predictnl phat = normal(xb()), se(phat_se)
. list phat phat_se lwt smoke ptl ht in -10/l
```

	phat	phat_se	lwt	smoke	ptl	ht
180.	.2363556	.042707	120	0	0	0
181.	.6577712	.1580714	154	0	1	1
182.	.2793261	.0519958	106	0	0	0
183.	.1502118	.0676338	190	1	0	0
184.	.5702871	.0819911	101	1	1	0
185.	.4477045	.079889	95	1	0	0
186.	.2988379	.0576306	100	0	0	0
187.	.4514706	.080815	94	1	0	0
188.	.5615571	.1551051	142	0	0	1
189.	.7316517	.1361469	130	1	0	1

which yields the same results. This approach is easier, produces more readable code, and is less prone to error, such as forgetting to include a term in the sum.

Here we used `xb()` without arguments because we have only one equation in our model. In multiple-equation (ME) settings, `xb()` (or equivalently `xb(#1)`) yields the linear predictor from the first equation, `xb(#2)` from the second, etc. You can also refer to equations by their names, for example, `xb(income)`.

❑ Technical note

Most estimation commands in Stata allow the postestimation calculation of linear predictors and their standard errors via `predict`. For example, to obtain these for the first (or only) equation in the model, you could type

```
predict xbvar, xb
predict stdpvar, stdp
```

Equivalently, you could type

```
predictnl xbvar = xb(), se(stdpvar)
```

but we recommend the first method, as it is faster. As we demonstrated above, however, `predictnl` is more general.

❑

Returning to our probit example, we can further simplify the calculation by using the inline function `predict()`. `predict(`*pred_options*`)` works by substituting, within our `predictnl` expression, the calculation performed by

```
predict ..., pred_options
```

In our example, we are interested in the predicted probabilities after a probit regression, normally obtained via

```
predict ..., p
```

We can obtain these predictions (and standard errors) by using

```
. drop phat phat_se
. predictnl phat = predict(p), se(phat_se)
. list phat phat_se lwt smoke ptl ht in -10/l
```

	phat	phat_se	lwt	smoke	ptl	ht
180.	.2363556	.042707	120	0	0	0
181.	.6577712	.1580714	154	0	1	1
182.	.2793261	.0519958	106	0	0	0
183.	.1502118	.0676338	190	1	0	0
184.	.5702871	.0819911	101	1	1	0
185.	.4477045	.079889	95	1	0	0
186.	.2988379	.0576306	100	0	0	0
187.	.4514706	.080815	94	1	0	0
188.	.5615571	.1551051	142	0	0	1
189.	.7316517	.1361469	130	1	0	1

which again replicates what we have already done by other means. However, this version did not require knowledge of the formula for the predicted probabilities after a probit regression—`predict(p)` took care of that for us.

Because the predicted probability is the default prediction after `probit`, we could have just used `predict()` without arguments, namely,

```
. predictnl phat = predict(), se(phat_se)
```

Also, the expression *pnl_exp* can be inordinately complicated, with multiple calls to `predict()` and `xb()`. For example,

```
. predictnl phat = normal(invnormal(predict()) + predict(xb)/xb() - 1),
> se(phat_se)
```

is perfectly valid and will give the same result as before, albeit a bit inefficiently.

❑ Technical note

When using `predict()` and `xb()`, the *formula* for the calculation is substituted within *pnl_exp*, not the values that result from the application of that formula. To see this, note the subtle difference between

```
. predict xbeta, xb
. predictnl phat = normal(xbeta), se(phat_se)
```

and

```
. predictnl phat = normal(xb()), se(phat_se)
```

Both sequences will yield the same phat, yet for the first sequence, phat_se will equal zero for all observations. The reason is that, once evaluated, xbeta will contain the values of the linear predictor, yet these values are treated as fixed and nonstochastic as far as predictnl is concerned. By contrast, because xb() is shorthand for the formula used to calculate the linear predictor, it contains not values, but references to the estimated regression coefficients and corresponding variables. Thus the second method produces the desired result.

❑

Multiple-equation (ME) estimators

In [R] **mlogit**, data on insurance choice (Tarlov et al. 1989; Wells et al. 1989) were examined, and a multinomial logit was used to assess the effects of age, gender, race, and site of study (one of three sites) on the type of insurance:

```
. use http://www.stata-press.com/data/r12/sysdsn1, clear
(Health insurance data)
. mlogit insure age male nonwhite i.site, nolog
Multinomial logistic regression                   Number of obs   =        615
                                                  LR chi2(10)     =      42.99
                                                  Prob > chi2     =     0.0000
Log likelihood = -534.36165                       Pseudo R2       =     0.0387
```

insure	Coef.	Std. Err.	z	P>|z|	[95% Conf.	Interval]
Indemnity	(base outcome)					
Prepaid						
age	-.011745	.0061946	-1.90	0.058	-.0238862	.0003962
male	.5616934	.2027465	2.77	0.006	.1643175	.9590693
nonwhite	.9747768	.2363213	4.12	0.000	.5115955	1.437958
site						
2	.1130359	.2101903	0.54	0.591	-.2989296	.5250013
3	-.5879879	.2279351	-2.58	0.010	-1.034733	-.1412433
_cons	.2697127	.3284422	0.82	0.412	-.3740222	.9134476
Uninsure						
age	-.0077961	.0114418	-0.68	0.496	-.0302217	.0146294
male	.4518496	.3674867	1.23	0.219	-.268411	1.17211
nonwhite	.2170589	.4256361	0.51	0.610	-.6171725	1.05129
site						
2	-1.211563	.4705127	-2.57	0.010	-2.133751	-.2893747
3	-.2078123	.3662926	-0.57	0.570	-.9257327	.510108
_cons	-1.286943	.5923219	-2.17	0.030	-2.447872	-.1260134

Of particular interest is the estimation of the relative risk, which, for a given selection, is the ratio of the probability of making that selection to the probability of selecting the base category (Indemnity here), given a set of covariate values. In a multinomial logit model, the relative risk (when comparing to the base category) simplifies to the exponentiated linear predictor for that selection.

Using this example, we can estimate the observation-specific relative risks of selecting a prepaid plan over the base category (with standard errors) by either referring to the Prepaid equation by name or number,

```
. predictnl RRppaid = exp(xb(Prepaid)), se(SERRppaid)
```

or

```
. predictnl RRppaid = exp(xb(#1)), se(SERRppaid)
```

because Prepaid is the first equation in the model.

Those of us for whom the simplified formula for the relative risk does not immediately come to mind may prefer to calculate the relative risk directly from its definition, that is, as a ratio of two predicted probabilities. After mlogit, the predicted probability for a category may be obtained using predict, but we must specify the category as the outcome:

```
. predictnl RRppaid = predict(outcome(Prepaid))/predict(outcome(Indemnity)),
> se(SERRppaid)
(1 missing value generated)
. list RRppaid SERRppaid age male nonwhite site in 1/10
```

	RRppaid	SERRpp~d	age	male	nonwhite	site
1.	.6168578	.1503759	73.722107	0	0	2
2.	1.056658	.1790703	27.89595	0	0	2
3.	.8426442	.1511281	37.541397	0	0	1
4.	1.460581	.3671465	23.641327	0	1	3
5.	.9115747	.1324168	40.470901	0	0	2
6.	1.034701	.1696923	29.683777	0	0	2
7.	.9223664	.1344981	39.468857	0	0	2
8.	1.678312	.4216626	26.702255	1	0	1
9.	.9188519	.2256017	63.101974	0	1	3
10.	.5766296	.1334877	69.839828	0	0	1

The "(1 missing value generated)" message is not an error; further examination of the data would reveal that age is missing in one observation and that the offending observation (among others) is not in the estimation sample. Just as with predict, predictnl can generate predictions in or out of the estimation sample.

Thus we estimate (among other things) that a white, female, 73-year-old from site 2 is less likely to choose a prepaid plan over an indemnity plan—her relative risk is about 62% with standard error 15%.

Test statistics and significance levels

Often a standard error calculation is just a means to an end, and what is really desired is a test of the hypothesis,

$$H_0 : g(\boldsymbol{\theta}, \mathbf{x}_i) = 0$$

versus the two-sided alternative.

We can use predictnl to obtain the Wald test statistics or significance levels (or both) for the above tests, whether or not we want standard errors. To obtain the Wald test statistics, we use the wald() option; for significance levels, we use p().

Returning to our mlogit example, suppose that we wanted for each observation a test of whether the relative risk of choosing a prepaid plan over an indemnity plan is different from one. One way to do this would be to define $g()$ to be the relative risk minus one and then test whether $g()$ is different from zero.

```
. predictnl RRm1 = exp(xb(Prepaid)) - 1, wald(W_RRm1) p(sig_RRm1)
(1 missing value generated)
note: significance levels are with respect to the chi-squared(1) distribution.
. list RRm1 W_RRm1 sig_RRm1 age male nonwhite in 1/10
```

	RRm1	W_RRm1	sig_RRm1	age	male	nonwhite
1.	-.3831422	6.491778	.0108375	73.722107	0	0
2.	.0566578	.100109	.7516989	27.89595	0	0
3.	-.1573559	1.084116	.2977787	37.541397	0	0
4.	.4605812	1.573743	.2096643	23.641327	0	1
5.	-.0884253	.4459299	.5042742	40.470901	0	0
6.	.0347015	.0418188	.8379655	29.683777	0	0
7.	-.0776336	.3331707	.563798	39.468857	0	0
8.	.6783119	2.587788	.1076906	26.702255	1	0
9.	-.0811482	.1293816	.719074	63.101974	0	1
10.	-.4233705	10.05909	.001516	69.839828	0	0

The newly created variable W_RRm1 contains the Wald test statistic for each observation, and sig_RRm1 contains the level of significance. Thus our 73-year-old white female represented by the first observation would have a relative risk of choosing prepaid over indemnity that is significantly different from 1, at least at the 5% level. For this test, it was not necessary to generate a variable containing the standard error of the relative risk minus 1, but we could have done so had we wanted. We could have also omitted specifying wald(W_RRm1) if all we cared about were, say, the significance levels of the tests.

In this regard, predictnl acts as an observation-specific version of testnl, with the test results vectorized over the observations in the data. The significance levels are pointwise—they are not adjusted to reflect any simultaneous testing over the observations in the data.

Manipulability

There are many ways to specify $g(\boldsymbol{\theta}, \mathbf{x}_i)$ to yield tests such that, for multiple specifications of $g()$, the theoretical conditions for which

$$H_0\colon g(\boldsymbol{\theta}, \mathbf{x}_i) = 0$$

is true will be equivalent. However, this does not mean that the tests themselves will be equivalent. This is known as the manipulability of the Wald test for nonlinear hypotheses; also see [R] **boxcox**.

As an example, consider the previous section where we defined $g()$ to be the relative risk between choosing a prepaid plan over an indemnity plan, minus 1. We could also have defined $g()$ to be the risk difference—the probability of choosing a prepaid plan minus the probability of choosing an indemnity plan. Either specification of $g()$ yields a mathematically equivalent specification of $H_0\colon g() = 0$; that is, the risk difference will equal zero when the relative risk equals one. However, the tests themselves do not give the same results:

```
. predictnl RD = predict(outcome(Prepaid)) - predict(outcome(Indemnity)),
> wald(W_RD) p(sig_RD)
(1 missing value generated)
note: significance levels are with respect to the chi-squared(1) distribution.
. list RD W_RD sig_RD RRm1 W_RRm1 sig_RRm1 in 1/10
```

	RD	W_RD	sig_RD	RRm1	W_RRm1	sig_RRm1
1.	-.2303744	4.230243	.0397097	-.3831422	6.491778	.0108375
2.	.0266902	.1058542	.7449144	.0566578	.100109	.7516989
3.	-.0768078	.9187646	.3377995	-.1573559	1.084116	.2977787
4.	.1710702	2.366535	.1239619	.4605812	1.573743	.2096643
5.	-.0448509	.4072922	.5233471	-.0884253	.4459299	.5042742
6.	.0165251	.0432816	.835196	.0347015	.0418188	.8379655
7.	-.0391535	.3077611	.5790573	-.0776336	.3331707	.563798
8.	.22382	4.539085	.0331293	.6783119	2.587788	.1076906
9.	-.0388409	.1190183	.7301016	-.0811482	.1293816	.719074
10.	-.2437626	6.151558	.0131296	-.4233705	10.05909	.001516

In certain cases (such as subject 8), the difference can be severe enough to potentially change the conclusion. The reason for this inconsistency is that the nonlinear Wald test is actually a standard Wald test of a first-order Taylor approximation of $g()$, and this approximation can differ according to how $g()$ is specified.

As such, keep in mind the manipulability of nonlinear Wald tests when drawing scientific conclusions.

Confidence intervals

We can also use `predictnl` to obtain confidence intervals for the observation-specific $g(\boldsymbol{\theta}, \mathbf{x}_i)$ by using the `ci()` option to specify two new variables to contain the left and right endpoints of the confidence interval, respectively. For example, we could generate confidence intervals for the risk differences calculated previously:

```
. drop RD
. predictnl RD = predict(outcome(Prepaid)) - predict(outcome(Indemnity)),
> ci(RD_lcl RD_rcl)
(1 missing value generated)
note: Confidence intervals calculated using Z critical values.
. list RD RD_lcl RD_rcl age male nonwhite in 1/10
```

	RD	RD_lcl	RD_rcl	age	male	nonwhite
1.	-.2303744	-.4499073	-.0108415	73.722107	0	0
2.	.0266902	-.1340948	.1874752	27.89595	0	0
3.	-.0768078	-.2338625	.080247	37.541397	0	0
4.	.1710702	-.0468844	.3890248	23.641327	0	1
5.	-.0448509	-.1825929	.092891	40.470901	0	0
6.	.0165251	-.1391577	.1722078	29.683777	0	0
7.	-.0391535	-.177482	.099175	39.468857	0	0
8.	.22382	.0179169	.4297231	26.702255	1	0
9.	-.0388409	-.2595044	.1818226	63.101974	0	1
10.	-.2437626	-.4363919	-.0511332	69.839828	0	0

The confidence level, here, 95%, is either set using the `level()` option or obtained from the current default level, `c(level)`; see [U] **20.7 Specifying the width of confidence intervals**.

From the above output, we can see that, for subjects 1, 8, and 10, a 95% confidence interval for the risk difference does not contain zero, meaning that, for these subjects, there is some evidence of a significant difference in risks.

The confidence intervals calculated by `predictnl` are pointwise; there is no adjustment (such as a Bonferroni correction) made so that these confidence intervals may be considered jointly at the specified level.

Methods and formulas

`predictnl` is implemented as an ado-file.

For the ith observation, consider the transformation $g(\boldsymbol{\theta}, \mathbf{x}_i)$, estimated by $g(\widehat{\boldsymbol{\theta}}, \mathbf{x}_i)$, for the $1 \times k$ parameter vector $\boldsymbol{\theta}$ and data $\mathbf{x}_i$ ($\mathbf{x}_i$ is assumed fixed). The variance of $g(\widehat{\boldsymbol{\theta}}, \mathbf{x}_i)$ is estimated by

$$\widehat{\text{Var}}\left\{g(\widehat{\boldsymbol{\theta}}, \mathbf{x}_i)\right\} = \mathbf{GVG}'$$

where $\mathbf{G}$ is the vector of derivatives

$$\mathbf{G} = \left\{ \left. \frac{\partial g(\boldsymbol{\theta}, \mathbf{x}_i)}{\partial \boldsymbol{\theta}} \right|_{\boldsymbol{\theta}=\widehat{\boldsymbol{\theta}}} \right\}_{(1\times k)}$$

and $\mathbf{V}$ is the estimated variance–covariance matrix of $\widehat{\boldsymbol{\theta}}$. Standard errors, $\widehat{\text{se}}\{g(\widehat{\boldsymbol{\theta}}, \mathbf{x}_i)\}$, are obtained as the square roots of the variances.

The Wald test statistic for testing

$$H_0\colon g(\boldsymbol{\theta}, \mathbf{x}_i) = 0$$

versus the two-sided alternative is given by

$$W_i = \frac{\left\{g(\widehat{\boldsymbol{\theta}}, \mathbf{x}_i)\right\}^2}{\widehat{\text{Var}}\left\{g(\widehat{\boldsymbol{\theta}}, \mathbf{x}_i)\right\}}$$

When the variance–covariance matrix of $\widehat{\boldsymbol{\theta}}$ is an asymptotic covariance matrix, W_i is approximately distributed as χ^2 with 1 degree of freedom. For linear regression, W_i is taken to be approximately distributed as $F_{1,r}$, where r is the residual degrees of freedom from the original model fit. The levels of significance of the observation-by-observation tests of H_0 versus the two-sided alternative are given by

$$p_i = \Pr(T > W_i)$$

where T is either a χ^2- or F-distributed random variable, as described above.

A $(1-\alpha) \times 100\%$ confidence interval for $g(\boldsymbol{\theta}, \mathbf{x}_i)$ is given by

$$g(\widehat{\boldsymbol{\theta}}, \mathbf{x}_i) \pm z_{\alpha/2} \left[\widehat{\text{se}}\left\{g(\widehat{\boldsymbol{\theta}}, \mathbf{x}_i)\right\}\right]$$

when W_i is χ^2-distributed, and

$$g(\widehat{\boldsymbol{\theta}}, \mathbf{x}_i) \pm t_{\alpha/2,r} \left[\widehat{\text{se}}\left\{g(\widehat{\boldsymbol{\theta}}, \mathbf{x}_i)\right\}\right]$$

when W_i is F-distributed. z_p is the $1-p$ quantile of the standard normal distribution, and $t_{p,r}$ is the $1-p$ quantile of the t distribution with r degrees of freedom.

References

Gould, W. W. 1996. crc43: Wald test of nonlinear hypotheses after model estimation. *Stata Technical Bulletin* 29: 2–4. Reprinted in *Stata Technical Bulletin Reprints*, vol. 5, pp. 15–18. College Station, TX: Stata Press.

Hosmer, D. W., Jr., and S. Lemeshow. 2000. *Applied Logistic Regression*. 2nd ed. New York: Wiley.

Phillips, P. C. B., and J. Y. Park. 1988. On the formulation of Wald tests of nonlinear restrictions. *Econometrica* 56: 1065–1083.

Tarlov, A. R., J. E. Ware, Jr., S. Greenfield, E. C. Nelson, E. Perrin, and M. Zubkoff. 1989. The medical outcomes study. An application of methods for monitoring the results of medical care. *Journal of the American Medical Association* 262: 925–930.

Wells, K. B., R. D. Hays, M. A. Burnam, W. H. Rogers, S. Greenfield, and J. E. Ware, Jr. 1989. Detection of depressive disorder for patients receiving prepaid or fee-for-service care. Results from the Medical Outcomes Survey. *Journal of the American Medical Association* 262: 3298–3302.

Also see

[R] **lincom** — Linear combinations of estimators

[R] **nlcom** — Nonlinear combinations of estimators

[R] **predict** — Obtain predictions, residuals, etc., after estimation

[R] **test** — Test linear hypotheses after estimation

[R] **testnl** — Test nonlinear hypotheses after estimation

[U] **20 Estimation and postestimation commands**

Title

probit — Probit regression

Syntax

`probit` *depvar* [*indepvars*] [*if*] [*in*] [*weight*] [, *options*]

options	Description
Model	
`noconstant`	suppress constant term
`offset(`*varname*`)`	include *varname* in model with coefficient constrained to 1
`asis`	retain perfect predictor variables
`constraints(`*constraints*`)`	apply specified linear constraints
`collinear`	keep collinear variables
SE/Robust	
`vce(`*vcetype*`)`	*vcetype* may be `oim`, `robust`, `cluster` *clustvar*, `bootstrap`, or `jackknife`
Reporting	
`level(`*#*`)`	set confidence level; default is `level(95)`
`nocnsreport`	do not display constraints
display_options	control column formats, row spacing, line width, and display of omitted variables and base and empty cells
Maximization	
maximize_options	control the maximization process; seldom used
`nocoef`	do not display the coefficient table; seldom used
`coeflegend`	display legend instead of statistics

indepvars may contain factor variables; see [U] **11.4.3 Factor variables**.

depvar and *indepvars* may contain time-series operators; see [U] **11.4.4 Time-series varlists**.

`bootstrap`, `by`, `fracpoly`, `jackknife`, `mfp`, `mi estimate`, `nestreg`, `rolling`, `statsby`, `stepwise`, and `svy` are allowed; see [U] **11.1.10 Prefix commands**.

`vce(bootstrap)` and `vce(jackknife)` are not allowed with the `mi estimate` prefix; see [MI] **mi estimate**.

Weights are not allowed with the `bootstrap` prefix; see [R] **bootstrap**.

`vce()`, `nocoef`, and weights are not allowed with the `svy` prefix; see [SVY] **svy**.

`fweight`s, `iweight`s, and `pweight`s are allowed; see [U] **11.1.6 weight**.

`nocoef` and `coeflegend` do not appear in the dialog box.

See [U] **20 Estimation and postestimation commands** for more capabilities of estimation commands.

Menu

Statistics > Binary outcomes > Probit regression

Description

probit fits a maximum-likelihood probit model.

If estimating on grouped data, see the bprobit command described in [R] **glogit**.

Several auxiliary commands may be run after probit, logit, or logistic; see [R] **logistic postestimation** for a description of these commands.

See [R] **logistic** for a list of related estimation commands.

Options

Model

noconstant, offset(*varname*), constraints(*constraints*), collinear; see [R] **estimation options**.

asis specifies that all specified variables and observations be retained in the maximization process. This option is typically not specified and may introduce numerical instability. Normally probit drops variables that perfectly predict success or failure in the dependent variable along with their associated observations. In those cases, the effective coefficient on the dropped variables is infinity (negative infinity) for variables that completely determine a success (failure). Dropping the variable and perfectly predicted observations has no effect on the likelihood or estimates of the remaining coefficients and increases the numerical stability of the optimization process. Specifying this option forces retention of perfect predictor variables and their associated observations.

SE/Robust

vce(*vcetype*) specifies the type of standard error reported, which includes types that are derived from asymptotic theory, that are robust to some kinds of misspecification, that allow for intragroup correlation, and that use bootstrap or jackknife methods; see [R] ***vce_option***.

Reporting

level(#); see [R] **estimation options**.

nocnsreport; see [R] **estimation options**.

display_options: noomitted, vsquish, noemptycells, baselevels, allbaselevels, cformat(%*fmt*), pformat(%*fmt*), sformat(%*fmt*), and nolstretch; see [R] **estimation options**.

Maximization

maximize_options: difficult, technique(*algorithm_spec*), iterate(#), [no]log, trace, gradient, showstep, hessian, showtolerance, tolerance(#), ltolerance(#), nrtolerance(#), nonrtolerance, and from(*init_specs*); see [R] **maximize**. These options are seldom used.

The following options are available with probit but are not shown in the dialog box:

nocoef specifies that the coefficient table not be displayed. This option is sometimes used by programmers but is of no use interactively.

coeflegend; see [R] **estimation options**.

Remarks

Remarks are presented under the following headings:

Robust standard errors
Model identification

`probit` fits maximum likelihood models with dichotomous dependent (left-hand-side) variables coded as 0/1 (more precisely, coded as 0 and not 0).

▷ Example 1

We have data on the make, weight, and mileage rating of 22 foreign and 52 domestic automobiles. We wish to fit a probit model explaining whether a car is foreign based on its weight and mileage. Here is an overview of our data:

```
. use http://www.stata-press.com/data/r12/auto
(1978 Automobile Data)

. keep make mpg weight foreign

. describe

Contains data from http://www.stata-press.com/data/r12/auto.dta
  obs:            74                          1978 Automobile Data
 vars:             4                          13 Apr 2011 17:45
 size:         1,702                           (_dta has notes)
--------------------------------------------------------------------------------
              storage  display     value
variable name   type   format      label      variable label
--------------------------------------------------------------------------------
make            str18  %-18s                  Make and Model
mpg             int    %8.0g                  Mileage (mpg)
weight          int    %8.0gc                 Weight (lbs.)
foreign         byte   %8.0g       origin     Car type
--------------------------------------------------------------------------------
Sorted by:  foreign
     Note:  dataset has changed since last saved

. inspect foreign

foreign:  Car type                              Number of Observations
------------------                        --------------------------------
                                             Total    Integers   Nonintegers
|  #                         Negative            -           -             -
|  #                         Zero               52          52             -
|  #                         Positive           22          22             -
|  #                                         -----       -----         -----
|  #   #                     Total              74          74             -
|  #   #                     Missing             -
+----------------------                      -----
0                     1                         74
   (2 unique values)

      foreign is labeled and all values are documented in the label.
```

The `foreign` variable takes on two unique values, 0 and 1. The value 0 denotes a domestic car, and 1 denotes a foreign car.

The model that we wish to fit is

$$\Pr(\texttt{foreign} = 1) = \Phi(\beta_0 + \beta_1\texttt{weight} + \beta_2\texttt{mpg})$$

where Φ is the cumulative normal distribution.

To fit this model, we type

```
. probit foreign weight mpg
Iteration 0:   log likelihood =  -45.03321
Iteration 1:   log likelihood = -27.914626
  (output omitted)
Iteration 5:   log likelihood = -26.844189
Probit regression                                 Number of obs   =         74
                                                  LR chi2(2)      =      36.38
                                                  Prob > chi2     =     0.0000
Log likelihood = -26.844189                       Pseudo R2       =     0.4039
```

foreign	Coef.	Std. Err.	z	P>|z|	[95% Conf.	Interval]
weight	-.0023355	.0005661	-4.13	0.000	-.003445	-.0012261
mpg	-.1039503	.0515689	-2.02	0.044	-.2050235	-.0028772
_cons	8.275464	2.554142	3.24	0.001	3.269437	13.28149

We find that heavier cars are less likely to be foreign and that cars yielding better gas mileage are also less likely to be foreign, at least holding the weight of the car constant.

See [R] **maximize** for an explanation of the output.

◁

❑ Technical note

Stata interprets a value of 0 as a negative outcome (failure) and treats all other values (except missing) as positive outcomes (successes). Thus if your dependent variable takes on the values 0 and 1, then 0 is interpreted as failure and 1 as success. If your dependent variable takes on the values 0, 1, and 2, then 0 is still interpreted as failure, but both 1 and 2 are treated as successes.

If you prefer a more formal mathematical statement, when you type `probit` y x, Stata fits the model

$$\Pr(y_j \neq 0 \mid \mathbf{x}_j) = \Phi(\mathbf{x}_j\boldsymbol{\beta})$$

where Φ is the standard cumulative normal.

❑

Robust standard errors

If you specify the `vce(robust)` option, `probit` reports robust standard errors; see [U] **20.20 Obtaining robust variance estimates**.

▷ Example 2

For the model from example 1, the robust calculation increases the standard error of the coefficient on `mpg` by almost 15%:

```
. probit foreign weight mpg, vce(robust) nolog
Probit regression                                 Number of obs   =         74
                                                  Wald chi2(2)    =      30.26
                                                  Prob > chi2     =     0.0000
Log pseudolikelihood = -26.844189                 Pseudo R2       =     0.4039
```

foreign	Coef.	Robust Std. Err.	z	P>|z|	[95% Conf.	Interval]
weight	-.0023355	.0004934	-4.73	0.000	-.0033025	-.0013686
mpg	-.1039503	.0593548	-1.75	0.080	-.2202836	.0123829
_cons	8.275464	2.539177	3.26	0.001	3.298769	13.25216

Without vce(robust), the standard error for the coefficient on mpg was reported to be 0.052 with a resulting confidence interval of $[-0.21, -0.00]$.

◁

▷ Example 3

The vce(cluster *clustvar*) option can relax the independence assumption required by the probit estimator to independence between clusters. To demonstrate, we will switch to a different dataset.

We are studying unionization of women in the United States and have a dataset with 26,200 observations on 4,434 women between 1970 and 1988. We will use the variables age (the women were 14–26 in 1968, and our data span the age range of 16–46), grade (years of schooling completed, ranging from 0 to 18), not_smsa (28% of the person-time was spent living outside an SMSA—standard metropolitan statistical area), south (41% of the person-time was in the South), and year. Each of these variables is included in the regression as a covariate along with the interaction between south and year. This interaction, along with the south and year variables, is specified in the probit command using factor-variables notation, south##c.year. We also have variable union, indicating union membership. Overall, 22% of the person-time is marked as time under union membership, and 44% of these women have belonged to a union.

We fit the following model, ignoring that the women are observed an average of 5.9 times each in these data:

```
. use http://www.stata-press.com/data/r12/union, clear
(NLS Women 14-24 in 1968)
. probit union age grade not_smsa south##c.year
Iteration 0:   log likelihood =  -13864.23
Iteration 1:   log likelihood = -13545.541
Iteration 2:   log likelihood = -13544.385
Iteration 3:   log likelihood = -13544.385
Probit regression                                 Number of obs   =      26200
                                                  LR chi2(6)      =     639.69
                                                  Prob > chi2     =     0.0000
Log likelihood = -13544.385                       Pseudo R2       =     0.0231

------------------------------------------------------------------------------
       union |      Coef.   Std. Err.      z    P>|z|     [95% Conf. Interval]
-------------+----------------------------------------------------------------
         age |   .0118481   .0029072     4.08   0.000     .0061502     .017546
       grade |   .0267365   .0036689     7.29   0.000     .0195457    .0339273
    not_smsa |  -.1293525   .0202595    -6.38   0.000    -.1690604   -.0896445
     1.south |  -.8281077   .2472219    -3.35   0.001    -1.312654   -.3435618
        year |  -.0080931   .0033469    -2.42   0.016    -.0146529   -.0015333
             |
south#c.year |
           1 |   .0057369   .0030917     1.86   0.064    -.0003226    .0117965
             |
       _cons |  -.6542487   .2007777    -3.26   0.001    -1.047766   -.2607316
------------------------------------------------------------------------------
```

The reported standard errors in this model are probably meaningless. Women are observed repeatedly, and so the observations are not independent. Looking at the coefficients, we find a large southern effect against unionization and a time trend for the south that is almost significantly different from the overall downward trend. The vce(cluster *clustvar*) option provides a way to fit this model and obtains correct standard errors:

```
. probit union age grade not_smsa south##c.year, vce(cluster id)
Iteration 0:   log pseudolikelihood =  -13864.23
Iteration 1:   log pseudolikelihood = -13545.541
Iteration 2:   log pseudolikelihood = -13544.385
Iteration 3:   log pseudolikelihood = -13544.385
Probit regression                               Number of obs   =      26200
                                                Wald chi2(6)    =     166.53
                                                Prob > chi2     =     0.0000
Log pseudolikelihood = -13544.385               Pseudo R2       =     0.0231
                           (Std. Err. adjusted for 4434 clusters in idcode)
```

union	Coef.	Robust Std. Err.	z	P>|z|	[95% Conf.	Interval]
age	.0118481	.0056625	2.09	0.036	.0007499	.0229463
grade	.0267365	.0078124	3.42	0.001	.0114244	.0420486
not_smsa	-.1293525	.0403885	-3.20	0.001	-.2085125	-.0501925
1.south	-.8281077	.3201584	-2.59	0.010	-1.455607	-.2006089
year	-.0080931	.0060829	-1.33	0.183	-.0200153	.0038292
south#c.year						
1	.0057369	.0040133	1.43	0.153	-.002129	.0136029
_cons	-.6542487	.3485976	-1.88	0.061	-1.337487	.02899

These standard errors are larger than those reported by the inappropriate conventional calculation. By comparison, another model we could fit is an equal-correlation population-averaged probit model:

```
. xtprobit union age grade not_smsa south##c.year, pa
Iteration 1: tolerance = .12544249
Iteration 2: tolerance = .0034686
Iteration 3: tolerance = .00017448
Iteration 4: tolerance = 8.382e-06
Iteration 5: tolerance = 3.997e-07
GEE population-averaged model                 Number of obs      =     26200
Group variable:                     idcode    Number of groups   =      4434
Link:                               probit    Obs per group: min =         1
Family:                           binomial                   avg =       5.9
Correlation:                  exchangeable                   max =        12
                                              Wald chi2(6)       =    242.57
Scale parameter:                         1    Prob > chi2        =    0.0000
```

union	Coef.	Std. Err.	z	P>|z|	[95% Conf.	Interval]
age	.0089699	.0053208	1.69	0.092	-.0014586	.0193985
grade	.0333174	.0062352	5.34	0.000	.0210966	.0455382
not_smsa	-.0715717	.027543	-2.60	0.009	-.1255551	-.0175884
1.south	-1.017368	.207931	-4.89	0.000	-1.424905	-.6098308
year	-.0062708	.0055314	-1.13	0.257	-.0171122	.0045706
south#c.year						
1	.0086294	.00258	3.34	0.001	.0035727	.013686
_cons	-.8670997	.294771	-2.94	0.003	-1.44484	-.2893592

The coefficient estimates are similar, but these standard errors are smaller than those produced by `probit, vce(cluster` *clustvar*`)`, as we would expect. If the equal-correlation assumption is valid, the population-averaged probit estimator above should be more efficient.

Is the assumption valid? That is a difficult question to answer. The default population-averaged estimates correspond to an assumption of exchangeable correlation within person. It would not be unreasonable to assume an AR(1) correlation within person or to assume that the observations are correlated but that we do not wish to impose any structure. See [XT] **xtprobit** and [XT] **xtgee** for full details.

◁

`probit, vce(cluster` *clustvar*`)` is robust to assumptions about within-cluster correlation. That is, it inefficiently sums within cluster for the standard error calculation rather than attempting to exploit what might be assumed about the within-cluster correlation.

Model identification

The `probit` command has one more feature that is probably the most useful. It will automatically check the model for identification and, if the model is underidentified, drop whatever variables and observations are necessary for estimation to proceed.

▷ Example 4

Have you ever fit a probit model where one or more of your independent variables perfectly predicted one or the other outcome?

For instance, consider the following data:

Outcome y	Independent variable x
0	1
0	1
0	0
1	0

Say that we wish to predict the outcome on the basis of the independent variable. The outcome is always zero when the independent variable is one. In our data, $\Pr(y = 0 \mid x = 1) = 1$, which means that the probit coefficient on x must be minus infinity with a corresponding infinite standard error. At this point, you may suspect that we have a problem.

Unfortunately, not all such problems are so easily detected, especially if you have many independent variables in your model. If you have ever had such difficulties, then you have experienced one of the more unpleasant aspects of computer optimization. The computer has no idea that it is trying to solve for an infinite coefficient as it begins its iterative process. All it knows is that, at each step, making the coefficient a little bigger, or a little smaller, works wonders. It continues on its merry way until either 1) the whole thing comes crashing to the ground when a numerical overflow error occurs or 2) it reaches some predetermined cutoff that stops the process. Meanwhile, you have been waiting. And the estimates that you finally receive, if any, may be nothing more than numerical roundoff.

Stata watches for these sorts of problems, alerts you, fixes them, and then properly fits the model.

Let's return to our automobile data. Among the variables we have in the data is one called `repair` that takes on three values. A value of 1 indicates that the car has a poor repair record, 2 indicates an average record, and 3 indicates a better-than-average record. Here is a tabulation of our data:

```
. use http://www.stata-press.com/data/r12/repair
(1978 Automobile Data)

. tabulate foreign repair

           |              repair
  Car type |         1          2          3 |     Total
-----------+---------------------------------+----------
  Domestic |        10         27          9 |        46
   Foreign |         0          3          9 |        12
-----------+---------------------------------+----------
     Total |        10         30         18 |        58
```

All the cars with poor repair records (`repair` = 1) are domestic. If we were to attempt to predict `foreign` on the basis of the repair records, the predicted probability for the `repair` = 1 category would have to be zero. This in turn means that the probit coefficient must be minus infinity, and that would set most computer programs buzzing.

Let's try using Stata on this problem.

```
. probit foreign b3.repair
note: 1.repair != 0 predicts failure perfectly
      1.repair dropped and 10 obs not used

Iteration 0:   log likelihood = -26.992087
Iteration 1:   log likelihood = -22.276479
Iteration 2:   log likelihood = -22.229184
Iteration 3:   log likelihood = -22.229138
Iteration 4:   log likelihood = -22.229138

Probit regression                                 Number of obs   =         48
                                                  LR chi2(1)      =       9.53
                                                  Prob > chi2     =     0.0020
Log likelihood = -22.229138                       Pseudo R2       =     0.1765

------------------------------------------------------------------------------
     foreign |      Coef.   Std. Err.      z    P>|z|     [95% Conf. Interval]
-------------+----------------------------------------------------------------
      repair |
          1  |          0  (empty)
          2  |  -1.281552   .4297326    -2.98   0.003    -2.123812   -.4392911
             |
       _cons |   9.89e-17    .295409     0.00   1.000     -.578991     .578991
------------------------------------------------------------------------------
```

Remember that all the cars with poor repair records (`repair` = 1) are domestic, so the model cannot be fit, or at least it cannot be fit if we restrict ourselves to finite coefficients. Stata noted that fact "note: 1.repair != 0 predicts failure perfectly". This is Stata's mathematically precise way of saying what we said in English. When `repair` is 1, the car is domestic.

Stata then went on to say, "1.repair dropped and 10 obs not used". This is Stata eliminating the problem. First, `1.repair` had to be removed from the model because it would have an infinite coefficient. Then the 10 observations that led to the problem had to be eliminated, as well, so as not to bias the remaining coefficients in the model. The 10 observations that are not used are the 10 domestic cars that have poor repair records.

Stata then fit what was left of the model, using the remaining observations. Because no observations remained for cars with poor repair records, Stata reports "(empty)" in the row for `repair` = 1.

◁

❑ Technical note

Stata is pretty smart about catching these problems. It will catch "one-way causation by a dummy variable", as we demonstrated above.

Stata also watches for "two-way causation", that is, a variable that perfectly determines the outcome, both successes and failures. Here Stata says that the variable "predicts outcome perfectly" and stops. Statistics dictate that no model can be fit.

Stata also checks your data for collinear variables; it will say "so-and-so omitted because of collinearity". No observations need to be eliminated here and model fitting will proceed without the offending variable.

It will also catch a subtle problem that can arise with continuous data. For instance, if we were estimating the chances of surviving the first year after an operation, and if we included in our model `age`, and if all the persons over 65 died within the year, Stata will say, "age $>$ 65 predicts failure perfectly". It will then inform us about how it resolves the issue and fit what can be fit of our model.

`probit` (and `logit`, `logistic`, and `ivprobit`) will also occasionally fail to converge and then display messages such as

```
Note: 4 failures and 0 successes completely determined.
```

The cause of this message and what to do if you see it are described in [R] **logit**. ❑

Saved results

`probit` saves the following in `e()`:

Scalars

`e(N)`	number of observations
`e(N_cds)`	number of completely determined successes
`e(N_cdf)`	number of completely determined failures
`e(k)`	number of parameters
`e(k_eq)`	number of equations in `e(b)`
`e(k_eq_model)`	number of equations in overall model test
`e(k_dv)`	number of dependent variables
`e(df_m)`	model degrees of freedom
`e(r2_p)`	pseudo-R-squared
`e(ll)`	log likelihood
`e(ll_0)`	log likelihood, constant-only model
`e(N_clust)`	number of clusters
`e(chi2)`	χ^2
`e(p)`	significance of model test
`e(rank)`	rank of `e(V)`
`e(ic)`	number of iterations
`e(rc)`	return code
`e(converged)`	`1` if converged, `0` otherwise

Macros	
e(cmd)	probit
e(cmdline)	command as typed
e(depvar)	name of dependent variable
e(wtype)	weight type
e(wexp)	weight expression
e(title)	title in estimation output
e(clustvar)	name of cluster variable
e(offset)	linear offset variable
e(chi2type)	Wald or LR; type of model χ^2 test
e(vce)	*vcetype* specified in vce()
e(vcetype)	title used to label Std. Err.
e(opt)	type of optimization
e(which)	max or min; whether optimizer is to perform maximization or minimization
e(ml_method)	type of ml method
e(user)	name of likelihood-evaluator program
e(technique)	maximization technique
e(properties)	b V
e(estat_cmd)	program used to implement estat
e(predict)	program used to implement predict
e(asbalanced)	factor variables fvset as asbalanced
e(asobserved)	factor variables fvset as asobserved
Matrices	
e(b)	coefficient vector
e(Cns)	constraints matrix
e(ilog)	iteration log (up to 20 iterations)
e(gradient)	gradient vector
e(mns)	vector of means of the independent variables
e(rules)	information about perfect predictors
e(V)	variance–covariance matrix of the estimators
e(V_modelbased)	model-based variance
Functions	
e(sample)	marks estimation sample

Methods and formulas

probit is implemented as an ado-file.

Probit analysis originated in connection with bioassay, and the word probit, a contraction of "probability unit", was suggested by Bliss (1934a, 1934b). For an introduction to probit and logit, see, for example, Aldrich and Nelson (1984), Cameron and Trivedi (2010), Greene (2012), Long (1997), Pampel (2000), or Powers and Xie (2008). Long and Freese (2006, chap. 4) and Jones (2007, chap. 3) provide introductions to probit and logit, along with Stata examples.

The log-likelihood function for probit is

$$\ln L = \sum_{j \in S} w_j \ln\Phi(\mathbf{x}_j\boldsymbol{\beta}) + \sum_{j \notin S} w_j \ln\Big\{1 - \Phi(\mathbf{x}_j\boldsymbol{\beta})\Big\}$$

where Φ is the cumulative normal and w_j denotes the optional weights. $\ln L$ is maximized, as described in [R] **maximize**.

This command supports the Huber/White/sandwich estimator of the variance and its clustered version using `vce(robust)` and `vce(cluster` *clustvar*`)`, respectively. See [P] **_robust**, particularly *Maximum likelihood estimators* and *Methods and formulas*. The scores are calculated as $\mathbf{u}_j = \{\phi(\mathbf{x}_j\mathbf{b})/\Phi(\mathbf{x}_j\mathbf{b})\}\mathbf{x}_j$ for the positive outcomes and $-[\phi(\mathbf{x}_j\mathbf{b})/\{1-\Phi(\mathbf{x}_j\mathbf{b})\}]\mathbf{x}_j$ for the negative outcomes, where ϕ is the normal density.

`probit` also supports estimation with survey data. For details on VCEs with survey data, see [SVY] **variance estimation**.

Chester Ittner Bliss (1899–1979) was born in Ohio. He was educated as an entomologist, earning degrees from Ohio State and Columbia, and was employed by the United States Department of Agriculture until 1933. When he lost his job because of the Depression, Bliss then worked with R. A. Fisher in London and at the Institute of Plant Protection in Leningrad before returning to a post at the Connecticut Agricultural Experiment Station in 1938. He was also a lecturer at Yale for 25 years. Among many contributions to biostatistics, his development and application of probit methods to biological problems are outstanding.

References

Aldrich, J. H., and F. D. Nelson. 1984. *Linear Probability, Logit, and Probit Models.* Newbury Park, CA: Sage.

Berkson, J. 1944. Application of the logistic function to bio-assay. *Journal of the American Statistical Association* 39: 357–365.

Bliss, C. I. 1934a. The method of probits. *Science* 79: 38–39.

——. 1934b. The method of probits—a correction. *Science* 79: 409–410.

Cameron, A. C., and P. K. Trivedi. 2010. *Microeconometrics Using Stata.* Rev. ed. College Station, TX: Stata Press.

Cochran, W. G., and D. J. Finney. 1979. Chester Ittner Bliss 1899–1979. *Biometrics* 35: 715–717.

De Luca, G. 2008. SNP and SML estimation of univariate and bivariate binary-choice models. *Stata Journal* 8: 190–220.

Greene, W. H. 2012. *Econometric Analysis.* 7th ed. Upper Saddle River, NJ: Prentice Hall.

Hilbe, J. M. 1996. sg54: Extended probit regression. *Stata Technical Bulletin* 32: 20–21. Reprinted in *Stata Technical Bulletin Reprints*, vol. 6, pp. 131–132. College Station, TX: Stata Press.

Jones, A. 2007. *Applied Econometrics for Health Economists: A Practical Guide.* 2nd ed. Abingdon, UK: Radcliffe.

Judge, G. G., W. E. Griffiths, R. C. Hill, H. Lütkepohl, and T.-C. Lee. 1985. *The Theory and Practice of Econometrics.* 2nd ed. New York: Wiley.

Long, J. S. 1997. *Regression Models for Categorical and Limited Dependent Variables.* Thousand Oaks, CA: Sage.

Long, J. S., and J. Freese. 2006. *Regression Models for Categorical Dependent Variables Using Stata.* 2nd ed. College Station, TX: Stata Press.

Miranda, A., and S. Rabe-Hesketh. 2006. Maximum likelihood estimation of endogenous switching and sample selection models for binary, ordinal, and count variables. *Stata Journal* 6: 285–308.

Pampel, F. C. 2000. *Logistic Regression: A Primer.* Thousand Oaks, CA: Sage.

Powers, D. A., and Y. Xie. 2008. *Statistical Methods for Categorical Data Analysis.* 2nd ed. Bingley, UK: Emerald.

Xu, J., and J. S. Long. 2005. Confidence intervals for predicted outcomes in regression models for categorical outcomes. *Stata Journal* 5: 537–559.

Also see

[R] **probit postestimation** — Postestimation tools for probit

[R] **asmprobit** — Alternative-specific multinomial probit regression

[R] **biprobit** — Bivariate probit regression

[R] **brier** — Brier score decomposition

[R] **glm** — Generalized linear models

[R] **hetprob** — Heteroskedastic probit model

[R] **ivprobit** — Probit model with continuous endogenous regressors

[R] **logistic** — Logistic regression, reporting odds ratios

[R] **logit** — Logistic regression, reporting coefficients

[R] **mprobit** — Multinomial probit regression

[R] **roc** — Receiver operating characteristic (ROC) analysis

[R] **scobit** — Skewed logistic regression

[MI] **estimation** — Estimation commands for use with mi estimate

[SVY] **svy estimation** — Estimation commands for survey data

[XT] **xtprobit** — Random-effects and population-averaged probit models

[U] **20 Estimation and postestimation commands**

Title

probit postestimation — Postestimation tools for probit

Description

The following postestimation commands are of special interest after `probit`:

Command	Description
`estat classification`	report various summary statistics, including the classification table
`estat gof`	Pearson or Hosmer–Lemeshow goodness-of-fit test
`lroc`	compute area under ROC curve and graph the curve
`lsens`	graph sensitivity and specificity versus probability cutoff

These commands are not appropriate after the `svy` prefix.

For information about these commands, see [R] **logistic postestimation**.

The following standard postestimation commands are also available:

Command	Description
`contrast`	contrasts and ANOVA-style joint tests of estimates
`estat`	AIC, BIC, VCE, and estimation sample summary
`estat` (svy)	postestimation statistics for survey data
`estimates`	cataloging estimation results
`hausman`	Hausman's specification test
`lincom`	point estimates, standard errors, testing, and inference for linear combinations of coefficients
`linktest`	link test for model specification
`lrtest`[1]	likelihood-ratio test
`margins`	marginal means, predictive margins, marginal effects, and average marginal effects
`marginsplot`	graph the results from margins (profile plots, interaction plots, etc.)
`nlcom`	point estimates, standard errors, testing, and inference for nonlinear combinations of coefficients
`predict`	predictions, residuals, influence statistics, and other diagnostic measures
`predictnl`	point estimates, standard errors, testing, and inference for generalized predictions
`pwcompare`	pairwise comparisons of estimates
`suest`	seemingly unrelated estimation
`test`	Wald tests of simple and composite linear hypotheses
`testnl`	Wald tests of nonlinear hypotheses

[1] `lrtest` is not appropriate with `svy` estimation results.

See the corresponding entries in the *Base Reference Manual* for details, but see [SVY] **estat** for details about `estat` (svy).

Syntax for predict

`predict [type] newvar [if] [in] [, statistic nooffset rules asif]`

statistic	Description
Main	
pr	probability of a positive outcome; the default
xb	linear prediction
stdp	standard error of the linear prediction
*deviance	deviance residual
score	first derivative of the log likelihood with respect to $\mathbf{x}_j\boldsymbol{\beta}$

Unstarred statistics are available both in and out of sample; type `predict ... if e(sample) ...` if wanted only for the estimation sample. Starred statistics are calculated only for the estimation sample, even when `if e(sample)` is not specified.

Menu

Statistics > Postestimation > Predictions, residuals, etc.

Options for predict

Main

`pr`, the default, calculates the probability of a positive outcome.

`xb` calculates the linear prediction.

`stdp` calculates the standard error of the linear prediction.

`deviance` calculates the deviance residual.

`score` calculates the equation-level score, $\partial \ln L/\partial(\mathbf{x}_j\boldsymbol{\beta})$.

`nooffset` is relevant only if you specified `offset(`*varname*`)` for `probit`. It modifies the calculations made by `predict` so that they ignore the offset variable; the linear prediction is treated as $\mathbf{x}_j\mathbf{b}$ rather than as $\mathbf{x}_j\mathbf{b} + \text{offset}_j$.

`rules` requests that Stata use any rules that were used to identify the model when making the prediction. By default, Stata calculates missing for excluded observations.

`asif` requests that Stata ignore the rules and exclusion criteria and calculate predictions for all observations possible using the estimated parameter from the model.

Remarks

Remarks are presented under the following headings:

Obtaining predicted values
Performing hypothesis tests

Obtaining predicted values

Once you have fit a probit model, you can obtain the predicted probabilities by using the predict command for both the estimation sample and other samples; see [U] **20 Estimation and postestimation commands** and [R] **predict**. Here we will make only a few additional comments.

predict without arguments calculates the predicted probability of a positive outcome. With the xb option, predict calculates the linear combination $\mathbf{x}_j\mathbf{b}$, where $\mathbf{x}_j$ are the independent variables in the jth observation and $\mathbf{b}$ is the estimated parameter vector. This is known as the index function because the cumulative density indexed at this value is the probability of a positive outcome.

In both cases, Stata remembers any rules used to identify the model and calculates missing for excluded observations unless rules or asif is specified. This is covered in the following example.

With the stdp option, predict calculates the standard error of the prediction, which is *not* adjusted for replicated covariate patterns in the data.

You can calculate the unadjusted-for-replicated-covariate-patterns diagonal elements of the hat matrix, or leverage, by typing

```
. predict pred
. predict stdp, stdp
. generate hat = stdp^2*pred*(1-pred)
```

▷ Example 1

In example 4 of [R] **probit**, we fit the probit model probit foreign b3.repair. To obtain predicted probabilities, we type

```
. predict p
(option pr assumed; Pr(foreign))
(10 missing values generated)
. summarize foreign p
```

Variable	Obs	Mean	Std. Dev.	Min	Max
foreign	58	.2068966	.4086186	0	1
p	48	.25	.1956984	.1	.5

Stata remembers any rules used to identify the model and sets predictions to missing for any excluded observations. In the previous example, probit dropped the variable 1.repair from our model and excluded 10 observations. When we typed predict p, those same 10 observations were again excluded and their predictions set to missing.

predict's rules option uses the rules in the prediction. During estimation, we were told, "1.repair != 0 predicts failure perfectly", so the rule is that when 1.repair is not zero, we should predict 0 probability of success or a positive outcome:

```
. predict p2, rules
. summarize foreign p p2
```

Variable	Obs	Mean	Std. Dev.	Min	Max
foreign	58	.2068966	.4086186	0	1
p	48	.25	.1956984	.1	.5
p2	58	.2068966	.2016268	0	.5

predict's asif option ignores the rules and the exclusion criteria and calculates predictions for all observations possible using the estimated parameters from the model:

```
. predict p3, asif
. summarize for p p2 p3
```

Variable	Obs	Mean	Std. Dev.	Min	Max
foreign	58	.2068966	.4086186	0	1
p	48	.25	.1956984	.1	.5
p2	58	.2068966	.2016268	0	.5
p3	58	.2931034	.2016268	.1	.5

Which is right? By default, predict uses the most conservative approach. If many observations had been excluded due to a simple rule, we could be reasonably certain that the rules prediction is correct. The asif prediction is correct only if the exclusion is a fluke and we would be willing to exclude the variable from the analysis, anyway. Then, however, we should refit the model to include the excluded observations. ◁

Performing hypothesis tests

After estimation with probit, you can perform hypothesis tests by using the test or testnl command; see [U] **20 Estimation and postestimation commands**.

Methods and formulas

All postestimation commands listed above are implemented as ado-files.

predict after probit

Let index j be used to index observations, not covariate patterns. Define M_j for each observation as the total number of observations sharing j's covariate pattern. Define Y_j as the total number of positive responses among observations sharing j's covariate pattern. Define p_j as the predicted probability of a positive outcome for observation j.

For $M_j > 1$, the deviance residual d_j is defined as

$$d_j = \pm\left(2\left[Y_j \ln\left(\frac{Y_j}{M_j p_j}\right) + (M_j - Y_j)\ln\left\{\frac{M_j - Y_j}{M_j(1-p_j)}\right\}\right]\right)^{1/2}$$

where the sign is the same as the sign of $(Y_j - M_j p_j)$. In the limiting cases, the deviance residual is given by

$$d_j = \begin{cases} -\sqrt{2M_j|\ln(1-p_j)|} & \text{if } Y_j = 0 \\ \sqrt{2M_j|\ln p_j|} & \text{if } Y_j = M_j \end{cases}$$

Also see

[R] **probit** — Probit regression

[R] **logistic postestimation** — Postestimation tools for logistic

[U] **20 Estimation and postestimation commands**

Title

proportion — Estimate proportions

Syntax

`proportion` *varlist* [*if*] [*in*] [*weight*] [, *options*]

options	Description
Model	
`stdize(`*varname*`)`	variable identifying strata for standardization
`stdweight(`*varname*`)`	weight variable for standardization
`nostdrescale`	do not rescale the standard weight variable
`nolabel`	suppress value labels from *varlist*
`missing`	treat missing values like other values
if/in/over	
`over(`*varlist*[, `nolabel`]`)`	group over subpopulations defined by *varlist*; optionally, suppress group labels
SE/Cluster	
`vce(`*vcetype*`)`	*vcetype* may be `analytic`, `cluster` *clustvar*, `bootstrap`, or `jackknife`
Reporting	
`level(`*#*`)`	set confidence level; default is `level(95)`
`noheader`	suppress table header
`nolegend`	suppress table legend
display_options	control column formats and line width
`coeflegend`	display legend instead of statistics

`bootstrap`, `jackknife`, `mi estimate`, `rolling`, `statsby`, and `svy` are allowed; see [U] **11.1.10 Prefix commands**.
`vce(bootstrap)` and `vce(jackknife)` are not allowed with the `mi estimate` prefix; see [MI] **mi estimate**.
Weights are not allowed with the `bootstrap` prefix; see [R] **bootstrap**.
`vce()` and weights are not allowed with the `svy` prefix; see [SVY] **svy**.
`fweight`s, `iweight`s, and `pweight`s are allowed; see [U] **11.1.6 weight**.
`coeflegend` does not appear in the dialog box.
See [U] **20 Estimation and postestimation commands** for more capabilities of estimation commands.

Menu

Statistics > Summaries, tables, and tests > Summary and descriptive statistics > Proportions

Description

`proportion` produces estimates of proportions, along with standard errors, for the categories identified by the values in each variable of *varlist*.

Options

Model

`stdize(`*varname*`)` specifies that the point estimates be adjusted by direct standardization across the strata identified by *varname*. This option requires the `stdweight()` option.

`stdweight(`*varname*`)` specifies the weight variable associated with the standard strata identified in the `stdize()` option. The standardization weights must be constant within the standard strata.

`nostdrescale` prevents the standardization weights from being rescaled within the `over()` groups. This option requires `stdize()` but is ignored if the `over()` option is not specified.

`nolabel` specifies that value labels attached to the variables in *varlist* be ignored.

`missing` specifies that missing values in *varlist* be treated as valid categories, rather than omitted from the analysis (the default).

if/in/over

`over(`*varlist* [, `nolabel`]`)` specifies that estimates be computed for multiple subpopulations, which are identified by the different values of the variables in *varlist*.

When this option is supplied with one variable name, such as `over(`*varname*`)`, the value labels of *varname* are used to identify the subpopulations. If *varname* does not have labeled values (or there are unlabeled values), the values themselves are used, provided that they are nonnegative integers. Noninteger values, negative values, and labels that are not valid Stata names are substituted with a default identifier.

When `over()` is supplied with multiple variable names, each subpopulation is assigned a unique default identifier.

`nolabel` requests that value labels attached to the variables identifying the subpopulations be ignored.

SE/Cluster

`vce(`*vcetype*`)` specifies the type of standard error reported, which includes types that are derived from asymptotic theory, that allow for intragroup correlation, and that use bootstrap or jackknife methods; see [R] ***vce_option***.

`vce(analytic)`, the default, uses the analytically derived variance estimator associated with the sample proportion.

Reporting

`level(`*#*`)`; see [R] **estimation options**.

`noheader` prevents the table header from being displayed. This option implies `nolegend`.

`nolegend` prevents the table legend identifying the subpopulations from being displayed.

display_options: `cformat(%`*fmt*`)` and `nolstretch`; see [R] **estimation options**.

The following option is available with `proportion` but is not shown in the dialog box:

`coeflegend`; see [R] **estimation options**.

Remarks

▷ Example 1

We can estimate the proportion of each repair rating in the auto data:

```
. use http://www.stata-press.com/data/r12/auto
(1978 Automobile Data)
. proportion rep78
Proportion estimation                Number of obs    =      69
```

	Proportion	Std. Err.	[95% Conf.	Interval]
rep78				
1	.0289855	.0203446	-.0116115	.0695825
2	.115942	.0388245	.0384689	.1934152
3	.4347826	.0601159	.3148232	.554742
4	.2608696	.0532498	.1546113	.3671278
5	.1594203	.0443922	.070837	.2480036

Here we use the `missing` option to include missing values as a category of `rep78`:

```
. proportion rep78, missing
Proportion estimation                Number of obs    =      74
      _prop_6: rep78 = .
```

	Proportion	Std. Err.	[95% Conf.	Interval]
rep78				
1	.027027	.0189796	-.0107994	.0648534
2	.1081081	.0363433	.0356761	.1805401
3	.4054054	.0574637	.2908804	.5199305
4	.2432432	.0502154	.1431641	.3433224
5	.1486486	.0416364	.0656674	.2316299
_prop_6	.0675676	.0293776	.0090181	.1261171

◁

▷ Example 2

We can also estimate proportions over groups:

```
. proportion rep78, over(foreign)
Proportion estimation                 Number of obs    =      69

      _prop_1: rep78 = 1
      _prop_2: rep78 = 2
      _prop_3: rep78 = 3
      _prop_4: rep78 = 4
      _prop_5: rep78 = 5

     Domestic: foreign = Domestic
      Foreign: foreign = Foreign
```

Over	Proportion	Std. Err.	[95% Conf.	Interval]
_prop_1				
Domestic	.0416667	.0291477	-.0164966	.0998299
Foreign	.	(no observations)		
_prop_2				
Domestic	.1666667	.0543607	.0581916	.2751417
Foreign	.	(no observations)		
_prop_3				
Domestic	.5625	.0723605	.4181069	.7068931
Foreign	.1428571	.0782461	-.0132805	.2989948
_prop_4				
Domestic	.1875	.0569329	.0738921	.3011079
Foreign	.4285714	.1106567	.2077595	.6493834
_prop_5				
Domestic	.0416667	.0291477	-.0164966	.0998299
Foreign	.4285714	.1106567	.2077595	.6493834

◁

Saved results

`proportion` saves the following in `e()`:

Scalars

`e(N)`	number of observations
`e(N_over)`	number of subpopulations
`e(N_stdize)`	number of standard strata
`e(N_clust)`	number of clusters
`e(k_eq)`	number of equations in `e(b)`
`e(df_r)`	sample degrees of freedom
`e(rank)`	rank of `e(V)`

Macros

`e(cmd)`	`proportion`
`e(cmdline)`	command as typed
`e(varlist)`	*varlist*
`e(stdize)`	*varname* from `stdize()`
`e(stdweight)`	*varname* from `stdweight()`
`e(wtype)`	weight type
`e(wexp)`	weight expression
`e(title)`	title in estimation output
`e(cluster)`	name of cluster variable

e(over)	*varlist* from over()
e(over_labels)	labels from over() variables
e(over_namelist)	names from e(over_labels)
e(namelist)	proportion identifiers
e(label#)	labels from #th variable in *varlist*
e(vce)	*vcetype* specified in vce()
e(vcetype)	title used to label Std. Err.
e(properties)	b V
e(estat_cmd)	program used to implement estat
e(marginsnotok)	predictions disallowed by margins
Matrices	
e(b)	vector of proportion estimates
e(V)	(co)variance estimates
e(_N)	vector of numbers of nonmissing observations
e(_N_stdsum)	number of nonmissing observations within the standard strata
e(_p_stdize)	standardizing proportions
e(error)	error code corresponding to e(b)
Functions	
e(sample)	marks estimation sample

Methods and formulas

proportion is implemented as an ado-file.

Proportions are means of indicator variables; see [R] **mean**.

References

Cochran, W. G. 1977. *Sampling Techniques.* 3rd ed. New York: Wiley.

Stuart, A., and J. K. Ord. 1994. *Kendall's Advanced Theory of Statistics: Distribution Theory, Vol I.* 6th ed. London: Arnold.

Also see

[R] **proportion postestimation** — Postestimation tools for proportion

[R] **mean** — Estimate means

[R] **ratio** — Estimate ratios

[R] **total** — Estimate totals

[MI] **estimation** — Estimation commands for use with mi estimate

[SVY] **direct standardization** — Direct standardization of means, proportions, and ratios

[SVY] **poststratification** — Poststratification for survey data

[SVY] **subpopulation estimation** — Subpopulation estimation for survey data

[SVY] **svy estimation** — Estimation commands for survey data

[SVY] **variance estimation** — Variance estimation for survey data

[U] **20 Estimation and postestimation commands**

Title

proportion postestimation — Postestimation tools for proportion

Description

The following postestimation commands are available after `proportion`:

Command	Description
`estat`	VCE
`estat` (svy)	postestimation statistics for survey data
`estimates`	cataloging estimation results
`lincom`	point estimates, standard errors, testing, and inference for linear combinations of coefficients
`nlcom`	point estimates, standard errors, testing, and inference for nonlinear combinations of coefficients
`test`	Wald tests of simple and composite linear hypotheses
`testnl`	Wald tests of nonlinear hypotheses

See the corresponding entries in the *Base Reference Manual* for details, but see [SVY] **estat** for details about `estat` (svy).

Methods and formulas

All postestimation commands listed above are implemented as ado-files.

Also see

[R] **proportion** — Estimate proportions

[SVY] **svy postestimation** — Postestimation tools for svy

[U] **20 Estimation and postestimation commands**

Title

prtest — One- and two-sample tests of proportions

Syntax

One-sample test of proportion

`prtest` *varname* `==` $\#_p$ [*if*] [*in*] [, `level(#)`]

Two-sample test of proportions

`prtest` $varname_1$ `==` $varname_2$ [*if*] [*in*] [, `level(#)`]

Two-group test of proportions

`prtest` *varname* [*if*] [*in*] , `by(`*groupvar*`)` [`level(#)`]

Immediate form of one-sample test of proportion

`prtesti` $\#_{\text{obs1}}$ $\#_{p1}$ $\#_{p2}$ [, `level(#)` `count`]

Immediate form of two-sample test of proportions

`prtesti` $\#_{\text{obs1}}$ $\#_{p1}$ $\#_{\text{obs2}}$ $\#_{p2}$ [, `level(#)` `count`]

`by` is allowed with `prtest`; see [D] **by**.

Menu

one-sample

Statistics > Summaries, tables, and tests > Classical tests of hypotheses > One-sample proportion test

two-sample

Statistics > Summaries, tables, and tests > Classical tests of hypotheses > Two-sample proportion test

two-group

Statistics > Summaries, tables, and tests > Classical tests of hypotheses > Two-group proportion test

immediate command: one-sample

Statistics > Summaries, tables, and tests > Classical tests of hypotheses > One-sample proportion calculator

immediate command: two-sample

Statistics > Summaries, tables, and tests > Classical tests of hypotheses > Two-sample proportion calculator

Description

prtest performs tests on the equality of proportions using large-sample statistics.

In the first form, prtest tests that *varname* has a proportion of $\#_p$. In the second form, prtest tests that $varname_1$ and $varname_2$ have the same proportion. In the third form, prtest tests that *varname* has the same proportion within the two groups defined by *groupvar*.

prtesti is the immediate form of prtest; see [U] **19 Immediate commands**.

The bitest command is a better version of the first form of prtest in that it gives exact *p*-values. Researchers should use bitest when possible, especially for small samples; see [R] **bitest**.

Options

Main

by(*groupvar*) specifies a numeric variable that contains the group information for a given observation. This variable must have only two values. Do not confuse the by() option with the by prefix; both may be specified.

level(#) specifies the confidence level, as a percentage, for confidence intervals. The default is level(95) or as set by set level; see [U] **20.7 Specifying the width of confidence intervals**.

count specifies that integer counts instead of proportions be used in the immediate forms of prtest. In the first syntax, prtesti expects that $\#_{\text{obs1}}$ and $\#_{p1}$ are counts—$\#_{p1} \leq \#_{\text{obs1}}$—and $\#_{p2}$ is a proportion. In the second syntax, prtesti expects that all four numbers are integer counts, that $\#_{\text{obs1}} \geq \#_{p1}$, and that $\#_{\text{obs2}} \geq \#_{p2}$.

Remarks

The prtest output follows the output of ttest in providing a lot of information. Each proportion is presented along with a confidence interval. The appropriate one- or two-sample test is performed, and the two-sided and both one-sided results are included at the bottom of the output. For a two-sample test, the calculated difference is also presented with its confidence interval. This command may be used for both large-sample testing and large-sample interval estimation.

▷ Example 1: One-sample test of proportion

In the first form, prtest tests whether the mean of the sample is equal to a known constant. Assume that we have a sample of 74 automobiles. We wish to test whether the proportion of automobiles that are foreign is different from 40%.

```
. use http://www.stata-press.com/data/r12/auto
(1978 Automobile Data)

. prtest foreign == .4

One-sample test of proportion                  foreign: Number of obs =       74
------------------------------------------------------------------------------
    Variable |       Mean   Std. Err.                     [95% Conf. Interval]
-------------+----------------------------------------------------------------
     foreign |   .2972973   .0531331                      .1931583    .4014363
------------------------------------------------------------------------------
    p = proportion(foreign)                                       z =  -1.8034
Ho: p = 0.4

     Ha: p < 0.4                 Ha: p != 0.4                   Ha: p > 0.4
 Pr(Z < z) = 0.0357         Pr(|Z| > |z|) = 0.0713          Pr(Z > z) = 0.9643
```

The test indicates that we cannot reject the hypothesis that the proportion of foreign automobiles is 0.40 at the 5% significance level.

◁

▷ Example 2: Two-sample test of proportions

We have two headache remedies that we give to patients. Each remedy's effect is recorded as 0 for failing to relieve the headache and 1 for relieving the headache. We wish to test the equality of the proportion of people relieved by the two treatments.

```
. use http://www.stata-press.com/data/r12/cure
. prtest cure1 == cure2
Two-sample test of proportions                  cure1: Number of obs =        50
                                                cure2: Number of obs =        59

    Variable |       Mean   Std. Err.       z    P>|z|     [95% Conf. Interval]
-------------+----------------------------------------------------------------
       cure1 |        .52   .0706541                        .3815205    .6584795
       cure2 |   .7118644   .0589618                        .5963013    .8274275
-------------+----------------------------------------------------------------
        diff |  -.1918644   .0920245                        -.372229   -.0114998
             |  under Ho:   .0931155    -2.06   0.039
------------------------------------------------------------------------------
        diff = prop(cure1) - prop(cure2)                          z =  -2.0605
    Ho: diff = 0

    Ha: diff < 0                 Ha: diff != 0                 Ha: diff > 0
 Pr(Z < z) = 0.0197         Pr(|Z| < |z|) = 0.0394          Pr(Z > z) = 0.9803
```

We find that the proportions are statistically different from each other at any level greater than 3.9%.

◁

▷ Example 3: Immediate form of one-sample test of proportion

`prtesti` is like `prtest`, except that you specify summary statistics rather than variables as arguments. For instance, we are reading an article that reports the proportion of registered voters among 50 randomly selected eligible voters as 0.52. We wish to test whether the proportion is 0.7:

```
. prtesti 50 .52 .70
One-sample test of proportion                       x: Number of obs =        50

    Variable |       Mean   Std. Err.                      [95% Conf. Interval]
-------------+----------------------------------------------------------------
           x |        .52   .0706541                        .3815205    .6584795
------------------------------------------------------------------------------
    p = proportion(x)                                             z =  -2.7775
Ho: p = 0.7

     Ha: p < 0.7                 Ha: p != 0.7                   Ha: p > 0.7
 Pr(Z < z) = 0.0027         Pr(|Z| > |z|) = 0.0055          Pr(Z > z) = 0.9973
```

◁

▷ Example 4: Immediate form of two-sample test of proportions

To judge teacher effectiveness, we wish to test whether the same proportion of people from two classes will answer an advanced question correctly. In the first classroom of 30 students, 40% answered the question correctly, whereas in the second classroom of 45 students, 67% answered the question correctly.

```
. prtesti 30 .4 45 .67
Two-sample test of proportions                  x: Number of obs =        30
                                                y: Number of obs =        45

    Variable |       Mean   Std. Err.      z    P>|z|     [95% Conf. Interval]
-------------+----------------------------------------------------------------
           x |         .4   .0894427                      .2246955    .5753045
           y |        .67   .0700952                       .532616     .807384
-------------+----------------------------------------------------------------
        diff |       -.27   .1136368                     -.4927241   -.0472759
             |  under Ho:   .1169416   -2.31   0.021
------------------------------------------------------------------------------
        diff = prop(x) - prop(y)                                  z =  -2.3088
    Ho: diff = 0

    Ha: diff < 0                 Ha: diff != 0                 Ha: diff > 0
 Pr(Z < z) = 0.0105         Pr(|Z| < |z|) = 0.0210          Pr(Z > z) = 0.9895
```

◁

Saved results

prtest and prtesti save the following in r():

Scalars

r(z)	z statistic	r(N_#)	number of observations for variable #
r(P_#)	proportion for variable #		

Methods and formulas

prtest and prtesti are implemented as ado-files.

See Acock (2010, 149–155) for additional examples of tests of proportions using Stata.

A large-sample $100(1-\alpha)\%$ confidence interval for a proportion p is

$$\widehat{p} \pm z_{1-\alpha/2}\sqrt{\frac{\widehat{p}\,\widehat{q}}{n}}$$

and a $100(1-\alpha)\%$ confidence interval for the difference of two proportions is given by

$$(\widehat{p}_1 - \widehat{p}_2) \pm z_{1-\alpha/2}\sqrt{\frac{\widehat{p}_1\widehat{q}_1}{n_1} + \frac{\widehat{p}_2\widehat{q}_2}{n_2}}$$

where $\widehat{q} = 1 - \widehat{p}$ and z is calculated from the inverse cumulative standard normal distribution.

The one-tailed and two-tailed tests of a population proportion use a normally distributed test statistic calculated as

$$z = \frac{\widehat{p} - p_0}{\sqrt{p_0 q_0 / n}}$$

where p_0 is the hypothesized proportion. A test of the difference of two proportions also uses a normally distributed test statistic calculated as

$$z = \frac{\widehat{p}_1 - \widehat{p}_2}{\sqrt{\widehat{p}_p \widehat{q}_p (1/n_1 + 1/n_2)}}$$

where

$$\widehat{p}_p = \frac{x_1 + x_2}{n_1 + n_2}$$

and x_1 and x_2 are the total number of successes in the two populations.

References

Acock, A. C. 2010. *A Gentle Introduction to Stata*. 3rd ed. College Station, TX: Stata Press.

Wang, D. 2000. sg154: Confidence intervals for the ratio of two binomial proportions by Koopman's method. *Stata Technical Bulletin* 58: 16–19. Reprinted in *Stata Technical Bulletin Reprints*, vol. 10, pp. 244–247. College Station, TX: Stata Press.

Also see

[R] **bitest** — Binomial probability test

[R] **proportion** — Estimate proportions

[R] **ttest** — Mean-comparison tests

[MV] **hotelling** — Hotelling's T-squared generalized means test

Title

pwcompare — Pairwise comparisons

Syntax

`pwcompare` *marginlist* [, *options*]

where *marginlist* is a list of factor variables or interactions that appear in the current estimation results or `_eqns` to reference equations. The variables may be typed with or without the `i.` prefix, and you may use any factor-variable syntax:

```
. pwcompare i.sex i.group i.sex#i.group
. pwcompare sex group sex#group
. pwcompare sex##group
```

options	Description
Main	
`mcompare(`*method*`)`	adjust for multiple comparisons; default is `mcompare(noadjust)`
`asobserved`	treat all factor variables as observed
Equations	
`equation(`*eqspec*`)`	perform comparisons within equation *eqspec*
`atequations`	perform comparisons within each equation
Advanced	
`emptycells(`*empspec*`)`	treatment of empty cells for balanced factors
`noestimcheck`	suppress estimability checks
Reporting	
`level(`*#*`)`	confidence level; default is `level(95)`
`cieffects`	show effects table with confidence intervals; the default
`pveffects`	show effects table with *p*-values
`effects`	show effects table with confidence intervals and *p*-values
`cimargins`	show table of margins and confidence intervals
`groups`	show table of margins and group codes
`sort`	sort the margins or contrasts within each term
`post`	post margins and their VCEs as estimation results
display_options	control column formats, line width, and suppress blank lines between terms
eform_option	report exponentiated contrasts

method	Description
noadjust	do not adjust for multiple comparisons; the default
bonferroni [adjustall]	Bonferroni's method; adjust across all terms
sidak [adjustall]	Šidák's method; adjust across all terms
scheffe	Scheffé's method
*tukey	Tukey's method
*snk	Student–Newman–Keuls' method
*duncan	Duncan's method
*dunnett	Dunnett's method

* tukey, snk, duncan, and dunnett are only allowed with results from anova, manova, regress, and mvreg. tukey, snk, duncan, and dunnett are not allowed with results from svy.

Time-series operators are allowed if they were used in the estimation.

Menu

Statistics > Postestimation > Pairwise comparisons

Description

pwcompare performs pairwise comparisons across the levels of factor variables from the most recently fit model. pwcompare can compare estimated cell means, marginal means, intercepts, marginal intercepts, slopes, or marginal slopes—collectively called margins. pwcompare reports the comparisons as contrasts (differences) of margins along with significance tests or confidence intervals for the contrasts. The tests and confidence intervals can be adjusted for multiple comparisons.

pwcompare can be used with svy estimation results; see [SVY] **svy postestimation**.

See [R] **margins, pwcompare** for performing pairwise comparisons of margins of linear and nonlinear predictions.

Options

Main

mcompare(*method*) specifies the method for computing p-values and confidence intervals that account for multiple comparisons within a factor-variable term.

Most methods adjust the comparisonwise error rate, α_c, to achieve a prespecified experimentwise error rate, α_e.

mcompare(noadjust) is the default; it specifies no adjustment.

$$\alpha_c = \alpha_e$$

mcompare(bonferroni) adjusts the comparisonwise error rate based on the upper limit of the Bonferroni inequality:

$$\alpha_e \leq m\alpha_c$$

where *m* is the number of comparisons within the term.

The adjusted comparisonwise error rate is

$$\alpha_c = \alpha_e/m$$

`mcompare(sidak)` adjusts the comparisonwise error rate based on the upper limit of the probability inequality

$$\alpha_e \leq 1 - (1 - \alpha_c)^m$$

where *m* is the number of comparisons within the term.

The adjusted comparisonwise error rate is

$$\alpha_c = 1 - (1 - \alpha_e)^{1/m}$$

This adjustment is exact when the *m* comparisons are independent.

`mcompare(scheffe)` controls the experimentwise error rate using the F (or χ^2) distribution with degrees of freedom equal to the rank of the term.

For results from `anova`, `regress`, `manova`, and `mvreg` (see [R] **anova**, [R] **regress**, [MV] **manova**, and [R] **mvreg**), `pwcompare` allows the following additional methods. These methods are not allowed with results that used `vce(robust)` or `vce(cluster` *clustvar*`)`.

`mcompare(tukey)` uses what is commonly referred to as Tukey's honestly significant difference. This method uses the Studentized range distribution instead of the t distribution.

`mcompare(snk)` is a variation on `mcompare(tukey)` that counts only the number of margins in the range for a given comparison instead of the full number of margins.

`mcompare(duncan)` is a variation on `mcompare(snk)` with additional adjustment to the significance probabilities.

`mcompare(dunnett)` uses Dunnett's method for making comparisons with a reference category.

`mcompare(`*method* `adjustall)` specifies that the multiple-comparison adjustments count all comparisons across all terms rather than performing multiple comparisons term by term. This leads to more conservative adjustments when multiple variables or terms are specified in *marginlist*. This option is compatible only with the `bonferroni` and `sidak` methods.

`asobserved` specifies that factor covariates be evaluated using the cell frequencies observed when the model was fit. The default is to treat all factor covariates as though there were an equal number of observations at each level.

Equations

`equation(`*eqspec*`)` specifies the equation from which margins are to be computed. The default is to compute margins from the first equation.

`atequations` specifies that the margins be computed within each equation.

Advanced

`emptycells(`*empspec*`)` specifies how empty cells are handled in interactions involving factor variables that are being treated as balanced.

`emptycells(strict)` is the default; it specifies that margins involving empty cells be treated as not estimable.

`emptycells(reweight)` specifies that the effects of the observed cells be increased to accommodate any missing cells. This makes the margins estimable but changes their interpretation.

`noestimcheck` specifies that `pwcompare` not check for estimability. By default, the requested margins are checked and those found not estimable are reported as such. Nonestimability is usually caused by empty cells. If `noestimcheck` is specified, estimates are computed in the usual way and reported even though the resulting estimates are manipulable, which is to say they can differ across equivalent models having different parameterizations.

Reporting

`level(#)` specifies the confidence level, as a percentage, for confidence intervals. The default is `level(95)` or as set by `set level`; see [U] **20.7 Specifying the width of confidence intervals**. The significance level used by the `groups` option is $100 - \#$, expressed as a percentage.

`cieffects` specifies that a table of the pairwise comparisons with their standard errors and confidence intervals be reported. This is the default.

`pveffects` specifies that a table of the pairwise comparisons with their standard errors, test statistics, and p-values be reported.

`effects` specifies that a table of the pairwise comparisons with their standard errors, test statistics, p-values, and confidence intervals be reported.

`cimargins` specifies that a table of the margins with their standard errors and confidence intervals be reported.

`groups` specifies that a table of the margins with their standard errors and group codes be reported. Margins with the same letter in the group code are not significantly different at the specified significance level.

`sort` specifies that the reported tables be sorted on the margins or differences in each term.

`post` causes `pwcompare` to behave like a Stata estimation (e-class) command. `pwcompare` posts the vector of estimated margins along with the estimated variance–covariance matrix to `e()`, so you can treat the estimated margins just as you would results from any other estimation command. For example, you could use `test` to perform simultaneous tests of hypotheses on the margins, or you could use `lincom` to create linear combinations.

display_options: `vsquish`, `cformat(%`*fmt*`)`, `pformat(%`*fmt*`)`, `sformat(%`*fmt*`)`, and `nolstretch`.

`vsquish` specifies that the blank space separating factor-variable terms or time-series–operated variables from other variables in the model be suppressed.

`cformat(%`*fmt*`)` specifies how to format contrasts or margins, standard errors, and confidence limits in the table of pairwise comparisons.

`pformat(%`*fmt*`)` specifies how to format p-values in the table of pairwise comparisons.

`sformat(%`*fmt*`)` specifies how to format test statistics in the table of pairwise comparisons.

`nolstretch` specifies that the width of the table of pairwise comparisons not be automatically widened to accommodate longer variable names. The default, `lstretch`, is to automatically widen the table of pairwise comparisons up to the width of the Results window. To change the default, use `set lstretch off`. `nolstretch` is not shown in the dialog box.

eform_option specifies that the contrasts table be displayed in exponentiated form. e^{contrast} is displayed rather than contrast. Standard errors and confidence intervals are also transformed. See [R] ***eform_option*** for the list of available options.

Remarks

`pwcompare` performs pairwise comparisons of margins across the levels of factor variables from the most recently fit model. The margins can be estimated cell means, marginal means, intercepts, marginal intercepts, slopes, or marginal slopes. With the exception of slopes, we can also consider these margins to be marginal linear predictions.

The margins are calculated as linear combinations of the coefficients. Let k be the number of levels for a factor term in our model; then there are k margins for that term, and

$$m = \binom{k}{2} = \frac{k(k-1)}{2}$$

unique pairwise comparisons of those margins.

The confidence intervals and p-values for these pairwise comparisons can be adjusted to account for multiple comparisons. Bonferroni's, Šidák's, and Scheffé's adjustments can be made for multiple comparisons after fitting any type of model. In addition, Tukey's, Student–Newman–Keuls', Duncan's, and Dunnett's adjustments are available when fitting ANOVA, linear regression, MANOVA, or multivariate regression models.

Remarks are presented under the following headings:

Pairwise comparisons of means

Suppose we are interested in the effects of five different fertilizers on wheat yield. We could estimate the following linear regression model to determine the effect of each type of fertilizer on the yield.

```
. use http://www.stata-press.com/data/r12/yield
(Artificial wheat yield dataset)

. regress yield i.fertilizer

      Source |       SS       df       MS              Number of obs =     200
-------------+------------------------------           F(  4,   195) =    5.33
       Model |  1078.84207     4  269.710517           Prob > F      =  0.0004
    Residual |  9859.55334   195   50.561812           R-squared     =  0.0986
-------------+------------------------------           Adj R-squared =  0.0801
       Total |  10938.3954   199  54.9668111           Root MSE      =  7.1107

------------------------------------------------------------------------------
       yield |      Coef.   Std. Err.      t    P>|t|     [95% Conf. Interval]
-------------+----------------------------------------------------------------
  fertilizer |
          2  |    3.62272   1.589997     2.28   0.024      .4869212    6.758518
          3  |   .4906299   1.589997     0.31   0.758     -2.645169    3.626428
          4  |   4.922803   1.589997     3.10   0.002      1.787005    8.058602
          5  |  -1.238328   1.589997    -0.78   0.437     -4.374127     1.89747
             |
       _cons |   41.36243   1.124298    36.79   0.000      39.14509    43.57977
------------------------------------------------------------------------------
```

In this simple case, the coefficients for fertilizers 2 through 5 indicate the difference in the mean yield for that fertilizer versus the mean yield for fertilizer 1. That the standard errors of all four coefficients are identical results from having perfectly balanced data.

Marginal means

We can use `pwcompare` with the `cimargins` option to compute the mean yield for each of the fertilizers.

```
. pwcompare fertilizer, cimargins

Pairwise comparisons of marginal linear predictions

Margins      : asbalanced

---------------------------------------------------------------
             |                              Unadjusted
             |     Margin   Std. Err.     [95% Conf. Interval]
-------------+-------------------------------------------------
  fertilizer |
          1  |   41.36243   1.124298      39.14509    43.57977
          2  |   44.98515   1.124298       42.7678    47.20249
          3  |   41.85306   1.124298      39.63571     44.0704
          4  |   46.28523   1.124298      44.06789    48.50258
          5  |    40.1241   1.124298      37.90676    42.34145
---------------------------------------------------------------
```

Looking at the confidence intervals for fertilizers 1 and 2 in the table above, we might be tempted to conclude that these means are not significantly different because the intervals overlap. However, as discussed in *Interaction plots* of [R] **marginsplot**, we cannot draw conclusions about the differences in means by looking at confidence intervals for the means themselves. Instead, we would need to look at confidence intervals for the difference in means.

All pairwise comparisons

By default, pwcompare calculates all pairwise differences of the margins, in this case pairwise differences of the mean yields.

```
. pwcompare fertilizer
Pairwise comparisons of marginal linear predictions
Margins      : asbalanced
```

	Contrast	Std. Err.	Unadjusted [95% Conf.	Interval]
fertilizer				
2 vs 1	3.62272	1.589997	.4869212	6.758518
3 vs 1	.4906299	1.589997	-2.645169	3.626428
4 vs 1	4.922803	1.589997	1.787005	8.058602
5 vs 1	-1.238328	1.589997	-4.374127	1.89747
3 vs 2	-3.13209	1.589997	-6.267889	.0037086
4 vs 2	1.300083	1.589997	-1.835715	4.435882
5 vs 2	-4.861048	1.589997	-7.996847	-1.725249
4 vs 3	4.432173	1.589997	1.296375	7.567972
5 vs 3	-1.728958	1.589997	-4.864757	1.406841
5 vs 4	-6.161132	1.589997	-9.29693	-3.025333

If a confidence interval does not include zero, the means for the compared fertilizers are significantly different. Therefore, at the 5% significance level, we would reject the hypothesis that the means for fertilizers 1 and 2 are equivalent—as we would do for 4 vs 1, 5 vs 2, 4 vs 3, and 5 vs 4.

We may prefer to see the p-values instead of looking at confidence intervals to determine whether the pairwise differences are significantly different from zero. We could use the pveffects option to see the differences with standard errors and p-values, or we could use the effects option to see both p-values and confidence intervals in the same table. Here we specify effects as well as the sort option so that the differences are sorted from smallest to largest.

```
. pwcompare fertilizer, effects sort
Pairwise comparisons of marginal linear predictions
Margins      : asbalanced
```

	Contrast	Std. Err.	Unadjusted t	P>|t|	Unadjusted [95% Conf.	Interval]
fertilizer						
5 vs 4	-6.161132	1.589997	-3.87	0.000	-9.29693	-3.025333
5 vs 2	-4.861048	1.589997	-3.06	0.003	-7.996847	-1.725249
3 vs 2	-3.13209	1.589997	-1.97	0.050	-6.267889	.0037086
5 vs 3	-1.728958	1.589997	-1.09	0.278	-4.864757	1.406841
5 vs 1	-1.238328	1.589997	-0.78	0.437	-4.374127	1.89747
3 vs 1	.4906299	1.589997	0.31	0.758	-2.645169	3.626428
4 vs 2	1.300083	1.589997	0.82	0.415	-1.835715	4.435882
2 vs 1	3.62272	1.589997	2.28	0.024	.4869212	6.758518
4 vs 3	4.432173	1.589997	2.79	0.006	1.296375	7.567972
4 vs 1	4.922803	1.589997	3.10	0.002	1.787005	8.058602

We find that 5 of the 10 pairs of means are significantly different at the 5% significance level.

We can use the groups option to obtain a table that identifies groups whose means are not significantly different by assigning them the same letter.

```
. pwcompare fertilizer, groups sort
Pairwise comparisons of marginal linear predictions
Margins      : asbalanced

                                                Unadjusted
                     Margin   Std. Err.            Groups

  fertilizer
           5        40.1241   1.124298                 A
           1       41.36243   1.124298                 A
           3       41.85306   1.124298                AB
           2       44.98515   1.124298                 BC
           4       46.28523   1.124298                  C

Note: Margins sharing a letter in the group label
      are not significantly different at the 5%
      level.
```

The letter A that is assigned to fertilizers 5, 1, and 3 designates that the mean yields for these fertilizers are not different at the 5% level.

Overview of multiple-comparison methods

For a single test, if we choose a 5% significance level, we would have a 5% chance of concluding that two margins are different when the population values are actually equal. This is known as making a type I error. When we perform $m = k(k-1)/2$ pairwise comparisons of the k margins, we have m opportunities to make a type I error.

pwcompare with the mcompare() option allows us to adjust the confidence intervals and p-values for each comparison to account for the increased probability of making a type I error when making multiple comparisons. Bonferroni's adjustment, Šidák's adjustment, and Scheffé's adjustment can be used when making pairwise comparisons of the margins after any estimation command. Tukey's honestly significant difference, Student–Newman–Keuls' method, Duncan's method, and Dunnett's method are only available when fitting linear models after anova, manova, regress, or mvreg.

Fisher's protected least-significant difference (LSD)

pwcompare does not offer an mcompare() option specifically for Fisher's protected least-significant difference (LSD). In this methodology, no adjustment is made to the confidence intervals or p-values. However, it is protected in the sense that no pairwise comparisons are tested unless the joint test for the corresponding term in the model is significant. Therefore, the default mcompare(noadjust) corresponds to Fisher's protected LSD assuming that the corresponding joint test was performed before using pwcompare.

Milliken and Johnson (2009) recommend using this methodology for planned comparisons, assuming the corresponding joint test is significant.

Bonferroni's adjustment

mcompare(bonferroni) adjusts significance levels based on the Bonferroni inequality, which, in the case of multiple testing, tells us that the maximum error rate for all comparisons is the sum of the error rates for the individual comparisons. Assuming that we are using the same significance level for all tests, the experimentwise error rate is the error rate for a single test multiplied by the number of comparisons. Therefore, a p-value for each comparison can be computed by multiplying the unadjusted p-value by the total number of comparisons. If the adjusted p-value is greater than 1, then `pwcompare` will report a p-value of 1.

Bonferroni's adjustment is popular because it is easy to compute manually and because it can be applied to any set of tests, not only the pairwise comparisons available in `pwcompare`. In addition, this method does not require equal sample sizes.

Because Bonferroni's adjustment is so general, it is more conservative than many of the other adjustments. It is especially conservative when a large number of tests is being performed.

Šidák's adjustment

`mcompare(sidak)` performs an adjustment using Šidák's method. This adjustment, like Bonferroni's adjustment, is derived from an inequality. However, in this case, the inequality is based on the probability of not making a type I error. For a single test, the probability that we do not make a type I error is $1-\alpha$. For two independent tests, both using α as a significance level, the probability is $(1-\alpha)(1-\alpha)$. Likewise, for m independent tests, the probability of not making a type I error is $(1-\alpha)^m$. Therefore, the probability of making one or more type I errors is $1-(1-\alpha)^m$. When tests are not independent, the probability of making at least one error is less than $1-(1-\alpha)^m$. Therefore, we can compute an adjusted p-value as $1-(1-{}_up)^m$, where ${}_up$ is the unadjusted p-value for a single comparison.

Šidák's method is also conservative although slightly less so than Bonferroni's method. Like Bonferroni's method, this method does not require equal sample sizes.

Scheffé's adjustment

Scheffé's adjustment is used when `mcompare(scheffe)` is specified. This adjustment is derived from the joint F test and its correspondence to the maximum normalized comparison. To adjust for multiple comparisons, the absolute value of the t statistic for a particular comparison can be compared with a critical value of $\sqrt{(k-1)F_{k-1,\nu}}$, where ν is the residual degrees of freedom. $F_{k-1,\nu}$ is the distribution of the joint F test for the corresponding term in a one-way ANOVA model. Winer, Brown, and Michels (1991, 191–195) discuss this in detail. For estimation commands that report z statistics instead of t statistics for the tests on coefficients, a χ^2 distribution is used instead of an F distribution.

Scheffé's method allows for making all possible comparisons of the k margins, not just the pairwise comparisons. Unlike the methods described above, it does not take into account the number of comparisons that are currently being made. Therefore, this method is even more conservative than the others. Because this method adjusts for all possible comparisons of the levels of the term, Milliken and Johnson (2009) recommend using this procedure when making unplanned contrasts that are suggested by the data. As Winer, Brown, and Michels (1991, 191) put it, this method is often used to adjust for "unfettered data snooping". When using this adjustment, a contrast will never be significant if the joint F or χ^2 test for the term is not also significant.

This is another method that does not require equal sample sizes.

Tukey's HSD adjustment

Tukey's adjustment is also referred to as Tukey's honestly significant difference (HSD) and is used when `mcompare(tukey)` is specified. It is often applied to all pairwise comparisons of means. Tukey's HSD is commonly used as a post hoc test although this is not a requirement.

To adjust for multiple comparisons, Tukey's method compares the absolute value of the t statistic from the individual comparison with a critical value based on a Studentized range distribution with parameter equal to the number of levels in the term. When applied to pairwise comparisons of means,

$$q = \frac{\text{mean}_{\text{max}} - \text{mean}_{\text{min}}}{s}$$

follows a Studentized range distribution with parameter k and ν degrees of freedom. Here mean_{max} and mean_{min} are the largest and smallest marginal means, and s is an estimate of the standard error of the means.

Now for the comparison of the smallest and largest means, we can say that the probability of not making a type I error is

$$\Pr\left(\frac{\text{mean}_{\text{max}} - \text{mean}_{\text{min}}}{s} \leq q_{k,\nu}\right) = 1 - \alpha$$

Then the following inquality holds for all pairs of means simultaneously:

$$\Pr\left(\frac{|\text{mean}_i - \text{mean}_j|}{s} \leq q_{k,\nu}\right) \geq 1 - \alpha$$

Based on this procedure, Tukey's HSD computes the p-value for each of the individual comparisons using the Studentized range distribution. However, because the equality holds only for the difference in the largest and smallest means, this procedure produces conservative tests for the remaining comparisons. Winer, Brown, and Michels (1991, 172–182) discuss this in further detail.

Tukey's HSD requires equal sample sizes.

Student–Newman–Keuls' adjustment

The Student–Newman–Keuls (SNK) method is used when `mcompare(snk)` is specified. It is a modification to Tukey's method and is less conservative. In this procedure, we first order the means. We then test the difference in the smallest and largest means using a critical value from the Studentized range distribution with parameter k, where k is the number of levels in the term. This step uses the same methodology as in Tukey's procedure. However, in the next step, we will then test for differences in the two sets of means that are the endpoints of the two ranges including $k - 1$ means. Specifically, we test the difference in the smallest mean and the second-largest mean using a critical value from the Studentized range distribution with parameter $k - 1$. We would also test the difference in the second-smallest mean and the largest mean using this critical value. Likewise, the means that are the endpoints of ranges including $k - 2$ means when ordered are tested using the Studentized range distribution with parameter $k - 2$, and so on.

As with Tukey's method, equal sample sizes are required.

Duncan's adjustment

When mcompare(duncan) is specified, tests are adjusted for multiple comparisons using Duncan's method, which is sometimes referred to as Duncan's new multiple range method. This adjustment produces tests that are less conservative than both Tukey's HSD and SNK. This procedure is performed in the same manner as SNK except that the p-values for the individual comparisons are adjusted as $1-(1-{}_{\rm snk}p_i)^{1/(r+1)}$, where ${}_{\rm snk}p$ is the p-value computed using the SNK method and r represents the number of means that, when ordered, fall between the two that are being compared.

Again equal sample sizes are required for this adjustment.

Dunnett's adjustment

Dunnett's adjustment is obtained by specifying mcompare(dunnett). It is used when one of the levels of a factor can be considered a control or reference level with which each of the other levels is being compared. When Dunnett's adjustment is requested, $k-1$ instead of $k(k-1)/2$ pairwise comparisons are made. Dunnett (1955, 1964) developed tables of critical values for what Miller (1981, 76) refers to as the "many-one t statistic". The t statistics for individual comparisons are compared with these critical values when making many comparisons to a single reference level.

This method also requires equal sample sizes.

Example adjustments using one-way models

Fisher's protected LSD

Fisher's protected LSD requires that we first verify that the joint test for a term in our model is significant before proceeding with pairwise comparisons. Using our previous example, we could have first used the contrast command to obtain a joint test for the effects of fertilizer.

```
. contrast fertilizer
Contrasts of marginal linear predictions
Margins      : asbalanced
```

	df	F	P>F
fertilizer	4	5.33	0.0004
Residual	195		

This test for the effects of fertilizer is highly significant. Now we can say we are using Fisher's protected LSD when looking at the unadjusted p-values that were obtained from our previous command,

```
. pwcompare fertilizer, effects sort
```

Tukey's HSD

Because we fit a linear regression model and are interested in all pairwise comparisons of the marginal means, we may instead choose to use Tukey's HSD.

```
. pwcompare fertilizer, effects sort mcompare(tukey)
Pairwise comparisons of marginal linear predictions
Margins      : asbalanced
```

	Number of Comparisons
fertilizer	10

	Contrast	Std. Err.	Tukey t	Tukey P>|t|	Tukey [95% Conf.	Interval]
fertilizer						
5 vs 4	-6.161132	1.589997	-3.87	0.001	-10.53914	-1.78312
5 vs 2	-4.861048	1.589997	-3.06	0.021	-9.239059	-.4830368
3 vs 2	-3.13209	1.589997	-1.97	0.285	-7.510101	1.245921
5 vs 3	-1.728958	1.589997	-1.09	0.813	-6.106969	2.649053
5 vs 1	-1.238328	1.589997	-0.78	0.936	-5.616339	3.139683
3 vs 1	.4906299	1.589997	0.31	0.998	-3.887381	4.868641
4 vs 2	1.300083	1.589997	0.82	0.925	-3.077928	5.678095
2 vs 1	3.62272	1.589997	2.28	0.156	-.7552913	8.000731
4 vs 3	4.432173	1.589997	2.79	0.046	.0541623	8.810185
4 vs 1	4.922803	1.589997	3.10	0.019	.5447922	9.300815

This time, our p-values have been modified, and we find that only four of the pairwise differences are considered significantly different from zero at the 5% level.

If we only are interested in performing pairwise comparisons of a subset of our means, we can use factor-variable operators to select the levels of the factor that we want to compare. Here we exclude all comparisons involving fertilizer 1.

```
. pwcompare i(2/5).fertilizer, effects sort mcompare(tukey)
Pairwise comparisons of marginal linear predictions
Margins      : asbalanced
```

	Number of Comparisons
fertilizer	6

	Contrast	Std. Err.	Tukey t	Tukey P>|t|	Tukey [95% Conf.	Interval]
fertilizer						
5 vs 4	-6.161132	1.589997	-3.87	0.001	-10.28133	-2.040937
5 vs 2	-4.861048	1.589997	-3.06	0.013	-8.981242	-.7408538
3 vs 2	-3.13209	1.589997	-1.97	0.203	-7.252284	.9881042
5 vs 3	-1.728958	1.589997	-1.09	0.698	-5.849152	2.391236
4 vs 2	1.300083	1.589997	0.82	0.846	-2.820111	5.420278
4 vs 3	4.432173	1.589997	2.79	0.030	.3119792	8.552368

The adjusted p-values and confidence intervals differ from those in the previous output because Tukey's adjustment takes into account the total number of comparisons being made when determining the appropriate degrees of freedom to use for the Studentized range distribution.

Dunnett's method for comparisons to a control

If one of our five fertilizer groups represents fields where no fertilizer was applied, we may want to use Dunnett's method to compare each of the four fertilizers with the control group. In this case, we make only $k - 1$ comparisons for k groups.

```
. pwcompare fertilizer, effects mcompare(dunnett)
Pairwise comparisons of marginal linear predictions
Margins      : asbalanced
```

	Number of Comparisons
fertilizer	4

	Contrast	Std. Err.	Dunnett t	Dunnett P>|t|	Dunnett [95% Conf.	Interval]
fertilizer						
2 vs 1	3.62272	1.589997	2.28	0.079	-.2918331	7.537273
3 vs 1	.4906299	1.589997	0.31	0.994	-3.423923	4.405183
4 vs 1	4.922803	1.589997	3.10	0.008	1.00825	8.837356
5 vs 1	-1.238328	1.589997	-0.78	0.852	-5.152881	2.676225

In our previous `regress` command, fertilizer 1 was treated as the base. Therefore, by default, it was treated as the control when using Dunnett's adjustment, and the pairwise comparisons are equivalent to the coefficients reported by `regress`. Based on our `regress` output, we would conclude that fertilizers 2 and 4 are different from fertilizer 1 at the 5% level. However, using Dunnett's adjustment, we find only fertilizer 4 to be different from fertilizer 1 at this same significance level.

If the model is fit without a base level for a factor variable, then `pwcompare` will choose the first level as the reference level. If we want to make comparisons with a different level than the one `mcompare(dunnett)` chooses by default, we can use the `b.` operator to override the default. Here we use fertilizer 5 as the reference level.

```
. pwcompare b5.fertilizer, effects sort mcompare(dunnett)
Pairwise comparisons of marginal linear predictions
Margins      : asbalanced
```

	Number of Comparisons
fertilizer	4

	Contrast	Std. Err.	Dunnett t	Dunnett P>|t|	Dunnett [95% Conf.	Interval]
fertilizer						
1 vs 5	1.238328	1.589997	0.78	0.852	-2.676225	5.152881
3 vs 5	1.728958	1.589997	1.09	0.649	-2.185595	5.643511
2 vs 5	4.861048	1.589997	3.06	0.009	.9464951	8.775601
4 vs 5	6.161132	1.589997	3.87	0.001	2.246579	10.07568

Two-way models

In the previous examples, we have performed pairwise comparisons after fitting a model with a single factor. Now we include two factors and their interaction in our model.

```
. regress yield fertilizer##irrigation
```

Source	SS	df	MS
Model	6200.81605	9	688.979561
Residual	4737.57936	190	24.9346282
Total	10938.3954	199	54.9668111

Number of obs	200
F(9, 190)	27.63
Prob > F	0.0000
R-squared	0.5669
Adj R-squared	0.5464
Root MSE	4.9935

yield	Coef.	Std. Err.	t	P>|t|	[95% Conf.	Interval]
fertilizer						
2	1.882256	1.57907	1.19	0.235	-1.232505	4.997016
3	-.5687418	1.57907	-0.36	0.719	-3.683502	2.546019
4	4.904999	1.57907	3.11	0.002	1.790239	8.01976
5	-1.217496	1.57907	-0.77	0.442	-4.332257	1.897264
1.irrigation	8.899721	1.57907	5.64	0.000	5.784961	12.01448
fertilizer# irrigation						
2 1	3.480928	2.233143	1.56	0.121	-.9240084	7.885865
3 1	2.118743	2.233143	0.95	0.344	-2.286193	6.52368
4 1	.0356082	2.233143	0.02	0.987	-4.369328	4.440545
5 1	-.0416636	2.233143	-0.02	0.985	-4.4466	4.363273
_cons	36.91257	1.116571	33.06	0.000	34.7101	39.11504

We can perform pairwise comparisons of the cell means defined by the fertilizer and irrigation interaction.

```
. pwcompare fertilizer#irrigation, sort groups mcompare(tukey)
Pairwise comparisons of marginal linear predictions
Margins      : asbalanced
```

	Number of Comparisons
fertilizer#irrigation	45

	Margin	Std. Err.	Tukey Groups
fertilizer#irrigation			
5 0	35.69507	1.116571	A
3 0	36.34383	1.116571	A
1 0	36.91257	1.116571	AB
2 0	38.79482	1.116571	AB
4 0	41.81757	1.116571	BC
5 1	44.55313	1.116571	CD
1 1	45.81229	1.116571	CDE
3 1	47.36229	1.116571	DEF
4 1	50.7529	1.116571	EF
2 1	51.17547	1.116571	F

```
Note: Margins sharing a letter in the group label are
      not significantly different at the 5% level.
```

Based on Tukey's HSD and a 5% significance level, we would conclude that the mean yield for fertilizer 5 without irrigation is not significantly different from the mean yields for fertilizers 1, 2, and 3 when used without irrigation but is significantly different from the remaining means.

Up to this point, most of the pairwise comparisons that we have performed could have also been obtained with `pwmean` (see [R] **pwmean**) if we had not been interested in examining the results from the estimation command before making pairwise comparisons of the means. For instance, we could reproduce the results from the above `pwcompare` command by typing

```
. pwmean yield, over(fertilizer irrigation) sort group mcompare(tukey)
```

However, `pwcompare` extends the capabilities of `pwmean` in many ways. For instance, `pwmean` only allows for pairwise comparisons of the cell means determined by the highest level interaction of the variables specified in the `over()` option. However, `pwcompare` allows us to fit a single model, such as the two-way model that we fit above,

```
. regress yield fertilizer##irrigation
```

and compute pairwise comparisons of the marginal means for only one of the variables in the model:

```
. pwcompare fertilizer, sort effects mcompare(tukey)
Pairwise comparisons of marginal linear predictions
Margins      : asbalanced
```

	Number of Comparisons
fertilizer	10

	Contrast	Std. Err.	Tukey t	Tukey P>|t|	Tukey [95% Conf.	Interval]
fertilizer						
5 vs 4	-6.161132	1.116571	-5.52	0.000	-9.236338	-3.085925
5 vs 2	-4.861048	1.116571	-4.35	0.000	-7.936255	-1.785841
3 vs 2	-3.13209	1.116571	-2.81	0.044	-6.207297	-.0568832
5 vs 3	-1.728958	1.116571	-1.55	0.532	-4.804165	1.346249
5 vs 1	-1.238328	1.116571	-1.11	0.802	-4.313535	1.836879
3 vs 1	.4906299	1.116571	0.44	0.992	-2.584577	3.565837
4 vs 2	1.300083	1.116571	1.16	0.772	-1.775123	4.37529
2 vs 1	3.62272	1.116571	3.24	0.012	.5475131	6.697927
4 vs 3	4.432173	1.116571	3.97	0.001	1.356967	7.50738
4 vs 1	4.922803	1.116571	4.41	0.000	1.847597	7.99801

Here the standard errors for the differences in marginal means and the residual degrees of freedom are based on the full model. Therefore, the results will differ from those obtained from pwcompare after fitting the one-way model with only fertilizer (or equivalently using pwmean).

Pairwise comparisons of slopes

If we fit a model with a factor variable that is interacted with a continuous variable, pwcompare will even allow us to make pairwise comparisons of the slopes of the continuous variable for the levels of the factor variable.

In this case, we have a continuous variable, NO3_N, indicating the amount of nitrate nitrogen already existing in the soil, based on a sample taken from each field.

```
. regress yield fertilizer##c.NO3_N

      Source |       SS       df       MS              Number of obs =     200
-------------+------------------------------           F(  9,   190) =   37.61
       Model |  7005.69932     9  778.411035           Prob > F      =  0.0000
    Residual |  3932.69609   190  20.6984005           R-squared     =  0.6405
-------------+------------------------------           Adj R-squared =  0.6234
       Total |  10938.3954   199  54.9668111           Root MSE      =  4.5495

------------------------------------------------------------------------------
       yield |      Coef.   Std. Err.      t    P>|t|     [95% Conf. Interval]
-------------+----------------------------------------------------------------
  fertilizer |
          2  |   18.65019   8.452061     2.21   0.029      1.97826    35.32212
          3  |  -13.34076   10.07595    -1.32   0.187    -33.21585    6.534327
          4  |   24.35061   9.911463     2.46   0.015     4.799973    43.90125
          5  |   17.58529   8.446736     2.08   0.039     .9238646    34.24671
             |
       NO3_N |   4.915653   .7983509     6.16   0.000     3.340884    6.490423
             |
 fertilizer# |
     c.NO3_N |
          2  |  -1.282039   .8953419    -1.43   0.154    -3.048126    .4840487
          3  |   -1.00571   .9025862    -1.11   0.267    -2.786087    .7746662
          4  |   -2.97627   .9136338    -3.26   0.001    -4.778438   -1.174102
          5  |  -3.275947   .8247385    -3.97   0.000    -4.902767   -1.649127
             |
       _cons |  -5.459168   7.638241    -0.71   0.476    -20.52581    9.607477
------------------------------------------------------------------------------
```

These are the pairwise differences of the slopes of NO3_N for each pair of fertilizers:

```
. pwcompare fertilizer#c.NO3_N, pveffects sort mcompare(scheffe)

Pairwise comparisons of marginal linear predictions

Margins      : asbalanced

---------------------------
                 |    Number of
                 |  Comparisons
-----------------+---------------
fertilizer#c.NO3_N |         10
---------------------------

-------------------------------------------------------
                 |                           Scheffe
                 |   Contrast   Std. Err.      t    P>|t|
-----------------+-------------------------------------
fertilizer#c.NO3_N |
         5 vs 1  |  -3.275947   .8247385    -3.97   0.004
         4 vs 1  |   -2.97627   .9136338    -3.26   0.034
         5 vs 3  |  -2.270237   .4691771    -4.84   0.000
         5 vs 2  |  -1.993909   .4550851    -4.38   0.001
         4 vs 3  |   -1.97056    .612095    -3.22   0.038
         4 vs 2  |  -1.694232   .6013615    -2.82   0.099
         2 vs 1  |  -1.282039   .8953419    -1.43   0.727
         3 vs 1  |   -1.00571   .9025862    -1.11   0.871
         5 vs 4  |  -.2996772   .4900939    -0.61   0.984
         3 vs 2  |    .276328   .5844405     0.47   0.994
-------------------------------------------------------
```

Using Scheffé's adjustment, we find that five of the pairs have significantly different slopes at the 5% level.

Nonlinear models

pwcompare can also perform pairwise comparisons of the marginal linear predictions after fitting a nonlinear model. For instance, we can use the dataset from *Beyond linear models* in [R] **contrast** and fit the following logistic regression model of patient satisfaction on hospital:

```
. use http://www.stata-press.com/data/r12/hospital
(Artificial hospital satisfaction data)

. logit satisfied i.hospital

Iteration 0:   log likelihood = -393.72216
Iteration 1:   log likelihood = -387.55736
Iteration 2:   log likelihood =  -387.4768
Iteration 3:   log likelihood = -387.47679

Logistic regression                               Number of obs   =        802
                                                  LR chi2(2)      =      12.49
                                                  Prob > chi2     =     0.0019
Log likelihood = -387.47679                       Pseudo R2       =     0.0159

------------------------------------------------------------------------------
   satisfied |      Coef.   Std. Err.      z    P>|z|     [95% Conf. Interval]
-------------+----------------------------------------------------------------
    hospital |
          2  |   .5348129   .2136021     2.50   0.012     .1161604    .9534654
          3  |   .7354519   .2221929     3.31   0.001     .2999618    1.170942
             |
       _cons |   1.034708   .1391469     7.44   0.000     .7619855    1.307431
------------------------------------------------------------------------------
```

For this model, the marginal linear predictions are the predicted log odds for each hospital and can be obtained with the cimargins option:

```
. pwcompare hospital, cimargins

Pairwise comparisons of marginal linear predictions

Margins      : asbalanced

--------------------------------------------------------------
             |                              Unadjusted
             |     Margin   Std. Err.     [95% Conf. Interval]
-------------+------------------------------------------------
    hospital |
          1  |   1.034708   .1391469      .7619855    1.307431
          2  |   1.569521   .1620618      1.251886    1.887157
          3  |    1.77016   .1732277       1.43064     2.10968
--------------------------------------------------------------
```

The pairwise comparisons are, therefore, differences in the log odds. We can specify mcompare(bonferroni) and effects to request Bonferroni-adjusted p-values and confidence intervals.

```
. pwcompare hospital, effects mcompare(bonferroni)

Pairwise comparisons of marginal linear predictions

Margins      : asbalanced

---------------------------
             |    Number of
             |  Comparisons
-------------+-------------
satisfied    |
    hospital |            3
---------------------------
```

	Contrast	Std. Err.	Bonferroni z	Bonferroni P>|z|	Bonferroni [95% Conf.	Interval]
satisfied						
hospital						
2 vs 1	.5348129	.2136021	2.50	0.037	.0234537	1.046172
3 vs 1	.7354519	.2221929	3.31	0.003	.2035265	1.267377
3 vs 2	.200639	.2372169	0.85	1.000	-.3672535	.7685314

For nonlinear models, only Bonferroni's adjustment, Šidák's adjustment, and Scheffé's adjustment are available.

If we want pairwise comparisons reported as odds ratios, we can specify the `or` option.

```
. pwcompare hospital, effects mcompare(bonferroni) or
Pairwise comparisons of marginal linear predictions
Margins      : asbalanced
```

	Number of Comparisons
satisfied	
hospital	3

	Odds Ratio	Std. Err.	Bonferroni z	Bonferroni P>|z|	Bonferroni [95% Conf.	Interval]
satisfied						
hospital						
2 vs 1	1.707129	.3646464	2.50	0.037	1.023731	2.846733
3 vs 1	2.086425	.4635888	3.31	0.003	1.225718	3.551525
3 vs 2	1.222183	.2899226	0.85	1.000	.6926341	2.156597

Notice that these tests are still performed on the marginal linear predictions. The odds ratios reported here are the exponentiated versions of the pairwise differences of log odds in the previous output. For further discussion, see [R] **contrast**.

Multiple-equation models

`pwcompare` works with models containing multiple equations. Commands such as `intreg` and `gnbreg` allow their ancillary parameters to be modeled as a function of independent variables, and `pwcompare` can compare the margins within these equations. The `equation()` option can be used to specify the equation for which pairwise comparisons of the margins should be made. The `atequations` option specifies that pairwise comparisons be computed for each equation. In addition, `pwcompare` allows a special pseudofactor for equation—called `_eqns`—when working with results from `manova`, `mvreg`, `mlogit`, and `mprobit`.

Here we use the jaw fracture dataset described in example 4 of [MV] **manova**. We fit a multivariate regression model including one independent factor variable, `fracture`.

```
. use http://www.stata-press.com/data/r12/jaw
(Table 4.6 Two-Way Unbalanced Data for Fractures of the Jaw -- Rencher (1998))
. mvreg y1 y2 y3 = i.fracture
```

Equation	Obs	Parms	RMSE	"R-sq"	F	P
y1	27	3	10.42366	0.2966	5.060804	0.0147
y2	27	3	6.325398	0.1341	1.858342	0.1777
y3	27	3	5.976973	0.1024	1.368879	0.2735

	Coef.	Std. Err.	t	P>|t|	[95% Conf.	Interval]
y1						
fracture						
2	-8.833333	4.957441	-1.78	0.087	-19.06499	1.398322
3	6	5.394759	1.11	0.277	-5.134235	17.13423
_cons	37	3.939775	9.39	0.000	28.8687	45.1313
y2						
fracture						
2	-5.761905	3.008327	-1.92	0.067	-11.97079	.446977
3	-3.053571	3.273705	-0.93	0.360	-9.810166	3.703023
_cons	38.42857	2.390776	16.07	0.000	33.49425	43.36289
y3						
fracture						
2	4.261905	2.842618	1.50	0.147	-1.60497	10.12878
3	.9285714	3.093377	0.30	0.767	-5.455846	7.312989
_cons	58.57143	2.259083	25.93	0.000	53.90891	63.23395

pwcompare performs pairwise comparisons of the margins using the coefficients from the first equation by default:

```
. pwcompare fracture, mcompare(bonferroni)
Pairwise comparisons of marginal linear predictions
Margins      : asbalanced
```

	Number of Comparisons
y1	
fracture	3

	Contrast	Std. Err.	Bonferroni [95% Conf.	Interval]
y1				
fracture				
2 vs 1	-8.833333	4.957441	-21.59201	3.925341
3 vs 1	6	5.394759	-7.884173	19.88417
3 vs 2	14.83333	4.75773	2.588644	27.07802

We can use the equation() option to get pwcompare to perform comparisons in the y2 equation:

```
. pwcompare fracture, equation(y2) mcompare(bonferroni)
Pairwise comparisons of marginal linear predictions
Margins      : asbalanced
```

	Number of Comparisons
y2	
fracture	3

	Contrast	Std. Err.	Bonferroni [95% Conf.	Interval]
y2				
fracture				
2 vs 1	-5.761905	3.008327	-13.50426	1.980449
3 vs 1	-3.053571	3.273705	-11.47891	5.371769
3 vs 2	2.708333	2.887136	-4.722119	10.13879

Because we are working with mvreg results, we can use the _eqns pseudofactor to compare the margins between the three dependent variables. The levels of _eqns index the equations: 1 for the first equation, 2 for the second, and 3 for the third.

```
. pwcompare _eqns, mcompare(bonferroni)
Pairwise comparisons of marginal linear predictions
Margins      : asbalanced
```

	Number of Comparisons
_eqns	3

	Contrast	Std. Err.	Bonferroni [95% Conf.	Interval]
_eqns				
2 vs 1	-.5654762	2.545923	-7.117768	5.986815
3 vs 1	24.24603	2.320677	18.27344	30.21862
3 vs 2	24.81151	2.368188	18.71664	30.90637

For the previous command, the only methods available are mcompare(bonferroni), mcompare(sidak), or mcompare(scheffe). Methods that use the Studentized range are not appropriate for making comparisons across equations.

Unbalanced data

pwcompare treats all factors as balanced when it computes the marginal means. By "balanced", we mean that the number of observations in each combination of factor levels (in each cell mean) is equal. We can alternatively specify the asobserved option when we have unbalanced data to obtain marginal means that are based on the observed cell frequencies from the model fit. For more details on the difference in these two types of marginal means and a discussion of when each may be appropriate, see [R] **margins** and [R] **contrast**.

In addition, when our data are not balanced, some of the multiple-comparison adjustments are no longer appropriate. Tukey's method, Student–Newman–Keuls' method, Duncan's method, and Dunnett's method assume equal numbers of observations per group.

Here we use an unbalanced dataset and fit a two-way ANOVA model for cholesterol levels on race and age group. Then we perform pairwise comparisons of the mean cholesterol levels for each race, requesting Šidák's adjustment as well as marginal means that are computed using the observed cell frequencies.

```
. use http://www.stata-press.com/data/r12/cholesterol3
(Artificial cholesterol data, unbalanced)

. anova chol race##agegrp

                           Number of obs =      67     R-squared     =  0.8179
                           Root MSE      = 8.37496     Adj R-squared =  0.7689

                  Source |  Partial SS    df       MS           F     Prob > F
             ------------+----------------------------------------------------
                   Model |  16379.9926    14  1169.99947      16.68     0.0000
                         |
                    race |  230.754396     2  115.377198       1.64     0.2029
                  agegrp |  13857.9877     4  3464.49693      49.39     0.0000
             race#agegrp |  857.815209     8  107.226901       1.53     0.1701
                         |
                Residual |   3647.2774    52    70.13995
             ------------+----------------------------------------------------
                   Total |    20027.27    66  303.443485

. pwcompare race, asobserved mcompare(sidak)

Pairwise comparisons of marginal linear predictions

Margins      : asobserved

---------------------------
             |    Number of
             |  Comparisons
-------------+-------------
        race |            3
---------------------------

--------------------------------------------------------------
             |                                 Sidak
             |   Contrast   Std. Err.     [95% Conf. Interval]
-------------+------------------------------------------------
        race |
     2 vs 1  |  -7.232433   2.686089     -13.85924   -.6056277
     3 vs 1  |  -5.231198   2.651203     -11.77194    1.309541
     3 vs 2  |   2.001235   2.414964     -3.956682    7.959152
--------------------------------------------------------------
```

Empty cells

An empty cell is a combination of the levels of factor variables that is not observed in the estimation sample. When we have empty cells in our data, the marginal means involving those empty cells are not estimable as described in [R] **margins**. In addition, all pairwise comparisons involving a marginal mean that is not estimable are themselves not estimable. Here we use a dataset where we do not have any observations for white individuals in the 20–29 age group. We can use the `emptycells(reweight)` option to reweight the nonempty cells so that we can estimate the marginal mean for whites and compute pairwise comparisons involving that marginal mean.

```
. use http://www.stata-press.com/data/r12/cholesterol2
(Artificial cholesterol data, empty cells)
. tabulate race agegrp

                            agegrp
      race      10-19      20-29      30-39      40-59      60-79      Total

     black          5          5          5          5          5         25
     white          5          0          5          5          5         20
     other          5          5          5          5          5         25

     Total         15         10         15         15         15         70

. anova chol race##agegrp

                          Number of obs =      70     R-squared     =  0.7582
                          Root MSE      = 9.47055     Adj R-squared =  0.7021

                 Source   Partial SS    df       MS           F     Prob > F

                  Model   15751.6113    13  1211.66241      13.51     0.0000

                   race    305.49046     2   152.74523       1.70     0.1914
                 agegrp   14387.8559     4  3596.96397      40.10     0.0000
            race#agegrp   795.807574     7  113.686796       1.27     0.2831

               Residual   5022.71559    56  89.6913498

                  Total   20774.3269    69  301.077201

. pwcompare race, emptycells(reweight)
Pairwise comparisons of marginal linear predictions
Margins      : asbalanced
Empty cells  : reweight

                                                  Unadjusted
                     Contrast   Std. Err.     [95% Conf. Interval]

              race
            2 vs 1   2.922769   2.841166     -2.768769    8.614308
            3 vs 1   -4.12621   2.678677     -9.492244    1.239824
            3 vs 2  -7.048979   2.841166     -12.74052    -1.35744
```

For further details on the `emptycells(reweight)` option, see [R] **margins** and [R] **contrast**.

Saved results

pwcompare saves the following in r():

Scalars

r(df_r)	variance degrees of freedom, from e(df_r)
r(k_terms)	number of terms in *marginlist*
r(level)	confidence level of confidence intervals
r(balanced)	1 if fully balanced data; 0 otherwise

Macros

r(cmd)	pwcompare
r(cmdline)	command as typed
r(est_cmd)	e(cmd) from original estimation results
r(est_cmdline)	e(cmdline) from original estimation results
r(title)	title in output
r(emptycells)	*empspec* from emptycells()
r(groups#)	group codes for the #th margin in r(b)
r(mcmethod_vs)	*method* from mcompare()
r(mctitle_vs)	title for *method* from mcompare()
r(mcadjustall_vs)	adjustall or empty
r(margin_method)	asbalanced or asobserved
r(vce)	*vcetype* specified in vce() in original estimation command

Matrices

r(b)	margin estimates
r(V)	variance–covariance matrix of the margin estimates
r(error)	margin estimability codes; 0 means estimable, 8 means not estimable
r(table)	matrix containing the margins with their standard errors, test statistics, p-values, and confidence intervals
r(M)	matrix that produces the margins from the model coefficients
r(b_vs)	margin difference estimates
r(V_vs)	variance–covariance matrix of the margin difference estimates
r(error_vs)	margin difference estimability codes; 0 means estimable, 8 means not estimable
r(table_vs)	matrix containing the margin differences with their standard errors, test statistics, p-values, and confidence intervals
r(L)	matrix that produces the margin differences from the model coefficients
r(k_groups)	number of significance groups for each term

pwcompare with the post option also saves the following in e():

Scalars

e(df_r)	variance degrees of freedom, from e(df_r)
e(k_terms)	number of terms in *marginlist*
e(balanced)	1 if fully balanced data; 0 otherwise

Macros

e(cmd)	pwcompare
e(cmdline)	command as typed
e(est_cmd)	e(cmd) from original estimation results
e(est_cmdline)	e(cmdline) from original estimation results
e(title)	title in output
e(emptycells)	*empspec* from emptycells()
e(margin_method)	asbalanced or asobserved
e(vce)	*vcetype* specified in vce() in original estimation command
e(properties)	b V

Matrices

e(b)	margin estimates
e(V)	variance–covariance matrix of the margin estimates
e(error)	margin estimability codes; 0 means estimable, 8 means not estimable
e(M)	matrix that produces the margins from the model coefficients
e(b_vs)	margin difference estimates
e(V_vs)	variance–covariance matrix of the margin difference estimates
e(error_vs)	margin difference estimability codes; 0 means estimable, 8 means not estimable
e(L)	matrix that produces the margin differences from the model coefficients
e(k_groups)	number of significance groups for each term

Methods and formulas

pwcompare is implemented as an ado-file.

Methods and formulas are presented under the following headings:

Notation

pwcompare performs comparisons of margins; see *Methods and formulas* in [R] **contrast**.

If there are k margins for a given factor term, then there are

$$m = \binom{k}{2} = \frac{k(k-1)}{2}$$

unique pairwise comparisons. Let the ith pairwise comparison be denoted by

$$\widehat{\delta}_i = l_i'\mathbf{b}$$

where $\mathbf{b}$ is a column vector of coefficients from the fitted model and l_i is a column vector that forms the corresponding linear combination. If $\widehat{\mathbf{V}}$ denotes the estimated variance matrix for $\mathbf{b}$, then the standard error for $\widehat{\delta}_i$ is given by

$$\widehat{\text{se}}(\widehat{\delta}_i) = \sqrt{l_i'\widehat{\mathbf{V}}l_i}$$

The corresponding test statistic is then

$$t_i = \frac{\widehat{\delta}_i}{\widehat{\text{se}}(\widehat{\delta}_i)}$$

and the limits for a $100(1-\alpha)\%$ confidence interval for the expected value of $\widehat{\delta}_i$ are

$$\widehat{\delta}_i \pm c_i(\alpha)\,\widehat{\text{se}}(\widehat{\delta}_i)$$

where $c_i(\alpha)$ is the critical value corresponding to the chosen multiple-comparison method.

Unadjusted comparisons

`pwcompare` computes unadjusted p-values and confidence intervals by default. `pwcompare` uses the t distribution with $\nu =$ `e(df_r)` degrees of freedom when `e(df_r)` is posted by the estimation command. The unadjusted two-sided p-value is

$${}_up_i = 2\,\Pr(t_\nu > |t_i|)$$

and the unadjusted critical value ${}_uc_i(\alpha)$ satisfies the following probability statement:

$$\alpha = 2\,\Pr\left\{t_\nu > {}_uc_i(\alpha)\right\}$$

`pwcompare` uses the standard normal distribution when `e(df_r)` is not posted.

Bonferroni's method

For `mcompare(bonferroni)`, the adjusted p-value is

$${}_bp_i = \min(1, m\,{}_up_i)$$

and the adjusted critical value is

$${}_bc_i(\alpha) = {}_uc_i(\alpha/m)$$

Šidák's method

For mcompare(sidak), the adjusted p-value is

$$_{\text{si}}p_i = 1 - (1 - {_u}p_i)^m$$

and the adjusted critical value is

$$_{\text{si}}c_i(\alpha) = {_u}c_i\left\{1 - (1-\alpha)^{1/m}\right\}$$

Scheffé's method

For mcompare(scheffe), the adjusted p-value is

$$_{\text{sc}}p_i = \Pr\left(F_{d,\nu} > t_i^2/d\right)$$

where $F_{d,\nu}$ is distributed as an F with d numerator and ν denominator degrees of freedom and d is the rank of the VCE for the term. The adjusted critical value satisfies the following probability statement:

$$\alpha = \Pr\left[F_{d,\nu} > \{{_{\text{sc}}}c_i(\alpha)\}^2/d\right]$$

pwcompare uses the χ^2 distribution when e(df_r) is not posted.

Tukey's method

For mcompare(tukey), the adjusted p-value is

$$_tp_i = \Pr\left(q_{k,\nu} > |t_i|\sqrt{2}\right)$$

where $q_{k,\nu}$ is distributed as the Studentized range statistic for k means and ν residual degrees of freedom (Miller 1981). The adjusted critical value satisfies the following probability statement:

$$\alpha = \Pr\left\{q_{k,\nu} > {_t}c_i(\alpha)\sqrt{2}\right\}$$

Student–Newman–Keuls' method

For mcompare(snk), suppose t_i is comparing two margins that have r other margins between them. Then the adjusted p-value is

$$_{\text{snk}}p_i = \Pr\left(q_{r+2,\nu} > |t_i|\sqrt{2}\right)$$

where r ranges from 0 to $k-2$. The adjusted critical value $_{\text{snk}}c_i(\alpha)$ satisfies the following probability statement:

$$\alpha = \Pr\left\{q_{r+2,\nu} > {_{\text{snk}}}c_i(\alpha)\sqrt{2}\right\}$$

Duncan's method

For mcompare(duncan), the adjusted p-value is

$$_{\mathrm{dunc}}p_i = 1 - (1 - {}_{\mathrm{snk}}p_i)^{1/(r+1)}$$

and the adjusted critical value is

$$_{\mathrm{dunc}}c_i(\alpha) = {}_{\mathrm{snk}}c_i\left\{1 - (1 - \alpha)^{r+1}\right\}$$

Dunnett's method

For mcompare(dunnett), the margins are compared with a reference category, resulting in only $k - 1$ pairwise comparisons. The adjusted p-value is

$$_{\mathrm{dunn}}p_i = \Pr(d_{k-1,\nu} > |t_i|)$$

where $d_{k-1,\nu}$ is distributed as the many-one t statistic (Miller 1981, 76). The adjusted critical value $_{\mathrm{dunn}}c_i(\alpha)$ satisfies the following probability statement:

$$\alpha = \Pr\left\{d_{k-1,\nu} > {}_{\mathrm{dunn}}c_i(\alpha)\right\}$$

The multiple-comparison methods for mcompare(tukey), mcompare(snk), mcompare(duncan), and mcompare(dunnett) assume the normal distribution with equal variance and sample size for each marginal mean; thus these methods are allowed only with results from anova, regress, manova, and mvreg. These options will cause pwcompare to report a footnote if unbalanced factors are detected.

References

Dunnett, C. W. 1955. A multiple comparison for comparing several treatments with a control. *Journal of the American Statistical Association* 50: 1096–1121.

——. 1964. New tables for multiple comparisons with a control. *Biometrics* 20: 482–491.

Miller, R. G., Jr. 1981. *Simultaneous Statistical Inference*. 2nd ed. New York: Springer.

Milliken, G. A., and D. E. Johnson. 2009. *Analysis of Messy Data, Volume 1: Designed Experiments*. 2nd ed. Boca Raton, FL: CRC Press.

Searle, S. R. 1997. *Linear Models for Unbalanced Data*. New York: Wiley.

Winer, B. J., D. R. Brown, and K. M. Michels. 1991. *Statistical Principles in Experimental Design*. 3rd ed. New York: McGraw–Hill.

Also see

[R] **contrast** — Contrasts and linear hypothesis tests after estimation

[R] **pwcompare postestimation** — Postestimation tools for pwcompare

[R] **lincom** — Linear combinations of estimators

[R] **margins** — Marginal means, predictive margins, and marginal effects

[R] **margins, pwcompare** — Pairwise comparisons of margins

[R] **test** — Test linear hypotheses after estimation

Title

pwcompare postestimation — Postestimation tools for pwcompare

Description

The following postestimation commands are available after `pwcompare, post`:

Command	Description
`estat`	VCE; `estat vce` only
`estat` (svy)	postestimation statistics for survey data
`estimates`	cataloging estimation results
`lincom`	point estimates, standard errors, testing, and inference for linear combinations of coefficients
`nlcom`	point estimates, standard errors, testing, and inference for nonlinear combinations of coefficients
`test`	Wald tests of simple and composite linear hypotheses
`testnl`	Wald tests of nonlinear hypotheses

See the corresponding entries in the *Base Reference Manual* for details, but see [SVY] **estat** for details about `estat` (svy).

Remarks

When we use the `post` option with `pwcompare`, the marginal linear predictions are posted as estimation results, and we can use postestimation commands to perform further analysis on them.

In *Pairwise comparisons of means* of [R] **pwcompare**, we fit a regression of wheat yield on types of fertilizers.

```
. use http://www.stata-press.com/data/r12/yield
(Artificial wheat yield dataset)
. regress yield i.fertilizer
  (output omitted)
```

We also used `pwcompare` with the `cimargins` option to obtain the marginal mean yield for each fertilizer. We can add the `post` option to this command to post these marginal means and their VCEs as estimation results.

```
. pwcompare fertilizer, cimargins post
Pairwise comparisons of marginal linear predictions
Margins      : asbalanced
```

	Margin	Std. Err.	Unadjusted [95% Conf.	Interval]
fertilizer				
1	41.36243	1.124298	39.14509	43.57977
2	44.98515	1.124298	42.7678	47.20249
3	41.85306	1.124298	39.63571	44.0704
4	46.28523	1.124298	44.06789	48.50258
5	40.1241	1.124298	37.90676	42.34145

Now we can use nlcom to compute a percentage improvement in the mean yield for fertilizer 2 when compared with fertilizer 1.

```
. nlcom (pct_chg: 100*(_b[2.fertilizer] - _b[1.fertilizer])/_b[1.fertilizer])
     pct_chg:  100*(_b[2.fertilizer] - _b[1.fertilizer])/_b[1.fertilizer]
```

	Coef.	Std. Err.	t	P>|t|	[95% Conf.	Interval]
pct_chg	8.758479	4.015932	2.18	0.030	.838243	16.67872

The mean yield for fertilizer 2 is about 9% higher than that of fertilizer 1, with a standard error of 4%.

Also see

[R] **pwcompare** — Pairwise comparisons

Title

pwmean — Pairwise comparisons of means

Syntax

`pwmean` *varname*, `over(`*varlist*`)` [*options*]

options	Description
Main	
*`over(`*varlist*`)`	compare means across each combination of the levels in *varlist*
`mcompare(`*method*`)`	adjust for multiple comparisons; default is `mcompare(noadjust)`
Reporting	
`level(`*#*`)`	confidence level; default is `level(95)`
`cieffects`	display a table of mean differences and confidence intervals; the default
`pveffects`	display a table of mean differences and *p*-values
`effects`	display a table of mean differences with *p*-values and confidence intervals
`cimeans`	display a table of means and confidence intervals
`groups`	display a table of means with codes that group them with other means that are not significantly different
`sort`	sort results tables by displayed mean or difference
display_options	control column formats and line width

*`over(`*varlist*`)` is required.

method	Description
`noadjust`	do not adjust for multiple comparisons; the default
`bonferroni`	Bonferroni's method
`sidak`	Šidák's method
`scheffe`	Scheffé's method
`tukey`	Tukey's method
`snk`	Student–Newman–Keuls' method
`duncan`	Duncan's method
`dunnett`	Dunnett's method

Menu

Statistics > Summaries, tables, and tests > Summary and descriptive statistics > Pairwise comparisons of means

Description

pwmean performs pairwise comparisons of means. It computes all pairwise differences of the means of *varname* over the combination of the levels of the variables in *varlist*. The tests and confidence intervals for the pairwise comparisons assume equal variances across groups. pwmean also allows for adjusting the confidence intervals and p-values to account for multiple comparisons using Bonferroni's method, Scheffé's method, Tukey's method, Dunnett's method, and others.

See [R] **pwcompare** for performing pairwise comparisons of means, estimated marginal means, and other types of marginal linear predictions after anova, regress, and most other estimation commands.

See [R] **margins, pwcompare** for performing pairwise comparisons of marginal probabilities and other linear and nonlinear predictions after estimation commands.

Options

Main

over(*varlist*) is required and specifies that means are computed for each combination of the levels of the variables in *varlist*.

mcompare(*method*) specifies the method for computing p-values and confidence intervals that account for multiple comparisons.

Most methods adjust the comparisonwise error rate, α_c, to achieve a prespecified experimentwise error rate, α_e.

mcompare(noadjust) is the default; it specifies no adjustment.

$$\alpha_c = \alpha_e$$

mcompare(bonferroni) adjusts the comparisonwise error rate based on the upper limit of the Bonferroni inequality:

$$\alpha_e \leq m\alpha_c$$

where m is the number of comparisons within the term.

The adjusted comparisonwise error rate is

$$\alpha_c = \alpha_e/m$$

mcompare(sidak) adjusts the comparisonwise error rate based on the upper limit of the probability inequality

$$\alpha_e \leq 1 - (1 - \alpha_c)^m$$

where m is the number of comparisons within the term.

The adjusted comparisonwise error rate is

$$\alpha_c = 1 - (1 - \alpha_e)^{1/m}$$

This adjustment is exact when the m comparisons are independent.

mcompare(scheffe) controls the experimentwise error rate using the F (or χ^2) distribution with degrees of freedom equal to $k - 1$ where k is the number of means being compared.

mcompare(tukey) uses what is commonly referred to as Tukey's honestly significant difference. This method uses the Studentized range distribution instead of the t distribution.

mcompare(snk) is a variation on mcompare(tukey) that counts only the number of means participating in the range for a given comparison instead of the full number of means.

mcompare(duncan) is a variation on mcompare(snk) with additional adjustment to the significance probabilities.

mcompare(dunnett) uses Dunnett's method for making comparisons with a reference category.

Reporting

level(#) specifies the confidence level, as a percentage, for confidence intervals. The default is level(95) or as set by set level; see [U] **20.7 Specifying the width of confidence intervals**. The significance level used by the groups option is $100 - \#$, expressed as a percentage.

cieffects specifies that a table of the pairwise comparisons of means with their standard errors and confidence intervals be reported. This is the default.

pveffects specifies that a table of the pairwise comparisons of means with their standard errors, test statistics, and p-values be reported.

effects specifies that a table of the pairwise comparisons of means with their standard errors, test statistics, p-values, and confidence intervals be reported.

cimeans specifies that a table of the means with their standard errors and confidence intervals be reported.

groups specifies that a table of the means with their standard errors and group codes be reported. Means with the same letter in the group code are not significantly different at the specified significance level.

sort specifies that the reported tables be sorted by the mean or difference that is displayed in the table.

display_options: cformat(%*fmt*), pformat(%*fmt*), sformat(%*fmt*), and nolstretch.

cformat(%*fmt*) specifies how to format means, standard errors, and confidence limits in the table of pairwise comparison of means.

pformat(%*fmt*) specifies how to format p-values in the table of pairwise comparison of means.

sformat(%*fmt*) specifies how to format test statistics in the table of pairwise comparison of means.

nolstretch specifies that the width of the table of estimated comparisons not be automatically widened to accommodate longer variable names. The default, lstretch, is to automatically widen the table of estimated comparisons up to the width of the Results window. To change the default, use set lstretch off. nolstretch is not shown in the dialog box.

Remarks

pwmean performs pairwise comparisons (differences) of means, assuming a common variance among groups. It can easily adjust the p-values and confidence intervals for the differences to account for the elevated type I error rate due to multiple comparisons. Adjustments for multiple comparisons can be made using Bonferroni's method, Scheffé's method, Tukey's method, Dunnett's method, and others.

Remarks are presented under the following headings:

Group means

Suppose we have data on the wheat yield of fields that were each randomly assigned an application of one of five types of fertilizers. Let's first look at the mean yield for each type of fertilizer.

```
. use http://www.stata-press.com/data/r12/yield
(Artificial wheat yield dataset)

. pwmean yield, over(fertilizer) cimeans

Pairwise comparisons of means with equal variances

over         : fertilizer
```

yield	Mean	Std. Err.	Unadjusted [95% Conf.	Interval]
fertilizer				
1	41.36243	1.124298	39.14509	43.57977
2	44.98515	1.124298	42.7678	47.20249
3	41.85306	1.124298	39.63571	44.0704
4	46.28523	1.124298	44.06789	48.50258
5	40.1241	1.124298	37.90676	42.34145

Pairwise differences of means

We can compute all pairwise differences in mean wheat yields for the types of fertilizers.

```
. pwmean yield, over(fertilizer) effects

Pairwise comparisons of means with equal variances

over         : fertilizer
```

yield	Contrast	Std. Err.	Unadjusted t	P>|t|	Unadjusted [95% Conf.	Interval]
fertilizer						
2 vs 1	3.62272	1.589997	2.28	0.024	.4869212	6.758518
3 vs 1	.4906299	1.589997	0.31	0.758	-2.645169	3.626428
4 vs 1	4.922803	1.589997	3.10	0.002	1.787005	8.058602
5 vs 1	-1.238328	1.589997	-0.78	0.437	-4.374127	1.89747
3 vs 2	-3.13209	1.589997	-1.97	0.050	-6.267889	.0037086
4 vs 2	1.300083	1.589997	0.82	0.415	-1.835715	4.435882
5 vs 2	-4.861048	1.589997	-3.06	0.003	-7.996847	-1.725249
4 vs 3	4.432173	1.589997	2.79	0.006	1.296375	7.567972
5 vs 3	-1.728958	1.589997	-1.09	0.278	-4.864757	1.406841
5 vs 4	-6.161132	1.589997	-3.87	0.000	-9.29693	-3.025333

The contrast in the row labeled (2 vs 1) is the difference in the mean wheat yield for fertilizer 2 and fertilizer 1. At a 5% significance level, we conclude that there is a difference in the means for these two fertilizers. Likewise, the rows labeled (4 vs 1), (5 vs 2), (4 vs 3) and (5 vs 4) show differences in these pairs of means. In all, we find that 5 of the 10 mean differences are significantly different from zero at a 5% significance level.

We can specify the sort option to order the differences from smallest to largest in the table.

```
. pwmean yield, over(fertilizer) effects sort
Pairwise comparisons of means with equal variances
over         : fertilizer

─────────────┬──────────────────────────────────────────────────────────────
             │                            Unadjusted           Unadjusted
       yield │   Contrast   Std. Err.      t    P>|t|     [95% Conf. Interval]
─────────────┼──────────────────────────────────────────────────────────────
  fertilizer │
      5 vs 4 │  -6.161132   1.589997    -3.87   0.000     -9.29693   -3.025333
      5 vs 2 │  -4.861048   1.589997    -3.06   0.003    -7.996847   -1.725249
      3 vs 2 │   -3.13209   1.589997    -1.97   0.050    -6.267889    .0037086
      5 vs 3 │  -1.728958   1.589997    -1.09   0.278    -4.864757    1.406841
      5 vs 1 │  -1.238328   1.589997    -0.78   0.437    -4.374127     1.89747
      3 vs 1 │   .4906299   1.589997     0.31   0.758    -2.645169    3.626428
      4 vs 2 │   1.300083   1.589997     0.82   0.415    -1.835715    4.435882
      2 vs 1 │    3.62272   1.589997     2.28   0.024     .4869212    6.758518
      4 vs 3 │   4.432173   1.589997     2.79   0.006     1.296375    7.567972
      4 vs 1 │   4.922803   1.589997     3.10   0.002     1.787005    8.058602
─────────────┴──────────────────────────────────────────────────────────────
```

Ordering the pairwise differences is particularly convenient when we are comparing means for a large number of groups.

Group output

We can use the group option to see the mean of each group and a visual representation of the tests for differences.

```
. pwmean yield, over(fertilizer) group sort
Pairwise comparisons of means with equal variances
over         : fertilizer

─────────────┬─────────────────────────────────────
             │                           Unadjusted
       yield │       Mean   Std. Err.        Groups
─────────────┼─────────────────────────────────────
  fertilizer │
           5 │    40.1241   1.124298            A
           1 │   41.36243   1.124298            A
           3 │   41.85306   1.124298            AB
           2 │   44.98515   1.124298             BC
           4 │   46.28523   1.124298              C
─────────────┴─────────────────────────────────────
Note: Means sharing a letter in the group label
      are not significantly different at the 5%
      level.
```

Fertilizers 5, 1, and 3 are all in group A. This means that at our 5% level of significance, we have insufficient information to distinguish their means. Likewise, fertilizers 3 and 2 are in group B and cannot be distinguished at the 5% level. The same is true for fertilizers 2 and 4 in group C.

Fertilizer 5 and fertilizer 2 have no letters in common, indicating that the mean yields of these two groups are significantly different at the 5% level. We can conclude that any other fertilizers without a letter in common have significantly different means as well.

Adjusting for multiple comparisons

The statistics in the examples above take no account that we are performing 10 comparisons. With our 5% significance level and assuming the comparisons are independent, we expect 1 in 20 tests of comparisons to be significant, even if all the population means are truly the same. If we are performing many comparisons, then we should account for the fact that some tests will be found significant by chance alone. More formally, the test for each pairwise comparison is made without adjusting for the elevated type I experimentwise error rate that is introduced when performing multiple tests. We can use the `mcompare()` option to adjust the confidence intervals and p-values for multiple comparisons.

Tukey's method

Of the available adjustments for multiple comparisons, Tukey's honestly significant difference, Student–Newman–Keuls' method, and Duncan's method are most often used when performing all pairwise comparisons of means. Of these, Tukey's method is the most conservative and Duncan's method is the least conservative. For further discussion of each of the multiple-comparison adjustments, see [R] **pwcompare**.

Here we use Tukey's adjustment to compute p-values and confidence intervals for the pairwise differences.

```
. pwmean yield, over(fertilizer) effects sort mcompare(tukey)
Pairwise comparisons of means with equal variances
over         : fertilizer
```

	Number of Comparisons
fertilizer	10

yield	Contrast	Std. Err.	Tukey t	Tukey P>|t|	Tukey [95% Conf.	Interval]
fertilizer						
5 vs 4	-6.161132	1.589997	-3.87	0.001	-10.53914	-1.78312
5 vs 2	-4.861048	1.589997	-3.06	0.021	-9.239059	-.4830368
3 vs 2	-3.13209	1.589997	-1.97	0.285	-7.510101	1.245921
5 vs 3	-1.728958	1.589997	-1.09	0.813	-6.106969	2.649053
5 vs 1	-1.238328	1.589997	-0.78	0.936	-5.616339	3.139683
3 vs 1	.4906299	1.589997	0.31	0.998	-3.887381	4.868641
4 vs 2	1.300083	1.589997	0.82	0.925	-3.077928	5.678095
2 vs 1	3.62272	1.589997	2.28	0.156	-.7552913	8.000731
4 vs 3	4.432173	1.589997	2.79	0.046	.0541623	8.810185
4 vs 1	4.922803	1.589997	3.10	0.019	.5447922	9.300815

When using a 5% significance level, Tukey's adjustment indicates that four pairs of means are different. With the adjustment, we no longer conclude that the difference in the mean yields for fertilizers 2 and 1 is significantly different from zero.

Dunnett's method

Now let's suppose that fertilizer 1 actually represents fields on which no fertilizer was applied. In this case, we can use Dunnett's method for comparing each of the fertilizers to the control.

```
. pwmean yield, over(fertilizer) effects mcompare(dunnett)
Pairwise comparisons of means with equal variances
over         : fertilizer
```

	Number of Comparisons
fertilizer	4

yield	Contrast	Std. Err.	Dunnett t	P>|t|	Dunnett [95% Conf.	Interval]
fertilizer						
2 vs 1	3.62272	1.589997	2.28	0.079	-.2918331	7.537273
3 vs 1	.4906299	1.589997	0.31	0.994	-3.423923	4.405183
4 vs 1	4.922803	1.589997	3.10	0.008	1.00825	8.837356
5 vs 1	-1.238328	1.589997	-0.78	0.852	-5.152881	2.676225

Using Dunnett's adjustment, we conclude that only fertilizer 4 produces a mean yield that is significantly different from the mean yield of the field with no fertilizer applied.

By default, `pwmean` treats the lowest level of the group variable as the control. If, for instance, fertilizer 3 was our control group, we could type

```
. pwmean yield, over(b3.fertilizer) effects mcompare(dunnett)
```

using the `b3.` factor-variable operator to specify this level as the reference level.

Multiple over() variables

When we specify more than one variable in the `over()` option, pairwise comparisons are performed for the means defined by each combination of levels of these variables.

```
. pwmean yield, over(fertilizer irrigation) group
Pairwise comparisons of means with equal variances
over         : fertilizer irrigation
```

yield	Mean	Std. Err.	Unadjusted Groups
fertilizer#irrigation			
1 0	36.91257	1.116571	A
1 1	45.81229	1.116571	B
2 0	38.79482	1.116571	A C
2 1	51.17547	1.116571	E
3 0	36.34383	1.116571	A
3 1	47.36229	1.116571	B
4 0	41.81757	1.116571	CD
4 1	50.7529	1.116571	E
5 0	35.69507	1.116571	A
5 1	44.55313	1.116571	B D

```
Note: Means sharing a letter in the group label are not
      significantly different at the 5% level.
```

Here the row labeled `1 0` is the mean for the fields treated with fertilizer 1 and without irrigation. This mean is significantly different from the mean of all fertilizer/irrigation pairings that do not have an A in the "Unadjusted Groups" column. These include all pairings where the fields were irrigated as well as the fields treated with fertilizer 4 but without irrigation.

Equal variance assumption

`pwmean` performs multiple comparisons assuming that there is a common variance for all groups. In the case of two groups, this is equivalent to performing the familiar two-sample t test when equal variances are assumed.

```
. ttest yield, by(irrigation)
Two-sample t test with equal variances
```

Group	Obs	Mean	Std. Err.	Std. Dev.	[95% Conf.	Interval]
0	100	37.91277	.5300607	5.300607	36.86102	38.96453
1	100	47.93122	.5630353	5.630353	46.81403	49.0484
combined	200	42.92199	.5242462	7.413961	41.8882	43.95579
diff		-10.01844	.7732872		-11.54338	-8.493509

```
    diff = mean(0) - mean(1)                                      t = -12.9557
Ho: diff = 0                                     degrees of freedom =      198

    Ha: diff < 0                 Ha: diff != 0                 Ha: diff > 0
 Pr(T < t) = 0.0000         Pr(|T| > |t|) = 0.0000          Pr(T > t) = 1.0000
```

```
. pwmean yield, over(irrigation) effects
Pairwise comparisons of means with equal variances
over         : irrigation

─────────────┬────────────────────────────────────────────────────────────────
             │                               Unadjusted           Unadjusted
       yield │   Contrast   Std. Err.      t    P>|t|     [95% Conf. Interval]
─────────────┼────────────────────────────────────────────────────────────────
  irrigation │
      1 vs 0 │   10.01844   .7732872    12.96   0.000     8.493509    11.54338
─────────────┴────────────────────────────────────────────────────────────────
```

The signs for the difference, the test statistic, and the confidence intervals are reversed because the difference is taken in the opposite direction. The p-value from pwmean is equivalent to the one for the two-sided test in the ttest output.

pwmean extends the capabilities of ttest to allow for simultaneously comparing all pairs of means and to allow for using one common variance estimate for all the tests instead of computing a separate pooled variance for each pair of means when using multiple ttest commands. In addition, pwmean allows adjustments for multiple comparisons, many of which rely on an assumption of equal variances among groups.

Saved results

pwmean saves the following in e():

Scalars

e(df_r)	variance degrees of freedom
e(balanced)	1 if fully balanced data; 0 otherwise

Macros

e(cmd)	pwmean
e(cmdline)	command as typed
e(title)	title in output
e(depvar)	name of variable from which the means are computed
e(over)	*varlist* from over()
e(properties)	b V

Matrices

e(b)	mean estimates
e(V)	variance–covariance matrix of the mean estimates
e(error)	mean estimability codes; 0 means estimable, 8 means not estimable
e(b_vs)	mean difference estimates
e(V_vs)	variance–covariance matrix of the mean difference estimates
e(error_vs)	mean difference estimability codes; 0 means estimable, 8 means not estimable
e(k_groups)	number of significance groups for each term

Methods and formulas

pwmean is implemented as an ado-file.

pwmean is a convenience command that uses pwcompare after fitting a fully factorial linear model. See *Methods and formulas* described in [R] **pwcompare**.

Reference

Searle, S. R. 1997. *Linear Models for Unbalanced Data.* New York: Wiley.

Also see

[R] **contrast** — Contrasts and linear hypothesis tests after estimation

[R] **pwcompare** — Pairwise comparisons

[R] **margins** — Marginal means, predictive margins, and marginal effects

[R] **margins, pwcompare** — Pairwise comparisons of margins

[R] **ttest** — Mean-comparison tests

Title

pwmean postestimation — Postestimation tools for pwmean

Description

The following postestimation commands are available after `pwmean`:

Command	Description
`estat`	VCE; `estat vce` only
`estimates`	cataloging estimation results
`lincom`	point estimates, standard errors, testing, and inference for linear combinations of coefficients
`nlcom`	point estimates, standard errors, testing, and inference for nonlinear combinations of coefficients
`test`	Wald tests of simple and composite linear hypotheses
`testnl`	Wald tests of nonlinear hypotheses

Remarks

In *Pairwise differences of means* of [R] **pwmean**, we computed all pairwise differences in mean wheat yields for five fertilizers.

```
. use http://www.stata-press.com/data/r12/yield
(Artificial wheat yield dataset)
. pwmean yield, over(fertilizer)
Pairwise comparisons of means with equal variances
over         : fertilizer
```

yield	Contrast	Std. Err.	Unadjusted [95% Conf.	Interval]
fertilizer				
2 vs 1	3.62272	1.589997	.4869212	6.758518
3 vs 1	.4906299	1.589997	-2.645169	3.626428
4 vs 1	4.922803	1.589997	1.787005	8.058602
5 vs 1	-1.238328	1.589997	-4.374127	1.89747
3 vs 2	-3.13209	1.589997	-6.267889	.0037086
4 vs 2	1.300083	1.589997	-1.835715	4.435882
5 vs 2	-4.861048	1.589997	-7.996847	-1.725249
4 vs 3	4.432173	1.589997	1.296375	7.567972
5 vs 3	-1.728958	1.589997	-4.864757	1.406841
5 vs 4	-6.161132	1.589997	-9.29693	-3.025333

After `pwmean`, we can use `testnl` to test whether the improvement in mean wheat yield when using fertilizer 4 instead of fertilizer 5 is significantly different from 10%.

```
. testnl (_b[4.fertilizer] - _b[5.fertilizer])/_b[5.fertilizer] = 0.1
  (1)  (_b[4.fertilizer] - _b[5.fertilizer])/_b[5.fertilizer] = 0.1
               F(1, 195) =        1.57
                Prob > F =        0.2121
```

The improvement is not significantly different from 10%.

Also see

[R] **pwmean** — Pairwise comparisons of means

Title

qc — Quality control charts

Syntax

Draw a c chart

`cchart` *defect_var unit_var* [, *cchart_options*]

Draw a p (fraction-defective) chart

`pchart` *reject_var unit_var ssize_var* [, *pchart_options*]

Draw an R (range or dispersion) chart

`rchart` *varlist* [*if*] [*in*] [, *rchart_options*]

Draw an $\overline{X}$ (control line) chart

`xchart` *varlist* [*if*] [*in*] [, *xchart_options*]

Draw vertically aligned $\overline{X}$ and R charts

`shewhart` *varlist* [*if*] [*in*] [, *shewhart_options*]

cchart_options	Description
Main	
`nograph`	suppress graph
Plot	
connect_options	affect rendition of the plotted points
marker_options	change look of markers (color, size, etc.)
marker_label_options	add marker labels; change look or position
Control limits	
`clopts(`*cline_options*`)`	affect rendition of the control limits
Add plots	
`addplot(`*plot*`)`	add other plots to the generated graph
Y axis, X axis, Titles, Legend, Overall	
twoway_options	any options other than `by()` documented in [G-3] ***twoway_options***

pchart_options	Description
Main	
stabilized	stabilize the p chart when sample sizes are unequal
nograph	suppress graph
generate(*newvar*$_f$ *newvar*$_{lcl}$ *newvar*$_{ucl}$)	store the fractions of defective elements and the lower and upper control limits
Plot	
connect_options	affect rendition of the plotted points
marker_options	change look of markers (color, size, etc.)
marker_label_options	add marker labels; change look or position
Control limits	
clopts(*cline_options*)	affect rendition of the control limits
Add plots	
addplot(*plot*)	add other plots to the generated graph
Y axis, X axis, Titles, Legend, Overall	
twoway_options	any options other than by() documented in [G-3] ***twoway_options***

rchart_options	Description
Main	
std(*#*)	user-specified standard deviation
nograph	suppress graph
Plot	
connect_options	affect rendition of the plotted points
marker_options	change look of markers (color, size, etc.)
marker_label_options	add marker labels; change look or position
Control limits	
clopts(*cline_options*)	affect rendition of the control limits
Add plots	
addplot(*plot*)	add other plots to the generated graph
Y axis, X axis, Titles, Legend, Overall	
twoway_options	any options other than by() documented in [G-3] ***twoway_options***

xchart_options	Description
Main	
std(#)	user-specified standard deviation
mean(#)	user-specified mean
lower(#) upper(#)	lower and upper limits of the X-bar limits
nograph	suppress graph
Plot	
connect_options	affect rendition of the plotted points
marker_options	change look of markers (color, size, etc.)
marker_label_options	add marker labels; change look or position
Control limits	
clopts(*cline_options*)	affect rendition of the control limits
Add plots	
addplot(*plot*)	add other plots to the generated graph
Y axis, X axis, Titles, Legend, Overall	
twoway_options	any options other than by() documented in [G-3] ***twoway_options***

shewhart_options	Description
Main	
std(#)	user-specified standard deviation
mean(#)	user-specified mean
nograph	suppress graph
Plot	
connect_options	affect rendition of the plotted points
marker_options	change look of markers (color, size, etc.)
marker_label_options	add marker labels; change look or position
Control limits	
clopts(*cline_options*)	affect rendition of the control limits
Y axis, X axis, Titles, Legend, Overall	
combine_options	any options documented in [G-2] **graph combine**

Menu

cchart

Statistics > Other > Quality control > C chart

pchart

Statistics > Other > Quality control > P chart

rchart

Statistics > Other > Quality control > R chart

xchart

Statistics > Other > Quality control > X-bar chart

shewhart

Statistics > Other > Quality control > Vertically aligned X-bar and R chart

Description

These commands provide standard quality-control charts. `cchart` draws a c chart; `pchart`, a p (fraction-defective) chart; `rchart`, an R (range or dispersion) chart; `xchart`, an $\overline{X}$ (control line) chart; and `shewhart`, vertically aligned $\overline{X}$ and R charts.

Options

Main

`stabilized` stabilizes the p chart when sample sizes are unequal.

`std(`*#*`)` specifies the standard deviation of the process. The R chart is calculated (based on the range) if this option is not specified.

`mean(`*#*`)` specifies the grand mean, which is calculated if not specified.

`lower(`*#*`)` and `upper(`*#*`)` must be specified together or not at all. They specify the lower and upper limits of the $\overline{X}$ chart. Calculations based on the mean and standard deviation (whether specified by option or calculated) are used otherwise.

`nograph` suppresses the graph.

`generate(`$newvar_f$ $newvar_{\rm lcl}$ $newvar_{\rm ucl}$`)` stores the plotted values in the p chart. $newvar_f$ will contain the fractions of defective elements; $newvar_{\rm lcl}$ and $newvar_{\rm ucl}$ will contain the lower and upper control limits, respectively.

Plot

connect_options affect whether lines connect the plotted points and the rendition of those lines; see [G-3] ***connect_options***.

marker_options affect the rendition of markers drawn at the plotted points, including their shape, size, color, and outline; see [G-3] ***marker_options***.

marker_label_options specify if and how the markers are to be labeled; see [G-3] ***marker_label_options***.

Control limits

`clopts(`*cline_options*`)` affects the rendition of the control limits; see [G-3] ***cline_options***.

Add plots

`addplot(`*plot*`)` provides a way to add other plots to the generated graph. See [G-3] ***addplot_option***.

Y axis, X axis, Titles, Legend, Overall

twoway_options are any of the options documented in [G-3] ***twoway_options***, excluding by(). These include options for titling the graph (see [G-3] ***title_options***) and for saving the graph to disk (see [G-3] ***saving_option***).

combine_options (shewhart only) are any of the options documented in [G-2] **graph combine**. These include options for titling the graph (see [G-3] ***title_options***) and for saving the graph to disk (see [G-3] ***saving_option***).

Remarks

Control charts may be used to define the goal of a repetitive process, to control that process, and to determine if the goal has been achieved. Walter A. Shewhart of Bell Telephone Laboratories devised the first control chart in 1924. In 1931, Shewhart published *Economic Control of Quality of Manufactured Product.* According to Burr, "Few fields of knowledge have ever been so completely explored and charted in the first exposition" (1976, 29). Shewhart states that "a phenomenon will be said to be controlled when, through the use of past experience, we can predict, at least within limits, how the phenomenon may be expected to vary in the future. Here it is understood that prediction within limits means that we can state, at least approximately, the probability that the observed phenomenon will fall within given limits" (1931, 6).

For more information on quality-control charts, see Burr (1976), Duncan (1986), Harris (1999), or Ryan (2000).

▷ Example 1: cchart

cchart graphs a c chart showing the number of nonconformities in a unit, where *defect_var* records the number of defects in each inspection unit and *unit_var* records the unit number. The unit numbers need not be in order. For instance, consider the following example dataset from Ryan (2000, 156):

```
. use http://www.stata-press.com/data/r12/ncu

. describe

Contains data from http://www.stata-press.com/data/r12/ncu.dta
  obs:            30
 vars:             2                          31 Mar 2011 03:56
 size:           240
--------------------------------------------------------------------------------
              storage  display     value
variable name   type   format      label      variable label
--------------------------------------------------------------------------------
day             float  %9.0g                  Days in April
defects         float  %9.0g                  Numbers of Nonconforming Units
--------------------------------------------------------------------------------
Sorted by:

. list in 1/5

     +---------------+
     | day   defects |
     |---------------|
  1. |   1         7 |
  2. |   2         5 |
  3. |   3        11 |
  4. |   4        13 |
  5. |   5         9 |
     +---------------+
```

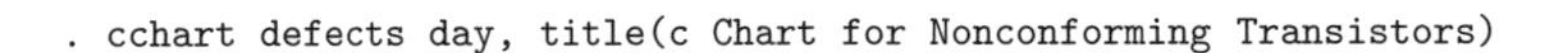

```
. cchart defects day, title(c Chart for Nonconforming Transistors)
```

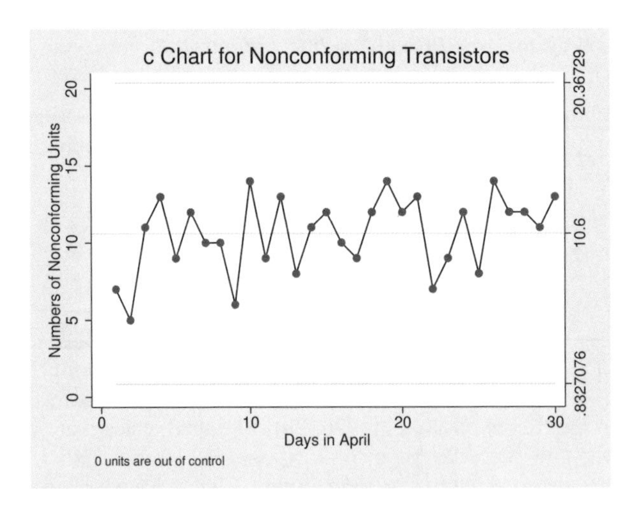

The expected number of defects is 10.6, with lower and upper control limits of 0.8327 and 20.37, respectively. No units are out of control.

◁

▷ Example 2: pchart

`pchart` graphs a p chart, which shows the fraction of nonconforming items in a subgroup, where *reject_var* records the number rejected in each inspection unit, *unit_var* records the inspection unit number, and *ssize_var* records the number inspected in each unit.

Consider the example dataset from Ryan (2000, 156) of the number of nonconforming transistors out of 1,000 inspected each day during the month of April:

```
. use http://www.stata-press.com/data/r12/ncu2

. describe

Contains data from http://www.stata-press.com/data/r12/ncu2.dta
  obs:            30
 vars:             3                          31 Mar 2011 14:13
 size:           360

              storage  display     value
variable name   type   format      label      variable label

day             float  %9.0g                  Days in April
rejects         float  %9.0g                  Numbers of Nonconforming Units
ssize           float  %9.0g                  Sample size

Sorted by:
```

```
. list in 1/5
```

	day	rejects	ssize
1.	1	7	1000
2.	2	5	1000
3.	3	11	1000
4.	4	13	1000
5.	5	9	1000

```
. pchart rejects day ssize
```

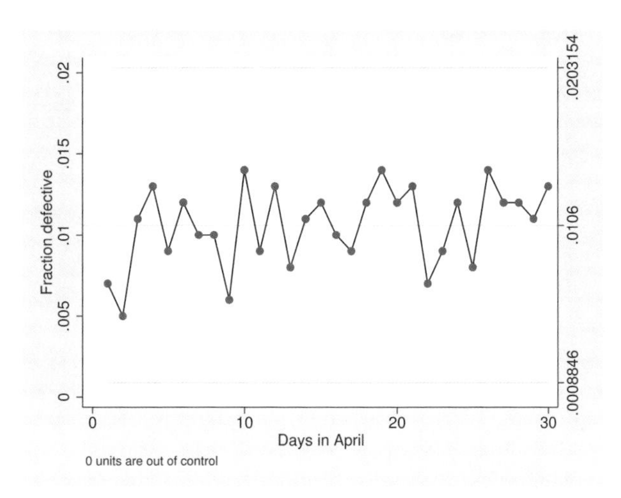

All the points are within the control limits, which are 0.0009 for the lower limit and 0.0203 for the upper limit.

Here the sample sizes are fixed at 1,000, so the `ssize` variable contains 1,000 for each observation. Sample sizes need not be fixed, however. Say that our data were slightly different:

```
. use http://www.stata-press.com/data/r12/ncu3
. list in 1/5
```

	day	rejects	ssize
1.	1	7	920
2.	2	5	920
3.	3	11	920
4.	4	13	950
5.	5	9	950

```
. pchart rejects day ssize
```

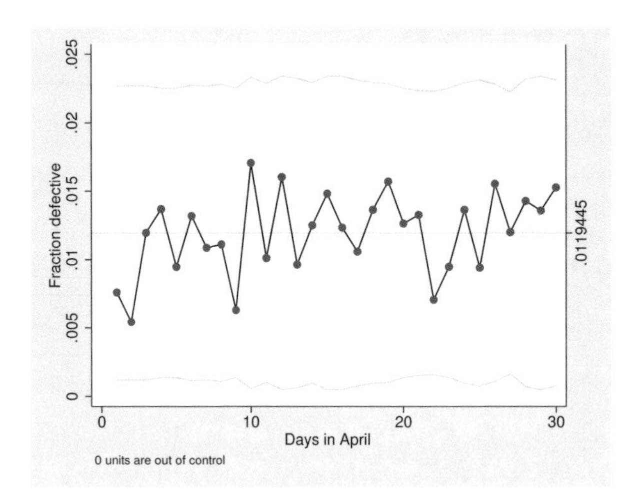

Here the control limits are, like the sample size, no longer constant. The `stabilize` option will stabilize the control chart:

```
. pchart rejects day ssize, stabilize
```

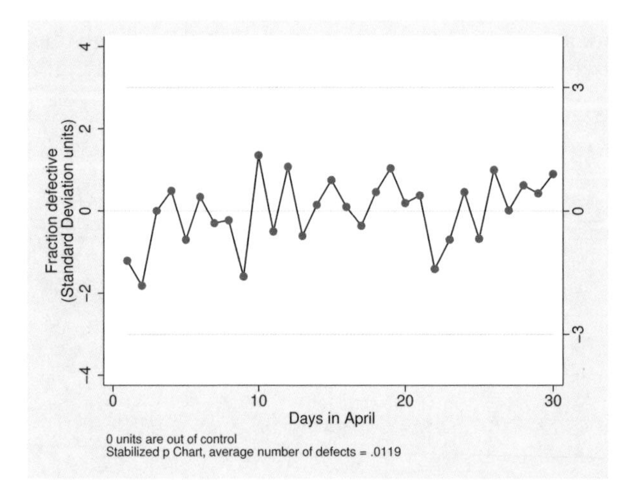

◁

▷ Example 3: rchart

`rchart` displays an R chart showing the range for repeated measurements at various times. Variables within observations record measurements. Observations represent different samples.

For instance, say that we take five samples of 5 observations each. In our first sample, our measurements are 10, 11, 10, 11, and 12. The data are

```
. list

     +------------------------+
     | m1   m2   m3   m4   m5 |
     |------------------------|
  1. | 10   11   10   11   12 |
  2. | 12   10    9   10    9 |
  3. | 10   11   10   12   10 |
  4. |  9    9    9   10   11 |
  5. | 12   12   12   12   13 |
     +------------------------+

. rchart m1-m5, connect(l)
```

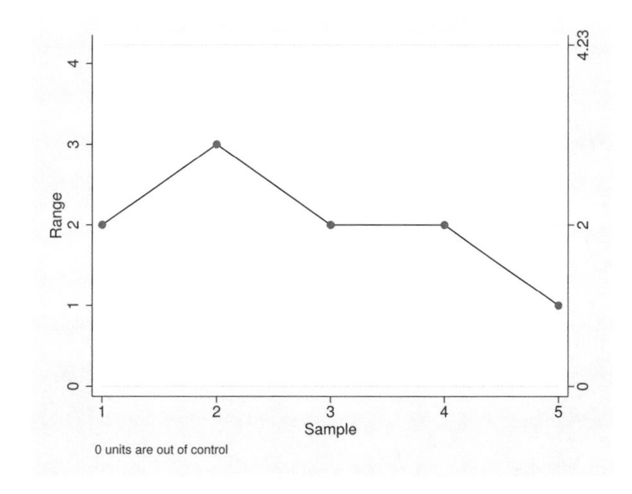

The expected range in each sample is 2 with lower and upper control limits of 0 and 4.23, respectively. If we know that the process standard deviation is 0.3, we could specify

```
. rchart m1-m5, connect(l) std(.3)
```

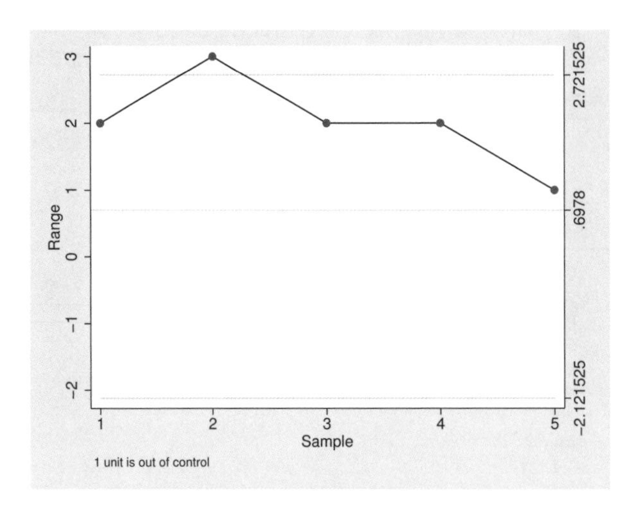

◁

▷ Example 4: xchart

`xchart` graphs an $\overline{X}$ chart for repeated measurements at various times. Variables within observations record measurements, and observations represent different samples. Using the same data as in the previous example, we type

```
. xchart m1-m5, connect(l)
```

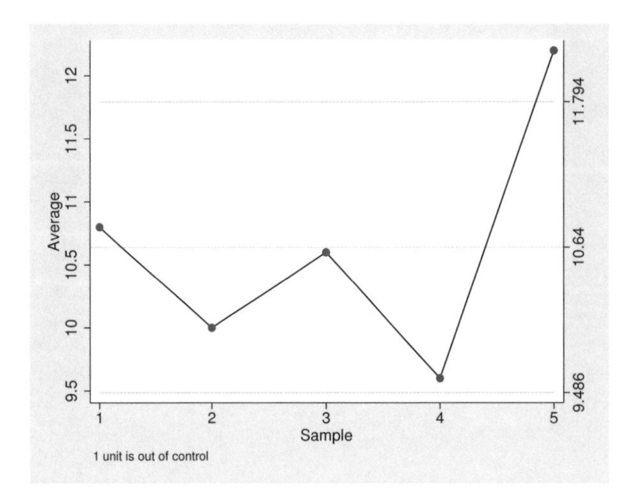

The average measurement in the sample is 10.64, and the lower and upper control limits are 9.486 and 11.794, respectively. Suppose that we knew from prior information that the mean of the process is 11. Then we would type

```
. xchart m1-m5, connect(l) mean(11)
```

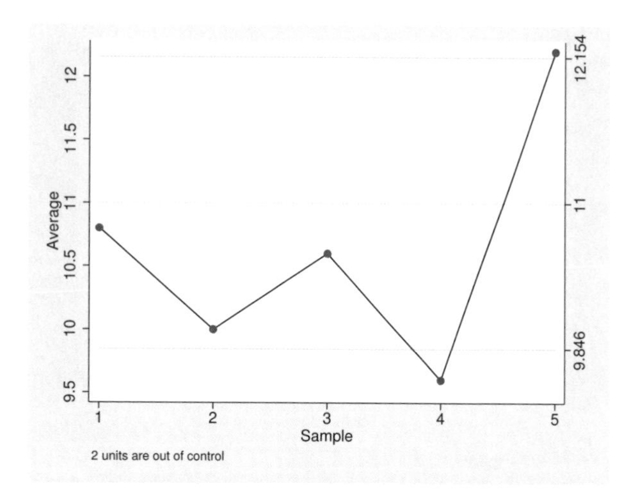

If we also know that the standard deviation of the process is 0.3, we could type

```
. xchart m1-m5, connect(l) mean(11) std(.3)
```

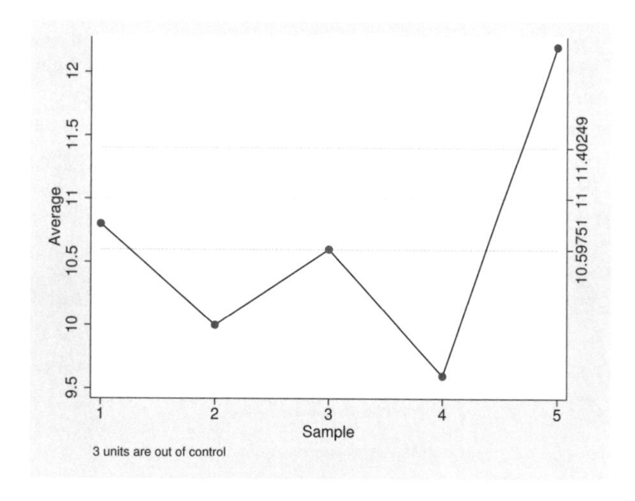

Finally, `xchart` allows us to specify our own control limits:

```
. xchart m1-m5, connect(l) mean(11) lower(10) upper(12)
```

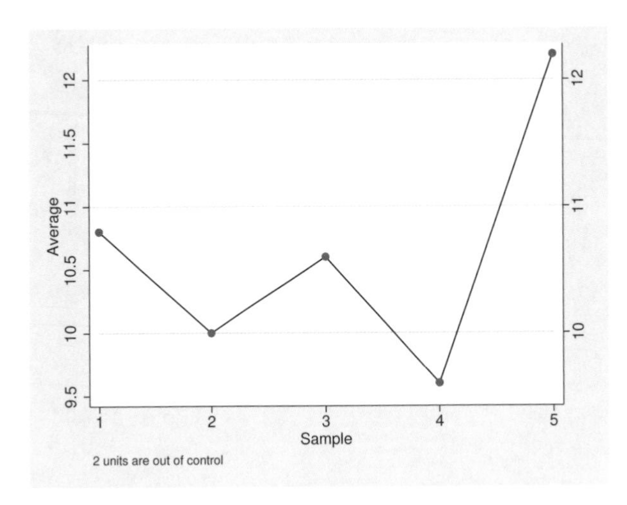

◁

> Walter Andrew Shewhart (1891–1967) was born in Illinois and educated as a physicist, with degrees from the Universities of Illinois and California. After a brief period teaching physics, he worked for the Western Electric Company and (from 1925) the Bell Telephone Laboratories. His name is most associated with control charts used in quality controls, but his many other interests ranged generally from quality assurance to the philosophy of science.

▷ Example 5: shewhart

`shewhart` displays a vertically aligned $\overline{X}$ and R chart in the same image. To produce the best-looking combined image possible, you will want to use the `xchart` and `rchart` commands separately and then combine the graphs. `shewhart`, however, is more convenient.

Using the same data as previously, but realizing that the standard deviation should have been 0.4, we type

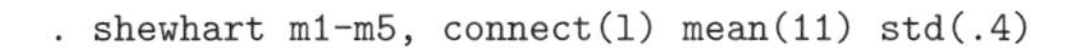

```
. shewhart m1-m5, connect(l) mean(11) std(.4)
```

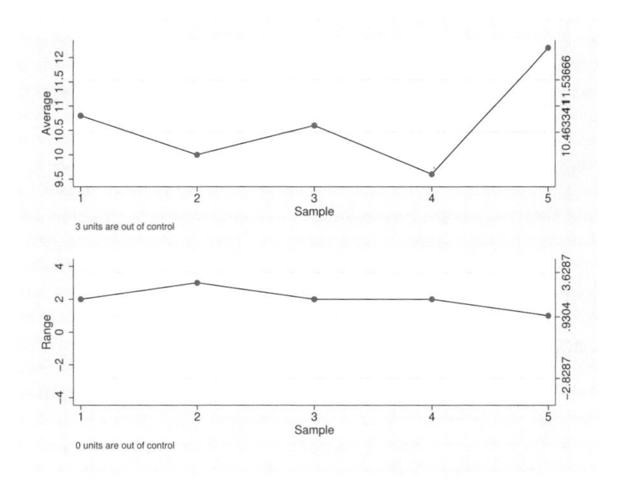

◁

Saved results

cchart saves the following in r():

Scalars

r(cbar)	expected number of nonconformities
r(lcl_c)	lower control limit
r(ucl_c)	upper control limit
r(N)	number of observations
r(out_c)	number of units out of control
r(below_c)	number of units below the lower limit
r(above_c)	number of units above the upper limit

pchart saves the following in r():

Scalars

r(pbar)	average fraction of nonconformities
r(lcl_p)	lower control limit
r(ucl_p)	upper control limit
r(N)	number of observations
r(out_p)	number of units out of control
r(below_p)	number of units below the lower limit
r(above_p)	number of units above the upper limit

rchart saves the following in r():

Scalars

r(central_line)	ordinate of the central line
r(lcl_r)	lower control limit
r(ucl_r)	upper control limit
r(N)	number of observations
r(out_r)	number of units out of control
r(below_r)	number of units below the lower limit
r(above_r)	number of units above the upper limit

xchart saves the following in r():

Scalars

r(xbar)	grand mean
r(lcl_x)	lower control limit
r(ucl_x)	upper control limit
r(N)	number of observations
r(out_x)	number of units out of control
r(below_x)	number of units below the lower limit

shewhart saves in r() the combination of saved results from xchart and rchart.

Methods and formulas

cchart, pchart, rchart, xchart, and shewhart are implemented as ado-files.

For the c chart, the number of defects per unit, C, is taken to be a value of a random variable having a Poisson distribution. If k is the number of units available for estimating λ, the parameter of the Poisson distribution, and if C_i is the number of defects in the ith unit, then λ is estimated by $\overline{C} = \sum_i C_i/k$. Then

$$\text{central line} = \overline{C}$$

$$\text{UCL} = \overline{C} + 3\sqrt{\overline{C}}$$

$$\text{LCL} = \overline{C} - 3\sqrt{\overline{C}}$$

Control limits for the p chart are based on the sampling theory for proportions, using the normal approximation to the binomial. If k samples are taken, the estimator of p is given by $\overline{p} = \sum_i \widehat{p}_i/k$, where $\widehat{p}_i = x_i/n_i$, and x_i is the number of defects in the ith sample of size n_i. The central line and the control limits are given by

$$\text{central line} = \overline{p}$$

$$\text{UCL} = \overline{p} + 3\sqrt{\overline{p}(1-\overline{p})/n_i}$$

$$\text{LCL} = \overline{p} - 3\sqrt{\overline{p}(1-\overline{p})/n_i}$$

Control limits for the R chart are based on the distribution of the range of samples of size n from a normal population. If the standard deviation of the process, σ, is known,

$$\text{central line} = d_2\sigma$$

$$\text{UCL} = D_2\sigma$$

$$\text{LCL} = D_1\sigma$$

where d_2, D_1, and D_2 are functions of the number of observations in the sample and are obtained from the table published in Beyer (1976).

When σ is unknown,

$$\text{central line} = \overline{R}$$

$$\text{UCL} = (D_2/d_2)\overline{R}$$

$$\text{LCL} = (D_1/d_2)\overline{R}$$

where $\overline{R} = \sum_i R_i/k$ is the range of the k sample ranges R_i.

Control limits for the $\overline{\mathrm{X}}$ chart are given by

$$\begin{aligned} \text{central line} &= \overline{x} \\ \text{UCL} &= \overline{x} + (3/\sqrt{n})\sigma \\ \text{LCL} &= \overline{x} - (3/\sqrt{n})\sigma \end{aligned}$$

if σ is known. If σ is unknown,

$$\begin{aligned} \text{central line} &= \overline{x} \\ \text{UCL} &= \overline{x} + A_2\overline{R} \\ \text{LCL} &= \overline{x} - A_2\overline{R} \end{aligned}$$

where $\overline{R}$ is the average range as defined above and A_2 is a function (op. cit.) of the number of observations in the sample.

References

Bayart, D. 2001. Walter Andrew Shewhart. In *Statisticians of the Centuries*, ed. C. C. Heyde and E. Seneta, 398–401. New York: Springer.

Beyer, W. H. 1976. Factors for computing control limits. In Vol. 2 of *Handbook of Tables for Probability and Statistics*, ed. W. H. Beyer, 451–465. Cleveland, OH: The Chemical Rubber Company.

Burr, I. W. 1976. *Statistical Quality Control Methods*. New York: Dekker.

Caulcutt, R. 2004. Control charts in practice. *Significance* 1: 81–84.

Duncan, A. J. 1986. *Quality Control and Industrial Statistics*. 5th ed. Homewood, IL: Irwin.

Harris, R. L. 1999. *Information Graphics: A Comprehensive Illustrated Reference*. New York: Oxford University Press.

Ryan, T. P. 2000. *Statistical Methods for Quality Improvement*. 2nd ed. New York: Wiley.

Saw, S. L. C., and T. W. Soon. 1994. sqc1: Estimating process capability indices with Stata. *Stata Technical Bulletin* 17: 18–19. Reprinted in *Stata Technical Bulletin Reprints*, vol. 3, pp. 174–175. College Station, TX: Stata Press.

Shewhart, W. A. 1931. *Economic Control of Quality of Manufactured Product*. New York: Van Nostrand.

Also see

[R] **serrbar** — Graph standard error bar chart

Title

qreg — Quantile regression

Syntax

Quantile regression

qreg *depvar* [*indepvars*] [*if*] [*in*] [*weight*] [, *qreg_options*]

Interquantile range regression

iqreg *depvar* [*indepvars*] [*if*] [*in*] [, *iqreg_options*]

Simultaneous-quantile regression

sqreg *depvar* [*indepvars*] [*if*] [*in*] [, *sqreg_options*]

Bootstrapped quantile regression

bsqreg *depvar* [*indepvars*] [*if*] [*in*] [, *bsqreg_options*]

Internal estimation command for quantile regression

_qreg [*depvar* [*indepvars*] [*if*] [*in*] [*weight*]] [, *_qreg_options*]

qreg_options	Description
Model	
quantile(*#*)	estimate # quantile; default is quantile(.5)
Reporting	
level(*#*)	set confidence level; default is level(95)
display_options	control column formats and line width
Optimization	
optimization_options	control the optimization process; seldom used
wlsiter(*#*)	attempt # weighted least-squares iterations before doing linear programming iterations

iqreg_options	Description
Model	
quantiles(*# #*)	interquantile range; default is quantiles(.25 .75)
reps(*#*)	perform # bootstrap replications; default is reps(20)
Reporting	
level(*#*)	set confidence level; default is level(95)
nodots	suppress display of the replication dots
display_options	control column formats and line width

sqreg_options	Description
Model	
quantiles(*#* [# [# ...]])	estimate # quantiles; default is quantiles(.5)
reps(*#*)	perform # bootstrap replications; default is reps(20)
Reporting	
level(*#*)	set confidence level; default is level(95)
nodots	suppress display of the replication dots
display_options	control column formats and line width

bsqreg_options	Description
Model	
quantile(*#*)	estimate # quantile; default is quantile(.5)
reps(*#*)	perform # bootstrap replications; default is reps(20)
Reporting	
level(*#*)	set confidence level; default is level(95)
display_options	control column formats and line width

_qreg_options	Description
quantile(*#*)	estimate # quantile; default is quantile(.5)
level(*#*)	set confidence level; default is level(95)
accuracy(*#*)	relative accuracy required for linear programming algorithm; should not be specified
optimization_options	control the optimization process; seldom used

by, mi estimate, rolling, statsby, and xi are allowed by qreg, iqreg, sqreg, and bsqreg; fracpoly, mfp, nestreg, and stepwise are allowed only with qreg; see [U] **11.1.10 Prefix commands**.

qreg and _qreg allow aweights and fweights; see [U] **11.1.6 weight**.

See [U] **20 Estimation and postestimation commands** for more capabilities of estimation commands.

Menu

qreg

Statistics > Nonparametric analysis > Quantile regression

iqreg

Statistics > Nonparametric analysis > Interquantile regression

sqreg

Statistics > Nonparametric analysis > Simultaneous-quantile regression

bsqreg

Statistics > Nonparametric analysis > Bootstrapped quantile regression

Description

`qreg` fits quantile (including median) regression models, also known as least–absolute-value models (LAV or MAD) and minimum L1-norm models.

`iqreg` estimates interquantile range regressions, regressions of the difference in quantiles. The estimated variance–covariance matrix of the estimators (VCE) is obtained via bootstrapping.

`sqreg` estimates simultaneous-quantile regression. It produces the same coefficients as `qreg` for each quantile. Reported standard errors will be similar, but `sqreg` obtains an estimate of the VCE via bootstrapping, and the VCE includes between-quantile blocks. Thus you can test and construct confidence intervals comparing coefficients describing different quantiles.

`bsqreg` is equivalent to `sqreg` with one quantile.

`_qreg` is the internal estimation command for quantile regression. `_qreg` is not intended to be used directly; see *Methods and formulas* below.

Options for qreg

Model

`quantile(`*#*`)` specifies the quantile to be estimated and should be a number between 0 and 1, exclusive. Numbers larger than 1 are interpreted as percentages. The default value of 0.5 corresponds to the median.

Reporting

`level(`*#*`)`; see [R] **estimation options**.

display_options: `cformat(`*%fmt*`)`, `pformat(`*%fmt*`)`, `sformat(`*%fmt*`)`, and `nolstretch`; see [R] **estimation options**.

Optimization

optimization_options: `iterate(`*#*`)`, [`no`]`log`, `trace`. `iterate()` specifies the maximum number of iterations; `log`/`nolog` specifies whether to show the iteration log; and `trace` specifies that the iteration log should include the current parameter vector. These options are seldom used.

`wlsiter(`*#*`)` specifies the number of weighted least-squares iterations that will be attempted before the linear programming iterations are started. The default value is 1. If there are convergence problems, increasing this number should help.

Options for iqreg

Model

`quantiles(`*# #*`)` specifies the quantiles to be compared. The first number must be less than the second, and both should be between 0 and 1, exclusive. Numbers larger than 1 are interpreted as percentages. Not specifying this option is equivalent to specifying `quantiles(.25 .75)`, meaning the interquartile range.

`reps(`*#*`)` specifies the number of bootstrap replications to be used to obtain an estimate of the variance–covariance matrix of the estimators (standard errors). `reps(20)` is the default and is arguably too small. `reps(100)` would perform 100 bootstrap replications. `reps(1000)` would perform 1,000 replications.

Reporting

level(#); see [R] **estimation options**.

nodots suppresses display of the replication dots.

display_options: cformat(%*fmt*), pformat(%*fmt*), sformat(%*fmt*), and nolstretch; see [R] **estimation options**.

Options for sqreg

Model

quantiles(# [# [# ...]]) specifies the quantiles to be estimated and should contain numbers between 0 and 1, exclusive. Numbers larger than 1 are interpreted as percentages. The default value of 0.5 corresponds to the median.

reps(#) specifies the number of bootstrap replications to be used to obtain an estimate of the variance–covariance matrix of the estimators (standard errors). reps(20) is the default and is arguably too small. reps(100) would perform 100 bootstrap replications. reps(1000) would perform 1,000 replications.

Reporting

level(#); see [R] **estimation options**.

nodots suppresses display of the replication dots.

display_options: cformat(%*fmt*), pformat(%*fmt*), sformat(%*fmt*), and nolstretch; see [R] **estimation options**.

Options for bsqreg

Model

quantile(#) specifies the quantile to be estimated and should be a number between 0 and 1, exclusive. Numbers larger than 1 are interpreted as percentages. The default value of 0.5 corresponds to the median.

reps(#) specifies the number of bootstrap replications to be used to obtain an estimate of the variance–covariance matrix of the estimators (standard errors). reps(20) is the default and is arguably too small. reps(100) would perform 100 bootstrap replications. reps(1000) would perform 1,000 replications.

Reporting

level(#); see [R] **estimation options**.

display_options: cformat(%*fmt*), pformat(%*fmt*), sformat(%*fmt*), and nolstretch; see [R] **estimation options**.

Options for _qreg

quantile(#) specifies the quantile to be estimated and should be a number between 0 and 1, exclusive. The default value of 0.5 corresponds to the median.

level(#); see [R] **estimation options**.

accuracy(#) should not be specified; it specifies the relative accuracy required for the linear programming algorithm. If the potential for improving the sum of weighted deviations by deleting an observation from the basis is less than this on a percentage basis, the algorithm will be said to have converged. The default value is 10^{-10}.

optimization_options: iterate(#), [no]log, trace. iterate() specifies the maximum number of iterations; log/nolog specifies whether to show the iteration log; and trace specifies that the iteration log should include the current parameter vector. These options are seldom used.

Remarks

Remarks are presented under the following headings:

Median regression
Generalized quantile regression
Estimated standard errors
Interquantile and simultaneous-quantile regression

Median regression

qreg without options fits quantile regression models. The most common form is median regression, where the object is to estimate the median of the dependent variable, conditional on the values of the independent variables. This method is similar to ordinary regression, where the objective is to estimate the mean of the dependent variable. Simply put, median regression finds a line through the data that minimizes the sum of the *absolute* residuals rather than the sum of the *squares* of the residuals, as in ordinary regression. Cameron and Trivedi (2010, chap. 7) provide a nice introduction to quantile regression using Stata.

▷ Example 1

Consider a two-group experimental design with 5 observations per group:

```
. use http://www.stata-press.com/data/r12/twogrp
. list

     +--------+
     | x    y |
     |--------|
  1. | 0    0 |
  2. | 0    1 |
  3. | 0    3 |
  4. | 0    4 |
  5. | 0   95 |
     |--------|
  6. | 1   14 |
  7. | 1   19 |
  8. | 1   20 |
  9. | 1   22 |
 10. | 1   23 |
     +--------+
```

```
. qreg y x
Iteration  1:  WLS sum of weighted deviations =  121.88268

Iteration  1: sum of abs. weighted deviations =        111
Iteration  2: sum of abs. weighted deviations =        110

Median regression                                   Number of obs =        10
  Raw sum of deviations       157 (about 14)
  Min sum of deviations       110                   Pseudo R2     =    0.2994

------------------------------------------------------------------------------
           y |      Coef.   Std. Err.      t    P>|t|     [95% Conf. Interval]
-------------+----------------------------------------------------------------
           x |         17    3.924233     4.33   0.003     7.950702    26.0493
       _cons |          3    2.774852     1.08   0.311     -3.39882    9.39882
------------------------------------------------------------------------------
```

We have estimated the equation

$$\text{y}_{\text{median}} = 3 + 17\,\text{x}$$

We look back at our data. x takes on the values 0 and 1, so the median for the $\text{x} = 0$ group is 3, whereas for $\text{x} = 1$ it is $3 + 17 = 20$. The output reports that the raw sum of absolute deviations about 14 is 157; that is, the sum of $|\text{y} - 14|$ is 157. Fourteen is the unconditional median of y, although in these data, any value between 14 and 19 could also be considered an unconditional median (we have an even number of observations, so the median is bracketed by those two values). In any case, the raw sum of deviations of y about the median would be the same no matter what number we choose between 14 and 19. (With a "median" of 14, the raw sum of deviations is 157. Now think of choosing a slightly larger number for the median and recalculating the sum. Half the observations will have larger negative residuals, but the other half will have smaller positive residuals, resulting in no net change.)

We turn now to the actual estimated equation. The sum of the absolute deviations about the solution $\text{y}_{\text{median}} = 3 + 17\text{x}$ is 110. The pseudo-R^2 is calculated as $1 - 110/157 \approx 0.2994$. This result is based on the idea that the median regression is the maximum likelihood estimate for the double-exponential distribution.

◁

❑ Technical note

qreg is an alternative to regular regression or robust regression—see [R] **regress** and [R] **rreg**. Let's compare the results:

```
. regress y x
      Source |       SS       df       MS              Number of obs =      10
-------------+------------------------------           F(  1,     8) =    0.00
       Model |        2.5      1         2.5           Prob > F      =  0.9586
    Residual |     6978.4      8       872.3           R-squared     =  0.0004
-------------+------------------------------           Adj R-squared = -0.1246
       Total |     6980.9      9  775.655556           Root MSE      =  29.535

------------------------------------------------------------------------------
           y |      Coef.   Std. Err.      t    P>|t|     [95% Conf. Interval]
-------------+----------------------------------------------------------------
           x |         -1     18.6794    -0.05   0.959    -44.07477    42.07477
       _cons |       20.6    13.20833     1.56   0.157    -9.858465    51.05847
------------------------------------------------------------------------------
```

Unlike qreg, regress fits ordinary linear regression and is concerned with predicting the mean rather than the median, so both results are, in a technical sense, correct. Putting aside those technicalities, however, we tend to use either regression to describe the central tendency of the data, of which the mean is one measure and the median another. Thus we can ask, "which method better describes the central tendency of these data?"

Means—and therefore ordinary linear regression—are sensitive to outliers, and our data were purposely designed to contain two such outliers: 95 for x = 0 and 14 for x = 1. These two outliers dominated the ordinary regression and produced results that do not reflect the central tendency well—you are invited to enter the data and graph y against x.

Robust regression attempts to correct the outlier-sensitivity deficiency in ordinary regression:

```
. rreg y x, genwt(wt)
   Huber iteration 1:  maximum difference in weights = .7311828
   Huber iteration 2:  maximum difference in weights = .17695779
   Huber iteration 3:  maximum difference in weights = .03149585
Biweight iteration 4:  maximum difference in weights = .1979335
Biweight iteration 5:  maximum difference in weights = .23332905
Biweight iteration 6:  maximum difference in weights = .09960067
Biweight iteration 7:  maximum difference in weights = .02691458
Biweight iteration 8:  maximum difference in weights = .0009113

Robust regression                                     Number of obs =      10
                                                      F(  1,      8) =   80.63
                                                      Prob > F      =  0.0000

------------------------------------------------------------------------------
           y |      Coef.   Std. Err.      t    P>|t|     [95% Conf. Interval]
-------------+----------------------------------------------------------------
           x |   18.16597   2.023114     8.98   0.000     13.50066    22.83128
       _cons |   2.000003   1.430558     1.40   0.200    -1.298869    5.298875
------------------------------------------------------------------------------
```

Here rreg discarded the first outlier completely. (We know this because we included the genwt() option on rreg and, after fitting the robust regression, examined the weights.) For the other "outlier", rreg produced a weight of 0.47.

In any case, the answers produced by qreg and rreg to describe the central tendency are similar, but the standard errors are different. In general, robust regression will have smaller standard errors because it is not as sensitive to the exact placement of observations near the median. Also, some authors (Rousseeuw and Leroy 1987, 11) have noted that quantile regression, unlike the median, may be sensitive to even one outlier, if its leverage is high enough. ❑

▷ Example 2

Let's now consider a less artificial example using the automobile data described in [U] **1.2.2 Example datasets**. Using median regression, we will regress each car's price on its weight and length and whether it is of foreign manufacture:

```
. use http://www.stata-press.com/data/r12/auto, clear
(1978 Automobile Data)

. qreg price weight length foreign
Iteration  1:  WLS sum of weighted deviations =  112795.66

Iteration  1: sum of abs. weighted deviations =     111901
Iteration  2: sum of abs. weighted deviations =  110529.43
  (output omitted)
Iteration  8: sum of abs. weighted deviations =  108822.59

Median regression                                   Number of obs =         74
  Raw sum of deviations    142205 (about 4934)
  Min sum of deviations 108822.6                    Pseudo R2     =     0.2347

------------------------------------------------------------------------------
       price |      Coef.   Std. Err.      t    P>|t|     [95% Conf. Interval]
-------------+----------------------------------------------------------------
      weight |   3.933588   .8602183     4.57   0.000     2.217937    5.649239
      length |  -41.25191    28.8693    -1.43   0.157    -98.82991    16.32609
     foreign |   3377.771   577.3391     5.85   0.000     2226.305    4529.237
       _cons |   344.6494   3260.244     0.11   0.916    -6157.702    6847.001
------------------------------------------------------------------------------
```

The estimated equation is

$$\texttt{price}_{\rm median} = 3.93\,\texttt{weight} - 41.25\,\texttt{length} + 3377.8\,\texttt{foreign} + 344.65$$

The output may be interpreted in the same way as linear regression output; see [R] **regress**. The variables `weight` and `foreign` are significant, but `length` is not significant. The median price of the cars in these data is $4,934. This value is a median (one of the two center observations), not *the* median, which would typically be defined as the midpoint of the two center observations.

◁

Generalized quantile regression

Generalized quantile regression is similar to median regression in that it estimates an equation describing a quantile other than the 0.5 (median) quantile. For example, specifying `quant(.25)` estimates the 25th percentile or the first quartile.

▷ Example 3

Again we will begin with the 10-observation artificial dataset we used at the beginning of *Median regression* above. We will estimate the 0.6667 quantile:

```
. use http://www.stata-press.com/data/r12/twogrp

. qreg y x, quant(0.6667)
Iteration  1:  WLS sum of weighted deviations =  152.32472

Iteration  1: sum of abs. weighted deviations =   138.0054
Iteration  2: sum of abs. weighted deviations =   136.6714

.6667 Quantile regression                           Number of obs =         10
  Raw sum of deviations 159.3334 (about 20)
  Min sum of deviations 136.6714                    Pseudo R2     =     0.1422

------------------------------------------------------------------------------
           y |      Coef.   Std. Err.      t    P>|t|     [95% Conf. Interval]
-------------+----------------------------------------------------------------
           x |         18   54.21918     0.33   0.748    -107.0297    143.0297
       _cons |          4   38.33875     0.10   0.919    -84.40932    92.40932
------------------------------------------------------------------------------
```

The 0.6667 quantile in the data is 20. The estimated values are 4 for $x = 0$ and 22 for $x = 1$. These values are appropriate because the usual convention is to "count in" $(n + 1) \times$ quantile observations.

◁

▷ Example 4

Returning to real data, the equation for the 25th percentile of `price` based on `weight`, `length`, and `foreign` in our automobile data is

```
. use http://www.stata-press.com/data/r12/auto
(1978 Automobile Data)

. qreg price weight length foreign, quant(.25)
Iteration  1:  WLS sum of weighted deviations =  98938.466

Iteration  1: sum of abs. weighted deviations =  99457.766
Iteration  2: sum of abs. weighted deviations =  91339.779
  (output omitted)
Iteration 10: sum of abs. weighted deviations =  69603.554

.25 Quantile regression                             Number of obs =         74
  Raw sum of deviations  83825.5 (about 4187)
  Min sum of deviations 69603.55                    Pseudo R2     =     0.1697
```

price	Coef.	Std. Err.	t	P>|t|	[95% Conf.	Interval]
weight	1.831789	.668093	2.74	0.008	.4993194	3.164258
length	2.845558	24.78057	0.11	0.909	-46.57773	52.26885
foreign	2209.925	434.631	5.08	0.000	1343.081	3076.769
_cons	-1879.775	2808.067	-0.67	0.505	-7480.287	3720.737

Compared with our previous median regression, the coefficient on `length` now has a positive sign, and the coefficients on `foreign` and `weight` are reduced. The actual lower quantile is $4,187, substantially less than the median $4,934. It appears that the factors are weaker in this part of the distribution.

We can also estimate the upper quartile as a function of the same three variables:

```
. qreg price weight length foreign, quant(.75)
Iteration  1:  WLS sum of weighted deviations =  110931.48

Iteration  1: sum of abs. weighted deviations =  111305.91
Iteration  2: sum of abs. weighted deviations =  105989.57
  (output omitted)
Iteration  7: sum of abs. weighted deviations =  98395.935

.75 Quantile regression                             Number of obs =         74
  Raw sum of deviations 159721.5 (about 6342)
  Min sum of deviations 98395.94                    Pseudo R2     =     0.3840
```

price	Coef.	Std. Err.	t	P>|t|	[95% Conf.	Interval]
weight	9.22291	2.653579	3.48	0.001	3.930515	14.51531
length	-220.7833	80.13907	-2.76	0.007	-380.6156	-60.95096
foreign	3595.133	1727.704	2.08	0.041	149.3355	7040.931
_cons	20242.9	8534.529	2.37	0.020	3221.323	37264.49

This result tells a different story: `weight` is much more important, and `length` is now significant—with a negative coefficient! The prices of high-priced cars seem to be determined by factors different from those affecting the prices of low-priced cars.

◁

❑ Technical note

One explanation for having substantially different regression functions for different quantiles is that the data are heteroskedastic, as we will demonstrate below. The following statements create a sharply heteroskedastic set of data:

```
. drop _all
. set obs 10000
obs was 0, now 10000
. set seed 50550
. gen x = .1 + .9 * runiform()
. gen y = x * runiform()^2
```

Let's now fit the regressions for the 5th and 95th quantiles:

```
. qreg y x, quant(.05)
Iteration  1:  WLS sum of weighted deviations =  1080.7273
Iteration  1: sum of abs. weighted deviations =  1078.3192
Iteration  2: sum of abs. weighted deviations =  282.73545
  (output omitted )
Iteration  9: sum of abs. weighted deviations =  182.25244
.05 Quantile regression                             Number of obs =      10000
  Raw sum of deviations  182.357 (about .0009234)
  Min sum of deviations 182.2524                    Pseudo R2     =     0.0006

------------------------------------------------------------------------------
           y |      Coef.   Std. Err.      t    P>|t|     [95% Conf. Interval]
-------------+----------------------------------------------------------------
           x |    .002601   .0002737     9.50   0.000     .0020646    .0031374
       _cons |  -.0001393   .0001666    -0.84   0.403     -.000466    .0001874
------------------------------------------------------------------------------

. qreg y x, quant(.95)
Iteration  1:  WLS sum of weighted deviations =  1237.5569
Iteration  1: sum of abs. weighted deviations =  1238.0014
Iteration  2: sum of abs. weighted deviations =  456.65044
  (output omitted )
Iteration  5: sum of abs. weighted deviations =   338.4389
.95 Quantile regression                             Number of obs =      10000
  Raw sum of deviations 554.6889 (about .61326343)
  Min sum of deviations 338.4389                    Pseudo R2     =     0.3899

------------------------------------------------------------------------------
           y |      Coef.   Std. Err.      t    P>|t|     [95% Conf. Interval]
-------------+----------------------------------------------------------------
           x |   .8898259   .0060398   147.33   0.000     .8779867     .901665
       _cons |   .0021514   .0036623     0.59   0.557    -.0050275    .0093302
------------------------------------------------------------------------------
```

The coefficient on x, in particular, differs markedly between the two estimates. For the mathematically inclined, it is not too difficult to show that the theoretical lines are $y = 0.0025\,x$ for the 5th percentile and $y = 0.9025\,x$ for the 95th, numbers in close agreement with our numerical results.

❑

Estimated standard errors

qreg estimates the variance–covariance matrix of the coefficients by using a method of Koenker and Bassett (1982) and Rogers (1993). This approach is described in *Methods and formulas* below. Rogers (1992) reports that, although this method seems adequate for homoskedastic errors, it appears to understate the standard errors for heteroskedastic errors. The irony is that exploring heteroskedastic errors is one of the major benefits of quantile regression. Gould (1992, 1997b) introduced generalized versions of qreg that obtain estimates of the standard errors by using bootstrap resampling (see Efron and Tibshirani [1993] or Wu [1986] for an introduction to bootstrap standard errors). The iqreg, sqreg, and bsqreg commands provide a bootstrapped estimate of the entire variance–covariance matrix of the estimators.

▷ Example 5

The first example of qreg on real data above was a median regression of price on weight, length, and foreign using the automobile data. Here is the result of repeating the estimation using bootstrap standard errors:

```
. set seed 1001
. use http://www.stata-press.com/data/r12/auto, clear
(1978 Automobile Data)
. bsqreg price weight length foreign
(fitting base model)
(bootstrapping ....................)
Median regression, bootstrap(20) SEs              Number of obs =        74
  Raw sum of deviations   142205 (about 4934)
  Min sum of deviations 108822.6                  Pseudo R2     =    0.2347
```

price	Coef.	Std. Err.	t	P>|t|	[95% Conf.	Interval]
weight	3.933588	3.12446	1.26	0.212	-2.297951	10.16513
length	-41.25191	83.71266	-0.49	0.624	-208.2116	125.7077
foreign	3377.771	1057.281	3.19	0.002	1269.09	5486.452
_cons	344.6494	7053.301	0.05	0.961	-13722.72	14412.01

The coefficient estimates are the same—indeed, they are obtained using the same technique. Only the standard errors differ. Therefore, the t statistics, significance levels, and confidence intervals also differ.

Because bsqreg (as well as sqreg and iqreg) obtains standard errors by randomly resampling the data, the standard errors it produces will not be the same from run to run, unless we first set the random-number seed to the same number; see [R] **set seed**.

By default, bsqreg, sqreg, and iqreg use 20 replications. We can control the number of replications by specifying the reps() option:

```
. bsqreg price weight length foreign, reps(1000)
(fitting base model)
(bootstrapping ...................(output omitted)...)
Median regression, bootstrap(1000) SEs              Number of obs =        74
  Raw sum of deviations    142205 (about 4934)
  Min sum of deviations 108822.6                    Pseudo R2     =    0.2347

------------------------------------------------------------------------------
       price |      Coef.   Std. Err.      t    P>|t|     [95% Conf. Interval]
-------------+----------------------------------------------------------------
      weight |   3.933588   2.659381     1.48   0.144    -1.370379    9.237555
      length |  -41.25191   69.29771    -0.60   0.554    -179.4618    96.95802
     foreign |   3377.771   1094.947     3.08   0.003     1193.967    5561.575
       _cons |   344.6494   5916.906     0.06   0.954    -11456.25    12145.55
------------------------------------------------------------------------------
```

A comparison of the standard errors is informative:

Variable	qreg	bsqreg reps(20)	bsqreg reps(1000)
weight	.8602	3.124	2.670
length	28.87	83.71	69.65
foreign	577.3	1057.	1094.
_cons	3260.	7053.	5945.

The results shown above are typical for models with heteroskedastic errors. (Our dependent variable is `price`; if our model had been in terms of ln(`price`), the standard errors estimated by `qreg` and `bsqreg` would have been nearly identical.) Also, even for heteroskedastic errors, 20 replications is generally sufficient for hypothesis tests against 0. ◁

Interquantile and simultaneous-quantile regression

Consider a quantile-regression model where the qth quantile is given by

$$Q_q(y) = a_q + b_{q,1}x_1 + b_{q,2}x_2$$

For instance, the 75th and 25th quantiles are given by

$$Q_{0.75}(y) = a_{0.75} + b_{0.75,1}x_1 + b_{0.75,2}x_2$$
$$Q_{0.25}(y) = a_{0.25} + b_{0.25,1}x_1 + b_{0.25,2}x_2$$

The difference in the quantiles is then

$$Q_{0.75}(y) - Q_{0.25}(y) = (a_{0.75} - a_{0.25}) + (b_{0.75,1} - b_{0.25,1})x_1 + (b_{0.75,2} - b_{0.25,2})x_2$$

`qreg` fits models such as $Q_{0.75}(y)$ and $Q_{0.25}(y)$. `iqreg` fits interquantile models, such as $Q_{0.75}(y) - Q_{0.25}(y)$. The relationships of the coefficients estimated by `qreg` and `iqreg` are exactly as shown: `iqreg` reports coefficients that are the difference in coefficients of two `qreg` models, and, of course, `iqreg` reports the appropriate standard errors, which it obtains by bootstrapping.

`sqreg` is like `qreg` in that it estimates the equations for the quantiles

$$Q_{0.75}(y) = a_{0.75} + b_{0.75,1}x_1 + b_{0.75,2}x_2$$
$$Q_{0.25}(y) = a_{0.25} + b_{0.25,1}x_1 + b_{0.25,2}x_2$$

The coefficients it obtains are the same that would be obtained by estimating each equation separately using qreg. sqreg differs from qreg in that it estimates the equations simultaneously and obtains an estimate of the entire variance–covariance matrix of the estimators by bootstrapping. Thus you can perform hypothesis tests concerning coefficients both within and across equations.

For example, to fit the above model, you could type

```
. qreg y x1 x2, q(.25)
. qreg y x1 x2, q(.75)
```

Doing this, you would obtain estimates of the parameters, but you could not test whether $b_{0.25,1} = b_{0.75,1}$ or, equivalently, $b_{0.75,1} - b_{0.25,1} = 0$. If your interest really is in the difference of coefficients, you could type

```
. iqreg y x1 x2, q(.25 .75)
```

The "coefficients" reported would be the difference in quantile coefficients. You could also estimate both quantiles simultaneously and then test the equality of the coefficients:

```
. sqreg y x1 x2, q(.25 .75)
. test [q25]x1 = [q75]x1
```

Whether you use iqreg or sqreg makes no difference for this test. sqreg, however, because it estimates the quantiles simultaneously, allows you to test other hypotheses. iqreg, by focusing on quantile differences, presents results in a way that is easier to read.

Finally, sqreg can estimate quantiles singly,

```
. sqreg y x1 x2, q(.5)
```

and can thereby be used as a substitute for the slower bsqreg. (Gould [1997b] presents timings demonstrating that sqreg is faster than bsqreg.) sqreg can also estimate more than two quantiles simultaneously:

```
. sqreg y x1 x2, q(.25 .5 .75)
```

▷ Example 6

In demonstrating qreg, we performed quantile regressions using the automobile data. We discovered that the regression of price on weight, length, and foreign produced vastly different coefficients for the 0.25, 0.5, and 0.75 quantile regressions. Here are the coefficients that we obtained:

Variable	25th percentile	50th percentile	75th percentile
weight	1.83	3.93	9.22
length	2.85	−41.25	−220.8
foreign	2209.9	3377.8	3595.1
_cons	−1879.8	344.6	20242.9

All we can say, having estimated these equations separately, is that price seems to depend differently on the weight, length, and foreign variables depending on the portion of the price distribution we examine. We cannot be more precise because the estimates have been made separately. With sqreg, however, we can estimate all the effects simultaneously:

```
. sqreg price weight length foreign, q(.25 .5 .75) reps(100)
(fitting base model)
(bootstrapping ................. (output omitted ) .....)
Simultaneous quantile regression                    Number of obs =         74
  bootstrap(100) SEs                                .25 Pseudo R2 =     0.1697
                                                    .50 Pseudo R2 =     0.2347
                                                    .75 Pseudo R2 =     0.3840
```

price	Coef.	Bootstrap Std. Err.	t	P>|t|	[95% Conf.	Interval]
q25						
weight	1.831789	1.244947	1.47	0.146	-.6511803	4.314758
length	2.845558	27.91648	0.10	0.919	-52.8321	58.52322
foreign	2209.925	911.6566	2.42	0.018	391.6836	4028.167
_cons	-1879.775	2756.871	-0.68	0.498	-7378.18	3618.63
q50						
weight	3.933588	2.732408	1.44	0.154	-1.516029	9.383205
length	-41.25191	75.8087	-0.54	0.588	-192.4476	109.9438
foreign	3377.771	921.578	3.67	0.000	1539.742	5215.8
_cons	344.6494	6810.32	0.05	0.960	-13238.1	13927.4
q75						
weight	9.22291	2.732795	3.37	0.001	3.772523	14.6733
length	-220.7833	87.38042	-2.53	0.014	-395.058	-46.50854
foreign	3595.133	1153.239	3.12	0.003	1295.07	5895.196
_cons	20242.9	9000.697	2.25	0.028	2291.579	38194.23

The coefficient estimates above are the same as those previously estimated, although the standard error estimates are a little different. `sqreg` obtains estimates of variance by bootstrapping. Rogers (1992) provides evidence that, for quantile regression, the bootstrap standard errors are better than those calculated analytically by Stata.

The important thing here, however, is that the full covariance matrix of the estimators has been estimated and stored, and thus it is now possible to perform hypothesis tests. Are the effects of `weight` the same at the 25th and 75th percentiles?

```
. test [q25]weight = [q75]weight
 ( 1)  [q25]weight - [q75]weight = 0
       F(  1,    70) =    8.29
            Prob > F =    0.0053
```

It appears that they are not. We can obtain a confidence interval for the difference by using `lincom`:

```
. lincom [q75]weight-[q25]weight
 ( 1)  - [q25]weight + [q75]weight = 0
```

price	Coef.	Std. Err.	t	P>|t|	[95% Conf.	Interval]
(1)	7.391121	2.567684	2.88	0.005	2.270036	12.51221

Indeed, we could test whether the `weight` and `length` sets of coefficients are equal at the three quantiles estimated:

```
. quietly test [q25]weight = [q50]weight
. quietly test [q25]weight = [q75]weight, accum
```

```
. quietly test [q25]length = [q50]length, accum
. test [q25]length = [q75]length, accum
 ( 1)  [q25]weight - [q50]weight = 0
 ( 2)  [q25]weight - [q75]weight = 0
 ( 3)  [q25]length - [q50]length = 0
 ( 4)  [q25]length - [q75]length = 0
       F(  4,    70) =    2.21
            Prob > F =    0.0767
```

iqreg focuses on one quantile comparison but presents results that are more easily interpreted:

```
. set seed 1001
. iqreg price weight length foreign, q(.25 .75) reps(100) nodots
.75-.25 Interquantile regression                    Number of obs =        74
  bootstrap(100) SEs                                .75 Pseudo R2 =    0.3840
                                                    .25 Pseudo R2 =    0.1697
```

price	Coef.	Bootstrap Std. Err.	t	P>|t|	[95% Conf.	Interval]
weight	7.391121	2.467548	3.00	0.004	2.469752	12.31249
length	-223.6288	83.09868	-2.69	0.009	-389.3639	-57.89376
foreign	1385.208	1193.557	1.16	0.250	-995.2672	3765.683
_cons	22122.68	9009.159	2.46	0.017	4154.478	40090.88

Looking only at the 0.25 and 0.75 quantiles (the interquartile range), the iqreg command output is easily interpreted. Increases in weight correspond significantly to increases in price dispersion. Increases in length correspond to decreases in price dispersion. The foreign variable does not significantly change price dispersion.

Do not make too much of these results; the purpose of this example is simply to illustrate the sqreg and iqreg commands and to do so in a context that suggests why analyzing dispersion might be of interest.

lincom after sqreg produced the same t statistic for the interquartile range of weight, as did the iqreg command above. In general, they will not agree exactly because of the randomness of bootstrapping, unless the random-number seed is set to the same value before estimation (as was done here).

◁

Gould (1997a) presents simulation results showing that the coverage—the actual percentage of confidence intervals containing the true value—for iqreg is appropriate.

Saved results

qreg saves the following in e():

Scalars

e(N)	number of observations
e(df_m)	model degrees of freedom
e(df_r)	residual degrees of freedom
e(q)	quantile requested
e(q_v)	value of the quantile
e(sum_adev)	sum of absolute deviations
e(sum_rdev)	sum of raw deviations
e(f_r)	residual density estimate
e(rank)	rank of e(V)
e(convcode)	0 if converged; otherwise, return code for why nonconvergence

Macros

e(cmd)	qreg
e(cmdline)	command as typed
e(depvar)	name of dependent variable
e(properties)	b V
e(predict)	program used to implement predict
e(marginsnotok)	predictions disallowed by margins

Matrices

e(b)	coefficient vector
e(V)	variance–covariance matrix of the estimators

Functions

e(sample)	marks estimation sample

iqreg saves the following in e():

Scalars

e(N)	number of observations
e(df_r)	residual degrees of freedom
e(q0)	lower quantile requested
e(q1)	upper quantile requested
e(reps)	number of replications
e(sumrdev0)	lower quantile sum of raw deviations
e(sumrdev1)	upper quantile sum of raw deviations
e(sumadev0)	lower quantile sum of absolute deviations
e(sumadev1)	upper quantile sum of absolute deviations
e(rank)	rank of e(V)
e(convcode)	0 if converged; otherwise, return code for why nonconvergence

Macros

e(cmd)	iqreg
e(cmdline)	command as typed
e(depvar)	name of dependent variable
e(vcetype)	title used to label Std. Err.
e(properties)	b V
e(predict)	program used to implement predict
e(marginsnotok)	predictions disallowed by margins

Matrices

e(b)	coefficient vector
e(V)	variance–covariance matrix of the estimators

Functions

e(sample)	marks estimation sample

sqreg saves the following in e():

Scalars

e(N)	number of observations
e(df_r)	residual degrees of freedom
e(n_q)	number of quantiles requested
e(q#)	the quantiles requested
e(reps)	number of replications
e(sumrdv#)	sum of raw deviations for q#
e(sumadv#)	sum of absolute deviations for q#
e(rank)	rank of e(V)
e(convcode)	0 if converged; otherwise, return code for why nonconvergence

Macros

e(cmd)	sqreg
e(cmdline)	command as typed
e(depvar)	name of dependent variable
e(eqnames)	names of equations
e(vcetype)	title used to label Std. Err.
e(properties)	b V
e(predict)	program used to implement predict
e(marginsnotok)	predictions disallowed by margins

Matrices

e(b)	coefficient vector
e(V)	variance–covariance matrix of the estimators

Functions

e(sample)	marks estimation sample

bsqreg saves the following in e():

Scalars

e(N)	number of observations
e(df_r)	residual degrees of freedom
e(q)	quantile requested
e(q_v)	value of the quantile
e(reps)	number of replications
e(sum_adev)	sum of absolute deviations
e(sum_rdev)	sum of raw deviations
e(rank)	rank of e(V)
e(convcode)	0 if converged; otherwise, return code for why nonconvergence

Macros

e(cmd)	bsqreg
e(cmdline)	command as typed
e(depvar)	name of dependent variable
e(properties)	b V
e(predict)	program used to implement predict
e(marginsnotok)	predictions disallowed by margins

Matrices

e(b)	coefficient vector
e(V)	variance–covariance matrix of the estimators

Functions

e(sample)	marks estimation sample

_qreg saves the following in r():

Scalars

r(N)	number of observations
r(df_m)	model degrees of freedom
r(q)	quantile requested
r(q_v)	value of the quantile
r(sum_w)	sum of the weights
r(sum_adev)	sum of absolute deviations
r(sum_rdev)	sum of raw deviations
r(f_r)	residual density estimate
r(ic)	number of iterations
r(convcode)	1 if converged, 0 otherwise

Methods and formulas

qreg, iqreg, sqreg, and bsqreg are implemented as ado-files.

According to Stuart and Ord (1991, 1084), the method of minimum absolute deviations was first proposed by Boscovich in 1757 and was later developed by Laplace; Stigler (1986, 39–55) and Hald (1998, 97–103, 112–116) provide historical details. According to Bloomfield and Steiger (1980), Harris (1950) later observed that the problem of minimum absolute deviations could be turned into the linear programming problem that was first implemented by Wagner (1959). Interest has grown in this method because of interest in robust methods. Statistical and computational properties of minimum absolute deviation estimators are surveyed by Narula and Wellington (1982). Hao and Naiman (2007) provide an excellent introduction to quantile-regression methods.

Define q as the quantile to be estimated; the median is $q = 0.5$. For each observation i, let r_i be the residual

$$r_i = y_i - \sum_j \beta_j x_{ij}$$

Define the multiplier h_i

$$h_i = \begin{cases} 2q & \text{if } r_i > 0 \\ 2(1-q) & \text{otherwise} \end{cases}$$

The quantity being minimized with respect to β_j is $\sum_i |r_i| h_i$, so quantiles other than the median are estimated by weighting the residuals. For example, if we want to estimate the 75th percentile, we weight the negative residuals by 0.50 and the positive residuals by 1.50. It can be shown that the criterion is minimized when 75% of the residuals are negative.

This is set up as a linear programming problem and is solved via linear programming techniques, as suggested by Armstrong, Frome, and Kung (1979) and used by courtesy of Marcel Dekker, Inc. The definition of convergence is exact in the sense that no amount of added iterations could improve the solution. Each step is described by a set of observations through which the regression plane passes, called the *basis*. A step is taken by replacing a point in the basis if the sum of weighted absolute deviations can be improved. If this occurs, a line is printed in the iteration log. The linear programming method is started by doing a weighted least-squares (WLS) regression to identify a good set of observations to use as a starting basis. The WLS algorithm for $q = 0.5$ is taken from Schlossmacher (1973) with a generalization for $0 < q < 1$ implied from Hunter and Lange (2000).

The variances are estimated using a method suggested by Koenker and Bassett (1982). This method can be put into a form recommended by Huber (1967) for M estimates, where

$$\text{cov}(\boldsymbol{\beta}) = \mathbf{R}_2^{-1}\mathbf{R}_1\mathbf{R}_2^{-1}$$

$\mathbf{R}_1 = \mathbf{X}'\mathbf{W}\mathbf{W}'\mathbf{X}$ (in the Huber formulation), $\mathbf{W}$ is a diagonal matrix with elements

$$W_{ii} = \begin{cases} q/f_{\text{residuals}}(0) & \text{if } r > 0 \\ (1-q)/f_{\text{residuals}}(0) & \text{if } r < 0 \\ 0 & \text{otherwise} \end{cases}$$

and $\mathbf{R}_2$ is the design matrix $\mathbf{X}'\mathbf{X}$. This is derived from formula 3.11 in Koenker and Bassett, although their notation is much different. $f_{\text{residuals}}()$ refers to the density of the true residuals. Koenker and Bassett leave much unspecified, including how to obtain a density estimate for the errors in real data. At this point, we offer our contribution (Rogers 1993).

We first sort the residuals and locate the observation in the residuals corresponding to the quantile in question, taking into account weights if they are applied. We then calculate w_n, the square root of the sum of the weights. Unweighted data are equivalent to weighted data in which each observation has weight 1, resulting in $w_n = \sqrt{n}$. For analytically weighted data, the weights are rescaled so that the sum of the weights is the number of observations, resulting in $\sqrt{n}$ again. For frequency-weighted data, w_n literally is the square root of the sum of the weights.

We locate the closest observation in each direction, such that the sum of weights for all closer observations is w_n. If we run off the end of the dataset, we stop. We calculate w_s, the sum of weights for all observations in this middle space. Typically, w_s is slightly greater than w_n.

If there are k parameters, then exactly k of the residuals must be zero. Thus we calculate an adjusted weight $w_a = w_s - k$. The density estimate is the distance spanned by these observations divided by w_a. Because the distance spanned by this mechanism converges toward zero, this estimate of density converges in probability to the true density.

The pseudo-R^2 is calculated as

$$1 - \frac{\text{sum of weighted deviations about estimated quantile}}{\text{sum of weighted deviations about raw quantile}}$$

This is based on the likelihood for a double-exponential distribution $e^{h_i |r_i|}$.

References

Angrist, J. D., and J.-S. Pischke. 2009. *Mostly Harmless Econometrics: An Empiricist's Companion*. Princeton, NJ: Princeton University Press.

Armstrong, R. D., E. L. Frome, and D. S. Kung. 1979. Algorithm 79-01: A revised simplex algorithm for the absolute deviation curve fitting problem. *Communications in Statistics, Simulation and Computation* 8: 175–190.

Bloomfield, P., and W. Steiger. 1980. Least absolute deviations curve-fitting. *SIAM Journal on Scientific Computing* 1: 290–301.

Cameron, A. C., and P. K. Trivedi. 2010. *Microeconometrics Using Stata*. Rev. ed. College Station, TX: Stata Press.

Efron, B., and R. J. Tibshirani. 1993. *An Introduction to the Bootstrap*. New York: Chapman & Hall/CRC.

Frölich, M., and B. Melly. 2010. Estimation of quantile treatment effects with Stata. *Stata Journal* 10: 423–457.

Gould, W. W. 1992. sg11.1: Quantile regression with bootstrapped standard errors. *Stata Technical Bulletin* 9: 19–21. Reprinted in *Stata Technical Bulletin Reprints*, vol. 2, pp. 137–139. College Station, TX: Stata Press.

——. 1997a. crc46: Better numerical derivatives and integrals. *Stata Technical Bulletin* 35: 3–5. Reprinted in *Stata Technical Bulletin Reprints*, vol. 6, pp. 8–12. College Station, TX: Stata Press.

——. 1997b. sg70: Interquantile and simultaneous-quantile regression. *Stata Technical Bulletin* 38: 14–22. Reprinted in *Stata Technical Bulletin Reprints*, vol. 7, pp. 167–176. College Station, TX: Stata Press.

Gould, W. W., and W. H. Rogers. 1994. Quantile regression as an alternative to robust regression. In *1994 Proceedings of the Statistical Computing Section*. Alexandria, VA: American Statistical Association.

Hald, A. 1998. *A History of Mathematical Statistics from 1750 to 1930*. New York: Wiley.

Hao, L., and D. Q. Naiman. 2007. *Quantile Regression*. Thousand Oaks, CA: Sage.

Harris, T. 1950. Regression using minimum absolute deviations. *American Statistician* 4: 14–15.

Huber, P. J. 1967. The behavior of maximum likelihood estimates under nonstandard conditions. In Vol. 1 of *Proceedings of the Fifth Berkeley Symposium on Mathematical Statistics and Probability*, 221–233. Berkeley: University of California Press.

——. 1981. *Robust Statistics*. New York: Wiley.

Hunter, D. R., and K. Lange. 2000. Quantile regression via an MM algorithm. *Journal of Computational and Graphical Statistics* 9: 60–77.

Jolliffe, D., B. Krushelnytskyy, and A. Semykina. 2000. sg153: Censored least absolute deviations estimator: CLAD. *Stata Technical Bulletin* 58: 13–16. Reprinted in *Stata Technical Bulletin Reprints*, vol. 10, pp. 240–244. College Station, TX: Stata Press.

Koenker, R., and G. Bassett, Jr. 1982. Robust tests for heteroscedasticity based on regression quantiles. *Econometrica* 50: 43–61.

Koenker, R., and K. Hallock. 2001. Quantile regression. *Journal of Economic Perspectives* 15: 143–156.

Narula, S. C., and J. F. Wellington. 1982. The minimum sum of absolute errors regression: A state of the art survey. *International Statistical Review* 50: 317–326.

Rogers, W. H. 1992. sg11: Quantile regression standard errors. *Stata Technical Bulletin* 9: 16–19. Reprinted in *Stata Technical Bulletin Reprints*, vol. 2, pp. 133–137. College Station, TX: Stata Press.

——. 1993. sg11.2: Calculation of quantile regression standard errors. *Stata Technical Bulletin* 13: 18–19. Reprinted in *Stata Technical Bulletin Reprints*, vol. 3, pp. 77–78. College Station, TX: Stata Press.

Rousseeuw, P. J., and A. M. Leroy. 1987. *Robust Regression and Outlier Detection*. New York: Wiley.

Schlossmacher, E. J. 1973. An iterative technique for absolute deviations curve fitting. *Journal of the American Statistical Association* 68: 857–859.

Stigler, S. M. 1986. *The History of Statistics: The Measurement of Uncertainty before 1900*. Cambridge, MA: Belknap Press.

Stuart, A., and J. K. Ord. 1991. *Kendall's Advanced Theory of Statistics: Distribution Theory, Vol I*. 5th ed. New York: Oxford University Press.

Wagner, H. M. 1959. Linear programming techniques for regression analysis. *Journal of the American Statistical Association* 54: 206–212.

Wu, C. F. J. 1986. Jackknife, bootstrap and other resampling methods in regression analysis. *Annals of Statistics* 14: 1261–1350 (including discussions and rejoinder).

Also see

[R] **qreg postestimation** — Postestimation tools for qreg, iqreg, sqreg, and bsqreg

[R] **bootstrap** — Bootstrap sampling and estimation

[R] **regress** — Linear regression

[R] **rreg** — Robust regression

[MI] **estimation** — Estimation commands for use with mi estimate

[U] **20 Estimation and postestimation commands**

Title

qreg postestimation — Postestimation tools for qreg, iqreg, sqreg, and bsqreg

Description

The following postestimation commands are available after `qreg`, `iqreg`, `bsqreg`, and `sqreg`:

Command	Description
`estat`	VCE and estimation sample summary
`estimates`	cataloging estimation results
`lincom`	point estimates, standard errors, testing, and inference for linear combinations of coefficients
`linktest`	link test for model specification
`margins`	marginal means, predictive margins, marginal effects, and average marginal effects
`marginsplot`	graph the results from margins (profile plots, interaction plots, etc.)
`nlcom`	point estimates, standard errors, testing, and inference for nonlinear combinations of coefficients
`predict`	predictions, residuals, influence statistics, and other diagnostic measures
`predictnl`	point estimates, standard errors, testing, and inference for generalized predictions
`test`	Wald tests of simple and composite linear hypotheses
`testnl`	Wald tests of nonlinear hypotheses

See the corresponding entries in the *Base Reference Manual* for details.

Syntax for predict

For qreg, iqreg, and bsqreg

`predict` [*type*] *newvar* [*if*] [*in*] [, [`xb` | `stdp` | `residuals`]]

For sqreg

`predict` [*type*] *newvar* [*if*] [*in*] [, `equation(`*eqno*[`,`*eqno*]`)` *statistic*]

statistic	Description
Main	
`xb`	linear prediction; the default
`stdp`	standard error of the linear prediction
`stddp`	standard error of the difference in linear predictions
`residuals`	residuals

These statistics are available both in and out of sample; type `predict ... if e(sample) ...` if wanted only for the estimation sample.

Menu

Statistics > Postestimation > Predictions, residuals, etc.

Options for predict

Main

`xb`, the default, calculates the linear prediction.

`stdp` calculates the standard error of the linear prediction.

`stddp` is allowed only after you have fit a model using `sqreg`. The standard error of the difference in linear predictions ($\mathbf{x}_{1j}\mathbf{b} - \mathbf{x}_{2j}\mathbf{b}$) between equations 1 and 2 is calculated.

`residuals` calculates the residuals, that is, $y_j - \mathbf{x}_j\mathbf{b}$.

`equation(`*eqno* [`,`*eqno*]`)` specifies the equation to which you are making the calculation.

`equation()` is filled in with one *eqno* for the `xb`, `stdp`, and `residuals` options. `equation(#1)` would mean that the calculation is to be made for the first equation, `equation(#2)` would mean the second, and so on. You could also refer to the equations by their names. `equation(income)` would refer to the equation named income and `equation(hours)` to the equation named hours.

If you do not specify `equation()`, results are the same as if you had specified `equation(#1)`.

To use `stddp`, you must specify two equations. You might specify `equation(#1, #2)` or `equation(q80, q20)` to indicate the 80th and 20th quantiles.

Methods and formulas

All postestimation commands listed above are implemented as ado-files.

Also see

[R] **qreg** — Quantile regression

[U] **20 Estimation and postestimation commands**

Title

query — Display system parameters

Syntax

`query [ memory | output | interface | graphics | efficiency | network | update | trace | mata | other ]`

Description

query displays the settings of various Stata parameters.

Remarks

query provides more system information than you will ever want to know. You do not need to understand every line of output that query produces if all you need is one piece of information. Here is what happens when you type query:

```
. query

    Memory settings
        set maxvar          5000          2048-32767; max. vars allowed
        set matsize         400           10-11000; max. # vars in models
        set niceness        5             0-10
        set min_memory      0             0-1600gc
        set max_memory      .             32mc-1600gc or .
        set segmentsize     32mc          1mc-32gc

    Output settings
        set more            on
        set rmsg            off
        set dp              period        may be period or comma
        set linesize        80            characters
        set pagesize        27            lines

        set level           95            percent confidence intervals

        set showbaselevels                may be empty, off, on, or all
        set showemptycells                may be empty, off, or on
        set showomitted                   may be empty, off, or on
        set lstretch                      may be empty, off, or on

        set cformat                       may be empty or a numerical format
        set pformat                       may be empty or a numerical format
        set sformat                       may be empty or a numerical format

        set logtype         smcl          may be smcl or text
```

```
    Interface settings
        set dockable        on
        set dockingguides   on
        set floatresults    off
        set floatwindows    off
        set locksplitters   off
        set pinnable        on
        set doublebuffer    on
        ---------------------------------------------------------------------
        set linegap         1          pixels
        set scrollbufsize   204800     characters
        set fastscroll      on
        set reventries      5000       lines
        ---------------------------------------------------------------------
        set maxdb           50         dialog boxes
-----------------------------------------------------------------------------
    Graphics settings
        set graphics        on
        set autotabgraphs   off
        set scheme          s2color
        set printcolor      automatic  may be automatic, asis, gs1, gs2, gs3
        set copycolor       automatic  may be automatic, asis, gs1, gs2, gs3
-----------------------------------------------------------------------------
    Efficiency settings
        set adosize         1000       kilobytes
-----------------------------------------------------------------------------
    Network settings
        set checksum        off
        set timeout1         30        seconds
        set timeout2        180        seconds
        ---------------------------------------------------------------------
        set httpproxy       off
        set httpproxyhost
        set httpproxyport   80
        ---------------------------------------------------------------------
        set httpproxyauth   off
        set httpproxyuser
        set httpproxypw
-----------------------------------------------------------------------------
    Update settings
        set update_query    on
        set update_interval 7
        set update_prompt   on
-----------------------------------------------------------------------------
    Trace (programming debugging) settings
        set trace           off
        set tracedepth      32000
        set traceexpand     on
        set tracesep        on
        set traceindent     on
        set tracenumber     off
        set tracehilite
-----------------------------------------------------------------------------
```

```
    Mata settings
        set matastrict      off
        set matalnum        off
        set mataoptimize    on
        set matafavor       space       may be space or speed
        set matacache       400         kilobytes
        set matalibs        lmatabase;lmataado;lmataopt
        set matamofirst     off
--------------------------------------------------------------------------
    Other settings
        set type            float       may be float or double
        set maxiter         16000       max iterations for estimation commands
        set searchdefault   local       may be local, net, or all
        set seed            X075bcd151f123bb5159a55e50022865746ad
        set varabbrev       on
        set emptycells      keep        may be keep or drop
        set processors      1
--------------------------------------------------------------------------
```

The output is broken into several divisions: memory, output, interface, graphics, efficiency, network, update, trace, mata, and other settings. We will discuss each one in turn.

We generated the output above using Stata/SE for Windows. Here is what happens when we type `query` and we are running Stata/IC for Mac:

```
. query
--------------------------------------------------------------------------
    Memory settings
        set maxvar          2048        (not settable in this version of Stata)
        set matsize         400         10-800; max. # vars in model
        set niceness        5           0-10
        set min_memory      0           0-1600g
        set max_memory      .           32m-1600g or .
        set segmentsize     32m         1m-32g
--------------------------------------------------------------------------
    Output settings
        set more            on
        set rmsg            off
        set dp              period      may be period or comma
        set linesize        80          characters
        set pagesize        25          lines
        ------------------------------------------------------------------
        set level           95          percent confidence intervals
        ------------------------------------------------------------------
        set showbaselevels              may be empty, off, on, or all
        set showemptycells              may be empty, off, or on
        set showomitted                 may be empty, off, or on
        set lstretch                    may be empty, off, or on
        ------------------------------------------------------------------
        set cformat                     may be empty or a numerical format
        set pformat                     may be empty or a numerical format
        set sformat                     may be empty or a numerical format
        ------------------------------------------------------------------
        set logtype         smcl        may be smcl or text
        ------------------------------------------------------------------
        set eolchar         unix        may be max or unix
        set notifyuser      on
        set playsnd         off
        set include_bitmap  on
--------------------------------------------------------------------------
```

```
    Interface settings
        set revkeyboard     on
        set varkeyboard     on
        set smoothfonts     on
        ---------------------------------------------------------------------
        set linegap         1           pixels
        set scrollbufsize   204800      characters
        set reventries      5000        lines
        ---------------------------------------------------------------------
        set maxdb           50          dialog boxes
-------------------------------------------------------------------------------
    Graphics settings
        set graphics        on
        set scheme          s2color
        set printcolor      automatic   may be automatic, asis, gs1, gs2, gs3
        set copycolor       automatic   may be automatic, asis, gs1, gs2, gs3
-------------------------------------------------------------------------------
    Efficiency settings
        set adosize         1000        kilobytes
-------------------------------------------------------------------------------
    Network settings
        set checksum        off
        set timeout1        30          seconds
        set timeout2        180         seconds
        ---------------------------------------------------------------------
        set httpproxy       off
        set httpproxyhost
        set httpproxyport   80
        ---------------------------------------------------------------------
        set httpproxyauth   off
        set httpproxyuser
        set httpproxypw
-------------------------------------------------------------------------------
    Update settings
        set update_query    on
        set update_interval 7
        set update_prompt   on
-------------------------------------------------------------------------------
    Trace (programming debugging) settings
        set trace           off
        set tracedepth      32000
        set traceexpand     on
        set tracesep        on
        set traceindent     on
        set tracenumber     off
        set tracehilite
-------------------------------------------------------------------------------
    Mata settings
        set matastrict      off
        set matalnum        off
        set mataoptimize    on
        set matafavor       space       may be space or speed
        set matacache       400         kilobytes
        set matalibs        lmatabase;lmataado;lmataopt
        set matamofirst     off
-------------------------------------------------------------------------------
```

```
Other settings
    set type            float       may be float or double
    set maxiter         16000       max iterations for estimation commands
    set searchdefault   local       may be local, net, or all
    set seed            X075bcd151f123bb5159a55e50022865746ad
    set varabbrev       on
    set emptycells      keep        may be keep or drop
    set processors      1
```

Memory settings

Memory settings indicate how memory is allocated, the maximum number of variables, and the maximum size of a matrix.

For more information, see

maxvar	[D] **memory**
matsize	[R] **matsize**
niceness	[D] **memory**
min_memory	[D] **memory**
max_memory	[D] **memory**
segmentsize	[D] **memory**

Output settings

Output settings show how Stata displays output on the screen and in log files.

For more information, see

more	[R] **more**
rmsg	[P] **rmsg**
dp	[D] **format**
linesize	[R] **log**
pagesize	[R] **more**
level	[R] **level**
showbaselevels	[R] **set showbaselevels**
showemptycells	[R] **set showbaselevels**
showomitted	[R] **set showbaselevels**
cformat	[R] **set cformat**
pformat	[R] **set cformat**
sformat	[R] **set cformat**
lstretch	[R] **set**
logtype	[R] **log**
eolchar	[R] **set**
notifyuser	[R] **set**
playsnd	[R] **set**
include_bitmap	[R] **set**

Interface settings

Interface settings control how Stata's interface works.

For more information, see

dockable	[R] **set**
dockingguides	[R] **set**
floatresults	[R] **set**
floatwindows	[R] **set**
locksplitters	[R] **set**
pinnable	[R] **set**
doublebuffer	[R] **set**
revkeyboard	[R] **set**
varkeyboard	[R] **set**
smoothfonts	[R] **set**
linegap	[R] **set**
scrollbufsize	[R] **set**
fastscroll	[R] **set**
reventries	[R] **set**
maxdb	[R] **db**

Graphics settings

Graphics settings indicate how Stata's graphics are displayed.

For more information, see

graphics	[G-2] **set graphics**
autotabgraphs	[R] **set**
scheme	[G-2] **set scheme**
printcolor	[G-2] **set printcolor**
copycolor	[G-2] **set printcolor**

Efficiency settings

The efficiency settings set the maximum amount of memory allocated to automatically loaded do-files, the maximum number of remembered-contents dialog boxes, and the use of virtual memory.

For more information, see

adosize	[P] **sysdir**

Network settings

Network settings determine how Stata interacts with the Internet.

For more information, see [R] **netio**.

Update settings

Update settings determine how Stata performs updates.

For more information, see [R] **update**.

Trace settings

Trace settings adjust Stata's behavior and are particularly useful in debugging code.

For more information, see [P] **trace**.

Mata settings

Mata settings affect Mata's system parameters.

For more information, see [M-3] **mata set**.

Other settings

The other settings are a miscellaneous collection.

For more information, see

`type`	[D] **generate**
`maxiter`	[R] **maximize**
`searchdefault`	[R] **search**
`seed`	[R] **set seed**
`varabbrev`	[R] **set**
`emptycells`	[R] **set**
`processors`	[R] **set**
`odbcmgr`	[D] **odbc**

In general, the parameters displayed by `query` can be changed by `set`; see [R] **set**.

Also see

[R] **set** — Overview of system parameters

[P] **creturn** — Return c-class values

[M-3] **mata set** — Set and display Mata system parameters

Title

ranksum — Equality tests on unmatched data

Syntax

Wilcoxon rank-sum test

`ranksum` *varname* [*if*] [*in*] , `by(`*groupvar*`)` [`porder`]

Nonparametric equality-of-medians test

`median` *varname* [*if*] [*in*] [*weight*] , `by(`*groupvar*`)` [*median_options*]

ranksum_options	Description
Main	
* `by(`*groupvar*`)`	grouping variable
`porder`	probability that variable for first group is larger than variable for second group

median_options	Description
Main	
* `by(`*groupvar*`)`	grouping variable
`exact`	perform Fisher's exact test
`medianties(below)`	assign values equal to the median to below group
`medianties(above)`	assign values equal to the median to above group
`medianties(drop)`	drop values equal to the median from the analysis
`medianties(split)`	split values equal to the median equally between the two groups

* `by(`*groupvar*`)` is required.

`by` is allowed with `ranksum` and `median`; see [D] **by**.

`fweight`s are allowed with `median`; see [U] **11.1.6 weight**.

Menu

ranksum

Statistics > Nonparametric analysis > Tests of hypotheses > Wilcoxon rank-sum test

median

Statistics > Nonparametric analysis > Tests of hypotheses > K-sample equality-of-medians test

Description

`ranksum` tests the hypothesis that two independent samples (that is, *unmatched* data) are from populations with the same distribution by using the Wilcoxon rank-sum test, which is also known as the Mann–Whitney two-sample statistic (Wilcoxon 1945; Mann and Whitney 1947).

`median` performs a nonparametric *k*-sample test on the equality of medians. It tests the null hypothesis that the *k* samples were drawn from populations with the same median. For two samples, the chi-squared test statistic is computed both with and without a continuity correction.

`ranksum` and `median` are for use with *unmatched* data. For equality tests on matched data, see [R] **signrank**.

Options for ranksum

Main

`by(`*groupvar*`)` is required. It specifies the name of the grouping variable.

`porder` displays an estimate of the probability that a random draw from the first population is larger than a random draw from the second population.

Options for median

Main

`by(`*groupvar*`)` is required. It specifies the name of the grouping variable.

`exact` displays the significance calculated by Fisher's exact test. For two samples, both one- and two-sided probabilities are displayed.

`medianties(below|above|drop|split)` specifies how values equal to the overall median are to be handled. The median test computes the median for *varname* by using all observations and then divides the observations into those falling above the median and those falling below the median. When values for an observation are equal to the sample median, they can be dropped from the analysis by specifying `medianties(drop)`; added to the group above or below the median by specifying `medianties(above)` or `medianties(below)`, respectively; or if there is more than 1 observation with values equal to the median, they can be equally divided into the two groups by specifying `medianties(split)`. If this option is not specified, `medianties(below)` is assumed.

Remarks

▷ Example 1

We are testing the effectiveness of a new fuel additive. We run an experiment with 24 cars: 12 cars with the fuel treatment and 12 cars without. We input these data by creating a dataset with 24 observations. `mpg` records the mileage rating, and `treat` records 0 if the mileage corresponds to untreated fuel and 1 if it corresponds to treated fuel.

```
. use http://www.stata-press.com/data/r12/fuel2
. ranksum mpg, by(treat)
Two-sample Wilcoxon rank-sum (Mann-Whitney) test
```

treat	obs	rank sum	expected
0	12	128	150
1	12	172	150
combined	24	300	300

```
unadjusted variance       300.00
adjustment for ties        -4.04
                        ----------
adjusted variance         295.96

Ho: mpg(treat==0) = mpg(treat==1)
             z =  -1.279
    Prob > |z| =   0.2010
```

These results indicate that the medians are not statistically different at any level smaller than 20.1%. Similarly, the median test,

```
. median mpg, by(treat) exact
Median test

   Greater |
  than the |        treat
    median |         0          1 |     Total
-----------+----------------------+----------
        no |         7          5 |        12
       yes |         5          7 |        12
-----------+----------------------+----------
     Total |        12         12 |        24

          Pearson chi2(1) =   0.6667   Pr = 0.414
           Fisher's exact =                 0.684
   1-sided Fisher's exact =                 0.342

   Continuity corrected:
          Pearson chi2(1) =   0.1667   Pr = 0.683
```

fails to reject the null hypothesis that there is no difference between the two fuel additives.

Compare these results from these two tests with those obtained from the signrank and signtest where we found significant differences; see [R] **signrank**. An experiment run on 24 different cars is not as powerful as a before-and-after comparison using the same 12 cars.

◁

Saved results

ranksum saves the following in r():

Scalars

r(N_1)	sample size n_1
r(N_2)	sample size n_2
r(z)	z statistic
r(Var_a)	adjusted variance
r(group1)	value of variable for first group
r(sum_obs)	actual sum of ranks for first group
r(sum_exp)	expected sum of ranks for first group
r(porder)	probability that draw from first population is larger than draw from second population

median saves the following in r():

Scalars

r(N)	sample size
r(chi2)	Pearson's χ^2
r(p)	significance of Pearson's χ^2
r(p_exact)	Fisher's exact p
r(groups)	number of groups compared
r(chi2_cc)	continuity-corrected Pearson's χ^2
r(p_cc)	continuity-corrected significance
r(p1_exact)	one-sided Fisher's exact p

Methods and formulas

ranksum and median are implemented as ado-files.

For a practical introduction to these techniques with an emphasis on examples rather than theory, see Acock (2010), Bland (2000), or Sprent and Smeeton (2007). For a summary of these tests, see Snedecor and Cochran (1989).

Methods and formulas are presented under the following headings:

ranksum
median

ranksum

For the Wilcoxon rank-sum test, there are two independent random variables, X_1 and X_2, and we test the null hypothesis that $X_1 \sim X_2$. We have a sample of size n_1 from X_1 and another of size n_2 from X_2.

The data are then ranked without regard to the sample to which they belong. If the data are tied, averaged ranks are used. Wilcoxon's test statistic (1945) is the sum of the ranks for the observations in the first sample:

$$T = \sum_{i=1}^{n_1} R_{1i}$$

Mann and Whitney's U statistic (1947) is the number of pairs (X_{1i}, X_{2j}) such that $X_{1i} > X_{2j}$. These statistics differ only by a constant:

$$U = T - \frac{n_1(n_1+1)}{2}$$

Again Fisher's principle of randomization provides a method for calculating the distribution of the test statistic, ties or not. The randomization distribution consists of the $\binom{n}{n_1}$ ways to choose n_1 ranks from the set of all $n = n_1 + n_2$ ranks and assign them to the first sample.

It is a straightforward exercise to verify that

$$E(T) = \frac{n_1(n+1)}{2} \qquad \text{and} \qquad \text{Var}(T) = \frac{n_1 n_2 s^2}{n}$$

where s is the standard deviation of the combined ranks, r_i, for both groups:

$$s^2 = \frac{1}{n-1}\sum_{i=1}^{n}(r_i - \overline{r})^2$$

This formula for the variance is exact and holds both when there are no ties and when there are ties and we use averaged ranks. (Indeed, the variance formula holds for the randomization distribution of choosing n_1 numbers from any set of n numbers.)

Using a normal approximation, we calculate

$$z = \frac{T - E(T)}{\sqrt{\text{Var}(T)}}$$

When the porder option is specified, the probability

$$p = \frac{U}{n_1 n_2}$$

is computed.

median

The median test examines whether it is likely that two or more samples came from populations with the same median. The null hypothesis is that the samples were drawn from populations with the same median. The alternative hypothesis is that at least one sample was drawn from a population with a different median. The test should be used only with ordinal or interval data.

Assume that there are score values for k independent samples to be compared. The median test is performed by first computing the median score for all observations combined, regardless of the sample group. Each score is compared with this computed grand median and is classified as being above the grand median, below the grand median, or equal to the grand median. Observations with scores equal to the grand median can be dropped, added to the "above" group, added to the "below" group, or split between the two groups.

Once all observations are classified, the data are cast into a $2 \times k$ contingency table, and a Pearson's chi-squared test or Fisher's exact test is performed.

Henry Berthold Mann (1905–2000) was born in Vienna, Austria, where he completed a doctorate in algebraic number theory. He moved to the United States in 1938 and for several years made his livelihood by tutoring in New York. During this time, he proved a celebrated conjecture in number theory and studied statistics at Columbia with Abraham Wald, with whom he wrote three papers. After the war, he taught at Ohio State and the Universities of Wisconsin and Arizona. In addition to his work in number theory and statistics, he made major contributions to algebra and combinatorics.

Donald Ransom Whitney (1915–2007) studied at Oberlin, Princeton, and Ohio State Universities and worked at the latter throughout his career. His PhD thesis under Henry Mann was on nonparametric statistics. It was this work that produced the test that bears their names.

References

Acock, A. C. 2010. *A Gentle Introduction to Stata*. 3rd ed. College Station, TX: Stata Press.

Bland, M. 2000. *An Introduction to Medical Statistics*. 3rd ed. Oxford: Oxford University Press.

Feiveson, A. H. 2002. Power by simulation. *Stata Journal* 2: 107–124.

Fisher, R. A. 1935. *The Design of Experiments*. Edinburgh: Oliver & Boyd.

Goldstein, R. 1997. sg69: Immediate Mann–Whitney and binomial effect-size display. *Stata Technical Bulletin* 36: 29–31. Reprinted in *Stata Technical Bulletin Reprints*, vol. 6, pp. 187–189. College Station, TX: Stata Press.

Kruskal, W. H. 1957. Historical notes on the Wilcoxon unpaired two-sample test. *Journal of the American Statistical Association* 52: 356–360.

Mann, H. B., and D. R. Whitney. 1947. On a test of whether one of two random variables is stochastically larger than the other. *Annals of Mathematical Statistics* 18: 50–60.

Newson, R. 2000a. snp15: somersd—Confidence intervals for nonparametric statistics and their differences. *Stata Technical Bulletin* 55: 47–55. Reprinted in *Stata Technical Bulletin Reprints*, vol. 10, pp. 312–322. College Station, TX: Stata Press.

——. 2000b. snp15.1: Update to somersd. *Stata Technical Bulletin* 57: 35. Reprinted in *Stata Technical Bulletin Reprints*, vol. 10, pp. 322–323. College Station, TX: Stata Press.

——. 2000c. snp15.2: Update to somersd. *Stata Technical Bulletin* 58: 30. Reprinted in *Stata Technical Bulletin Reprints*, vol. 10, p. 323. College Station, TX: Stata Press.

——. 2001. snp15.3: Update to somersd. *Stata Technical Bulletin* 61: 22. Reprinted in *Stata Technical Bulletin Reprints*, vol. 10, p. 324. College Station, TX: Stata Press.

——. 2003. snp15_4: Software update for somersd. *Stata Journal* 3: 325.

——. 2005. snp15_5: Software update for somersd. *Stata Journal* 5: 470.

Perkins, A. M. 1998. snp14: A two-sample multivariate nonparametric test. *Stata Technical Bulletin* 42: 47–49. Reprinted in *Stata Technical Bulletin Reprints*, vol. 7, pp. 243–245. College Station, TX: Stata Press.

Snedecor, G. W., and W. G. Cochran. 1989. *Statistical Methods*. 8th ed. Ames, IA: Iowa State University Press.

Sprent, P., and N. C. Smeeton. 2007. *Applied Nonparametric Statistical Methods*. 4th ed. Boca Raton, FL: Chapman & Hall/CRC.

Sribney, W. M. 1995. crc40: Correcting for ties and zeros in sign and rank tests. *Stata Technical Bulletin* 26: 2–4. Reprinted in *Stata Technical Bulletin Reprints*, vol. 5, pp. 5–8. College Station, TX: Stata Press.

Wilcoxon, F. 1945. Individual comparisons by ranking methods. *Biometrics* 1: 80–83.

Also see

[R] **signrank** — Equality tests on matched data

[R] **ttest** — Mean-comparison tests

Title

ratio — Estimate ratios

Syntax

Basic syntax

`ratio` [*name*:] *varname* [/] *varname*

Full syntax

`ratio` ([*name*:] *varname* [/] *varname*)
    [([*name*:] *varname* [/] *varname*) ...] [*if*] [*in*] [*weight*] [, *options*]

options	Description
Model	
`stdize(`*varname*`)`	variable identifying strata for standardization
`stdweight(`*varname*`)`	weight variable for standardization
`nostdrescale`	do not rescale the standard weight variable
if/in/over	
`over(`*varlist*[, `nolabel`]`)`	group over subpopulations defined by *varlist*; optionally, suppress group labels
SE/Cluster	
`vce(`*vcetype*`)`	*vcetype* may be `linearized`, `cluster` *clustvar*, `bootstrap`, or `jackknife`
Reporting	
`level(`*#*`)`	set confidence level; default is `level(95)`
`noheader`	suppress table header
`nolegend`	suppress table legend
display_options	control column formats and line width
`coeflegend`	display legend instead of statistics

`bootstrap`, `jackknife`, `mi estimate`, `rolling`, `statsby`, and `svy` are allowed; see [U] **11.1.10 Prefix commands**.
`vce(bootstrap)` and `vce(jackknife)` are not allowed with the `mi estimate` prefix; see [MI] **mi estimate**.
Weights are not allowed with the `bootstrap` prefix; see [R] **bootstrap**.
`vce()` and weights are not allowed with the `svy` prefix; see [SVY] **svy**.
`fweight`s, `iweight`s, and `pweight`s are allowed; see [U] **11.1.6 weight**.
`coeflegend` does not appear in the dialog box.
See [U] **20 Estimation and postestimation commands** for more capabilities of estimation commands.

Menu

Statistics > Summaries, tables, and tests > Summary and descriptive statistics > Ratios

Description

ratio produces estimates of ratios, along with standard errors.

Options

Model

stdize(*varname*) specifies that the point estimates be adjusted by direct standardization across the strata identified by *varname*. This option requires the stdweight() option.

stdweight(*varname*) specifies the weight variable associated with the standard strata identified in the stdize() option. The standardization weights must be constant within the standard strata.

nostdrescale prevents the standardization weights from being rescaled within the over() groups. This option requires stdize() but is ignored if the over() option is not specified.

if/in/over

over(*varlist* [, nolabel]) specifies that estimates be computed for multiple subpopulations, which are identified by the different values of the variables in *varlist*.

When this option is supplied with one variable name, such as over(*varname*), the value labels of *varname* are used to identify the subpopulations. If *varname* does not have labeled values (or there are unlabeled values), the values themselves are used, provided that they are nonnegative integers. Noninteger values, negative values, and labels that are not valid Stata names are substituted with a default identifier.

When over() is supplied with multiple variable names, each subpopulation is assigned a unique default identifier.

nolabel requests that value labels attached to the variables identifying the subpopulations be ignored.

SE/Cluster

vce(*vcetype*) specifies the type of standard error reported, which includes types that are derived from asymptotic theory, that allow for intragroup correlation, and that use bootstrap or jackknife methods; see [R] ***vce_option***.

vce(linearized), the default, uses the linearized or sandwich estimator of variance.

Reporting

level(*#*); see [R] **estimation options**.

noheader prevents the table header from being displayed. This option implies nolegend.

nolegend prevents the table legend identifying the subpopulations from being displayed.

display_options: cformat(%*fmt*) and nolstretch; see [R] **estimation options**.

The following option is available with ratio but is not shown in the dialog box:

coeflegend; see [R] **estimation options**.

Remarks

▷ Example 1

Using the fuel data from example 2 of [R] **ttest**, we estimate the ratio of mileage for the cars without the fuel treatment (mpg1) to those with the fuel treatment (mpg2).

```
. use http://www.stata-press.com/data/r12/fuel
. ratio myratio: mpg1/mpg2
Ratio estimation                  Number of obs    =      12
      myratio: mpg1/mpg2

                          Linearized
                 Ratio    Std. Err.     [95% Conf. Interval]

     myratio   .9230769    .032493     .8515603    .9945936
```

Using these results, we can test to see if this ratio is significantly different from one.

```
. test _b[myratio] = 1
 ( 1)  myratio = 1
       F(  1,    11) =    5.60
            Prob > F =    0.0373
```

We find that the ratio is different from one at the 5% significance level but not at the 1% significance level.

◁

▷ Example 2

Using state-level census data, we want to test whether the marriage rate is equal to the death rate.

```
. use http://www.stata-press.com/data/r12/census2
(1980 Census data by state)
. ratio (deathrate: death/pop) (marrate: marriage/pop)
Ratio estimation                  Number of obs    =      50
    deathrate: death/pop
      marrate: marriage/pop

                          Linearized
                 Ratio    Std. Err.     [95% Conf. Interval]

   deathrate   .0087368   .0002052     .0083244    .0091492
     marrate   .0105577   .0006184      .009315    .0118005

. test _b[deathrate]  =  _b[marrate]
 ( 1)  deathrate - marrate = 0
       F(  1,    49) =    6.93
            Prob > F =    0.0113
```

◁

Saved results

`ratio` saves the following in `e()`:

Scalars

`e(N)`	number of observations
`e(N_over)`	number of subpopulations
`e(N_stdize)`	number of standard strata
`e(N_clust)`	number of clusters
`e(k_eq)`	number of equations in `e(b)`
`e(df_r)`	sample degrees of freedom
`e(rank)`	rank of `e(V)`

Macros

`e(cmd)`	`ratio`
`e(cmdline)`	command as typed
`e(varlist)`	*varlist*
`e(stdize)`	*varname* from `stdize()`
`e(stdweight)`	*varname* from `stdweight()`
`e(wtype)`	weight type
`e(wexp)`	weight expression
`e(title)`	title in estimation output
`e(cluster)`	name of cluster variable
`e(over)`	*varlist* from `over()`
`e(over_labels)`	labels from `over()` variables
`e(over_namelist)`	names from `e(over_labels)`
`e(namelist)`	ratio identifiers
`e(vce)`	*vcetype* specified in `vce()`
`e(vcetype)`	title used to label Std. Err.
`e(properties)`	`b V`
`e(estat_cmd)`	program used to implement `estat`
`e(marginsnotok)`	predictions disallowed by `margins`

Matrices

`e(b)`	vector of mean estimates
`e(V)`	(co)variance estimates
`e(_N)`	vector of numbers of nonmissing observations
`e(_N_stdsum)`	number of nonmissing observations within the standard strata
`e(_p_stdize)`	standardizing proportions
`e(error)`	error code corresponding to `e(b)`

Functions

`e(sample)`	marks estimation sample

Methods and formulas

`ratio` is implemented as an ado-file.

Methods and formulas are presented under the following headings:

The ratio estimator
Survey data
The survey ratio estimator
The standardized ratio estimator
The poststratified ratio estimator
The standardized poststratified ratio estimator
Subpopulation estimation

The ratio estimator

Let $R = Y/X$ be the ratio to be estimated, where Y and X are totals; see [R] **total**. The estimate for R is $\widehat{R} = \widehat{Y}/\widehat{X}$ (the ratio of the sample totals). From the delta method (that is, a first-order Taylor expansion), the approximate variance of the sampling distribution of the linearized $\widehat{R}$ is

$$V(\widehat{R}) \approx \frac{1}{X^2}\left\{V(\widehat{Y}) - 2R\mathrm{Cov}(\widehat{Y},\widehat{X}) + R^2V(\widehat{X})\right\}$$

Direct substitution of $\widehat{X}$, $\widehat{R}$, and the estimated variances and covariance of $\widehat{X}$ and $\widehat{Y}$ leads to the following variance estimator:

$$\widehat{V}(\widehat{R}) = \frac{1}{\widehat{X}^2}\left\{\widehat{V}(\widehat{Y}) - 2\widehat{R}\widehat{\mathrm{Cov}}(\widehat{Y},\widehat{X}) + \widehat{R}^2\widehat{V}(\widehat{X})\right\} \tag{1}$$

Survey data

See [SVY] **variance estimation**, [SVY] **direct standardization**, and [SVY] **poststratification** for discussions that provide background information for the following formulas.

The survey ratio estimator

Let Y_j and X_j be survey items for the jth individual in the population, where $j = 1,\dots,M$ and M is the size of the population. The associated population ratio for the items of interest is $R = Y/X$ where

$$Y = \sum_{j=1}^{M} Y_j \qquad \text{and} \qquad X = \sum_{j=1}^{M} X_j$$

Let y_j and x_j be the corresponding survey items for the jth sampled individual from the population, where $j = 1,\dots,m$ and m is the number of observations in the sample.

The estimator $\widehat{R}$ for the population ratio R is $\widehat{R} = \widehat{Y}/\widehat{X}$, where

$$\widehat{Y} = \sum_{j=1}^{m} w_j y_j \qquad \text{and} \qquad \widehat{X} = \sum_{j=1}^{m} w_j x_j$$

and w_j is a sampling weight. The score variable for the ratio estimator is

$$z_j(\widehat{R}) = \frac{y_j - \widehat{R}x_j}{\widehat{X}} = \frac{\widehat{X}y_j - \widehat{Y}x_j}{\widehat{X}^2}$$

The standardized ratio estimator

Let D_g denote the set of sampled observations that belong to the gth standard stratum and define $I_{D_g}(j)$ to indicate if the jth observation is a member of the gth standard stratum; where $g = 1, \dots, L_D$ and L_D is the number of standard strata. Also, let π_g denote the fraction of the population that belongs to the gth standard stratum, thus $\pi_1 + \cdots + \pi_{L_D} = 1$. Note that π_g is derived from the `stdweight()` option.

The estimator for the standardized ratio is

$$\widehat{R}^D = \sum_{g=1}^{L_D} \pi_g \frac{\widehat{Y}_g}{\widehat{X}_g}$$

where

$$\widehat{Y}_g = \sum_{j=1}^{m} I_{D_g}(j)\, w_j y_j$$

and $\widehat{X}_g$ is similarly defined. The score variable for the standardized ratio is

$$z_j(\widehat{R}^D) = \sum_{g=1}^{L_D} \pi_g I_{D_g}(j) \frac{\widehat{X}_g y_j - \widehat{Y}_g x_j}{\widehat{X}_g^2}$$

The poststratified ratio estimator

Let P_k denote the set of sampled observations that belong to poststratum k, and define $I_{P_k}(j)$ to indicate if the jth observation is a member of poststratum k, where $k = 1, \dots, L_P$ and L_P is the number of poststrata. Also, let M_k denote the population size for poststratum k. P_k and M_k are identified by specifying the `poststrata()` and `postweight()` options on `svyset`; see [SVY] **svyset**.

The estimator for the poststratified ratio is

$$\widehat{R}^P = \frac{\widehat{Y}^P}{\widehat{X}^P}$$

where

$$\widehat{Y}^P = \sum_{k=1}^{L_P} \frac{M_k}{\widehat{M}_k} \widehat{Y}_k = \sum_{k=1}^{L_P} \frac{M_k}{\widehat{M}_k} \sum_{j=1}^{m} I_{P_k}(j)\, w_j y_j$$

and $\widehat{X}^P$ is similarly defined. The score variable for the poststratified ratio is

$$z_j(\widehat{R}^P) = \frac{z_j(\widehat{Y}^P) - \widehat{R}^P z_j(\widehat{X}^P)}{\widehat{X}^P} = \frac{\widehat{X}^P z_j(\widehat{Y}^P) - \widehat{Y}^P z_j(\widehat{X}^P)}{(\widehat{X}^P)^2}$$

where

$$z_j(\widehat{Y}^P) = \sum_{k=1}^{L_P} I_{P_k}(j) \frac{M_k}{\widehat{M}_k} \left(y_j - \frac{\widehat{Y}_k}{\widehat{M}_k} \right)$$

and $z_j(\widehat{X}^P)$ is similarly defined.

The standardized poststratified ratio estimator

The estimator for the standardized poststratified ratio is

$$\widehat{R}^{DP} = \sum_{g=1}^{L_D} \pi_g \frac{\widehat{Y}_g^P}{\widehat{X}_g^P}$$

where

$$\widehat{Y}_g^P = \sum_{k=1}^{L_p} \frac{M_k}{\widehat{M}_k} \widehat{Y}_{g,k} = \sum_{k=1}^{L_p} \frac{M_k}{\widehat{M}_k} \sum_{j=1}^{m} I_{D_g}(j) I_{P_k}(j)\, w_j y_j$$

and $\widehat{X}_g^P$ is similarly defined. The score variable for the standardized poststratified ratio is

$$z_j(\widehat{R}^{DP}) = \sum_{g=1}^{L_D} \pi_g \frac{\widehat{X}_g^P z_j(\widehat{Y}_g^P) - \widehat{Y}_g^P z_j(\widehat{X}_g^P)}{(\widehat{X}_g^P)^2}$$

where

$$z_j(\widehat{Y}_g^P) = \sum_{k=1}^{L_P} I_{P_k}(j) \frac{M_k}{\widehat{M}_k} \left\{ I_{D_g}(j) y_j - \frac{\widehat{Y}_{g,k}}{\widehat{M}_k} \right\}$$

and $z_j(\widehat{X}_g^P)$ is similarly defined.

Subpopulation estimation

Let S denote the set of sampled observations that belong to the subpopulation of interest, and define $I_S(j)$ to indicate if the jth observation falls within the subpopulation.

The estimator for the subpopulation ratio is $\widehat{R}^S = \widehat{Y}^S / \widehat{X}^S$, where

$$\widehat{Y}^S = \sum_{j=1}^{m} I_S(j)\, w_j y_j \qquad \text{and} \qquad \widehat{X}^S = \sum_{j=1}^{m} I_S(j)\, w_j x_j$$

Its score variable is

$$z_j(\widehat{R}^S) = I_S(j) \frac{y_j - \widehat{R}^S x_j}{\widehat{X}^S} = I_S(j) \frac{\widehat{X}^S y_j - \widehat{Y}^S x_j}{(\widehat{X}^S)^2}$$

The estimator for the standardized subpopulation ratio is

$$\widehat{R}^{DS} = \sum_{g=1}^{L_D} \pi_g \frac{\widehat{Y}_g^S}{\widehat{X}_g^S}$$

where

$$\widehat{Y}_g^S = \sum_{j=1}^{m} I_{D_g}(j) I_S(j)\, w_j y_j$$

and $\widehat{X}_g^S$ is similarly defined. Its score variable is

$$z_j(\widehat{R}^{DS}) = \sum_{g=1}^{L_D} \pi_g I_{D_g}(j) I_S(j) \frac{\widehat{X}_g^S y_j - \widehat{Y}_g^S x_j}{(\widehat{X}_g^S)^2}$$

The estimator for the poststratified subpopulation ratio is

$$\widehat{R}^{PS} = \frac{\widehat{Y}^{PS}}{\widehat{X}^{PS}}$$

where

$$\widehat{Y}^{PS} = \sum_{k=1}^{L_P} \frac{M_k}{\widehat{M}_k} \widehat{Y}_k^S = \sum_{k=1}^{L_P} \frac{M_k}{\widehat{M}_k} \sum_{j=1}^{m} I_{P_k}(j) I_S(j)\, w_j y_j$$

and $\widehat{X}^{PS}$ is similarly defined. Its score variable is

$$z_j(\widehat{R}^{PS}) = \frac{\widehat{X}^{PS} z_j(\widehat{Y}^{PS}) - \widehat{Y}^{PS} z_j(\widehat{X}^{PS})}{(\widehat{X}^{PS})^2}$$

where

$$z_j(\widehat{Y}^{PS}) = \sum_{k=1}^{L_P} I_{P_k}(j) \frac{M_k}{\widehat{M}_k} \left\{ I_S(j)\, y_j - \frac{\widehat{Y}_k^S}{\widehat{M}_k} \right\}$$

and $z_j(\widehat{X}^{PS})$ is similarly defined.

The estimator for the standardized poststratified subpopulation ratio is

$$\widehat{R}^{DPS} = \sum_{g=1}^{L_D} \pi_g \frac{\widehat{Y}_g^{PS}}{\widehat{X}_g^{PS}}$$

where

$$\widehat{Y}_g^{PS} = \sum_{k=1}^{L_p} \frac{M_k}{\widehat{M}_k} \widehat{Y}_{g,k}^S = \sum_{k=1}^{L_p} \frac{M_k}{\widehat{M}_k} \sum_{j=1}^{m} I_{D_g}(j) I_{P_k}(j) I_S(j)\, w_j y_j$$

and $\widehat{X}_g^{PS}$ is similarly defined. Its score variable is

$$z_j(\widehat{R}^{DPS}) = \sum_{g=1}^{L_D} \pi_g \frac{\widehat{X}_g^{PS} z_j(\widehat{Y}_g^{PS}) - \widehat{Y}_g^{PS} z_j(\widehat{X}_g^{PS})}{(\widehat{X}_g^{PS})^2}$$

where

$$z_j(\widehat{Y}_g^{PS}) = \sum_{k=1}^{L_P} I_{P_k}(j) \frac{M_k}{\widehat{M}_k} \left\{ I_{D_g}(j) I_S(j)\, y_j - \frac{\widehat{Y}_{g,k}^S}{\widehat{M}_k} \right\}$$

and $z_j(\widehat{X}_g^{PS})$ is similarly defined.

References

Cochran, W. G. 1977. *Sampling Techniques.* 3rd ed. New York: Wiley.

Stuart, A., and J. K. Ord. 1994. *Kendall's Advanced Theory of Statistics: Distribution Theory, Vol I.* 6th ed. London: Arnold.

Also see

[R] **ratio postestimation** — Postestimation tools for ratio

[R] **mean** — Estimate means

[R] **proportion** — Estimate proportions

[R] **total** — Estimate totals

[MI] **estimation** — Estimation commands for use with mi estimate

[SVY] **direct standardization** — Direct standardization of means, proportions, and ratios

[SVY] **poststratification** — Poststratification for survey data

[SVY] **subpopulation estimation** — Subpopulation estimation for survey data

[SVY] **svy estimation** — Estimation commands for survey data

[SVY] **variance estimation** — Variance estimation for survey data

[U] **20 Estimation and postestimation commands**

Title

ratio postestimation — Postestimation tools for ratio

Description

The following postestimation commands are available after `ratio`:

Command	Description
`estat`	VCE
`estat` (svy)	postestimation statistics for survey data
`estimates`	cataloging estimation results
`lincom`	point estimates, standard errors, testing, and inference for linear combinations of coefficients
`nlcom`	point estimates, standard errors, testing, and inference for nonlinear combinations of coefficients
`test`	Wald tests of simple and composite linear hypotheses
`testnl`	Wald tests of nonlinear hypotheses

See the corresponding entries in the *Base Reference Manual* for details, but see [SVY] **estat** for details about `estat` (svy).

Remarks

For examples of the use of `test` after `ratio`, see [R] **ratio**.

Methods and formulas

All postestimation commands listed above are implemented as ado-files.

Also see

[R] **ratio** — Estimate ratios

[SVY] **svy postestimation** — Postestimation tools for svy

[U] **20 Estimation and postestimation commands**

Title

reg3 — Three-stage estimation for systems of simultaneous equations

Syntax

Basic syntax

`reg3` (*depvar*$_1$ *varlist*$_1$) (*depvar*$_2$ *varlist*$_2$) ... (*depvar*$_N$ *varlist*$_N$) [*if*] [*in*] [*weight*]

Full syntax

`reg3` ([*eqname*$_1$:]*depvar*$_{1a}$ [*depvar*$_{1b}$ ...=]*varlist*$_1$ [, `noconstant`])

([*eqname*$_2$:]*depvar*$_{2a}$ [*depvar*$_{2b}$ ...=]*varlist*$_2$ [, `noconstant`])

...

([*eqname*$_N$:]*depvar*$_{Na}$ [*depvar*$_{Nb}$ ...=]*varlist*$_N$ [, `noconstant`])

[*if*] [*in*] [*weight*] [, *options*]

options	Description
Model	
`ireg3`	iterate until estimates converge
`constraints(`*constraints*`)`	apply specified linear constraints
Model 2	
`exog(`*varlist*`)`	exogenous variables not specified in system equations
`endog(`*varlist*`)`	additional right-hand-side endogenous variables
`inst(`*varlist*`)`	full list of exogenous variables
`allexog`	all right-hand-side variables are exogenous
`noconstant`	suppress constant from instrument list
Est. method	
`3sls`	three-stage least squares; the default
`2sls`	two-stage least squares
`ols`	ordinary least squares (OLS)
`sure`	seemingly unrelated regression estimation (SURE)
`mvreg`	`sure` with OLS degrees-of-freedom adjustment
`corr(`*correlation*`)`	`unstructured` or `independent` correlation structure; default is `unstructured`
df adj.	
`small`	report small-sample statistics
`dfk`	use small-sample adjustment
`dfk2`	use alternate adjustment

Reporting	
level(#)	set confidence level; default is level(95)
first	report first-stage regression
nocnsreport	do not display constraints
display_options	control column formats, row spacing, line width, and display of omitted variables and base and empty cells
Optimization	
optimization_options	control the optimization process; seldom used
noheader	suppress display of header
notable	suppress display of coefficient table
nofooter	suppress display of footer
coeflegend	display legend instead of statistics

varlist$_1$, ..., *varlist*$_N$ and the exog() and the inst() varlist may contain factor variables; see [U] **11.4.3 Factor variables**. You must have the same levels of factor variables in all equations that have factor variables.

depvar and *varlist* may contain time-series operators; see [U] **11.4.4 Time-series varlists**.

bootstrap, by, jackknife, rolling, and statsby are allowed; see [U] **11.1.10 Prefix commands**.

Weights are not allowed with the bootstrap prefix; see [R] **bootstrap**.

aweights are not allowed with the jackknife prefix; see [R] **jackknife**.

aweights and fweights are allowed; see [U] **11.1.6 weight**.

noheader, notable, nofooter, and coeflegend do not appear in the dialog box.

See [U] **20 Estimation and postestimation commands** for more capabilities of estimation commands.

Explicit equation naming (*eqname*:) cannot be combined with multiple dependent variables in an equation specification.

Menu

Statistics > Endogenous covariates > Three-stage least squares

Description

reg3 estimates a system of structural equations, where some equations contain endogenous variables among the explanatory variables. Estimation is via three-stage least squares (3SLS); see Zellner and Theil (1962). Typically, the endogenous explanatory variables are dependent variables from other equations in the system. reg3 supports iterated GLS estimation and linear constraints.

reg3 can also estimate systems of equations by seemingly unrelated regression estimation (SURE), multivariate regression (MVREG), and equation-by-equation ordinary least squares (OLS) or two-stage least squares (2SLS).

Nomenclature

Under 3SLS or 2SLS estimation, a *structural equation* is defined as one of the equations specified in the system. A *dependent variable* will have its usual interpretation as the left-hand-side variable in an equation with an associated disturbance term. All dependent variables are explicitly taken to be *endogenous* to the system and are treated as correlated with the disturbances in the system's equations. Unless specified in an endog() option, all other variables in the system are treated as *exogenous* to the system and uncorrelated with the disturbances. The exogenous variables are taken to be *instruments* for the endogenous variables.

Options

Model

`ireg3` causes `reg3` to iterate over the estimated disturbance covariance matrix and parameter estimates until the parameter estimates converge. Although the iteration is usually successful, there is no guarantee that it will converge to a stable point. Under SURE, this iteration converges to the maximum likelihood estimates.

`constraints(`*constraints*`)`; see [R] **estimation options**.

Model 2

`exog(`*varlist*`)` specifies additional exogenous variables that are included in none of the system equations. This can occur when the system contains identities that are not estimated. If implicitly exogenous variables from the equations are listed here, `reg3` will just ignore the additional information. Specified variables will be added to the exogenous variables in the system and used in the first stage as instruments for the endogenous variables. By specifying dependent variables from the structural equations, you can use `exog()` to override their endogeneity.

`endog(`*varlist*`)` identifies variables in the system that are not dependent variables but are endogenous to the system. These variables must appear in the variable list of at least one equation in the system. Again the need for this identification often occurs when the system contains identities. For example, a variable that is the sum of an exogenous variable and a dependent variable may appear as an explanatory variable in some equations.

`inst(`*varlist*`)` specifies a full list of all exogenous variables and may not be used with the `endog()` or `exog()` options. It must contain a full list of variables to be used as instruments for the endogenous regressors. Like `exog()`, the list may contain variables not specified in the system of equations. This option can be used to achieve the same results as the `endog()` and `exog()` options, and the choice is a matter of convenience. Any variable not specified in the *varlist* of the `inst()` option is assumed to be endogenous to the system. As with `exog()`, including the dependent variables from the structural equations will override their endogeneity.

`allexog` indicates that all right-hand-side variables are to be treated as exogenous—even if they appear as the dependent variable of another equation in the system. This option can be used to enforce a SURE or MVREG estimation even when some dependent variables appear as regressors.

`noconstant`; see [R] **estimation options**.

Est. method

`3sls` specifies the full 3SLS estimation of the system and is the default for `reg3`.

`2sls` causes `reg3` to perform equation-by-equation 2SLS on the full system of equations. This option implies `dfk`, `small`, and `corr(independent)`.

Cross-equation testing should not be performed after estimation with this option. With `2sls`, no covariance is estimated between the parameters of the equations. For cross-equation testing, use `3sls`.

`ols` causes `reg3` to perform equation-by-equation OLS on the system—even if dependent variables appear as regressors or the regressors differ for each equation; see [R] **mvreg**. `ols` implies `allexog`, `dfk`, `small`, and `corr(independent)`; `nodfk` and `nosmall` may be specified to override `dfk` and `small`.

The covariance of the coefficients between equations is not estimated under this option, and cross-equation tests should not be performed after estimation with `ols`. For cross-equation testing, use `sure` or `3sls` (the default).

sure causes reg3 to perform a SURE of the system—even if dependent variables from some equations appear as regressors in other equations; see [R] **sureg**. sure is a synonym for allexog.

mvreg is identical to sure, except that the disturbance covariance matrix is estimated with an OLS degrees-of-freedom adjustment—the dfk option. If the regressors are identical for all equations, the parameter point estimates will be the standard MVREG results. If any of the regressors differ, the point estimates are those for SURE with an OLS degrees-of-freedom adjustment in computing the covariance matrix. nodfk and nosmall may be specified to override dfk and small.

corr(*correlation*) specifies the assumed form of the correlation structure of the equation disturbances and is rarely requested explicitly. For the family of models fit by reg3, the only two allowable correlation structures are unstructured and independent. The default is unstructured.

This option is used almost exclusively to estimate a system of equations by 2SLS or to perform OLS regression with reg3 on multiple equations. In these cases, the correlation is set to independent, forcing reg3 to treat the covariance matrix of equation disturbances as diagonal in estimating model parameters. Thus a set of two-stage coefficient estimates can be obtained if the system contains endogenous right-hand-side variables, or OLS regression can be imposed, even if the regressors differ across equations. Without imposing independent disturbances, reg3 would estimate the former by 3SLS and the latter by SURE.

Any tests performed after estimation with the independent option will treat coefficients in different equations as having no covariance; cross-equation tests should not be used after specifying corr(independent).

df adj.

small specifies that small-sample statistics be computed. It shifts the test statistics from χ^2 and z statistics to F statistics and t statistics. This option is intended primarily to support MVREG. Although the standard errors from each equation are computed using the degrees of freedom for the equation, the degrees of freedom for the t statistics are all taken to be those for the first equation. This approach poses no problem under MVREG because the regressors are the same across equations.

dfk specifies the use of an alternative divisor in computing the covariance matrix for the equation residuals. As an asymptotically justified estimator, reg3 by default uses the number of sample observations n as a divisor. When the dfk option is set, a small-sample adjustment is made, and the divisor is taken to be $\sqrt{(n-k_i)(n-k_j)}$, where k_i and k_j are the numbers of parameters in equations i and j, respectively.

dfk2 specifies the use of an alternative divisor in computing the covariance matrix for the equation errors. When the dfk2 option is set, the divisor is taken to be the mean of the residual degrees of freedom from the individual equations.

Reporting

level(*#*); see [R] **estimation options**.

first requests that the first-stage regression results be displayed during estimation.

nocnsreport; see [R] **estimation options**.

display_options: noomitted, vsquish, noemptycells, baselevels, allbaselevels, cformat(%*fmt*), pformat(%*fmt*), sformat(%*fmt*), and nolstretch; see [R] **estimation options**.

Optimization

optimization_options control the iterative process that minimizes the sum of squared errors when ireg3 is specified. These options are seldom used.

iterate(#) specifies the maximum number of iterations. When the number of iterations equals #, the optimizer stops and presents the current results, even if the convergence tolerance has not been reached. The default value of iterate() is the current value of set maxiter (see [R] **maximize**), which is iterate(16000) if maxiter has not been changed.

trace adds to the iteration log a display of the current parameter vector.

nolog suppresses the display of the iteration log.

tolerance(#) specifies the tolerance for the coefficient vector. When the relative change in the coefficient vector from one iteration to the next is less than or equal to #, the optimization process is stopped. tolerance(1e-6) is the default.

The following options are available with reg3 but are not shown in the dialog box:

noheader suppresses display of the header reporting the estimation method and the table of equation summary statistics.

notable suppresses display of the coefficient table.

nofooter suppresses display of the footer reporting the list of endogenous and exogenous variables in the model.

coeflegend; see [R] **estimation options**.

Remarks

reg3 estimates systems of structural equations where some equations contain endogenous variables among the explanatory variables. Generally, these endogenous variables are the dependent variables of other equations in the system, though not always. The disturbance is correlated with the endogenous variables—violating the assumptions of OLS. Further, because some of the explanatory variables are the dependent variables of other equations in the system, the error terms among the equations are expected to be correlated. reg3 uses an instrumental-variables approach to produce consistent estimates and generalized least squares (GLS) to account for the correlation structure in the disturbances across the equations. Good general references on three-stage estimation include Davidson and MacKinnon (1993, 651–661) and Greene (2012, 331–334).

Three-stage least squares can be thought of as producing estimates from a three-step process.

Step 1. Develop instrumented values for all endogenous variables. These instrumented values can simply be considered as the predicted values resulting from a regression of each endogenous variable on all exogenous variables in the system. This stage is identical to the first step in 2SLS and is critical for the consistency of the parameter estimates.

Step 2. Obtain a consistent estimate for the covariance matrix of the equation disturbances. These estimates are based on the residuals from a 2SLS estimation of each structural equation.

Step 3. Perform a GLS-type estimation using the covariance matrix estimated in the second stage and with the instrumented values in place of the right-hand-side endogenous variables.

❑ Technical note

The estimation and use of the covariance matrix of disturbances in three-stage estimation is almost identical to the SURE method—sureg. As with SURE, using this covariance matrix improves the efficiency of the three-stage estimator. Even without the covariance matrix, the estimates would be consistent. (They would be 2SLS estimates.) This improvement in efficiency comes with a caveat. All the parameter estimates now depend on the consistency of the covariance matrix estimates. If one equation in the system is misspecified, the disturbance covariance estimates will be inconsistent, and the resulting coefficients will be biased and inconsistent. Alternatively, if each equation is estimated separately by 2SLS ([R] **regress**), only the coefficients in the misspecified equation are affected.

❑

❑ Technical note

If an equation is just identified, the 3SLS point estimates for that equation are identical to the 2SLS estimates. However, as with sureg, even if all equations are just identified, fitting the model via reg3 has at least one advantage over fitting each equation separately via ivregress; by using reg3, tests involving coefficients in different equations can be performed easily using test or testnl.

❑

▷ Example 1

A simple macroeconomic model relates consumption (consump) to private and government wages paid (wagepriv and wagegovt). Simultaneously, private wages depend on consumption, total government expenditures (govt), and the lagged stock of capital in the economy (capital1). Although this is not a plausible model, it does meet the criterion of being simple. This model could be written as

$$\texttt{consump} = \beta_0 + \beta_1\,\texttt{wagepriv} + \beta_2\,\texttt{wagegovt} + \epsilon_1$$

$$\texttt{wagepriv} = \beta_3 + \beta_4\,\texttt{consump} + \beta_5\,\texttt{govt} + \beta_6\,\texttt{capital1} + \epsilon_2$$

If we assume that this is the full system, consump and wagepriv will be endogenous variables, with wagegovt, govt, and capital1 exogenous. Data for the U.S. economy on these variables are taken from Klein (1950). This model can be fit with reg3 by typing

```
. use http://www.stata-press.com/data/r12/klein
. reg3 (consump wagepriv wagegovt) (wagepriv consump govt capital1)
Three-stage least-squares regression
----------------------------------------------------------------------
Equation          Obs  Parms        RMSE    "R-sq"       chi2        P
----------------------------------------------------------------------
consump            22      2    1.776297    0.9388     208.02   0.0000
wagepriv           22      3    2.372443    0.8542      80.04   0.0000
----------------------------------------------------------------------

------------------------------------------------------------------------------
             |      Coef.   Std. Err.      z    P>|z|     [95% Conf. Interval]
-------------+----------------------------------------------------------------
consump      |
    wagepriv |   .8012754   .1279329     6.26   0.000     .5505314    1.052019
    wagegovt |   1.029531   .3048424     3.38   0.001      .432051    1.627011
       _cons |    19.3559   3.583772     5.40   0.000     12.33184    26.37996
-------------+----------------------------------------------------------------
wagepriv     |
     consump |   .4026076   .2567312     1.57   0.117    -.1005764    .9057916
        govt |   1.177792   .5421253     2.17   0.030     .1152461    2.240338
    capital1 |  -.0281145   .0572111    -0.49   0.623    -.1402462    .0840173
       _cons |   14.63026   10.26693     1.42   0.154    -5.492552    34.75306
------------------------------------------------------------------------------
Endogenous variables:  consump wagepriv
Exogenous variables:   wagegovt govt capital1
------------------------------------------------------------------------------
```

Without showing the 2SLS results, we note that the consumption function in this system falls under the conditions noted earlier. That is, the 2SLS and 3SLS coefficients for the equation are identical. ◁

▷ Example 2

Some of the most common simultaneous systems encountered are supply-and-demand models. A simple system could be specified as

$$\texttt{qDemand} = \beta_0 + \beta_1\,\texttt{price} + \beta_2\,\texttt{pcompete} + \beta_3\,\texttt{income} + \epsilon_1$$

$$\texttt{qSupply} = \beta_4 + \beta_5\,\texttt{price} + \beta_6\,\texttt{praw} + \epsilon_2$$

$$\texttt{Equilibrium condition:}\;\; \texttt{quantity} = \texttt{qDemand} = \texttt{qSupply}$$

where

`quantity` is the quantity of a product produced and sold,
`price` is the price of the product,
`pcompete` is the price of a competing product,
`income` is the average income level of consumers, and
`praw` is the price of raw materials used to produce the product.

In this system, price is assumed to be determined simultaneously with demand. The important statistical implications are that price is not a predetermined variable and that it is correlated with the disturbances of both equations. The system is somewhat unusual: quantity is associated with two disturbances. This fact really poses no problem because the disturbances are specified on the behavioral demand and supply equations—two separate entities. Often one of the two equations is rewritten to place price on the left-hand side, making this endogeneity explicit in the specification.

To provide a concrete illustration of the effects of simultaneous equations, we can simulate data for the above system by using known coefficients and disturbance properties. Specifically, we will simulate the data as

$$\texttt{qDemand} = 40 - 1.0\,\texttt{price} + 0.25\,\texttt{pcompete} + 0.5\,\texttt{income} + \epsilon_1$$

$$\texttt{qSupply} = 0.5\,\texttt{price} - 0.75\,\texttt{praw} + \epsilon_2$$

where

$$\epsilon_1 \sim N(0, 2.4)$$

$$\epsilon_2 \sim N(0, 3.8)$$

For comparison, we can estimate the supply and demand equations separately by OLS. The estimates for the demand equation are

```
. use http://www.stata-press.com/data/r12/supDem
. regress quantity price pcompete income
      Source |       SS       df       MS              Number of obs =      49
-------------+------------------------------           F(  3,    45) =    1.00
       Model |  23.1579302     3  7.71931008           Prob > F      =  0.4004
    Residual |  346.459313    45  7.69909584           R-squared     =  0.0627
-------------+------------------------------           Adj R-squared =  0.0002
       Total |  369.617243    48  7.70035923           Root MSE      =  2.7747

------------------------------------------------------------------------------
    quantity |      Coef.   Std. Err.      t    P>|t|     [95% Conf. Interval]
-------------+----------------------------------------------------------------
       price |   .1186265   .1716014     0.69   0.493    -.2269965    .4642496
    pcompete |   .0946416   .1200815     0.79   0.435    -.1472149    .3364981
      income |   .0785339   .1159867     0.68   0.502    -.1550754    .3121432
       _cons |   7.563261   5.019479     1.51   0.139     -2.54649    17.67301
------------------------------------------------------------------------------
```

The OLS estimates for the supply equation are

```
. regress quantity price praw
      Source |       SS       df       MS              Number of obs =      49
-------------+------------------------------           F(  2,    46) =   35.71
       Model |  224.819549     2  112.409774           Prob > F      =  0.0000
    Residual |  144.797694    46  3.14777596           R-squared     =  0.6082
-------------+------------------------------           Adj R-squared =  0.5912
       Total |  369.617243    48  7.70035923           Root MSE      =  1.7742

------------------------------------------------------------------------------
    quantity |      Coef.   Std. Err.      t    P>|t|     [95% Conf. Interval]
-------------+----------------------------------------------------------------
       price |    .724675   .1095657     6.61   0.000     .5041307    .9452192
        praw |  -.8674796   .1066114    -8.14   0.000    -1.082077    -.652882
       _cons |   -6.97291   3.323105    -2.10   0.041    -13.66197    -.283847
------------------------------------------------------------------------------
```

Examining the coefficients from these regressions, we note that they are not close to the known parameters used to generate the simulated data. In particular, the positive coefficient on `price` in the demand equation stands out. We constructed our simulated data to be consistent with economic theory—people demand less of a product if its price rises and more if their personal income rises. Although the `price` coefficient is statistically insignificant, the positive value contrasts starkly with what is predicted from economic price theory and the -1.0 value that we used in the simulation. Likewise, we are disappointed with the insignificance and level of the coefficient on average `income`. The supply equation has correct signs on the two main parameters, but their levels are different from the known values. In fact, the coefficient on `price` (0.724675) is different from the simulated parameter (0.5) at the 5% level of significance.

All these problems are to be expected. We explicitly constructed a simultaneous system of equations that violated one of the assumptions of least squares. Specifically, the disturbances were correlated with one of the regressors—`price`.

Two-stage least squares can be used to address the correlation between regressors and disturbances. Using instruments for the endogenous variable, `price`, 2SLS will produce consistent estimates of the parameters in the system. Let's use `ivregress` (see [R] **ivregress**) to see how our simulated system behaves when fit using 2SLS.

```
. ivregress 2sls quantity (price = praw) pcompete income
Instrumental variables (2SLS) regression               Number of obs =      49
                                                       Wald chi2(3)  =    8.77
                                                       Prob > chi2   =  0.0326
                                                       R-squared     =       .
                                                       Root MSE      =  3.7333

------------------------------------------------------------------------------
    quantity |      Coef.   Std. Err.      z    P>|z|     [95% Conf. Interval]
-------------+----------------------------------------------------------------
       price |  -1.015817    .374209    -2.71   0.007    -1.749253    -.282381
    pcompete |   .3319504    .172912     1.92   0.055    -.0069508    .6708517
      income |   .5090607   .1919482     2.65   0.008     .1328491    .8852723
       _cons |   39.89988   10.77378     3.70   0.000     18.78366    61.01611
------------------------------------------------------------------------------
Instrumented:  price
Instruments:   pcompete income praw
. ivregress 2sls quantity (price = pcompete income) praw
Instrumental variables (2SLS) regression               Number of obs =      49
                                                       Wald chi2(2)  =   39.25
                                                       Prob > chi2   =  0.0000
                                                       R-squared     =  0.5928
                                                       Root MSE      =  1.7525

------------------------------------------------------------------------------
    quantity |      Coef.   Std. Err.      z    P>|z|     [95% Conf. Interval]
-------------+----------------------------------------------------------------
       price |   .5773133   .1749974     3.30   0.001     .2343247    .9203019
        praw |  -.7835496   .1312414    -5.97   0.000    -1.040778   -.5263213
       _cons |  -2.550694   5.273067    -0.48   0.629    -12.88571    7.784327
------------------------------------------------------------------------------
Instrumented:  price
Instruments:   praw pcompete income
```

We are now much happier with the estimation results. All the coefficients from both equations are close to the true parameter values for the system. In particular, the coefficients are all well within 95% confidence intervals for the parameters. The missing R-squared in the demand equation seems unusual; we will discuss that more later.

Finally, this system could be estimated using 3SLS. To demonstrate how large systems might be handled and to avoid multiline commands, we will use global macros (see [P] **macro**) to hold the specifications for our equations.

```
. global demand "(qDemand: quantity price pcompete income)"
. global supply "(qSupply: quantity price praw)"
. reg3 $demand $supply, endog(price)
```

We must specify `price` as endogenous because it does not appear as a dependent variable in either equation. Without this option, `reg3` would assume that there are no endogenous variables in the system and produce seemingly unrelated regression (`sureg`) estimates. The `reg3` output from our series of commands is

```
Three-stage least-squares regression
----------------------------------------------------------------------
Equation          Obs  Parms        RMSE    "R-sq"       chi2        P
----------------------------------------------------------------------
qDemand            49      3    3.739686   -0.8540       8.68   0.0338
qSupply            49      2    1.752501    0.5928      39.25   0.0000
----------------------------------------------------------------------

------------------------------------------------------------------------------
             |      Coef.   Std. Err.      z    P>|z|     [95% Conf. Interval]
-------------+----------------------------------------------------------------
qDemand      |
       price |  -1.014345   .3742036    -2.71   0.007     -1.74777   -.2809194
    pcompete |   .2647206   .1464194     1.81   0.071    -.0222561    .5516973
      income |   .5299146   .1898161     2.79   0.005     .1578819    .9019472
       _cons |   40.08749   10.77072     3.72   0.000     18.97726    61.19772
-------------+----------------------------------------------------------------
qSupply      |
       price |   .5773133   .1749974     3.30   0.001     .2343247    .9203019
        praw |  -.7835496   .1312414    -5.97   0.000    -1.040778   -.5263213
       _cons |  -2.550694   5.273067    -0.48   0.629    -12.88571    7.784327
------------------------------------------------------------------------------
Endogenous variables:  quantity price
Exogenous variables:   pcompete income praw
------------------------------------------------------------------------------
```

The use of 3SLS over 2SLS is essentially an efficiency issue. The coefficients of the demand equation from 3SLS are close to the coefficients from two-stage least squares, and those of the supply equation are identical. The latter case was mentioned earlier for systems with some exactly identified equations. However, even for the demand equation, we do not expect the coefficients to change systematically. What we do expect from three-stage least squares are more precise estimates of the parameters given the validity of our specification and `reg3`'s use of the covariances among the disturbances.

Let's summarize the results. With OLS, we got obviously biased estimates of the parameters. No amount of data would have improved the OLS estimates—they are inconsistent in the face of the violated OLS assumptions. With 2SLS, we obtained consistent estimates of the parameters, and these would have improved with more data. With 3SLS, we obtained consistent estimates of the parameters that are more efficient than those obtained by 2SLS.

◁

❑ Technical note

We noted earlier that the R-squared was missing from the two-stage estimates of the demand equation. Now we see that the R-squared is negative for the three-stage estimates of the same equation. How can we have a negative R-squared?

In most estimators, other than least squares, the R-squared is no more than a summary measure of the overall in-sample predictive power of the estimator. The computational formula for R-squared is $R\text{-squared} = 1 - \text{RSS}/\text{TSS}$, where RSS is the residual sum of squares (sum of squared residuals) and TSS is the total sum of squared deviations about the mean of the dependent variable. In a standard linear model with a constant, the model from which the TSS is computed is nested within the full model from which RSS is computed—they both have a constant term based on the same data. Thus it must be that $\text{TSS} \geq \text{RSS}$ and R-squared is constrained between 0 and 1.

For 2SLS and 3SLS, some of the regressors enter the model as instruments when the parameters are estimated. However, because our goal is to fit the structural model, the actual values, not the instruments for the endogenous right-hand-side variables, are used to determine R-squared. The model residuals are computed over a different set of regressors from those used to fit the model. The two-

or three-stage estimates are no longer nested within a constant-only model of the dependent variable, and the residual sum of squares is no longer constrained to be smaller than the total sum of squares.

A negative R-squared in 3SLS should be taken for exactly what it is—an indication that the structural model predicts the dependent variable worse than a constant-only model. Is this a problem? It depends on the application. Three-stage least squares applied to our contrived supply-and-demand example produced good estimates of the known true parameters. Still, the demand equation produced an R-squared of -0.854. How do we feel about our parameter estimates? This should be determined by the estimates themselves, their associated standard errors, and the overall model significance. On this basis, negative R-squared and all, we feel pretty good about all the parameter estimates for both the supply and demand equations. Would we want to make predictions about equilibrium quantity by using the demand equation alone? Probably not. Would we want to make these quantity predictions by using the supply equation? Possibly, because based on in-sample predictions, they seem better than those from the demand equations. However, both the supply and demand estimates are based on limited information. If we are interested in predicting quantity, a reduced-form equation containing all our independent variables would usually be preferred.

❑

❑ Technical note

As a matter of syntax, we could have specified the supply-and-demand model on one line without using global macros.

```
. reg3 (quantity price pcompete income) (quantity price praw), endog(price)
Three-stage least-squares regression
```

Equation	Obs	Parms	RMSE	"R-sq"	chi2	P
quantity	49	3	3.739686	-0.8540	8.68	0.0338
2quantity	49	2	1.752501	0.5928	39.25	0.0000

	Coef.	Std. Err.	z	P>|z|	[95% Conf.	Interval]
quantity						
price	-1.014345	.3742036	-2.71	0.007	-1.74777	-.2809194
pcompete	.2647206	.1464194	1.81	0.071	-.0222561	.5516973
income	.5299146	.1898161	2.79	0.005	.1578819	.9019472
_cons	40.08749	10.77072	3.72	0.000	18.97726	61.19772
2quantity						
price	.5773133	.1749974	3.30	0.001	.2343247	.9203019
praw	-.7835496	.1312414	-5.97	0.000	-1.040778	-.5263213
_cons	-2.550694	5.273067	-0.48	0.629	-12.88571	7.784327

```
Endogenous variables:  quantity price
Exogenous variables:   pcompete income praw
```

However, here `reg3` has been forced to create a unique equation name for the supply equation—`2quantity`. Both the supply and demand equations could not be designated as `quantity`, so a number was prefixed to the name for the supply equation.

We could have specified

```
. reg3 (qDemand: quantity price pcompete income) (qSupply: quantity price praw),
> endog(price)
```

and obtained the same results and equation labeling as when we used global macros to hold the equation specifications.

Without explicit equation names, reg3 always assumes that the dependent variable should be used to name equations. When each equation has a different dependent variable, this rule causes no problems and produces easily interpreted result tables. If the same dependent variable appears in more than one equation, however, reg3 will create a unique equation name based on the dependent variable name. Because equation names must be used for cross-equation tests, you have more control in this situation if explicit names are placed on the equations.

❑

▷ Example 3: Using the full syntax of reg3

Klein's (1950) model of the U.S. economy is often used to demonstrate system estimators. It contains several common features that will serve to demonstrate the full syntax of reg3. The Klein model is defined by the following seven relationships:

$$\mathtt{c} = \beta_0 + \beta_1\mathtt{p} + \beta_2\mathtt{p1} + \beta_3\mathtt{w} + \epsilon_1 \tag{1}$$

$$\mathtt{i} = \beta_4 + \beta_5\mathtt{p} + \beta_6\mathtt{p1} + \beta_7\mathtt{k1} + \epsilon_2 \tag{2}$$

$$\mathtt{wp} = \beta_8 + \beta_9\mathtt{y} + \beta_{10}\mathtt{y1} + \beta_{11}\mathtt{yr} + \epsilon_3 \tag{3}$$

$$\mathtt{y} = \mathtt{c} + \mathtt{i} + \mathtt{g} \tag{4}$$

$$\mathtt{p} = \mathtt{y} - \mathtt{t} - \mathtt{wp} \tag{5}$$

$$\mathtt{k} = \mathtt{k1} + \mathtt{i} \tag{6}$$

$$\mathtt{w} = \mathtt{wg} + \mathtt{wp} \tag{7}$$

The variables in the model are listed below. Two sets of variable names are shown. The concise first name uses traditional economics mnemonics, whereas the second name provides more guidance for everyone else. The concise names serve to keep the specification of the model small (and quite understandable to economists).

Short name	Long name	Variable definition	Type
c	consump	Consumption	endogenous
p	profits	Private industry profits	endogenous
p1	profits1	Last year's private industry profits	exogenous
wp	wagepriv	Private wage bill	endogenous
wg	wagegovt	Government wage bill	exogenous
w	wagetot	Total wage bill	endogenous
i	invest	Investment	endogenous
k1	capital1	Last year's level of capital stock	exogenous
y	totinc	Total income/demand	endogenous
y1	totinc1	Last year's total income	exogenous
g	govt	Government spending	exogenous
t	taxnetx	Indirect bus. taxes + net exports	exogenous
yr	year	Year—1931	exogenous

Equations (1)–(3) are behavioral and contain explicit disturbances (ϵ_1, ϵ_2, and ϵ_3). The remaining equations are identities that specify additional variables in the system and their accounting relationships with the variables in the behavioral equations. Some variables are explicitly endogenous by appearing as dependent variables in (1)–(3). Others are implicitly endogenous as linear combinations that contain other endogenous variables (for example, w and p). Still other variables are implicitly exogenous by appearing in the identities but not in the behavioral equations (for example, wg and g).

Using the concise names, we can fit Klein's model with the following command:

```
. use http://www.stata-press.com/data/r12/kleinAbr
. reg3 (c p p1 w) (i p p1 k1) (wp y y1 yr), endog(w p y) exog(t wg g)
Three-stage least-squares regression
```

Equation	Obs	Parms	RMSE	"R-sq"	chi2	P
c	21	3	.9443305	0.9801	864.59	0.0000
i	21	3	1.446736	0.8258	162.98	0.0000
wp	21	3	.7211282	0.9863	1594.75	0.0000

	Coef.	Std. Err.	z	P>|z|	[95% Conf.	Interval]
c						
p	.1248904	.1081291	1.16	0.248	-.0870387	.3368194
p1	.1631439	.1004382	1.62	0.104	-.0337113	.3599992
w	.790081	.0379379	20.83	0.000	.715724	.8644379
_cons	16.44079	1.304549	12.60	0.000	13.88392	18.99766
i						
p	-.0130791	.1618962	-0.08	0.936	-.3303898	.3042316
p1	.7557238	.1529331	4.94	0.000	.4559805	1.055467
k1	-.1948482	.0325307	-5.99	0.000	-.2586072	-.1310893
_cons	28.17785	6.793768	4.15	0.000	14.86231	41.49339
wp						
y	.4004919	.0318134	12.59	0.000	.3381388	.462845
y1	.181291	.0341588	5.31	0.000	.1143411	.2482409
yr	.149674	.0279352	5.36	0.000	.094922	.2044261
_cons	1.797216	1.115854	1.61	0.107	-.3898181	3.984251

```
Endogenous variables:  c i wp w p y
Exogenous variables:   p1 k1 y1 yr t wg g
```

We used the exog() option to identify t, wg, and g as exogenous variables in the system. These variables must be identified because they are part of the system but appear directly in none of the behavioral equations. Without this option, reg3 would not know they were part of the system. The endog() option specifying w, p, and y is also required. Without this information, reg3 would be unaware that these variables are linear combinations that include endogenous variables.

❑ Technical note

Rather than listing additional endogenous and exogenous variables, we could specify the full list of exogenous variables in an inst() option,

```
. reg3 (c p p1 w) (i p p1 k1) (wp y y1 yr), inst(g t wg yr p1 k1 y1)
```

or equivalently,

```
. global conseqn "(c p p1 w)"
. global inveqn  "(i p p1 k1)"
. global wageqn  "(wp y y1 yr)"
. global inlist  "g t wg yr p1 k1 y1"
. reg3 $conseqn $inveqn $wageqn, inst($inlist)
```

Macros and explicit equations can also be mixed in the specification

```
. reg3 $conseqn (i p p1 k1) $wageqn, endog(w p y) exog(t wg g)
```

or

```
. reg3 (c p p1 w) $inveqn (wp y y1 yr), endog(w p y) exog(t wg g)
```

Placing the equation-binding parentheses in the global macros was also arbitrary. We could have used

```
. global consump  "c p p1 w"
. global invest   "i p p1 k1"
. global wagepriv "wp y y1 yr"
. reg3 ($consump) ($invest) ($wagepriv), endog(w p y) exog(t wg g)
```

`reg3` is tolerant of all combinations, and these commands will produce identical output.

❑

Switching to the full variable names, we can fit Klein's model with the commands below. We will use global macros to store the lists of endogenous and exogenous variables. Again this is not necessary: these lists could have been typed directly on the command line. However, assigning the lists to local macros makes additional processing easier if alternative models are to be fit. We will also use the `ireg3` option to produce the iterated estimates.

```
. use http://www.stata-press.com/data/r12/klein
. global conseqn "(consump profits profits1 wagetot)"
. global inveqn  "(invest profits profits1 capital1)"
. global wageqn "(wagepriv totinc totinc1 year)"
. global enlist "wagetot profits totinc"
. global exlist "taxnetx wagegovt govt"
. reg3 $conseqn $inveqn $wageqn, endog($enlist) exog($exlist) ireg3
Iteration 1:   tolerance =  .3712549
Iteration 2:   tolerance =  .1894712
Iteration 3:   tolerance =  .1076401
  (output omitted)
Iteration 24:   tolerance =  7.049e-07
Three-stage least-squares regression, iterated
```

Equation	Obs	Parms	RMSE	"R-sq"	chi2	P
consump	21	3	.9565088	0.9796	970.31	0.0000
invest	21	3	2.134327	0.6209	56.78	0.0000
wagepriv	21	3	.7782334	0.9840	1312.19	0.0000

```
------------------------------------------------------------------------------
             |      Coef.   Std. Err.      z    P>|z|     [95% Conf. Interval]
-------------+----------------------------------------------------------------
consump      |
     profits |   .1645096   .0961979     1.71   0.087    -.0240348    .3530539
    profits1 |   .1765639   .0901001     1.96   0.050    -.0000291    .3531569
     wagetot |   .7658011   .0347599    22.03   0.000     .6976729    .8339294
       _cons |   16.55899   1.224401    13.52   0.000     14.15921    18.95877
-------------+----------------------------------------------------------------
invest       |
     profits |  -.3565316   .2601568    -1.37   0.171    -.8664296    .1533664
    profits1 |   1.011299   .2487745     4.07   0.000     .5237098    1.498888
    capital1 |     -.2602   .0508694    -5.12   0.000    -.3599022   -.1604978
       _cons |   42.89629   10.59386     4.05   0.000     22.13271    63.65987
-------------+----------------------------------------------------------------
wagepriv     |
      totinc |   .3747792   .0311027    12.05   0.000     .3138191    .4357394
     totinc1 |   .1936506   .0324018     5.98   0.000     .1301443     .257157
        year |   .1679262   .0289291     5.80   0.000     .1112263    .2246261
       _cons |   2.624766   1.195559     2.20   0.028     .2815124    4.968019
------------------------------------------------------------------------------
Endogenous variables:  consump invest wagepriv wagetot profits totinc
Exogenous variables:   profits1 capital1 totinc1 year taxnetx wagegovt govt
------------------------------------------------------------------------------
```

◁

▷ Example 4: Constraints with reg3

As a simple example of constraints, (1) above may be rewritten with both wages explicitly appearing (rather than as a variable containing the sum). Using the longer variable names, we have

$$\texttt{consump} = \beta_0 + \beta_1\,\texttt{profits} + \beta_2\,\texttt{profits1} + \beta_3\,\texttt{wagepriv} + \beta_{12}\,\texttt{wagegovt} + \epsilon_1$$

To retain the effect of the identity in (7), we need $\beta_3 = \beta_{12}$ as a constraint on the system. We obtain this result by defining the constraint in the usual way and then specifying its use in `reg3`. Because `reg3` is a system estimator, we will need to use the full equation syntax of `constraint`. The assumption that the following commands are entered after the model above has been estimated. We are simply changing the definition of the consumption equation (`consump`) and adding a constraint on two of its parameters. The rest of the model definition is carried forward.

```
. global conseqn "(consump profits profits1 wagepriv wagegovt)"
. constraint 1 [consump]wagepriv = [consump]wagegovt
. reg3 $conseqn $inveqn $wageqn, endog($enlist) exog($exlist) constr(1) ireg3
note:  additional endogenous variables not in the system have no effect
       and are ignored:  wagetot
Iteration 1:   tolerance =  .3712547
Iteration 2:   tolerance =   .189471
Iteration 3:   tolerance =    .10764
  (output omitted)
Iteration 24:   tolerance =  7.049e-07
```

```
Three-stage least-squares regression, iterated

Equation             Obs  Parms        RMSE    "R-sq"       chi2        P

consump               21      3    .9565086    0.9796     970.31   0.0000
invest                21      3    2.134326    0.6209      56.78   0.0000
wagepriv              21      3    .7782334    0.9840    1312.19   0.0000

 ( 1)  [consump]wagepriv - [consump]wagegovt = 0

                   Coef.   Std. Err.      z    P>|z|     [95% Conf. Interval]

consump
     profits    .1645097   .0961978     1.71   0.087    -.0240346     .353054
    profits1    .1765639   .0901001     1.96   0.050    -.0000291    .3531568
    wagepriv    .7658012   .0347599    22.03   0.000     .6976729    .8339294
    wagegovt    .7658012   .0347599    22.03   0.000     .6976729    .8339294
       _cons    16.55899   1.224401    13.52   0.000      14.1592    18.95877

invest
     profits   -.3565311   .2601567    -1.37   0.171    -.8664288    .1533666
    profits1    1.011298   .2487744     4.07   0.000     .5237096    1.498887
    capital1   -.2601999   .0508694    -5.12   0.000     -.359902   -.1604977
       _cons    42.89626   10.59386     4.05   0.000     22.13269    63.65984

wagepriv
      totinc    .3747792   .0311027    12.05   0.000      .313819    .4357394
     totinc1    .1936506   .0324018     5.98   0.000     .1301443     .257157
        year    .1679262   .0289291     5.80   0.000     .1112263    .2246261
       _cons    2.624766   1.195559     2.20   0.028      .281512    4.968019

Endogenous variables:  consump invest wagepriv wagetot profits totinc
Exogenous variables:   profits1 wagegovt capital1 totinc1 year taxnetx govt
```

As expected, none of the parameter or standard error estimates has changed from the previous estimates (before the seventh significant digit). We have simply decomposed the total wage variable into its two parts and constrained the coefficients on these parts. The warning about additional endogenous variables was just `reg3`'s way of letting us know that we had specified some information that was irrelevant to the estimation of the system. We had left the `wagetot` variable in our `endog` macro. It does not mean anything to the system to specify `wagetot` as endogenous because it is no longer in the system. That's fine with `reg3` and fine for our current purposes.

We can also impose constraints across the equations. For example, the admittedly meaningless constraint of requiring `profits` to have the same effect in both the consumption and investment equations could be imposed. Retaining the constraint on the wage coefficients, we would estimate this constrained system.

```
. constraint 2 [consump]profits = [invest]profits
. reg3 $conseqn $inveqn $wageqn, endog($enlist) exog($exlist) constr(1 2) ireg3
note:  additional endogenous variables not in the system have no effect
       and are ignored:  wagetot
Iteration 1:   tolerance =   .1427927
Iteration 2:   tolerance =    .032539
Iteration 3:   tolerance =  .00307811
Iteration 4:   tolerance =  .00016903
Iteration 5:   tolerance =  .00003409
Iteration 6:   tolerance =  7.763e-06
Iteration 7:   tolerance =  9.240e-07
```

```
Three-stage least-squares regression, iterated
----------------------------------------------------------------------
Equation          Obs  Parms        RMSE    "R-sq"       chi2        P
----------------------------------------------------------------------
consump            21      3    .9504669    0.9798    1019.54   0.0000
invest             21      3    1.247066    0.8706     144.57   0.0000
wagepriv           21      3    .7225276    0.9862    1537.45   0.0000
----------------------------------------------------------------------

 ( 1)  [consump]wagepriv - [consump]wagegovt = 0
 ( 2)  [consump]profits - [invest]profits = 0
------------------------------------------------------------------------------
             |      Coef.   Std. Err.      z    P>|z|     [95% Conf. Interval]
-------------+----------------------------------------------------------------
consump      |
     profits |   .1075413   .0957767     1.12   0.262    -.0801777    .2952602
    profits1 |   .1712756   .0912613     1.88   0.061    -.0075932    .3501444
    wagepriv |    .798484   .0340876    23.42   0.000     .7316734    .8652946
    wagegovt |    .798484   .0340876    23.42   0.000     .7316734    .8652946
       _cons |    16.2521   1.212157    13.41   0.000     13.87631    18.62788
-------------+----------------------------------------------------------------
invest       |
     profits |   .1075413   .0957767     1.12   0.262    -.0801777    .2952602
    profits1 |   .6443378   .1058682     6.09   0.000       .43684    .8518356
    capital1 |  -.1766669   .0261889    -6.75   0.000    -.2279962   -.1253375
       _cons |   24.31931   5.284325     4.60   0.000     13.96222     34.6764
-------------+----------------------------------------------------------------
wagepriv     |
      totinc |   .4014106   .0300552    13.36   0.000     .3425035    .4603177
     totinc1 |   .1775359   .0321583     5.52   0.000     .1145068     .240565
        year |   .1549211   .0282291     5.49   0.000      .099593    .2102492
       _cons |   1.959788    1.14467     1.71   0.087    -.2837242    4.203299
------------------------------------------------------------------------------
Endogenous variables:  consump invest wagepriv wagetot profits totinc
Exogenous variables:   profits1 wagegovt capital1 totinc1 year taxnetx govt
------------------------------------------------------------------------------
```

◁

❑ Technical note

Identification in a system of simultaneous equations involves the notion that there is enough information to estimate the parameters of the model given the specified functional form. Underidentification usually manifests itself as one matrix in the 3SLS computations. The most commonly violated order condition for 2SLS or 3SLS involves the number of endogenous and exogenous variables. There must be at least as many noncollinear exogenous variables in the remaining system as there are endogenous right-hand-side variables in an equation. This condition must hold for each structural equation in the system.

Put as a set of rules the following:

1. Count the number of right-hand-side endogenous variables in an equation and call this m_i.
2. Count the number of exogenous variables in the same equation and call this k_i.
3. Count the total number of exogenous variables in all the structural equations plus any additional variables specified in an `exog()` or `inst()` option and call this K.
4. If $m_i > (K - k_i)$ for any structural equation (i), then the system is underidentified and cannot be estimated by 3SLS.

We are also possibly in trouble if any of the exogenous variables are linearly dependent. We must have m_i linearly independent variables among the exogenous variables represented by $(K - k_i)$.

The complete conditions for identification involve rank-order conditions on several matrices. For a full treatment, see Theil (1971) or Greene (2012, 331–334).

❑

Saved results

reg3 saves the following in e():

Scalars	
e(N)	number of observations
e(k)	number of parameters
e(k_eq)	number of equations in e(b)
e(mss_#)	model sum of squares for equation #
e(df_m#)	model degrees of freedom for equation #
e(rss_#)	residual sum of squares for equation #
e(df_r)	residual degrees of freedom (small)
e(r2_#)	R-squared for equation #
e(F_#)	F statistic for equation # (small)
e(rmse_#)	root mean squared error for equation #
e(dfk2_adj)	divisor used with VCE when dfk2 specified
e(ll)	log likelihood
e(chi2_#)	χ^2 for equation #
e(p_#)	significance for equation #
e(cons_#)	1 when equation # has a constant, 0 otherwise
e(rank)	rank of e(V)
e(ic)	number of iterations
Macros	
e(cmd)	reg3
e(cmdline)	command as typed
e(depvar)	names of dependent variables
e(exog)	names of exogenous variables
e(endog)	names of endogenous variables
e(eqnames)	names of equations
e(corr)	correlation structure
e(wtype)	weight type
e(wexp)	weight expression
e(method)	3sls, 2sls, ols, sure, or mvreg
e(small)	small
e(dfk)	dfk, if specified
e(properties)	b V
e(predict)	program used to implement predict
e(marginsok)	predictions allowed by margins
e(marginsnotok)	predictions disallowed by margins
e(asbalanced)	factor variables fvset as asbalanced
e(asobserved)	factor variables fvset as asobserved
Matrices	
e(b)	coefficient vector
e(Cns)	constraints matrix
e(Sigma)	$\widehat{\Sigma}$ matrix
e(V)	variance–covariance matrix of the estimators
Functions	
e(sample)	marks estimation sample

Methods and formulas

reg3 is implemented as an ado-file.

The most concise way to represent a system of equations for 3SLS requires thinking of the individual equations and their associated data as being stacked. reg3 does not expect the data in this format, but it is a convenient shorthand. The system could then be formulated as

$$\begin{bmatrix} \mathbf{y}_1 \\ \mathbf{y}_2 \\ \vdots \\ \mathbf{y}_M \end{bmatrix} = \begin{bmatrix} \mathbf{Z}_1 & 0 & \dots & 0 \\ 0 & \mathbf{Z}_2 & \dots & 0 \\ \vdots & \vdots & \ddots & \vdots \\ 0 & 0 & \dots & \mathbf{Z}_M \end{bmatrix} \begin{bmatrix} \beta_1 \\ \beta_2 \\ \vdots \\ \beta_M \end{bmatrix} + \begin{bmatrix} \epsilon_1 \\ \epsilon_2 \\ \vdots \\ \epsilon_M \end{bmatrix}$$

In full matrix notation, this is just

$$\mathbf{y} = \mathbf{Z}\mathbf{B} + \epsilon$$

The $\mathbf{Z}$ elements in these matrices represent both the endogenous and the exogenous right-hand-side variables in the equations.

Also assume that there will be correlation between the disturbances of the equations so that

$$E(\epsilon\epsilon') = \mathbf{\Sigma}$$

where the disturbances are further assumed to have an expected value of 0; $E(\epsilon) = 0$.

The first stage of 3SLS regression requires developing instrumented values for the endogenous variables in the system. These values can be derived as the predictions from a linear regression of each endogenous regressor on all exogenous variables in the system or, more succinctly, as the projection of each regressor through the projection matrix of all exogenous variables onto the regressors. Designating the set of all exogenous variables as $\mathbf{X}$ results in

$$\widehat{\mathbf{z}}_i = \mathbf{X}(\mathbf{X}'\mathbf{X})^{-1}\mathbf{X}'\mathbf{z}_i \quad \text{for each } i$$

Taken collectively, these $\widehat{\mathbf{Z}}$ contain the instrumented values for all the regressors. They take on the actual values for the exogenous variables and first-stage predictions for the endogenous variables. Given these instrumented variables, a generalized least squares (GLS) or Aitken (1935) estimator can be formed for the parameters of the system

$$\widehat{\mathbf{B}} = \left\{\widehat{\mathbf{Z}}'(\mathbf{\Sigma}^{-1} \otimes \mathbf{I})\widehat{\mathbf{Z}}\right\}^{-1} \widehat{\mathbf{Z}}'(\mathbf{\Sigma}^{-1} \otimes \mathbf{I})\mathbf{y}$$

All that remains is to obtain a consistent estimator for $\mathbf{\Sigma}$. This estimate can be formed from the residuals of 2SLS estimates of each equation in the system. Alternately, and identically, the residuals can be computed from the estimates formed by taking $\mathbf{\Sigma}$ to be an identity matrix. This maintains the full system of coefficients and allows constraints to be applied when the residuals are computed.

If we take $\mathbf{E}$ to be the matrix of residuals from these estimates, a consistent estimate of $\mathbf{\Sigma}$ is

$$\widehat{\mathbf{\Sigma}} = \frac{\mathbf{E}'\mathbf{E}}{n}$$

where n is the number of observations in the sample. An alternative divisor for this estimate can be obtained with the dfk option as outlined under options.

With the estimate of $\widehat{\mathbf{\Sigma}}$ placed into the GLS estimating equation,

$$\widehat{\mathbf{B}} = \left\{\widehat{\mathbf{Z}}'(\widehat{\mathbf{\Sigma}}^{-1} \otimes \mathbf{I})\widehat{\mathbf{Z}}\right\}^{-1} \widehat{\mathbf{Z}}'(\widehat{\mathbf{\Sigma}}^{-1} \otimes \mathbf{I})\mathbf{y}$$

is the 3SLS estimates of the system parameters.

The asymptotic variance–covariance matrix of the estimator is just the standard formulation for a GLS estimator

$$\mathbf{V}_{\widehat{\mathbf{B}}} = \left\{\widehat{\mathbf{Z}}'(\widehat{\mathbf{\Sigma}}^{-1} \otimes \mathbf{I})\widehat{\mathbf{Z}}\right\}^{-1}$$

Iterated 3SLS estimates can be obtained by computing the residuals from the three-stage parameter estimates, using these to formulate a new $\widehat{\mathbf{\Sigma}}$, and recomputing the parameter estimates. This process is repeated until the estimates $\widehat{\mathbf{B}}$ converge—if they converge. Convergence is not guaranteed. When estimating a system by SURE, these iterated estimates will be the maximum likelihood estimates for the system. The iterated solution can also be used to produce estimates that are invariant to choice of system and restriction parameterization for many linear systems under full 3SLS.

The exposition above follows the parallel developments in Greene (2012) and Davidson and MacKinnon (1993).

Henri Theil (1924–2000) was born in Amsterdam and awarded a PhD in 1951 by the University of Amsterdam. He researched and taught econometric theory, statistics, microeconomics, macroeconomic modeling, and economic forecasting, and policy at (what is now) Erasmus University Rotterdam, the University of Chicago, and the University of Florida. Theil's many specific contributions include work on 2SLS and 3SLS, inequality and concentration, and consumer demand.

References

Aitken, A. C. 1935. On least squares and linear combination of observations. *Proceedings of the Royal Society of Edinburgh* 55: 42–48.

Bewley, R. 2000. Mr. Henri Theil: An interview with the International Journal of Forecasting. *International Journal of Forecasting* 16: 1–16.

Davidson, R., and J. G. MacKinnon. 1993. *Estimation and Inference in Econometrics*. New York: Oxford University Press.

Greene, W. H. 2012. *Econometric Analysis*. 7th ed. Upper Saddle River, NJ: Prentice Hall.

Klein, L. R. 1950. *Economic Fluctuations in the United States 1921-1941*. New York: Wiley.

Nichols, A. 2007. Causal inference with observational data. *Stata Journal* 7: 507–541.

Poi, B. P. 2006. Jackknife instrumental variables estimation in Stata. *Stata Journal* 6: 364–376.

Theil, H. 1971. *Principles of Econometrics*. New York: Wiley.

Weesie, J. 1999. sg121: Seemingly unrelated estimation and the cluster-adjusted sandwich estimator. *Stata Technical Bulletin* 52: 34–47. Reprinted in *Stata Technical Bulletin Reprints*, vol. 9, pp. 231–248. College Station, TX: Stata Press.

Zellner, A., and H. Theil. 1962. Three stage least squares: Simultaneous estimate of simultaneous equations. *Econometrica* 29: 54–78.

Also see

[R] **reg3 postestimation** — Postestimation tools for reg3

[R] **ivregress** — Single-equation instrumental-variables regression

[R] **mvreg** — Multivariate regression

[R] **nlsur** — Estimation of nonlinear systems of equations

[R] **regress** — Linear regression

[R] **sureg** — Zellner's seemingly unrelated regression

Stata Structural Equation Modeling Reference Manual

[U] **20 Estimation and postestimation commands**

Title

reg3 postestimation — Postestimation tools for reg3

Description

The following postestimation commands are available after `reg3`:

Command	Description
`contrast`	contrasts and ANOVA-style joint tests of estimates
*`estat`	AIC, BIC, VCE, and estimation sample summary
`estimates`	cataloging estimation results
`hausman`	Hausman's specification test
`lincom`	point estimates, standard errors, testing, and inference for linear combinations of coefficients
`margins`	marginal means, predictive margins, marginal effects, and average marginal effects
`marginsplot`	graph the results from margins (profile plots, interaction plots, etc.)
`nlcom`	point estimates, standard errors, testing, and inference for nonlinear combinations of coefficients
`predict`	predictions, residuals, influence statistics, and other diagnostic measures
`predictnl`	point estimates, standard errors, testing, and inference for generalized predictions
`pwcompare`	pairwise comparisons of estimates
`test`	Wald tests of simple and composite linear hypotheses
`testnl`	Wald tests of nonlinear hypotheses

*`estat ic` is not appropriate after `reg3, 2sls`.

See the corresponding entries in the *Base Reference Manual* for details.

Syntax for predict

`predict` [*type*] *newvar* [*if*] [*in*] [, `equation(`*eqno*[`,`*eqno*]`)` *statistic*]

statistic	Description
Main	
`xb`	linear prediction; the default
`stdp`	standard error of the linear prediction
`residuals`	residuals
`difference`	difference between the linear predictions of two equations
`stddp`	standard error of the difference in linear predictions

These statistics are available both in and out of sample; type `predict ... if e(sample) ...` if wanted only for the estimation sample.

Menu

Statistics > Postestimation > Predictions, residuals, etc.

Options for predict

Main

`equation(`*eqno*[`,`*eqno*]`)` specifies to which equation you are referring.

`equation()` is filled in with one *eqno* for the `xb`, `stdp`, and `residuals` options. `equation(#1)` would mean the calculation is to be made for the first equation, `equation(#2)` would mean the second, and so on. You could also refer to the equations by their names. `equation(income)` would refer to the equation named income and `equation(hours)` to the equation named hours.

If you do not specify `equation()`, results are the same as if you specified `equation(#1)`.

`difference` and `stddp` refer to between-equation concepts. To use these options, you must specify two equations, for example, `equation(#1,#2)` or `equation(income,hours)`. When two equations must be specified, `equation()` is required.

`xb`, the default, calculates the linear prediction (fitted values)—the prediction of $\mathbf{x}_j\mathbf{b}$ for the specified equation.

`stdp` calculates the standard error of the prediction for the specified equation. It can be thought of as the standard error of the predicted expected value or mean for the observation's covariate pattern. The standard error of the prediction is also referred to as the standard error of the fitted value.

`residuals` calculates the residuals.

`difference` calculates the difference between the linear predictions of two equations in the system. With `equation(#1,#2)`, `difference` computes the prediction of `equation(#1)` minus the prediction of `equation(#2)`.

`stddp` is allowed only after you have previously fit a multiple-equation model. The standard error of the difference in linear predictions ($\mathbf{x}_{1j}\mathbf{b} - \mathbf{x}_{2j}\mathbf{b}$) between equations 1 and 2 is calculated.

For more information on using `predict` after multiple-equation estimation commands, see [R] **predict**.

Remarks

▷ Example 1

In example 2 of [R] **reg3**, we fit a simple supply-and-demand model. Here we obtain the fitted supply and demand curves assuming that the exogenous regressors equal their sample means. We first `replace` each of the three exogenous regressors with their sample means, then we call `predict` to obtain the predictions.

```
. use http://www.stata-press.com/data/r12/supDem
. global demand "(qDemand: quantity price pcompete income)"
. global supply "(qSupply: quantity price praw)"
. reg3 $demand $supply, endog(price)
  (output omitted )
. summarize pcompete, meanonly
. replace pcompete = r(mean)
(49 real changes made)
```

```
. summarize income, meanonly
. replace income = r(mean)
(49 real changes made)
. summarize praw, meanonly
. replace praw = r(mean)
(49 real changes made)
. predict demand, equation(qDemand)
(option xb assumed; fitted values)
. predict supply, equation(qSupply)
(option xb assumed; fitted values)
. graph twoway line demand price, sort || line supply price
```

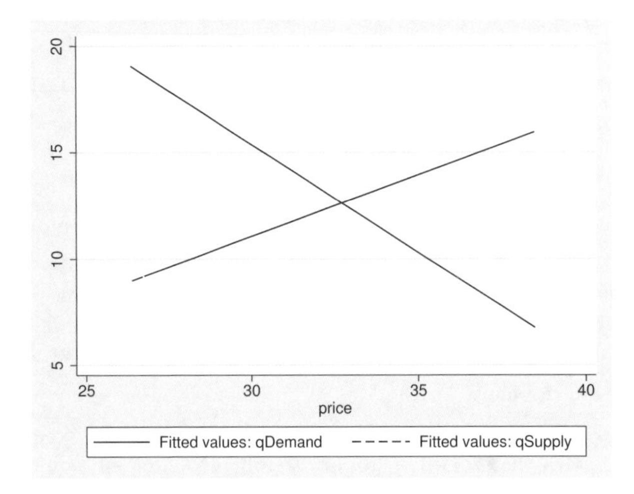

As we would expect based on economic theory, the demand curve slopes downward while the supply curve slopes upward. With the exogenous variables at their mean levels, the equilibrium price and quantity are slightly less than 33 and 13, respectively.

◁

Methods and formulas

All postestimation commands listed above are implemented as ado-files.

The computational formulas for the statistics produced by `predict` can be found in [R] **predict** and [R] **regress postestimation**.

Also see

[R] **reg3** — Three-stage estimation for systems of simultaneous equations

[U] **20 Estimation and postestimation commands**

Title

regress — Linear regression

Syntax

`regress` *depvar* [*indepvars*] [*if*] [*in*] [*weight*] [, *options*]

options	Description
Model	
`noconstant`	suppress constant term
`hascons`	has user-supplied constant
`tsscons`	compute total sum of squares with constant; seldom used
SE/Robust	
`vce(`*vcetype*`)`	*vcetype* may be `ols`, `robust`, `cluster` *clustvar*, `bootstrap`, `jackknife`, `hc2`, or `hc3`
Reporting	
`level(`*#*`)`	set confidence level; default is `level(95)`
`beta`	report standardized beta coefficients
`eform(`*string*`)`	report exponentiated coefficients and label as *string*
`depname(`*varname*`)`	substitute dependent variable name; programmer's option
display_options	control column formats, row spacing, line width, and display of omitted variables and base and empty cells
`noheader`	suppress table header
`notable`	suppress table header
`plus`	make table extendable
`mse1`	force mean squared error to 1
`coeflegend`	display legend instead of statistics

indepvars may contain factor variables; see [U] **11.4.3 Factor variables**.

depvar and *indepvars* may contain time-series operators; see [U] **11.4.4 Time-series varlists**.

`bootstrap`, `by`, `fracpoly`, `jackknife`, `mfp`, `mi estimate`, `nestreg`, `rolling`, `statsby`, `stepwise`, and `svy` are allowed; see [U] **11.1.10 Prefix commands**.

`vce(bootstrap)` and `vce(jackknife)` are not allowed with the `mi estimate` prefix; see [MI] **mi estimate**.

Weights are not allowed with the `bootstrap` prefix; see [R] **bootstrap**.

`aweight`s are not allowed with the `jackknife` prefix; see [R] **jackknife**.

`hascons`, `tsscons`, `vce()`, `beta`, `noheader`, `notable`, `plus`, `depname()`, `mse1`, and weights are not allowed with the `svy` prefix; see [SVY] **svy**.

`aweight`s, `fweight`s, `iweight`s, and `pweight`s are allowed; see [U] **11.1.6 weight**.

`noheader`, `notable`, `plus`, `mse1`, and `coeflegend` do not appear in the dialog box.

See [U] **20 Estimation and postestimation commands** for more capabilities of estimation commands.

Menu

Statistics > Linear models and related > Linear regression

Description

`regress` fits a model of *depvar* on *indepvars* using linear regression.

Here is a short list of other regression commands that may be of interest. See [I] **estimation commands** for a complete list.

Command	Entry	Description
areg	[R] **areg**	an easier way to fit regressions with many dummy variables
arch	[TS] **arch**	regression models with ARCH errors
arima	[TS] **arima**	ARIMA models
boxcox	[R] **boxcox**	Box–Cox regression models
cnsreg	[R] **cnsreg**	constrained linear regression
eivreg	[R] **eivreg**	errors-in-variables regression
frontier	[R] **frontier**	stochastic frontier models
gmm	[R] **gmm**	generalized method of moments estimation
heckman	[R] **heckman**	Heckman selection model
intreg	[R] **intreg**	interval regression
ivregress	[R] **ivregress**	single-equation instrumental-variables regression
ivtobit	[R] **ivtobit**	tobit regression with endogenous variables
newey	[TS] **newey**	regression with Newey–West standard errors
nl	[R] **nl**	nonlinear least-squares estimation
nlsur	[R] **nlsur**	estimation of nonlinear systems of equations
qreg	[R] **qreg**	quantile (including median) regression
reg3	[R] **reg3**	three-stage least-squares (3SLS) regression
rreg	[R] **rreg**	a type of robust regression
sureg	[R] **sureg**	seemingly unrelated regression
tobit	[R] **tobit**	tobit regression
treatreg	[R] **treatreg**	treatment-effects model
truncreg	[R] **truncreg**	truncated regression
xtabond	[XT] **xtabond**	Arellano–Bond linear dynamic panel-data estimation
xtdpd	[XT] **xtdpd**	linear dynamic panel-data estimation
xtfrontier	[XT] **xtfrontier**	panel-data stochastic frontier models
xtgls	[XT] **xtgls**	panel-data GLS models
xthtaylor	[XT] **xthtaylor**	Hausman–Taylor estimator for error-components models
xtintreg	[XT] **xtintreg**	panel-data interval regression models
xtivreg	[XT] **xtivreg**	panel-data instrumental-variables (2SLS) regression
xtpcse	[XT] **xtpcse**	linear regression with panel-corrected standard errors
xtreg	[XT] **xtreg**	fixed- and random-effects linear models
xtregar	[XT] **xtregar**	fixed- and random-effects linear models with an AR(1) disturbance
xttobit	[XT] **xttobit**	panel-data tobit models
	[SEM] *Stata Structural Equation Modeling Reference Manual*	

Options

Model

`noconstant`; see [R] **estimation options**.

`hascons` indicates that a user-defined constant or its equivalent is specified among the independent variables in *indepvars*. Some caution is recommended when specifying this option, as resulting estimates may not be as accurate as they otherwise would be. Use of this option requires "sweeping" the constant last, so the moment matrix must be accumulated in absolute rather than deviation form. This option may be safely specified when the means of the dependent and independent variables are all reasonable and there is not much collinearity between the independent variables. The best procedure is to view `hascons` as a reporting option—estimate with and without `hascons` and verify that the coefficients and standard errors of the variables not affected by the identity of the constant are unchanged.

`tsscons` forces the total sum of squares to be computed as though the model has a constant, that is, as deviations from the mean of the dependent variable. This is a rarely used option that has an effect only when specified with `noconstant`. It affects the total sum of squares and all results derived from the total sum of squares.

SE/Robust

`vce(`*vcetype*`)` specifies the type of standard error reported, which includes types that are derived from asymptotic theory, that are robust to some kinds of misspecification, that allow for intragroup correlation, and that use bootstrap or jackknife methods; see [R] ***vce_option***.

`vce(ols)`, the default, uses the standard variance estimator for ordinary least-squares regression.

`regress` also allows the following:

`vce(hc2)` and `vce(hc3)` specify an alternative bias correction for the robust variance calculation. `vce(hc2)` and `vce(hc3)` may not be specified with `svy` prefix. In the unclustered case, `vce(robust)` uses $\widehat{\sigma}_j^2 = \{n/(n-k)\}u_j^2$ as an estimate of the variance of the jth observation, where u_j is the calculated residual and $n/(n-k)$ is included to improve the overall estimate's small-sample properties.

`vce(hc2)` instead uses $u_j^2/(1-h_{jj})$ as the observation's variance estimate, where h_{jj} is the diagonal element of the hat (projection) matrix. This estimate is unbiased if the model really is homoskedastic. `vce(hc2)` tends to produce slightly more conservative confidence intervals.

`vce(hc3)` uses $u_j^2/(1-h_{jj})^2$ as suggested by Davidson and MacKinnon (1993), who report that this method tends to produce better results when the model really is heteroskedastic. `vce(hc3)` produces confidence intervals that tend to be even more conservative.

See Davidson and MacKinnon (1993, 554–556) and Angrist and Pischke (2009, 294–308) for more discussion on these two bias corrections.

Reporting

`level(`*#*`)`; see [R] **estimation options**.

`beta` asks that standardized beta coefficients be reported instead of confidence intervals. The beta coefficients are the regression coefficients obtained by first standardizing all variables to have a mean of 0 and a standard deviation of 1. `beta` may not be specified with `vce(cluster` *clustvar*`)` or the `svy` prefix.

`eform(`*string*`)` is used only in programs and ado-files that use `regress` to fit models other than linear regression. `eform()` specifies that the coefficient table be displayed in exponentiated form as defined in [R] **maximize** and that *string* be used to label the exponentiated coefficients in the table.

`depname(`*varname*`)` is used only in programs and ado-files that use `regress` to fit models other than linear regression. `depname()` may be specified only at estimation time. *varname* is recorded as the identity of the dependent variable, even though the estimates are calculated using *depvar*. This method affects the labeling of the output—not the results calculated—but could affect subsequent calculations made by `predict`, where the residual would be calculated as deviations from *varname* rather than *depvar*. `depname()` is most typically used when *depvar* is a temporary variable (see [P] **macro**) used as a proxy for *varname*.

`depname()` is not allowed with the `svy` prefix.

display_options: `noomitted`, `vsquish`, `noemptycells`, `baselevels`, `allbaselevels`, `cformat(%`*fmt*`)`, `pformat(%`*fmt*`)`, `sformat(%`*fmt*`)`, and `nolstretch`; see [R] **estimation options**.

The following options are available with `regress` but are not shown in the dialog box:

`noheader` suppresses the display of the ANOVA table and summary statistics at the top of the output; only the coefficient table is displayed. This option is often used in programs and ado-files.

`notable` suppresses display of the coefficient table.

`plus` specifies that the output table be made extendable. This option is often used in programs and ado-files.

`mse1` is used only in programs and ado-files that use `regress` to fit models other than linear regression and is not allowed with the `svy` prefix. `mse1` sets the mean squared error to `1`, forcing the variance–covariance matrix of the estimators to be $(\mathbf{X}'\mathbf{D}\mathbf{X})^{-1}$ (see *Methods and formulas* below) and affecting calculated standard errors. Degrees of freedom for t statistics are calculated as n rather than $n - k$.

`coeflegend`; see [R] **estimation options**.

Remarks

Remarks are presented under the following headings:

Ordinary least squares
Treatment of the constant
Robust standard errors
Weighted regression
Instrumental variables and two-stage least-squares regression

`regress` performs linear regression, including ordinary least squares and weighted least squares. For a general discussion of linear regression, see Draper and Smith (1998), Greene (2012), or Kmenta (1997).

See Wooldridge (2009) for an excellent treatment of estimation, inference, interpretation, and specification testing in linear regression models. This presentation stands out for its clarification of the statistical issues, as opposed to the algebraic issues. See Wooldridge (2010, chap. 4) for a more advanced discussion along the same lines.

See Hamilton (2009, chap. 6) and Cameron and Trivedi (2010, chap. 3) for an introduction to linear regression using Stata. Dohoo, Martin, and Stryhn (2010) discuss linear regression using examples from epidemiology, and Stata datasets and do-files used in the text are available. Cameron and Trivedi (2010) discuss linear regression using econometric examples with Stata.

Chatterjee and Hadi (2006) explain regression analysis by using examples containing typical problems that you might encounter when performing exploratory data analysis. We also recommend Weisberg (2005), who emphasizes the importance of the assumptions of linear regression and problems resulting from these assumptions. Angrist and Pischke (2009) approach regression as a tool for exploring relationships, estimating treatment effects, and providing answers to public policy questions. For a discussion of model-selection techniques and exploratory data analysis, see Mosteller and Tukey (1977). For a mathematically rigorous treatment, see Peracchi (2001, chap. 6). Finally, see Plackett (1972) if you are interested in the history of regression. Least squares, which dates back to the 1790s, was discovered independently by Legendre and Gauss.

Ordinary least squares

▷ Example 1

Suppose that we have data on the mileage rating and weight of 74 automobiles. The variables in our data are mpg, weight, and foreign. The last variable assumes the value 1 for foreign and 0 for domestic automobiles. We wish to fit the model

$$\texttt{mpg} = \beta_0 + \beta_1\texttt{weight} + \beta_2\texttt{weight}^2 + \beta_3\texttt{foreign} + \epsilon$$

We include c.weight#c.weight in our model for the weight-squared term (see [U] **11.4.3 Factor variables**):

```
. use http://www.stata-press.com/data/r12/auto
(1978 Automobile Data)

. regress mpg weight c.weight#c.weight foreign

      Source |       SS       df       MS              Number of obs =      74
-------------+------------------------------           F(  3,    70) =   52.25
       Model |  1689.15372     3   563.05124           Prob > F      =  0.0000
    Residual |   754.30574    70  10.7757963           R-squared     =  0.6913
-------------+------------------------------           Adj R-squared =  0.6781
       Total |  2443.45946    73  33.4720474           Root MSE      =  3.2827

------------------------------------------------------------------------------
         mpg |      Coef.   Std. Err.      t    P>|t|     [95% Conf. Interval]
-------------+----------------------------------------------------------------
      weight |  -.0165729   .0039692    -4.18   0.000    -.0244892   -.0086567
             |
   c.weight# |
    c.weight |   1.59e-06   6.25e-07     2.55   0.013     3.45e-07    2.84e-06
             |
     foreign |    -2.2035   1.059246    -2.08   0.041      -4.3161   -.0909002
       _cons |   56.53884   6.197383     9.12   0.000     44.17855    68.89913
------------------------------------------------------------------------------
```

◁

regress produces a variety of summary statistics along with the table of regression coefficients. At the upper left, regress reports an analysis-of-variance (ANOVA) table. The column headings SS, df, and MS stand for “sum of squares”, “degrees of freedom”, and “mean square”, respectively. In the previous example, the total sum of squares is 2,443.5: 1,689.2 accounted for by the model and 754.3 left unexplained. Because the regression included a constant, the total sum reflects the sum after removal of means, as does the sum of squares due to the model. The table also reveals that there are 73 total degrees of freedom (counted as 74 observations less 1 for the mean removal), of which 3 are consumed by the model, leaving 70 for the residual.

To the right of the ANOVA table are presented other summary statistics. The F statistic associated with the ANOVA table is 52.25. The statistic has 3 numerator and 70 denominator degrees of freedom. The F statistic tests the hypothesis that all coefficients *excluding the constant* are zero. The chance of observing an F statistic that large or larger is reported as 0.0000, which is Stata's way of indicating a number smaller than 0.00005. The R-squared (R^2) for the regression is 0.6913, and the R-squared adjusted for degrees of freedom (R_a^2) is 0.6781. The root mean squared error, labeled `Root MSE`, is 3.2827. It is the square root of the mean squared error reported for the residual in the ANOVA table.

Finally, Stata produces a table of the estimated coefficients. The first line of the table indicates that the left-hand-side variable is `mpg`. Thereafter follow the four estimated coefficients. Our fitted model is

$$\texttt{mpg\_hat} = 56.54 - 0.0166\,\texttt{weight} + 1.59 \times 10^{-6}\,\texttt{c.weight\#c.weight} - 2.20\,\texttt{foreign}$$

Reported to the right of the coefficients in the output are the standard errors. For instance, the standard error for the coefficient on `weight` is 0.0039692. The corresponding t statistic is -4.18, which has a two-sided significance level of 0.000. This number indicates that the significance is less than 0.0005. The 95% confidence interval for the coefficient is $[-0.024, -0.009]$.

▷ Example 2

`regress` shares the features of all estimation commands. Among other things, this means that after running a regression, we can use `test` to test hypotheses about the coefficients, `estat vce` to examine the covariance matrix of the estimators, and `predict` to obtain predicted values, residuals, and influence statistics. See [U] **20 Estimation and postestimation commands**. Options that affect how estimates are displayed, such as `beta` or `level()`, can be used when replaying results.

Suppose that we meant to specify the `beta` option to obtain beta coefficients (regression coefficients normalized by the ratio of the standard deviation of the regressor to the standard deviation of the dependent variable). Even though we forgot, we can specify the option now:

```
. regress, beta

      Source |       SS       df       MS              Number of obs =      74
-------------+------------------------------           F(  3,    70) =   52.25
       Model |  1689.15372     3   563.05124           Prob > F      =  0.0000
    Residual |   754.30574    70  10.7757963           R-squared     =  0.6913
-------------+------------------------------           Adj R-squared =  0.6781
       Total |  2443.45946    73  33.4720474           Root MSE      =  3.2827

------------------------------------------------------------------------------
         mpg |      Coef.   Std. Err.      t    P>|t|                     Beta
-------------+----------------------------------------------------------------
      weight |  -.0165729   .0039692    -4.18   0.000                -2.226321
             |
    c.weight#|
    c.weight |   1.59e-06   6.25e-07     2.55   0.013                  1.32654
             |
     foreign |    -2.2035   1.059246    -2.08   0.041                  -.17527
       _cons |   56.53884   6.197383     9.12   0.000                        .
------------------------------------------------------------------------------
```

◁

Treatment of the constant

By default, regress includes an intercept (constant) term in the model. The noconstant option suppresses it, and the hascons option tells regress that the model already has one.

▷ Example 3

We wish to fit a regression of the weight of an automobile against its length, and we wish to impose the constraint that the weight is zero when the length is zero.

If we simply type regress weight length, we are fitting the model

$$\texttt{weight} = \beta_0 + \beta_1 \,\texttt{length} + \epsilon$$

Here a length of zero corresponds to a weight of β_0. We want to force β_0 to be zero or, equivalently, estimate an equation that does not include an intercept:

$$\texttt{weight} = \beta_1 \,\texttt{length} + \epsilon$$

We do this by specifying the noconstant option:

```
. regress weight length, noconstant
      Source |       SS       df       MS              Number of obs =      74
-------------+------------------------------           F(  1,    73) = 3450.13
       Model |   703869302     1   703869302           Prob > F      =  0.0000
    Residual |  14892897.8    73  204012.299           R-squared     =  0.9793
-------------+------------------------------           Adj R-squared =  0.9790
       Total |   718762200    74   9713002.7           Root MSE      =  451.68

------------------------------------------------------------------------------
      weight |      Coef.   Std. Err.      t    P>|t|     [95% Conf. Interval]
-------------+----------------------------------------------------------------
      length |   16.29829   .2774752    58.74   0.000     15.74528     16.8513
------------------------------------------------------------------------------
```

In our data, length is measured in inches and weight in pounds. We discover that each inch of length adds 16 pounds to the weight.

◁

Sometimes there is no need for Stata to include a constant term in the model. Most commonly, this occurs when the model contains a set of mutually exclusive indicator variables. hascons is a variation of the noconstant option—it tells Stata not to add a constant to the regression because the regression specification already has one, either directly or indirectly.

For instance, we now refit our model of weight as a function of length and include separate constants for foreign and domestic cars by specifying bn.foreign. bn.foreign is factor-variable notation for "no base for foreign" or "include all levels of variable foreign in the model"; see [U] **11.4.3 Factor variables**.

```
. regress weight length bn.foreign, hascons

      Source |       SS           df       MS      Number of obs   =        74
-------------+----------------------------------   F(2, 71)        =    316.54
       Model |  39647744.7         2  19823872.3   Prob > F        =    0.0000
    Residual |   4446433.7        71  62625.8268   R-squared       =    0.8992
-------------+----------------------------------   Adj R-squared   =    0.8963
       Total |  44094178.4        73  604029.841   Root MSE        =    250.25

------------------------------------------------------------------------------
      weight |      Coef.   Std. Err.      t    P>|t|     [95% Conf. Interval]
-------------+----------------------------------------------------------------
      length |   31.44455   1.601234    19.64   0.000     28.25178    34.63732
             |
     foreign |
          0  |   -2850.25   315.9691    -9.02   0.000    -3480.274   -2220.225
          1  |  -2983.927   275.1041   -10.85   0.000    -3532.469   -2435.385
------------------------------------------------------------------------------
```

❑ Technical note

There is a subtle distinction between the `hascons` and `noconstant` options. We can most easily reveal it by refitting the last regression, specifying `noconstant` rather than `hascons`:

```
. regress weight length bn.foreign, noconstant

      Source |       SS           df       MS      Number of obs   =        74
-------------+----------------------------------   F(3, 71)        =   3802.03
       Model |   714315766         3   238105255   Prob > F        =    0.0000
    Residual |   4446433.7        71  62625.8268   R-squared       =    0.9938
-------------+----------------------------------   Adj R-squared   =    0.9936
       Total |   718762200        74   9713002.7   Root MSE        =    250.25

------------------------------------------------------------------------------
      weight |      Coef.   Std. Err.      t    P>|t|     [95% Conf. Interval]
-------------+----------------------------------------------------------------
      length |   31.44455   1.601234    19.64   0.000     28.25178    34.63732
             |
     foreign |
          0  |   -2850.25   315.9691    -9.02   0.000    -3480.274   -2220.225
          1  |  -2983.927   275.1041   -10.85   0.000    -3532.469   -2435.385
------------------------------------------------------------------------------
```

Comparing this output with that produced by the previous `regress` command, we see that they are almost, but not quite, identical. The parameter estimates and their associated statistics—the second half of the output—are identical. The overall summary statistics and the ANOVA table—the first half of the output—are different, however.

In the first case, the R^2 is shown as 0.8992; here it is shown as 0.9938. In the first case, the F statistic is 316.54; now it is 3,802.03. The numerator degrees of freedom are different as well. In the first case, the numerator degrees of freedom are 2; now they are 3. Which is correct?

Both are. Specifying the `hascons` option causes `regress` to adjust the ANOVA table and its associated statistics for the explanatory power of the constant. The regression in effect has a constant; it is just written in such a way that a separate constant is unnecessary. No such adjustment is made with the `noconstant` option.

❑

❑ Technical note

When the hascons option is specified, regress checks to make sure that the model does in fact have a constant term. If regress cannot find a constant term, it automatically adds one. Fitting a model of weight on length and specifying the hascons option, we obtain

```
. regress weight length, hascons
(note: hascons false)
      Source |       SS       df       MS              Number of obs =      74
-------------+------------------------------           F(  1,    72) =  613.27
       Model |  39461306.8     1  39461306.8           Prob > F      =  0.0000
    Residual |  4632871.55    72  64345.4382           R-squared     =  0.8949
-------------+------------------------------           Adj R-squared =  0.8935
       Total |  44094178.4    73  604029.841           Root MSE      =  253.66

------------------------------------------------------------------------------
      weight |      Coef.   Std. Err.      t    P>|t|     [95% Conf. Interval]
-------------+----------------------------------------------------------------
      length |   33.01988   1.333364    24.76   0.000     30.36187    35.67789
       _cons |  -3186.047   252.3113   -12.63   0.000     -3689.02   -2683.073
------------------------------------------------------------------------------
```

Even though we specified hascons, regress included a constant, anyway. It also added a note to our output: “note: hascons false”.

❑

❑ Technical note

Even if the model specification effectively includes a constant term, we need not specify the hascons option. regress is always on the lookout for collinear variables and omits them from the model. For instance,

```
. regress weight length bn.foreign
note: 1.foreign omitted because of collinearity
      Source |       SS       df       MS              Number of obs =      74
-------------+------------------------------           F(  2,    71) =  316.54
       Model |  39647744.7     2  19823872.3           Prob > F      =  0.0000
    Residual |   4446433.7    71  62625.8268           R-squared     =  0.8992
-------------+------------------------------           Adj R-squared =  0.8963
       Total |  44094178.4    73  604029.841           Root MSE      =  250.25

------------------------------------------------------------------------------
      weight |      Coef.   Std. Err.      t    P>|t|     [95% Conf. Interval]
-------------+----------------------------------------------------------------
      length |   31.44455   1.601234    19.64   0.000     28.25178    34.63732
             |
     foreign |
          0  |   133.6775   77.47615     1.73   0.089    -20.80555    288.1605
          1  |          0  (omitted)
             |
       _cons |  -2983.927   275.1041   -10.85   0.000    -3532.469   -2435.385
------------------------------------------------------------------------------
```

❑

Robust standard errors

`regress` with the `vce(robust)` option substitutes a robust variance matrix calculation for the conventional calculation, or if `vce(cluster` *clustvar*`)` is specified, allows relaxing the assumption of independence within groups. How this method works is explained in [U] **20.20 Obtaining robust variance estimates**. Below we show how well this approach works.

▷ Example 4

Specifying the `vce(robust)` option is equivalent to requesting White-corrected standard errors in the presence of heteroskedasticity. We use the automobile data and, in the process of looking at the energy efficiency of cars, analyze a variable with considerable heteroskedasticity.

We will examine the amount of energy—measured in gallons of gasoline—that the cars in the data need to move 1,000 pounds of their weight 100 miles. We are going to examine the relative efficiency of foreign and domestic cars.

```
. gen gpmw = ((1/mpg)/weight)*100*1000
. summarize gpmw
    Variable |       Obs        Mean    Std. Dev.       Min        Max
-------------+--------------------------------------------------------
        gpmw |        74    1.682184    .2426311    1.09553    2.30521
```

In these data, the engines consume between 1.10 and 2.31 gallons of gas to move 1,000 pounds of the car's weight 100 miles. If we ran a regression with conventional standard errors of `gpmw` on `foreign`, we would obtain

```
. regress gpmw foreign
      Source |       SS       df       MS              Number of obs =      74
-------------+------------------------------           F(  1,    72) =   20.07
       Model |  .936705572     1  .936705572           Prob > F      =  0.0000
    Residual |  3.36079459    72  .046677703           R-squared     =  0.2180
-------------+------------------------------           Adj R-squared =  0.2071
       Total |  4.29750017    73  .058869865           Root MSE      =  .21605

------------------------------------------------------------------------------
        gpmw |      Coef.   Std. Err.      t    P>|t|     [95% Conf. Interval]
-------------+----------------------------------------------------------------
     foreign |   .2461526   .0549487     4.48   0.000     .1366143    .3556909
       _cons |   1.609004   .0299608    53.70   0.000     1.549278     1.66873
------------------------------------------------------------------------------
```

`regress` with the `vce(robust)` option, on the other hand, reports

```
. regress gpmw foreign, vce(robust)
Linear regression                                      Number of obs =      74
                                                       F(  1,    72) =   13.13
                                                       Prob > F      =  0.0005
                                                       R-squared     =  0.2180
                                                       Root MSE      =  .21605

------------------------------------------------------------------------------
             |               Robust
        gpmw |      Coef.   Std. Err.      t    P>|t|     [95% Conf. Interval]
-------------+----------------------------------------------------------------
     foreign |   .2461526   .0679238     3.62   0.001     .1107489    .3815563
       _cons |   1.609004   .0234535    68.60   0.000      1.56225    1.655758
------------------------------------------------------------------------------
```

The point estimates are the same (foreign cars need one-quarter gallon more gas), but the standard errors differ by roughly 20%. Conventional regression reports the 95% confidence interval as $[0.14, 0.36]$, whereas the robust standard errors make the interval $[0.11, 0.38]$.

Which is right? Notice that `gpmw` is a variable with considerable heteroskedasticity:

```
. tabulate foreign, summarize(gpmw)

            |         Summary of gpmw
   Car type |        Mean   Std. Dev.       Freq.
------------+------------------------------------
   Domestic |   1.6090039   .16845182          52
    Foreign |   1.8551565   .30186861          22
------------+------------------------------------
      Total |   1.6821844   .24263113          74
```

Thus here we favor the robust standard errors. In [U] **20.20 Obtaining robust variance estimates**, we show another example using linear regression where it makes little difference whether we specify `vce(robust)`. The linear-regression assumptions were true, and we obtained nearly linear-regression results. The advantage of the robust estimate is that in neither case did we have to check assumptions.

◁

❑ Technical note

`regress` purposefully suppresses displaying the ANOVA table when `vce(robust)` is specified, as it is no longer appropriate in a statistical sense, even though, mechanically, the numbers would be unchanged. That is, sums of squares remain unchanged, but the meaning of those sums is no longer relevant. The F statistic, for instance, is no longer based on sums of squares; it becomes a Wald test based on the robustly estimated variance matrix. Nevertheless, `regress` continues to report the R^2 and the root MSE even though both numbers are based on sums of squares and are, strictly speaking, irrelevant. In this, the root MSE is more in violation of the spirit of the robust estimator than is R^2. As a goodness-of-fit statistic, R^2 is still fine; just do not use it in formulas to obtain F statistics because those formulas no longer apply. The root MSE is valid in a literal sense—it is the square root of the mean squared error, but it is no longer an estimate of σ because there is no single σ; the variance of the residual varies observation by observation.

❑

▷ Example 5

The `vce(hc2)` and `vce(hc3)` options modify the robust variance calculation. In the context of linear regression without clustering, the idea behind the robust calculation is somehow to measure σ_j^2, the variance of the residual associated with the jth observation, and then to use that estimate to improve the estimated variance of $\widehat{\beta}$. Because residuals have (theoretically and practically) mean 0, one estimate of σ_j^2 is the observation's squared residual itself—u_j^2. A finite-sample correction could improve that by multiplying u_j^2 by $n/(n-k)$, and, as a matter of fact, `vce(robust)` uses $\{n/(n-k)\}u_j^2$ as its estimate of the residual's variance.

`vce(hc2)` and `vce(hc3)` use alternative estimators of the observation-specific variances. For instance, if the residuals are homoskedastic, we can show that the expected value of u_j^2 is $\sigma^2(1-h_{jj})$, where h_{jj} is the jth diagonal element of the projection (hat) matrix. h_{jj} has average value k/n, so $1-h_{jj}$ has average value $1-k/n = (n-k)/n$. Thus the default robust estimator $\widehat{\sigma}_j = \{n/(n-k)\}u_j^2$ amounts to dividing u_j^2 by the average of the expectation.

vce(hc2) divides u_j^2 by $1-h_{jj}$ itself, so it should yield better estimates if the residuals really are homoskedastic. vce(hc3) divides u_j^2 by $(1-h_{jj})^2$ and has no such clean interpretation. Davidson and MacKinnon (1993) show that $u_j^2/(1-h_{jj})^2$ approximates a more complicated estimator that they obtain by jackknifing (MacKinnon and White 1985). Angrist and Pischke (2009) also illustrate the relative merits of these adjustments.

Here are the results of refitting our efficiency model using vce(hc2) and vce(hc3):

```
. regress gpmw foreign, vce(hc2)
Linear regression                                 Number of obs =         74
                                                  F(  1,     72) =      12.93
                                                  Prob > F      =     0.0006
                                                  R-squared     =     0.2180
                                                  Root MSE      =     .21605

------------------------------------------------------------------------------
             |             Robust HC2
        gpmw |      Coef.   Std. Err.      t    P>|t|     [95% Conf. Interval]
-------------+----------------------------------------------------------------
     foreign |   .2461526   .0684669     3.60   0.001     .1096662    .3826389
       _cons |   1.609004   .0233601    68.88   0.000     1.562437    1.655571
------------------------------------------------------------------------------

. regress gpmw foreign, vce(hc3)
Linear regression                                 Number of obs =         74
                                                  F(  1,     72) =      12.38
                                                  Prob > F      =     0.0008
                                                  R-squared     =     0.2180
                                                  Root MSE      =     .21605

------------------------------------------------------------------------------
             |             Robust HC3
        gpmw |      Coef.   Std. Err.      t    P>|t|     [95% Conf. Interval]
-------------+----------------------------------------------------------------
     foreign |   .2461526    .069969     3.52   0.001     .1066719    .3856332
       _cons |   1.609004    .023588    68.21   0.000     1.561982    1.656026
------------------------------------------------------------------------------
```

◁

▷ Example 6

The vce(cluster *clustvar*) option relaxes the assumption of independence. Below we have 28,534 observations on 4,711 women aged 14–46 years. Data were collected on these women between 1968 and 1988. We are going to fit a classic earnings model, and we begin by ignoring that each woman appears an average of 6.056 times in the data.

```
. use http://www.stata-press.com/data/r12/regsmpl, clear
(NLS Women 14-26 in 1968)

. regress ln_wage age c.age#c.age tenure

      Source |       SS       df       MS              Number of obs =   28101
-------------+------------------------------           F(  3, 28097) = 1842.45
       Model |  1054.52501     3  351.508335           Prob > F      =  0.0000
    Residual |  5360.43962 28097  .190783344           R-squared     =  0.1644
-------------+------------------------------           Adj R-squared =  0.1643
       Total |  6414.96462 28100  .228290556           Root MSE      =  .43679

------------------------------------------------------------------------------
     ln_wage |      Coef.   Std. Err.      t    P>|t|     [95% Conf. Interval]
-------------+----------------------------------------------------------------
         age |   .0752172   .0034736    21.65   0.000     .0684088    .0820257
             |
 c.age#c.age |  -.0010851   .0000575   -18.86   0.000    -.0011979   -.0009724
             |
      tenure |   .0390877   .0007743    50.48   0.000     .0375699    .0406054
       _cons |   .3339821   .0504413     6.62   0.000     .2351148    .4328495
------------------------------------------------------------------------------
```

The number of observations in our model is 28,101 because Stata drops observations that have a missing value for one or more of the variables in the model. We can be reasonably certain that the standard errors reported above are meaningless. Without a doubt, a woman with higher-than-average wages in one year typically has higher-than-average wages in other years, and so the residuals are not independent. One way to deal with this would be to fit a random-effects model—and we are going to do that—but first we fit the model using `regress` specifying `vce(cluster id)`, which treats only observations with different person `ids` as truly independent:

```
. regress ln_wage age c.age#c.age tenure, vce(cluster id)

Linear regression                                      Number of obs =   28101
                                                       F(  3,  4698) =  748.82
                                                       Prob > F      =  0.0000
                                                       R-squared     =  0.1644
                                                       Root MSE      =  .43679

                               (Std. Err. adjusted for 4699 clusters in idcode)
------------------------------------------------------------------------------
             |               Robust
     ln_wage |      Coef.   Std. Err.      t    P>|t|     [95% Conf. Interval]
-------------+----------------------------------------------------------------
         age |   .0752172   .0045711    16.45   0.000     .0662557    .0841788
             |
 c.age#c.age |  -.0010851   .0000778   -13.94   0.000    -.0012377   -.0009325
             |
      tenure |   .0390877   .0014425    27.10   0.000     .0362596    .0419157
       _cons |   .3339821   .0641918     5.20   0.000      .208136    .4598282
------------------------------------------------------------------------------
```

For comparison, we focus on the tenure coefficient, which in economics jargon can be interpreted as the rate of return for keeping your job. The 95% confidence interval we previously estimated—an interval we do not believe—is $[0.038, 0.041]$. The robust interval is twice as wide, being $[0.036, 0.042]$.

As we said, one correct way to fit this model is by random-effects regression. Here is the random-effects result:

```
. xtreg ln_wage age c.age#c.age tenure, re
Random-effects GLS regression                   Number of obs      =     28101
Group variable: idcode                          Number of groups   =      4699
R-sq:  within  = 0.1370                         Obs per group: min =         1
       between = 0.2154                                        avg =       6.0
       overall = 0.1608                                        max =        15
Random effects u_i ~ Gaussian                   Wald chi2(3)       =   4717.05
corr(u_i, X)       = 0 (assumed)                Prob > chi2        =    0.0000

------------------------------------------------------------------------------
     ln_wage |      Coef.   Std. Err.      z    P>|z|     [95% Conf. Interval]
-------------+----------------------------------------------------------------
         age |   .0568296   .0026958    21.08   0.000     .0515459    .0621132
             |
 c.age#c.age |  -.0007566   .0000447   -16.93   0.000    -.0008441    -.000669
             |
      tenure |   .0260135   .0007477    34.79   0.000     .0245481    .0274789
       _cons |   .6136792   .0394611    15.55   0.000     .5363368    .6910216
-------------+----------------------------------------------------------------
     sigma_u |  .33542449
     sigma_e |  .29674679
         rho |  .56095413   (fraction of variance due to u_i)
------------------------------------------------------------------------------
```

Robust regression estimated the 95% interval $[0.036, 0.042]$, and `xtreg` (see [XT] **xtreg**) estimates $[0.025, 0.027]$. Which is better? The random-effects regression estimator assumes a lot. We can check some of these assumptions by performing a Hausman test. Using `estimates` (see [R] **estimates store**), we save the random-effects estimation results, and then we run the required fixed-effects regression to perform the test.

```
. estimates store random
. xtreg ln_wage age c.age#c.age tenure, fe
Fixed-effects (within) regression               Number of obs      =     28101
Group variable: idcode                          Number of groups   =      4699
R-sq:  within  = 0.1375                         Obs per group: min =         1
       between = 0.2066                                        avg =       6.0
       overall = 0.1568                                        max =        15
                                                F(3,23399)         =   1243.00
corr(u_i, Xb)  = 0.1380                         Prob > F           =    0.0000

------------------------------------------------------------------------------
     ln_wage |      Coef.   Std. Err.      t    P>|t|     [95% Conf. Interval]
-------------+----------------------------------------------------------------
         age |   .0522751    .002783    18.78   0.000     .0468202      .05773
             |
 c.age#c.age |  -.0006717   .0000461   -14.56   0.000    -.0007621   -.0005813
             |
      tenure |    .021738    .000799    27.21   0.000      .020172     .023304
       _cons |    .687178   .0405944    16.93   0.000     .6076103    .7667456
-------------+----------------------------------------------------------------
     sigma_u |  .38743138
     sigma_e |  .29674679
         rho |   .6302569   (fraction of variance due to u_i)
------------------------------------------------------------------------------
F test that all u_i=0:     F(4698, 23399) =     7.98         Prob > F = 0.0000
```

```
. hausman . random

                 ---- Coefficients ----
             |      (b)          (B)            (b-B)     sqrt(diag(V_b-V_B))
             |       .          random       Difference          S.E.
-------------+----------------------------------------------------------------
         age |    .0522751     .0568296       -.0045545        .0006913
 c.age#c.age |   -.0006717    -.0007566        .0000849        .0000115
      tenure |     .021738     .0260135       -.0042756        .0002816
------------------------------------------------------------------------------
                           b = consistent under Ho and Ha; obtained from xtreg
            B = inconsistent under Ha, efficient under Ho; obtained from xtreg

    Test:  Ho:  difference in coefficients not systematic

                  chi2(3) = (b-B)'[(V_b-V_B)^(-1)](b-B)
                          =      336.62
                Prob>chi2 =      0.0000
```

The Hausman test casts grave suspicions on the random-effects model we just fit, so we should be careful in interpreting those results.

Meanwhile, our robust regression results still stand, as long as we are careful about the interpretation. The correct interpretation is that, if the data collection were repeated (on women sampled the same way as in the original sample), and if we were to refit the model, 95% of the time we would expect the estimated coefficient on tenure to be in the range $[0.036, 0.042]$.

Even with robust regression, we must be careful about going beyond that statement. Here the Hausman test is probably picking up something that differs within and between person, which would cast doubt on our robust regression model in terms of interpreting $[0.036, 0.042]$ to contain the rate of return for keeping a job, economywide, for all women, without exception.

◁

Weighted regression

`regress` can perform weighted and unweighted regression. We indicate the weight by specifying the [*weight*] qualifier. By default, `regress` assumes analytic weights; see the technical note below.

▷ Example 7

We have census data recording the death rate (`drate`) and median age (`medage`) for each state. The data also record the region of the country in which each state is located and the overall population of the state:

```
. use http://www.stata-press.com/data/r12/census9
(1980 Census data by state)

. describe

Contains data from http://www.stata-press.com/data/r12/census9.dta
  obs:            50                          1980 Census data by state
 vars:             6                          6 Apr 2011 15:43
 size:         1,450
-------------------------------------------------------------------------------
              storage  display     value
variable name   type   format      label      variable label
-------------------------------------------------------------------------------
state           str14  %-14s                  State
state2          str2   %-2s                   Two-letter state abbreviation
drate           float  %9.0g                  Death Rate
pop             long   %12.0gc                Population
medage          float  %9.2f                  Median age
region          byte   %-8.0g      cenreg     Census region
-------------------------------------------------------------------------------
Sorted by:
```

We can use factor variables to include dummy variables for region. Because the variables in the regression reflect means rather than individual observations, the appropriate method of estimation is analytically weighted least squares (Davidson and MacKinnon 2004, 261–262), where the weight is total population:

```
. regress drate medage i.region [w=pop]
(analytic weights assumed)
(sum of wgt is   2.2591e+08)
      Source |       SS       df       MS              Number of obs =      50
-------------+------------------------------           F(  4,    45) =   37.21
       Model |  4096.6093     4  1024.15232           Prob > F      =  0.0000
    Residual |  1238.40987    45  27.5202192           R-squared     =  0.7679
-------------+------------------------------           Adj R-squared =  0.7472
       Total |  5335.01916    49  108.877942           Root MSE      =   5.246

------------------------------------------------------------------------------
       drate |      Coef.   Std. Err.      t    P>|t|     [95% Conf. Interval]
-------------+----------------------------------------------------------------
      medage |   4.283183   .5393329     7.94   0.000     3.196911    5.369455
             |
      region |
          2  |   .3138738   2.456431     0.13   0.899    -4.633632     5.26138
          3  |  -1.438452   2.320244    -0.62   0.538    -6.111663    3.234758
          4  |  -10.90629   2.681349    -4.07   0.000    -16.30681   -5.505777
             |
       _cons |  -39.14727   17.23613    -2.27   0.028    -73.86262   -4.431915
------------------------------------------------------------------------------
```

To weight the regression by population, we added the qualifier `[w=pop]` to the end of the `regress` command. Our qualifier was vague (we did not say `[aweight=pop]`), but unless told otherwise, Stata assumes analytic weights for `regress`. Stata informed us that the sum of the weight is 2.2591×10^8; there were approximately 226 million people residing in the United States according to our 1980 data.

◁

❑ Technical note

Once we fit a weighted regression, we can obtain the appropriately weighted variance–covariance matrix of the estimators using `estat vce` and perform appropriately weighted hypothesis tests using `test`.

In the weighted regression in the previous example, we see that `4.region` is statistically significant but that `2.region` and `3.region` are not. We use `test` to test the joint significance of the region variables:

```
. test 2.region 3.region 4.region
 ( 1)  2.region = 0
 ( 2)  3.region = 0
 ( 3)  4.region = 0
       F(  3,    45) =    9.84
            Prob > F =    0.0000
```

The results indicate that the region variables are jointly significant.

❑

`regress` also accepts frequency weights (`fweights`). Frequency weights are appropriate when the data do not reflect cell means, but instead represent replicated observations. Specifying `aweights` or `fweights` will not change the parameter estimates, but it will change the corresponding significance levels.

For instance, if we specified [fweight=pop] in the weighted regression example above—which would be statistically incorrect—Stata would treat the data as if the data represented 226 million independent observations on death rates and median age. The data most certainly do not represent that—they represent 50 observations on state averages.

With aweights, Stata treats the number of observations on the process as the number of observations in the data. When we specify fweights, Stata treats the number of observations as if it were equal to the sum of the weights; see *Methods and formulas* below.

❑ Technical note

A popular request on the help line is to describe the effect of specifying [aweight=*exp*] with regress in terms of transformation of the dependent and independent variables. The mechanical answer is that typing

```
. regress y x1 x2 [aweight=n]
```

is equivalent to fitting the model

$$y_j\sqrt{n_j} = \beta_0\sqrt{n_j} + \beta_1 x_{1j}\sqrt{n_j} + \beta_2 x_{2j}\sqrt{n_j} + u_j\sqrt{n_j}$$

This regression will reproduce the coefficients and covariance matrix produced by the aweighted regression. The mean squared errors (estimates of the variance of the residuals) will, however, be different. The transformed regression reports s_t^2, an estimate of $\mathrm{Var}(u_j\sqrt{n_j})$. The aweighted regression reports s_a^2, an estimate of $\mathrm{Var}(u_j\sqrt{n_j}\sqrt{N/\sum_k n_k})$, where N is the number of observations. Thus

$$s_a^2 = \frac{N}{\sum_k n_k} s_t^2 = \frac{s_t^2}{\overline{n}} \tag{1}$$

The logic for this adjustment is as follows: Consider the model

$$y = \beta_0 + \beta_1 x_1 + \beta_2 x_2 + u$$

Assume that, were this model fit on individuals, $\mathrm{Var}(u) = \sigma_u^2$, a constant. Assume that individual data are not available; what is available are averages $(\overline{y}_j, \overline{x}_{1j}, \overline{x}_{2j})$ for $j = 1, \ldots, N$, and each average is calculated over n_j observations. Then it is still true that

$$\overline{y}_j = \beta_0 + \beta_1 \overline{x}_{1j} + \beta_2 \overline{x}_{2j} + \overline{u}_j$$

where $\overline{u}_j$ is the average of n_j mean 0, variance σ_u^2 deviates and has variance $\sigma_{\overline{u}}^2 = \sigma_u^2/n_j$. Thus multiplying through by $\sqrt{n_j}$ produces

$$\overline{y}_j\sqrt{n_j} = \beta_0\sqrt{n_j} + \beta_1 \overline{x}_{1j}\sqrt{n_j} + \beta_2 \overline{x}_{2j}\sqrt{n_j} + \overline{u}_j\sqrt{n_j}$$

and $\mathrm{Var}(\overline{u}_j\sqrt{n_j}) = \sigma_u^2$. The mean squared error, s_t^2, reported by fitting this transformed regression is an estimate of σ_u^2. The coefficients and covariance matrix could also be obtained by aweighted regress. The only difference would be in the reported mean squared error, which from (1) is $\sigma_u^2/\overline{n}$. On average, each observation in the data reflects the averages calculated over $\overline{n} = \sum_k n_k/N$ individuals, and thus this reported mean squared error is the average variance of an observation in the dataset. We can retrieve the estimate of σ_u^2 by multiplying the reported mean squared error by $\overline{n}$.

More generally, aweights are used to solve general heteroskedasticity problems. In these cases, we have the model

$$y_j = \beta_0 + \beta_1 x_{1j} + \beta_2 x_{2j} + u_j$$

and the variance of u_j is thought to be proportional to a_j. If the variance is proportional to a_j, it is also proportional to αa_j, where α is any positive constant. Not quite arbitrarily, but with no loss of generality, we could choose $\alpha = \sum_k (1/a_k)/N$, the average value of the inverse of a_j. We can then write $\text{Var}(u_j) = k\alpha a_j \sigma^2$, where k is the constant of proportionality that is no longer a function of the scale of the weights.

Dividing this regression through by the $\sqrt{a_j}$,

$$y_j/\sqrt{a_j} = \beta_0/\sqrt{a_j} + \beta_1 x_{1j}/\sqrt{a_j} + \beta_2 x_{2j}/\sqrt{a_j} + u_j/\sqrt{a_j}$$

produces a model with $\text{Var}(u_j/\sqrt{a_j}) = k\alpha\sigma^2$, which is the constant part of $\text{Var}(u_j)$. This variance is a function of α, the average of the reciprocal weights; if the weights are scaled arbitrarily, then so is this variance.

We can also fit this model by typing

```
. regress y x1 x2 [aweight=1/a]
```

This input will produce the same estimates of the coefficients and covariance matrix; the reported mean squared error is, from (1), $\{N/\sum_k (1/a_k)\} k\alpha\sigma^2 = k\sigma^2$. This variance is independent of the scale of a_j.

❑

Instrumental variables and two-stage least-squares regression

An alternate syntax for `regress` can be used to produce instrumental-variables (two-stage least squares) estimates.

<u>reg</u>ress *depvar* [*varlist*$_1$ [(*varlist*$_2$)]] [*if*] [*in*] [*weight*] [, *regress_options*]

This syntax is used mainly by programmers developing estimators using the instrumental-variables estimates as intermediate results. `ivregress` is normally used to directly fit these models; see [R] **ivregress**.

With this syntax, `regress` fits a structural equation of *depvar* on *varlist*$_1$ using instrumental variables regression; (*varlist*$_2$) indicates the list of instrumental variables. With the exception of `vce(hc2)` and `vce(hc3)`, all standard `regress` options are allowed.

Saved results

`regress` saves the following in `e()`:

Scalars

e(N)	number of observations
e(mss)	model sum of squares
e(df_m)	model degrees of freedom
e(rss)	residual sum of squares
e(df_r)	residual degrees of freedom
e(r2)	R-squared
e(r2_a)	adjusted R-squared
e(F)	F statistic
e(rmse)	root mean squared error
e(ll)	log likelihood under additional assumption of i.i.d. normal errors
e(ll_0)	log likelihood, constant-only model
e(N_clust)	number of clusters
e(rank)	rank of e(V)

Macros

e(cmd)	regress
e(cmdline)	command as typed
e(depvar)	name of dependent variable
e(model)	ols or iv
e(wtype)	weight type
e(wexp)	weight expression
e(title)	title in estimation output when vce() is not ols
e(clustvar)	name of cluster variable
e(vce)	*vcetype* specified in vce()
e(vcetype)	title used to label Std. Err.
e(properties)	b V
e(estat_cmd)	program used to implement estat
e(predict)	program used to implement predict
e(marginsok)	predictions allowed by margins
e(asbalanced)	factor variables fvset as asbalanced
e(asobserved)	factor variables fvset as asobserved

Matrices

e(b)	coefficient vector
e(V)	variance–covariance matrix of the estimators
e(V_modelbased)	model-based variance

Functions

e(sample)	marks estimation sample

Methods and formulas

`regress` is implemented as an ado-file.

Methods and formulas are presented under the following headings:

Coefficient estimation and ANOVA table
A general notation for the robust variance calculation
Robust calculation for regress

Coefficient estimation and ANOVA table

Variables printed in lowercase and not boldfaced (for example, x) are scalars. Variables printed in lowercase and boldfaced (for example, $\mathbf{x}$) are column vectors. Variables printed in uppercase and boldfaced (for example, $\mathbf{X}$) are matrices.

Let $\mathbf{v}$ be a column vector of weights specified by the user. If no weights are specified, $\mathbf{v} = \mathbf{1}$. Let $\mathbf{w}$ be a column vector of normalized weights. If no weights are specified or if the user specified fweights or iweights, $\mathbf{w} = \mathbf{v}$. Otherwise, $\mathbf{w} = \{\mathbf{v}/(\mathbf{1}'\mathbf{v})\}(\mathbf{1}'\mathbf{1})$.

The *number of observations*, n, is defined as $\mathbf{1}'\mathbf{w}$. For iweights, this is truncated to an integer. The *sum of the weights* is $\mathbf{1}'\mathbf{v}$. Define $c = 1$ if there is a constant in the regression and zero otherwise. Define k as the number of right-hand-side variables (including the constant).

Let $\mathbf{X}$ denote the matrix of observations on the right-hand-side variables, $\mathbf{y}$ the vector of observations on the left-hand-side variable, and $\mathbf{Z}$ the matrix of observations on the instruments. If the user specifies no instruments, then $\mathbf{Z} = \mathbf{X}$. In the following formulas, if the user specifies weights, then $\mathbf{X}'\mathbf{X}$, $\mathbf{X}'\mathbf{y}$, $\mathbf{y}'\mathbf{y}$, $\mathbf{Z}'\mathbf{Z}$, $\mathbf{Z}'\mathbf{X}$, and $\mathbf{Z}'\mathbf{y}$ are replaced by $\mathbf{X}'\mathbf{D}\mathbf{X}$, $\mathbf{X}'\mathbf{D}\mathbf{y}$, $\mathbf{y}'\mathbf{D}\mathbf{y}$, $\mathbf{Z}'\mathbf{D}\mathbf{Z}$, $\mathbf{Z}'\mathbf{D}\mathbf{X}$, and $\mathbf{Z}'\mathbf{D}\mathbf{y}$, respectively, where $\mathbf{D}$ is a diagonal matrix whose diagonal elements are the elements of $\mathbf{w}$. We suppress the $\mathbf{D}$ below to simplify the notation.

If no instruments are specified, define $\mathbf{A}$ as $\mathbf{X}'\mathbf{X}$ and $\mathbf{a}$ as $\mathbf{X}'\mathbf{y}$. Otherwise, define $\mathbf{A}$ as $\mathbf{X}'\mathbf{Z}(\mathbf{Z}'\mathbf{Z})^{-1}(\mathbf{X}'\mathbf{Z})'$ and $\mathbf{a}$ as $\mathbf{X}'\mathbf{Z}(\mathbf{Z}'\mathbf{Z})^{-1}\mathbf{Z}'\mathbf{y}$.

The coefficient vector $\mathbf{b}$ is defined as $\mathbf{A}^{-1}\mathbf{a}$. Although not shown in the notation, unless hascons is specified, $\mathbf{A}$ and $\mathbf{a}$ are accumulated in deviation form and the constant is calculated separately. This comment applies to all statistics listed below.

The *total sum of squares*, TSS, equals $\mathbf{y}'\mathbf{y}$ if there is no intercept and $\mathbf{y}'\mathbf{y} - \{(\mathbf{1}'\mathbf{y})^2/n\}$ otherwise. The *degrees of freedom* is $n - c$.

The *error sum of squares*, ESS, is defined as $\mathbf{y}'\mathbf{y} - 2\mathbf{b}\mathbf{X}'\mathbf{y} + \mathbf{b}'\mathbf{X}'\mathbf{X}\mathbf{b}$ if there are instruments and as $\mathbf{y}'\mathbf{y} - \mathbf{b}'\mathbf{X}'\mathbf{y}$ otherwise. The *degrees of freedom* is $n - k$.

The *model sum of squares*, MSS, equals TSS − ESS. The *degrees of freedom* is $k - c$.

The *mean squared error*, s^2, is defined as $\text{ESS}/(n-k)$. The *root mean squared error* is s, its square root.

The F statistic with $k - c$ and $n - k$ degrees of freedom is defined as

$$F = \frac{\text{MSS}}{(k-c)s^2}$$

if no instruments are specified. If instruments are specified and $c = 1$, then F is defined as

$$F = \frac{(\mathbf{b}-\mathbf{c})'\mathbf{A}(\mathbf{b}-\mathbf{c})}{(k-1)s^2}$$

where $\mathbf{c}$ is a vector of $k - 1$ zeros and kth element $\mathbf{1}'\mathbf{y}/n$. Otherwise, F is defined as *missing*. (Here you may use the test command to construct any F test that you wish.)

The *R-squared*, R^2, is defined as $R^2 = 1 - \text{ESS}/\text{TSS}$.

The *adjusted R-squared*, R_a^2, is $1 - (1 - R^2)(n - c)/(n - k)$.

If vce(robust) is not specified, the conventional estimate of variance is $s^2\mathbf{A}^{-1}$. The handling of vce(robust) is described below.

A general notation for the robust variance calculation

Put aside all context of linear regression and the notation that goes with it—we will return to it. First, we are going to establish a notation for describing robust variance calculations.

The calculation formula for the robust variance calculation is

$$\widehat{\mathcal{V}} = q_c \widehat{\mathbf{V}} \Big(\sum_{k=1}^{M} \mathbf{u}_k^{(G)\prime} \mathbf{u}_k^{(G)} \Big) \widehat{\mathbf{V}}$$

where

$$\mathbf{u}_k^{(G)} = \sum_{j \in G_k} w_j \mathbf{u}_j$$

$G_1, G_2, \ldots, G_M$ are the clusters specified by `vce(cluster` *clustvar*`)`, and w_j are the user-specified weights, normalized if `aweights` or `pweights` are specified and equal to 1 if no weights are specified.

For `fweights` without clusters, the variance formula is

$$\widehat{\mathcal{V}} = q_c \widehat{\mathbf{V}} \Big(\sum_{j=1}^{N} w_j \mathbf{u}_j' \mathbf{u}_j \Big) \widehat{\mathbf{V}}$$

which is the same as expanding the dataset and making the calculation on the unweighted data.

If `vce(cluster` *clustvar*`)` is not specified, $M = N$, and each cluster contains 1 observation. The inputs into this calculation are

- $\widehat{\mathbf{V}}$, which is typically a conventionally calculated variance matrix;
- $\mathbf{u}_j$, $j = 1, \ldots, N$, a row vector of scores; and
- $q_{\rm c}$, a constant finite-sample adjustment.

Thus we can now describe how estimators apply the robust calculation formula by defining $\widehat{\mathbf{V}}$, $\mathbf{u}_j$, and $q_{\rm c}$.

Two definitions are popular enough for $q_{\rm c}$ to deserve a name. The regression-like formula for $q_{\rm c}$ (Fuller et al. 1986) is

$$q_{\rm c} = \frac{N-1}{N-k} \frac{M}{M-1}$$

where M is the number of clusters and N is the number of observations. For weights, N refers to the sum of the weights if weights are frequency weights and the number of observations in the dataset (ignoring weights) in all other cases. Also note that, weighted or not, $M = N$ when `vce(cluster` *clustvar*`)` is not specified, and then $q_{\rm c} = N/(N-k)$.

The asymptotic-like formula for $q_{\rm c}$ is

$$q_{\rm c} = \frac{M}{M-1}$$

where $M = N$ if `vce(cluster` *clustvar*`)` is not specified.

See [U] **20.20 Obtaining robust variance estimates** and [P] **_robust** for a discussion of the robust variance estimator and a development of these formulas.

Robust calculation for regress

For `regress`, $\widehat{\mathbf{V}} = \mathbf{A}^{-1}$. The other terms are

No instruments, `vce(robust)`, but not `vce(hc2)` or `vce(hc3)`,

$$\mathbf{u}_j = (y_j - \mathbf{x}_j\mathbf{b})\mathbf{x}_j$$

and q_c is given by its regression-like definition.

No instruments, `vce(hc2)`,

$$\mathbf{u}_j = \frac{1}{\sqrt{1 - h_{jj}}}(y_j - \mathbf{x}_j\mathbf{b})\mathbf{x}_j$$

where $q_c = 1$ and $h_{jj} = \mathbf{x}_j(\mathbf{X}'\mathbf{X})^{-1}\mathbf{x}_j{}'$.

No instruments, `vce(hc3)`,

$$\mathbf{u}_j = \frac{1}{1 - h_{jj}}(y_j - \mathbf{x}_j\mathbf{b})\mathbf{x}_j$$

where $q_c = 1$ and $h_{jj} = \mathbf{x}_j(\mathbf{X}'\mathbf{X})^{-1}\mathbf{x}_j{}'$.

Instrumental variables,

$$\mathbf{u}_j = (y_j - \mathbf{x}_j\mathbf{b})\widehat{\mathbf{x}}_j$$

where q_c is given by its regression-like definition, and

$$\widehat{\mathbf{x}}'_j = \mathbf{P}\mathbf{z}_j{}'$$

where $\mathbf{P} = (\mathbf{X}'\mathbf{Z})(\mathbf{Z}'\mathbf{Z})^{-1}$.

Acknowledgments

The robust estimate of variance was first implemented in Stata by Mead Over, Center for Global Development; Dean Jolliffe, World Bank; and Andrew Foster, Department of Economics, Brown University (Over, Jolliffe, and Foster 1996).

The history of regression is long and complicated: the books by Stigler (1986) and Hald (1998) are devoted largely to the story. Legendre published first on least squares in 1805. Gauss published later in 1809, but he had the idea earlier. Gauss, and especially Laplace, tied least squares to a normal errors assumption. The idea of the normal distribution can itself be traced back to De Moivre in 1733. Laplace discussed a variety of other estimation methods and error assumptions over his long career, while linear models long predate either innovation. Most of this work was linked to problems in astronomy and geodesy.

A second wave of ideas started when Galton used graphical and descriptive methods on data bearing on heredity to develop what he called regression. His term reflects the common phenomenon that characteristics of offspring are positively correlated with those of parents but with regression slope such that offspring "regress toward the mean". Galton's work was rather intuitive: contributions from Pearson, Edgeworth, Yule, and others introduced more formal machinery, developed related ideas on correlation, and extended application into the biological and social sciences. So most of the elements of regression as we know it were in place by 1900.

Pierre-Simon Laplace (1749–1827) was born in Normandy and was early recognized as a remarkable mathematician. He weathered a changing political climate well enough to rise to Minister of the Interior under Napoleon in 1799 (although only for 6 weeks) and to be made a Marquis by Louis XVIII in 1817. He made many contributions to mathematics and physics, his two main interests being theoretical astronomy and probability theory (including statistics). Laplace transforms are named for him.

Adrien-Marie Legendre (1752–1833) was born in Paris (or possibly in Toulouse) and educated in mathematics and physics. He worked in number theory, geometry, differential equations, calculus, function theory, applied mathematics, and geodesy. The Legendre polynomials are named for him. His main contribution to statistics is as one of the discoverers of least squares. He died in poverty, having refused to bow to political pressures.

Johann Carl Friedrich Gauss (1777–1855) was born in Braunschweig (Brunswick), now in Germany. He studied there and at Göttingen. His doctoral dissertation at the University of Helmstedt was a discussion of the fundamental theorem of algebra. He made many fundamental contributions to geometry, number theory, algebra, real analysis, differential equations, numerical analysis, statistics, astronomy, optics, geodesy, mechanics, and magnetism. An outstanding genius, Gauss worked mostly in isolation in Göttingen.

Francis Galton (1822–1911) was born in Birmingham, England, into a well-to-do family with many connections: he and Charles Darwin were first cousins. After an unsuccessful foray into medicine, he became independently wealthy at the death of his father. Galton traveled widely in Europe, the Middle East, and Africa, and became celebrated as an explorer and geographer. His pioneering work on weather maps helped in the identification of anticyclones, which he named. From about 1865, most of his work was centered on quantitative problems in biology, anthropology, and psychology. In a sense, Galton (re)invented regression, and he certainly named it. Galton also promoted the normal distribution, correlation approaches, and the use of median and selected quantiles as descriptive statistics. He was knighted in 1909.

References

Adkins, L. C., and R. C. Hill. 2008. *Using Stata for Principles of Econometrics.* 3rd ed. Hoboken, NJ: Wiley.

Alexandersson, A. 1998. gr32: Confidence ellipses. *Stata Technical Bulletin* 46: 10–13. Reprinted in *Stata Technical Bulletin Reprints*, vol. 8, pp. 54–57. College Station, TX: Stata Press.

Angrist, J. D., and J.-S. Pischke. 2009. *Mostly Harmless Econometrics: An Empiricist's Companion*. Princeton, NJ: Princeton University Press.

Cameron, A. C., and P. K. Trivedi. 2010. *Microeconometrics Using Stata*. Rev. ed. College Station, TX: Stata Press.

Chatterjee, S., and A. S. Hadi. 2006. *Regression Analysis by Example*. 4th ed. New York: Wiley.

Davidson, R., and J. G. MacKinnon. 1993. *Estimation and Inference in Econometrics*. New York: Oxford University Press.

——. 2004. *Econometric Theory and Methods*. New York: Oxford University Press.

Dohoo, I., W. Martin, and H. Stryhn. 2010. *Veterinary Epidemiologic Research*. 2nd ed. Charlottetown, Prince Edward Island: VER Inc.

Draper, N., and H. Smith. 1998. *Applied Regression Analysis*. 3rd ed. New York: Wiley.

Dunnington, G. W. 1955. *Gauss: Titan of Science*. New York: Hafner Publishing.

Duren, P. 2009. Changing faces: The mistaken portrait of Legendre. *Notices of the American Mathematical Society* 56: 1440–1443.

Fuller, W. A., W. J. Kennedy, Jr., D. Schnell, G. Sullivan, and H. J. Park. 1986. *PC CARP.* Software package. Ames, IA: Statistical Laboratory, Iowa State University.

Gillham, N. W. 2001. *A Life of Sir Francis Galton: From African Exploration to the Birth of Eugenics*. New York: Oxford University Press.

Gillispie, C. C. 1997. *Pierre-Simon Laplace, 1749–1827: A Life in Exact Science*. Princeton: Princeton University Press.

Gould, W. W. 2011. Understanding matrices intuitively, part 1. The Stata Blog: Not Elsewhere Classified. http://blog.stata.com/2011/03/03/understanding-matrices-intuitively-part-1/

Greene, W. H. 2012. *Econometric Analysis*. 7th ed. Upper Saddle River, NJ: Prentice Hall.

Hald, A. 1998. *A History of Mathematical Statistics from 1750 to 1930*. New York: Wiley.

Hamilton, L. C. 2009. *Statistics with Stata (Updated for Version 10)*. Belmont, CA: Brooks/Cole.

Hill, R. C., W. E. Griffiths, and G. C. Lim. 2011. *Principles of Econometrics*. 4th ed. Hoboken, NJ: Wiley.

Kmenta, J. 1997. *Elements of Econometrics*. 2nd ed. Ann Arbor: University of Michigan Press.

Kohler, U., and F. Kreuter. 2009. *Data Analysis Using Stata*. 2nd ed. College Station, TX: Stata Press.

Long, J. S., and J. Freese. 2000. sg152: Listing and interpreting transformed coefficients from certain regression models. *Stata Technical Bulletin* 57: 27–34. Reprinted in *Stata Technical Bulletin Reprints*, vol. 10, pp. 231–240. College Station, TX: Stata Press.

MacKinnon, J. G., and H. White. 1985. Some heteroskedasticity-consistent covariance matrix estimators with improved finite sample properties. *Journal of Econometrics* 29: 305–325.

Mosteller, F., and J. W. Tukey. 1977. *Data Analysis and Regression: A Second Course in Statistics*. Reading, MA: Addison–Wesley.

Over, M., D. Jolliffe, and A. Foster. 1996. sg46: Huber correction for two-stage least squares estimates. *Stata Technical Bulletin* 29: 24–25. Reprinted in *Stata Technical Bulletin Reprints*, vol. 5, pp. 140–142. College Station, TX: Stata Press.

Peracchi, F. 2001. *Econometrics*. Chichester, UK: Wiley.

Plackett, R. L. 1972. Studies in the history of probability and statistics: XXIX. The discovery of the method of least squares. *Biometrika* 59: 239–251.

Rogers, W. H. 1991. smv2: Analyzing repeated measurements—some practical alternatives. *Stata Technical Bulletin* 4: 10–16. Reprinted in *Stata Technical Bulletin Reprints*, vol. 1, pp. 123–131. College Station, TX: Stata Press.

Royston, P., and G. Ambler. 1998. sg79: Generalized additive models. *Stata Technical Bulletin* 42: 38–43. Reprinted in *Stata Technical Bulletin Reprints*, vol. 7, pp. 217–224. College Station, TX: Stata Press.

Schonlau, M. 2005. Boosted regression (boosting): An introductory tutorial and a Stata plugin. *Stata Journal* 5: 330–354.

Stigler, S. M. 1986. *The History of Statistics: The Measurement of Uncertainty before 1900*. Cambridge, MA: Belknap Press.

Tyler, J. H. 1997. sg73: Table making programs. *Stata Technical Bulletin* 40: 18–23. Reprinted in *Stata Technical Bulletin Reprints*, vol. 7, pp. 186–192. College Station, TX: Stata Press.

Weesie, J. 1998. sg77: Regression analysis with multiplicative heteroscedasticity. *Stata Technical Bulletin* 42: 28–32. Reprinted in *Stata Technical Bulletin Reprints*, vol. 7, pp. 204–210. College Station, TX: Stata Press.

Weisberg, S. 2005. *Applied Linear Regression*. 3rd ed. New York: Wiley.

Wooldridge, J. M. 2009. *Introductory Econometrics: A Modern Approach*. 4th ed. Cincinnati, OH: South-Western.

——. 2010. *Econometric Analysis of Cross Section and Panel Data*. 2nd ed. Cambridge, MA: MIT Press.

Zimmerman, F. 1998. sg93: Switching regressions. *Stata Technical Bulletin* 45: 30–33. Reprinted in *Stata Technical Bulletin Reprints*, vol. 8, pp. 183–186. College Station, TX: Stata Press.

Also see

[R] **regress postestimation** — Postestimation tools for regress

[R] **regress postestimation time series** — Postestimation tools for regress with time series

[R] **anova** — Analysis of variance and covariance

[R] **contrast** — Contrasts and linear hypothesis tests after estimation

[MI] **estimation** — Estimation commands for use with mi estimate

[SVY] **svy estimation** — Estimation commands for survey data

Stata Structural Equation Modeling Reference Manual

[U] **20 Estimation and postestimation commands**

Title

regress postestimation — Postestimation tools for regress

Description

The following postestimation commands are of special interest after `regress`:

Command	Description
`dfbeta`	DFBETA influence statistics
`estat hettest`	tests for heteroskedasticity
`estat imtest`	information matrix test
`estat ovtest`	Ramsey regression specification-error test for omitted variables
`estat szroeter`	Szroeter's rank test for heteroskedasticity
`estat vif`	variance inflation factors for the independent variables
`acprplot`	augmented component-plus-residual plot
`avplot`	added-variable plot
`avplots`	all added-variables plots in one image
`cprplot`	component-plus-residual plot
`lvr2plot`	leverage-versus-squared-residual plot
`rvfplot`	residual-versus-fitted plot
`rvpplot`	residual-versus-predictor plot

These commands are not appropriate after the `svy` prefix.

For information about these commands, see below.

The following standard postestimation commands are also available:

Command	Description
`contrast`	contrasts and ANOVA-style joint tests of estimates
`estat`	AIC, BIC, VCE, and estimation sample summary
`estat` (svy)	postestimation statistics for survey data
`estimates`	cataloging estimation results
`hausman`	Hausman's specification test
`lincom`	point estimates, standard errors, testing, and inference for linear combinations of coefficients
`linktest`	link test for model specification
`lrtest`[1]	likelihood-ratio test
`margins`	marginal means, predictive margins, marginal effects, and average marginal effects
`marginsplot`	graph the results from margins (profile plots, interaction plots, etc.)
`nlcom`	point estimates, standard errors, testing, and inference for nonlinear combinations of coefficients
`predict`	predictions, residuals, influence statistics, and other diagnostic measures
`predictnl`	point estimates, standard errors, testing, and inference for generalized predictions
`pwcompare`	pairwise comparisons of estimates
`suest`	seemingly unrelated estimation
`test`	Wald tests of simple and composite linear hypotheses
`testnl`	Wald tests of nonlinear hypotheses

[1] `lrtest` is not appropriate with `svy` estimation results.

See the corresponding entries in the *Base Reference Manual* for details, but see [SVY] **estat** for details about `estat` (svy).

For postestimation tests specific to time series, see [R] **regress postestimation time series**.

Special-interest postestimation commands

These commands provide tools for diagnosing sensitivity to individual observations, analyzing residuals, and assessing specification.

`dfbeta` will calculate one, more than one, or all the DFBETAs after `regress`. Although `predict` will also calculate DFBETAs, `predict` can do this for only one variable at a time. `dfbeta` is a convenience tool for those who want to calculate DFBETAs for multiple variables. The names for the new variables created are chosen automatically and begin with the letters `_dfbeta_`.

`estat hettest` performs three versions of the Breusch–Pagan (1979) and Cook–Weisberg (1983) test for heteroskedasticity. All three versions of this test present evidence against the null hypothesis that $\mathbf{t} = \mathbf{0}$ in $\text{Var}(e) = \sigma^2 \exp(\mathbf{zt})$. In the `normal` version, performed by default, the null hypothesis also includes the assumption that the regression disturbances are independent-normal draws with variance σ^2. The normality assumption is dropped from the null hypothesis in the `iid` and `fstat` versions, which respectively produce the score and F tests discussed in *Methods and formulas*. If *varlist* is not specified, the fitted values are used for $\mathbf{z}$. If *varlist* or the `rhs` option is specified, the variables specified are used for $\mathbf{z}$.

`estat imtest` performs an information matrix test for the regression model and an orthogonal decomposition into tests for heteroskedasticity, skewness, and kurtosis due to Cameron and Trivedi (1990);

White's test for homoskedasticity against unrestricted forms of heteroskedasticity (1980) is available as an option. White's test is usually similar to the first term of the Cameron–Trivedi decomposition.

`estat ovtest` performs two versions of the Ramsey (1969) regression specification-error test (RESET) for omitted variables. This test amounts to fitting $y = \mathbf{xb} + \mathbf{zt} + u$ and then testing $\mathbf{t} = \mathbf{0}$. If the `rhs` option is not specified, powers of the fitted values are used for `z`. If `rhs` is specified, powers of the individual elements of **x** are used.

`estat szroeter` performs Szroeter's rank test for heteroskedasticity for each of the variables in *varlist* or for the explanatory variables of the regression if `rhs` is specified.

`estat vif` calculates the centered or uncentered variance inflation factors (VIFs) for the independent variables specified in a linear regression model.

`acprplot` graphs an augmented component-plus-residual plot (a.k.a. augmented partial residual plot) as described by Mallows (1986). This seems to work better than the component-plus-residual plot for identifying nonlinearities in the data.

`avplot` graphs an added-variable plot (a.k.a. partial-regression leverage plot, partial regression plot, or adjusted partial residual plot) after `regress`. *indepvar* may be an independent variable (a.k.a. predictor, carrier, or covariate) that is currently in the model or not.

`avplots` graphs all the added-variable plots in one image.

`cprplot` graphs a component-plus-residual plot (a.k.a. partial residual plot) after `regress`. *indepvar* must be an independent variable that is currently in the model.

`lvr2plot` graphs a leverage-versus-squared-residual plot (a.k.a. L-R plot).

`rvfplot` graphs a residual-versus-fitted plot, a graph of the residuals against the fitted values.

`rvpplot` graphs a residual-versus-predictor plot (a.k.a. independent variable plot or carrier plot), a graph of the residuals against the specified predictor.

Syntax for predict

`predict` [*type*] *newvar* [*if*] [*in*] [, *statistic*]

statistic	Description
Main	
`xb`	linear prediction; the default
`residuals`	residuals
`score`	score; equivalent to `residuals`
`rstandard`	standardized residuals
`rstudent`	Studentized (jackknifed) residuals
`cooksd`	Cook's distance
`leverage` | `hat`	leverage (diagonal elements of hat matrix)
`pr(`*a*`,`*b*`)`	$\Pr(y_j \mid a < y_j < b)$
`e(`*a*`,`*b*`)`	$E(y_j \mid a < y_j < b)$
`ystar(`*a*`,`*b*`)`	$E(y_j^*)$, $y_j^* = \max\{a, \min(y_j, b)\}$
* `dfbeta(`*varname*`)`	DFBETA for *varname*
`stdp`	standard error of the linear prediction
`stdf`	standard error of the forecast
`stdr`	standard error of the residual
* `covratio`	COVRATIO
* `dfits`	DFITS
* `welsch`	Welsch distance

Unstarred statistics are available both in and out of sample; type `predict ... if e(sample) ...` if wanted only for the estimation sample. Starred statistics are calculated only for the estimation sample, even when `if e(sample)` is not specified.

`rstandard`, `rstudent`, `cooksd`, `leverage`, `dfbeta()`, `stdf`, `stdr`, `covratio`, `dfits`, and `welsch` are not available if any `vce()` other than `vce(ols)` was specified with `regress`.

`xb`, `residuals`, `score`, and `stdp` are the only options allowed with `svy` estimation results.

where *a* and *b* may be numbers or variables; *a* missing ($a \geq .$) means $-\infty$, and *b* missing ($b \geq .$) means $+\infty$; see [U] **12.2.1 Missing values**.

Menu

Statistics > Postestimation > Predictions, residuals, etc.

Options for predict

Main

`xb`, the default, calculates the linear prediction.

`residuals` calculates the residuals.

`score` is equivalent to `residuals` in linear regression.

`rstandard` calculates the standardized residuals.

`rstudent` calculates the Studentized (jackknifed) residuals.

`cooksd` calculates the Cook's D influence statistic (Cook 1977).

`leverage` or `hat` calculates the diagonal elements of the projection hat matrix.

`pr(`*a*`,`*b*`)` calculates $\Pr(a < \mathbf{x}_j\mathbf{b} + u_j < b)$, the probability that $y_j|\mathbf{x}_j$ would be observed in the interval (a,b).

a and *b* may be specified as numbers or variable names; *lb* and *ub* are variable names;
`pr(20,30)` calculates $\Pr(20 < \mathbf{x}_j\mathbf{b} + u_j < 30)$;
`pr(`*lb*`,`*ub*`)` calculates $\Pr(lb < \mathbf{x}_j\mathbf{b} + u_j < ub)$; and
`pr(20,`*ub*`)` calculates $\Pr(20 < \mathbf{x}_j\mathbf{b} + u_j < ub)$.

a missing ($a \geq .$) means $-\infty$; `pr(.,30)` calculates $\Pr(-\infty < \mathbf{x}_j\mathbf{b} + u_j < 30)$;
`pr(`*lb*`,30)` calculates $\Pr(-\infty < \mathbf{x}_j\mathbf{b} + u_j < 30)$ in observations for which $lb \geq .$
and calculates $\Pr(lb < \mathbf{x}_j\mathbf{b} + u_j < 30)$ elsewhere.

b missing ($b \geq .$) means $+\infty$; `pr(20,.)` calculates $\Pr(+\infty > \mathbf{x}_j\mathbf{b} + u_j > 20)$;
`pr(20,`*ub*`)` calculates $\Pr(+\infty > \mathbf{x}_j\mathbf{b} + u_j > 20)$ in observations for which $ub \geq .$
and calculates $\Pr(20 < \mathbf{x}_j\mathbf{b} + u_j < ub)$ elsewhere.

`e(`*a*`,`*b*`)` calculates $E(\mathbf{x}_j\mathbf{b} + u_j \mid a < \mathbf{x}_j\mathbf{b} + u_j < b)$, the expected value of $y_j|\mathbf{x}_j$ conditional on $y_j|\mathbf{x}_j$ being in the interval (a,b), meaning that $y_j|\mathbf{x}_j$ is truncated.
a and *b* are specified as they are for `pr()`.

`ystar(`*a*`,`*b*`)` calculates $E(y_j^*)$, where $y_j^* = a$ if $\mathbf{x}_j\mathbf{b} + u_j \leq a$, $y_j^* = b$ if $\mathbf{x}_j\mathbf{b} + u_j \geq b$, and $y_j^* = \mathbf{x}_j\mathbf{b} + u_j$ otherwise, meaning that y_j^* is censored. *a* and *b* are specified as they are for `pr()`.

`dfbeta(`*varname*`)` calculates the DFBETA for *varname*, the difference between the regression coefficient when the jth observation is included and excluded, said difference being scaled by the estimated standard error of the coefficient. *varname* must have been included among the regressors in the previously fitted model. The calculation is automatically restricted to the estimation subsample.

`stdp` calculates the standard error of the prediction, which can be thought of as the standard error of the predicted expected value or mean for the observation's covariate pattern. The standard error of the prediction is also referred to as the standard error of the fitted value.

`stdf` calculates the standard error of the forecast, which is the standard error of the point prediction for 1 observation. It is commonly referred to as the standard error of the future or forecast value. By construction, the standard errors produced by `stdf` are always larger than those produced by `stdp`; see *Methods and formulas*.

`stdr` calculates the standard error of the residuals.

`covratio` calculates COVRATIO (Belsley, Kuh, and Welsch 1980), a measure of the influence of the jth observation based on considering the effect on the variance–covariance matrix of the estimates. The calculation is automatically restricted to the estimation subsample.

`dfits` calculates DFITS (Welsch and Kuh 1977) and attempts to summarize the information in the leverage versus residual-squared plot into one statistic. The calculation is automatically restricted to the estimation subsample.

`welsch` calculates Welsch distance (Welsch 1982) and is a variation on `dfits`. The calculation is automatically restricted to the estimation subsample.

Syntax for dfbeta

`dfbeta` [*indepvar* [*indepvar* [...]]] [, `stub(`*name*`)`]

Menu

Statistics > Linear models and related > Regression diagnostics > DFBETAs

Option for dfbeta

`stub(`*name*`)` specifies the leading characters `dfbeta` uses to name the new variables to be generated. The default is `stub(_dfbeta_)`.

Syntax for estat hettest

`estat hettest` [*varlist*] [, `rhs` [`normal`|`iid`|`fstat`] `mtest`[(*spec*)]]

Menu

Statistics > Postestimation > Reports and statistics

Options for estat hettest

`rhs` specifies that tests for heteroskedasticity be performed for the right-hand-side (explanatory) variables of the fitted regression model. The `rhs` option may be combined with a *varlist*.

`normal`, the default, causes `estat hettest` to compute the original Breusch–Pagan/Cook–Weisberg test, which assumes that the regression disturbances are normally distributed.

`iid` causes `estat hettest` to compute the $N * R^2$ version of the score test that drops the normality assumption.

`fstat` causes `estat hettest` to compute the F-statistic version that drops the normality assumption.

`mtest`[(*spec*)] specifies that multiple testing be performed. The argument specifies how p-values are adjusted. The following specifications, *spec*, are supported:

`bonferroni`	Bonferroni's multiple testing adjustment
`holm`	Holm's multiple testing adjustment
`sidak`	Šidák's multiple testing adjustment
`noadjust`	no adjustment is made for multiple testing

`mtest` may be specified without an argument. This is equivalent to specifying `mtest(noadjust)`; that is, tests for the individual variables should be performed with unadjusted p-values. By default, `estat hettest` does not perform multiple testing. `mtest` may not be specified with `iid` or `fstat`.

Syntax for estat imtest

`estat imtest` [, `preserve` `white`]

Menu

Statistics > Postestimation > Reports and statistics

Options for estat imtest

`preserve` specifies that the data in memory be preserved, all variables and cases that are not needed in the calculations be dropped, and at the conclusion the original data be restored. This option is costly for large datasets. However, because `estat imtest` has to perform an auxiliary regression on $k(k+1)/2$ temporary variables, where k is the number of regressors, it may not be able to perform the test otherwise.

`white` specifies that White's original heteroskedasticity test also be performed.

Syntax for estat ovtest

estat ovtest [, rhs]

Menu

Statistics > Postestimation > Reports and statistics

Option for estat ovtest

`rhs` specifies that powers of the right-hand-side (explanatory) variables be used in the test rather than powers of the fitted values.

Syntax for estat szroeter

estat szroeter [*varlist*] [, rhs mtest(*spec*)]

Either *varlist* or `rhs` must be specified.

Menu

Statistics > Postestimation > Reports and statistics

Options for estat szroeter

`rhs` specifies that tests for heteroskedasticity be performed for the right-hand-side (explanatory) variables of the fitted regression model. Option `rhs` may be combined with a *varlist*.

`mtest(`*spec*`)` specifies that multiple testing be performed. The argument specifies how p-values are adjusted. The following specifications, *spec*, are supported:

bonferroni	Bonferroni's multiple testing adjustment
holm	Holm's multiple testing adjustment
sidak	Šidák's multiple testing adjustment
noadjust	no adjustment is made for multiple testing

`estat szroeter` always performs multiple testing. By default, it does not adjust the p-values.

Syntax for estat vif

`estat vif [, uncentered]`

Menu

Statistics > Postestimation > Reports and statistics

Option for estat vif

`uncentered` requests that the computation of the uncentered variance inflation factors. This option is often used to detect the collinearity of the regressors with the constant. `estat vif, uncentered` may be used after regression models fit without the constant term.

Syntax for acprplot

`acprplot` *indepvar* [, *acprplot_options*]

acprplot_options	Description
Plot	
marker_options	change look of markers (color, size, etc.)
marker_label_options	add marker labels; change look or position
Reference line	
`rlopts(`*cline_options*`)`	affect rendition of the reference line
Options	
`lowess`	add a lowess smooth of the plotted points
`lsopts(`*lowess_options*`)`	affect rendition of the lowess smooth
`mspline`	add median spline of the plotted points
`msopts(`*mspline_options*`)`	affect rendition of the spline
Add plots	
`addplot(`*plot*`)`	add other plots to the generated graph
Y axis, X axis, Titles, Legend, Overall	
twoway_options	any options other than `by()` documented in [G-3] ***twoway_options***

Menu

Statistics > Linear models and related > Regression diagnostics > Augmented component-plus-residual plot

Options for acprplot

Plot

marker_options affect the rendition of markers drawn at the plotted points, including their shape, size, color, and outline; see [G-3] ***marker_options***.

marker_label_options specify if and how the markers are to be labeled; see [G-3] ***marker_label_options***.

Reference line

`rlopts(`*cline_options*`)` affects the rendition of the reference line. See [G-3] ***cline_options***.

Options

`lowess` adds a lowess smooth of the plotted points to assist in detecting nonlinearities.

`lsopts(`*lowess_options*`)` affects the rendition of the lowess smooth. For an explanation of these options, especially the `bwidth()` option, see [R] **lowess**. Specifying `lsopts()` implies the `lowess` option.

`mspline` adds a median spline of the plotted points to assist in detecting nonlinearities.

`msopts(`*mspline_options*`)` affects the rendition of the spline. For an explanation of these options, especially the `bands()` option, see [G-2] **graph twoway mspline**. Specifying `msopts()` implies the `mspline` option.

Add plots

`addplot(`*plot*`)` provides a way to add other plots to the generated graph. See [G-3] ***addplot_option***.

Y axis, X axis, Titles, Legend, Overall

twoway_options are any of the options documented in [G-3] ***twoway_options***, excluding `by()`. These include options for titling the graph (see [G-3] ***title_options***) and for saving the graph to disk (see [G-3] ***saving_option***).

Syntax for avplot

`avplot` *indepvar* [, *avplot_options*]

avplot_options	Description
Plot	
marker_options	change look of markers (color, size, etc.)
marker_label_options	add marker labels; change look or position
Reference line	
`rlopts(`*cline_options*`)`	affect rendition of the reference line
Add plots	
`addplot(`*plot*`)`	add other plots to the generated graph
Y axis, X axis, Titles, Legend, Overall	
twoway_options	any options other than `by()` documented in [G-3] ***twoway_options***

Menu

Statistics > Linear models and related > Regression diagnostics > Added-variable plot

Options for avplot

Plot

marker_options affect the rendition of markers drawn at the plotted points, including their shape, size, color, and outline; see [G-3] ***marker_options***.

marker_label_options specify if and how the markers are to be labeled; see [G-3] ***marker_label_options***.

Reference line

`rlopts(`*cline_options*`)` affects the rendition of the reference line. See [G-3] ***cline_options***.

Add plots

`addplot(`*plot*`)` provides a way to add other plots to the generated graph. See [G-3] ***addplot_option***.

Y axis, X axis, Titles, Legend, Overall

twoway_options are any of the options documented in [G-3] ***twoway_options***, excluding `by()`. These include options for titling the graph (see [G-3] ***title_options***) and for saving the graph to disk (see [G-3] ***saving_option***).

Syntax for avplots

`avplots` [, *avplots_options*]

avplots_options	Description
Plot	
marker_options	change look of markers (color, size, etc.)
marker_label_options	add marker labels; change look or position
combine_options	any of the options documented in [G-2] **graph combine**
Reference line	
`rlopts(`*cline_options*`)`	affect rendition of the reference line
Y axis, X axis, Titles, Legend, Overall	
twoway_options	any options other than `by()` documented in [G-3] ***twoway_options***

Menu

Statistics > Linear models and related > Regression diagnostics > Added-variable plot

Options for avplots

Plot

marker_options affect the rendition of markers drawn at the plotted points, including their shape, size, color, and outline; see [G-3] ***marker_options***.

marker_label_options specify if and how the markers are to be labeled; see [G-3] ***marker_label_options***.

combine_options are any of the options documented in [G-2] **graph combine**. These include options for titling the graph (see [G-3] ***title_options***) and for saving the graph to disk (see [G-3] ***saving_option***).

Reference line

`rlopts(`*cline_options*`)` affects the rendition of the reference line. See [G-3] ***cline_options***.

Y axis, X axis, Titles, Legend, Overall

twoway_options are any of the options documented in [G-3] ***twoway_options***, excluding `by()`. These include options for titling the graph (see [G-3] ***title_options***) and for saving the graph to disk (see [G-3] ***saving_option***).

Syntax for cprplot

`cprplot` *indepvar* [, *cprplot_options*]

cprplot_options	Description
Plot	
marker_options	change look of markers (color, size, etc.)
marker_label_options	add marker labels; change look or position
Reference line	
`rlopts(`*cline_options*`)`	affect rendition of the reference line
Options	
`lowess`	add a lowess smooth of the plotted points
`lsopts(`*lowess_options*`)`	affect rendition of the lowess smooth
`mspline`	add median spline of the plotted points
`msopts(`*mspline_options*`)`	affect rendition of the spline
Add plots	
`addplot(`*plot*`)`	add other plots to the generated graph
Y axis, X axis, Titles, Legend, Overall	
twoway_options	any options other than `by()` documented in [G-3] ***twoway_options***

Menu

Statistics > Linear models and related > Regression diagnostics > Component-plus-residual plot

Options for cprplot

Plot

marker_options affect the rendition of markers drawn at the plotted points, including their shape, size, color, and outline; see [G-3] ***marker_options***.

marker_label_options specify if and how the markers are to be labeled; see [G-3] ***marker_label_options***.

Reference line

`rlopts(`*cline_options*`)` affects the rendition of the reference line. See [G-3] ***cline_options***.

Options

`lowess` adds a lowess smooth of the plotted points to assist in detecting nonlinearities.

`lsopts(`*lowess_options*`)` affects the rendition of the lowess smooth. For an explanation of these options, especially the `bwidth()` option, see [R] **lowess**. Specifying `lsopts()` implies the `lowess` option.

`mspline` adds a median spline of the plotted points to assist in detecting nonlinearities.

`msopts(`*mspline_options*`)` affects the rendition of the spline. For an explanation of these options, especially the `bands()` option, see [G-2] **graph twoway mspline**. Specifying `msopts()` implies the `mspline` option.

Add plots

`addplot(`*plot*`)` provides a way to add other plots to the generated graph. See [G-3] ***addplot_option***.

Y axis, X axis, Titles, Legend, Overall

twoway_options are any of the options documented in [G-3] ***twoway_options***, excluding `by()`. These include options for titling the graph (see [G-3] ***title_options***) and for saving the graph to disk (see [G-3] ***saving_option***).

Syntax for lvr2plot

`lvr2plot` [, *lvr2plot_options*]

lvr2plot_options	Description
Plot	
marker_options	change look of markers (color, size, etc.)
marker_label_options	add marker labels; change look or position
Add plots	
`addplot(`*plot*`)`	add other plots to the generated graph
Y axis, X axis, Titles, Legend, Overall	
twoway_options	any options other than `by()` documented in [G-3] ***twoway_options***

Menu

Statistics > Linear models and related > Regression diagnostics > Leverage-versus-squared-residual plot

Options for lvr2plot

Plot

marker_options affect the rendition of markers drawn at the plotted points, including their shape, size, color, and outline; see [G-3] ***marker_options***.

marker_label_options specify if and how the markers are to be labeled; see [G-3] ***marker_label_options***.

Add plots

`addplot(`*plot*`)` provides a way to add other plots to the generated graph. See [G-3] ***addplot_option***.

Y axis, X axis, Titles, Legend, Overall

twoway_options are any of the options documented in [G-3] ***twoway_options***, excluding `by()`. These include options for titling the graph (see [G-3] ***title_options***) and for saving the graph to disk (see [G-3] ***saving_option***).

Syntax for rvfplot

`rvfplot` [, *rvfplot_options*]

rvfplot_options	Description
Plot	
marker_options	change look of markers (color, size, etc.)
marker_label_options	add marker labels; change look or position
Add plots	
`addplot(`*plot*`)`	add plots to the generated graph
Y axis, X axis, Titles, Legend, Overall	
twoway_options	any options other than `by()` documented in [G-3] ***twoway_options***

Menu

Statistics > Linear models and related > Regression diagnostics > Residual-versus-fitted plot

Options for rvfplot

Plot

marker_options affect the rendition of markers drawn at the plotted points, including their shape, size, color, and outline; see [G-3] ***marker_options***.

marker_label_options specify if and how the markers are to be labeled; see [G-3] ***marker_label_options***.

Add plots

addplot(*plot*) provides a way to add plots to the generated graph. See [G-3] ***addplot_option***.

Y axis, X axis, Titles, Legend, Overall

twoway_options are any of the options documented in [G-3] ***twoway_options***, excluding by(). These include options for titling the graph (see [G-3] ***title_options***) and for saving the graph to disk (see [G-3] ***saving_option***).

Syntax for rvpplot

rvpplot *indepvar* [, *rvpplot_options*]

rvpplot_options	Description
Plot	
marker_options	change look of markers (color, size, etc.)
marker_label_options	add marker labels; change look or position
Add plots	
addplot(*plot*)	add other plots to the generated graph
Y axis, X axis, Titles, Legend, Overall	
twoway_options	any options other than by() documented in [G-3] ***twoway_options***

Menu

Statistics > Linear models and related > Regression diagnostics > Residual-versus-predictor plot

Options for rvpplot

Plot

marker_options affect the rendition of markers drawn at the plotted points, including their shape, size, color, and outline; see [G-3] ***marker_options***.

marker_label_options specify if and how the markers are to be labeled; see [G-3] ***marker_label_options***.

Add plots

addplot(*plot*) provides a way to add other plots to the generated graph; see [G-3] ***addplot_option***.

Y axis, X axis, Titles, Legend, Overall

twoway_options are any of the options documented in [G-3] ***twoway_options***, excluding by(). These include options for titling the graph (see [G-3] ***title_options***) and for saving the graph to disk (see [G-3] ***saving_option***).

Remarks

Remarks are presented under the following headings:

Many of these commands concern identifying influential data in linear regression. This is, unfortunately, a field that is dominated by jargon, codified and partially begun by Belsley, Kuh, and Welsch (1980). In the words of Chatterjee and Hadi (1986, 416), "Belsley, Kuh, and Welsch's book, *Regression Diagnostics*, was a very valuable contribution to the statistical literature, but it unleashed on an unsuspecting statistical community a computer speak (à la Orwell), the likes of which we have never seen." Things have only gotten worse since then. Chatterjee and Hadi's (1986, 1988) own attempts to clean up the jargon did not improve matters (see Hoaglin and Kempthorne [1986], Velleman [1986], and Welsch [1986]). We apologize for the jargon, and for our contribution to the jargon in the form of inelegant command names, we apologize most of all.

Model *sensitivity* refers to how estimates are affected by subsets of our data. Imagine data on y and x, and assume that the data are to be fit by the regression $y_i = \alpha + \beta x_i + \epsilon_i$. The regression estimates of α and β are a and b, respectively. Now imagine that the estimated a and b would be different if a small portion of the dataset, perhaps even one observation, were deleted. As a data analyst, you would like to think that you are summarizing tendencies that apply to all the data, but you have just been told that the model you fit is unduly influenced by one point or just a few points and that, as a matter of fact, there is another model that applies to the rest of the data—a model that you have ignored. The search for subsets of the data that, if deleted, would change the results markedly is a predominant theme of this entry.

There are three key issues in identifying model sensitivity to individual observations, which go by the names *residuals*, *leverage*, and *influence*. In our $y_i = a + bx_i + e_i$ regression, the residuals are, of course, e_i—they reveal how much our fitted value $\widehat{y}_i = a + bx_i$ differs from the observed y_i. A point (x_i, y_i) with a corresponding large residual is called an outlier. Say that you are interested in outliers because you somehow think that such points will exert undue influence on your estimates. Your feelings are generally right, but there are exceptions. A point might have a huge residual and yet not affect the estimated b at all. Nevertheless, studying observations with large residuals almost always pays off.

(x_i, y_i) can be an outlier in another way—just as y_i can be far from $\widehat{y}_i$, x_i can be far from the center of mass of the other x's. Such an "outlier" should interest you just as much as the more traditional outliers. Picture a scatterplot of y against x with thousands of points in some sort of mass at the lower left of the graph and one point at the upper right of the graph. Now run a regression line through the points—the regression line will come close to the point at the upper right of the graph and may in fact, go through it. That is, this isolated point will not appear as an outlier as measured by residuals because its residual will be small. Yet this point might have a dramatic effect on our resulting estimates in the sense that, were you to delete the point, the estimates would change markedly. Such a point is said to have high leverage. Just as with traditional outliers, a high leverage point does not necessarily have an undue effect on regression estimates, but if it does not, it is more the exception than the rule.

Now all this is a most unsatisfactory state of affairs. Points with large residuals may, but need not, have a large effect on our results, and points with small residuals may still have a large effect. Points with high leverage may, but need not, have a large effect on our results, and points with low leverage may still have a large effect. Can you not identify the influential points and simply have the computer list them for you? You can, but you will have to define what you mean by "influential".

"Influential" is defined with respect to some statistic. For instance, you might ask which points in your data have a large effect on your estimated a, which points have a large effect on your estimated b, which points have a large effect on your estimated standard error of b, and so on, but do not be surprised when the answers to these questions are different. In any case, obtaining such measures is not difficult—all you have to do is fit the regression excluding each observation one at a time and record the statistic of interest which, in the day of the modern computer, is not too onerous. Moreover, you can save considerable computer time by doing algebra ahead of time and working out formulas that will calculate the same answers as if you ran each of the regressions. (Ignore the question of pairs of observations that, together, exert undue influence, and triples, and so on, which remains largely unsolved and for which the brute force fit-every-possible-regression procedure is not a viable alternative.)

Fitted values and residuals

Typing `predict` *newvar* with no options creates *newvar* containing the fitted values. Typing `predict` *newvar*`, resid` creates *newvar* containing the residuals.

▷ Example 1

Continuing with example 1 from [R] **regress**, we wish to fit the following model:

$$\texttt{mpg} = \beta_0 + \beta_1\texttt{weight} + \beta_2\texttt{weight}^2 + \beta_3\texttt{foreign} + \epsilon$$

```
. use http://www.stata-press.com/data/r12/auto
(1978 Automobile Data)

. regress mpg weight c.weight#c.weight foreign
      Source |       SS       df       MS              Number of obs =      74
-------------+------------------------------           F(  3,    70) =   52.25
       Model |  1689.15372     3   563.05124           Prob > F      =  0.0000
    Residual |   754.30574    70  10.7757963           R-squared     =  0.6913
-------------+------------------------------           Adj R-squared =  0.6781
       Total |  2443.45946    73  33.4720474           Root MSE      =  3.2827

------------------------------------------------------------------------------
         mpg |      Coef.   Std. Err.      t    P>|t|     [95% Conf. Interval]
-------------+----------------------------------------------------------------
      weight |  -.0165729   .0039692    -4.18   0.000    -.0244892   -.0086567
             |
   c.weight#|
    c.weight |   1.59e-06   6.25e-07     2.55   0.013     3.45e-07    2.84e-06
             |
     foreign |    -2.2035   1.059246    -2.08   0.041      -4.3161   -.0909002
       _cons |   56.53884   6.197383     9.12   0.000     44.17855    68.89913
------------------------------------------------------------------------------
```

That done, we can now obtain the predicted values from the regression. We will store them in a new variable called pmpg by typing `predict pmpg`. Because `predict` produces no output, we will follow that by summarizing our predicted and observed values.

```
. predict pmpg
(option xb assumed; fitted values)

. summarize pmpg mpg

    Variable |       Obs        Mean    Std. Dev.       Min        Max
-------------+--------------------------------------------------------
        pmpg |        74     21.2973    4.810311   13.59953   31.86288
         mpg |        74     21.2973    5.785503         12         41
```

◁

▷ Example 2: Out-of-sample predictions

We can just as easily obtain predicted values from the model by using a wholly different dataset from the one on which the model was fit. The only requirement is that the data have the necessary variables, which here are `weight` and `foreign`.

Using the data on two new cars (the Pontiac Sunbird and the Volvo 260) from the `newautos.dta` dataset, we can obtain out-of-sample predictions (or forecasts) by typing

```
. use http://www.stata-press.com/data/r12/newautos, clear
(New Automobile Models)

. predict pmpg
(option xb assumed; fitted values)

. list, divider
```

	make	weight	foreign	pmpg
1.	Pont. Sunbird	2690	Domestic	23.47137
2.	Volvo 260	3170	Foreign	17.78846

The Pontiac Sunbird has a predicted mileage rating of 23.5 mpg, whereas the Volvo 260 has a predicted rating of 17.8 mpg. In comparison, the actual mileage ratings are 24 for the Pontiac and 17 for the Volvo.

◁

Prediction standard errors

`predict` can calculate the standard error of the forecast (`stdf` option), the standard error of the prediction (`stdp` option), and the standard error of the residual (`stdr` option). It is easy to confuse `stdf` and `stdp` because both are often called the prediction error. Consider the prediction $\widehat{y}_j = \mathbf{x}_j\mathbf{b}$, where $\mathbf{b}$ is the estimated coefficient (column) vector and $\mathbf{x}_j$ is a (row) vector of independent variables for which you want the prediction. First, $\widehat{y}_j$ has a variance due to the variance of the estimated coefficient vector $\mathbf{b}$,

$$\mathrm{Var}(\widehat{y}_j) = \mathrm{Var}(\mathbf{x}_j\mathbf{b}) = s^2 h_j$$

where $h_j = \mathbf{x}_j(\mathbf{X}'\mathbf{X})^{-1}\mathbf{x}_j'$ and s^2 is the mean squared error of the regression. Do not panic over the algebra—just remember that $\mathrm{Var}(\widehat{y}_j) = s^2 h_j$, whatever s^2 and h_j are. `stdp` calculates this quantity. This is the error in the prediction due to the uncertainty about $\mathbf{b}$.

If you are about to hand this number out as your forecast, however, there is another error. According to your model, the true value of y_j is given by

$$y_j = \mathbf{x}_j\mathbf{b} + \epsilon_j = \widehat{y}_j + \epsilon_j$$

and thus the $\mathrm{Var}(y_j) = \mathrm{Var}(\widehat{y}_j) + \mathrm{Var}(\epsilon_j) = s^2h_j + s^2$, which is the square of `stdf`. `stdf`, then, is the sum of the error in the prediction plus the residual error.

`stdr` has to do with an analysis-of-variance decomposition of s^2, the estimated variance of y. The standard error of the prediction is s^2h_j, and therefore $s^2h_j + s^2(1-h_j) = s^2$ decomposes s^2 into the prediction and residual variances.

▷ Example 3: standard error of the forecast

Returning to our model of `mpg` on `weight`, `weight`^{2}, and `foreign`, we previously predicted the mileage rating for the Pontiac Sunbird and Volvo 260 as 23.5 and 17.8 mpg, respectively. We now want to put a standard error around our forecast. Remember, the data for these two cars were in `newautos.dta`:

```
. use http://www.stata-press.com/data/r12/newautos, clear
(New Automobile Models)
. predict pmpg
(option xb assumed; fitted values)
. predict se_pmpg, stdf
. list, divider
```

	make	weight	foreign	pmpg	se_pmpg
1.	Pont. Sunbird	2690	Domestic	23.47137	3.341823
2.	Volvo 260	3170	Foreign	17.78846	3.438714

Thus an approximate 95% confidence interval for the mileage rating of the Volvo 260 is $17.8 \pm 2 \cdot 3.44 = [\,10.92, 24.68\,]$.

◁

Prediction with weighted data

`predict` can be used after frequency-weighted (`fweight`) estimation, just as it is used after unweighted estimation. The technical note below concerns the use of `predict` after analytically weighted (`aweight`) estimation.

❑ Technical note

After analytically weighted estimation, `predict` is willing to calculate only the prediction (no options), residual (`residual` option), standard error of the prediction (`stdp` option), and diagonal elements of the projection matrix (`hat` option). Moreover, the results produced by `hat` need to be adjusted, as will be described. For analytically weighted estimation, the standard error of the forecast and residuals, the standardized and Studentized residuals, and Cook's D are not statistically well-defined concepts.

To obtain the correct values of the diagonal elements of the hat matrix, you can use `predict` with the `hat` option to make a first, partially adjusted calculation, and then follow that by completing the adjustment. Assume that you are fitting a linear regression model weighting the data with the variable `w` (`[aweight=w]`). Begin by creating a new variable, `w0`:

```
. predict resid if e(sample), resid
. summarize w if resid < . & e(sample)
. gen w0=w/r(mean)
```

Some caution is necessary at this step—the `summarize w` must be performed on the same sample that was used to fit the model, which means that you must include `if e(sample)` to restrict the prediction to the estimation sample. You created the residual and then included the modifier '`if resid < .`' so that if the dependent variable or any of the independent variables is missing, the corresponding observations will be excluded from the calculation of the average value of the original weight.

To correct `predict`'s `hat` calculation, multiply the result by `w0`:

```
. predict myhat, hat
. replace myhat = w0 * myhat
```

❑

Residual-versus-fitted plots

▷ Example 4: rvfplot

Using the automobile dataset described in [U] **1.2.2 Example datasets**, we will use `regress` to fit a model of `price` on `weight`, `mpg`, `foreign`, and the interaction of `foreign` with `mpg`. We specify `foreign##c.mpg` to obtain the interaction of `foreign` with `mpg`; see [U] **11.4.3 Factor variables**.

```
. use http://www.stata-press.com/data/r12/auto, clear
(1978 Automobile Data)

. regress price weight foreign##c.mpg

      Source |       SS       df       MS              Number of obs =      74
-------------+------------------------------           F(  4,    69) =   21.22
       Model |   350319665     4  87579916.3           Prob > F      =  0.0000
    Residual |   284745731    69  4126749.72           R-squared     =  0.5516
-------------+------------------------------           Adj R-squared =  0.5256
       Total |   635065396    73  8699525.97           Root MSE      =  2031.4

------------------------------------------------------------------------------
       price |      Coef.   Std. Err.      t    P>|t|     [95% Conf. Interval]
-------------+----------------------------------------------------------------
      weight |   4.613589   .7254961     6.36   0.000     3.166263    6.060914
   1.foreign |   11240.33   2751.681     4.08   0.000     5750.878    16729.78
         mpg |   263.1875   110.7961     2.38   0.020     42.15527    484.2197
             |
    foreign# |
       c.mpg |
          1  |  -307.2166   108.5307    -2.83   0.006    -523.7294   -90.70368
             |
       _cons |  -14449.58    4425.72    -3.26   0.002    -23278.65    -5620.51
------------------------------------------------------------------------------
```

Once we have fit a model, we may use any of the regression diagnostics commands. `rvfplot` (read residual-versus-fitted plot) graphs the residuals against the fitted values:

```
. rvfplot, yline(0)
```

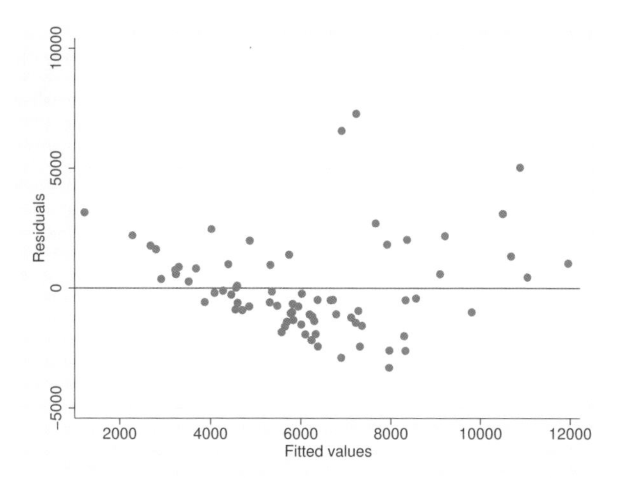

All the diagnostic plot commands allow the `graph twoway` and `graph twoway scatter` options; we specified a `yline(0)` to draw a line across the graph at $y = 0$; see [G-2] **graph twoway scatter**.

In a well-fitted model, there should be no pattern to the residuals plotted against the fitted values—something not true of our model. Ignoring the two outliers at the top center of the graph, we see curvature in the pattern of the residuals, suggesting a violation of the assumption that `price` is linear in our independent variables. We might also have seen increasing or decreasing variation in the residuals—heteroskedasticity. Any pattern whatsoever indicates a violation of the least-squares assumptions.

◁

Added-variable plots

▷ Example 5: avplot

We continue with our price model, and another diagnostic graph is provided by `avplot` (read added-variable plot, also known as the partial-regression leverage plot).

One of the wonderful features of one-regressor regressions (regressions of y on one x) is that we can graph the data and the regression line. There is no easier way to understand the regression than to examine such a graph. Unfortunately, we cannot do this when we have more than one regressor. With two regressors, it is still theoretically possible—the graph must be drawn in three dimensions, but with three or more regressors no graph is possible.

The added-variable plot is an attempt to project multidimensional data back to the two-dimensional world for each of the original regressors. This is, of course, impossible without making some concessions. Call the coordinates on an added-variable plot y and x. The added-variable plot has the following properties:

- There is a one-to-one correspondence between (x_i, y_i) and the ith observation used in the original regression.
- A regression of y on x has the same coefficient and standard error (up to a degree-of-freedom adjustment) as the estimated coefficient and standard error for the regressor in the original regression.

• The "outlierness" of each observation in determining the slope is in some sense preserved.

It is equally important to note the properties that are not listed. The y and x coordinates of the added-variable plot cannot be used to identify functional form, or, at least, not well (see Mallows [1986]). In the construction of the added-variable plot, the relationship between y and x is forced to be linear.

Let's examine the added-variable plot for mpg in our regression of price on weight and foreign##c.mpg:

```
. avplot mpg
```

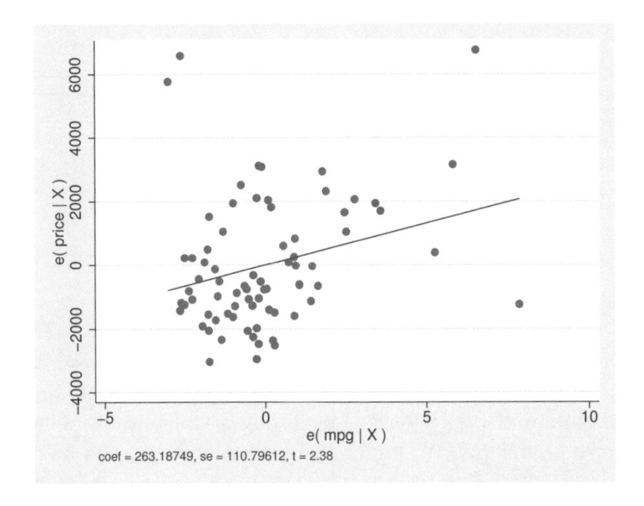

This graph suggests a problem in determining the coefficient on mpg. Were this a one-regressor regression, the two points at the top-left corner and the one at the top right would cause us concern, and so it does in our more complicated multiple-regressor case. To identify the problem points, we retyped our command, modifying it to read avplot mpg, mlabel(make), and discovered that the two cars at the top left are the Cadillac Eldorado and the Lincoln Versailles; the point at the top right is the Cadillac Seville. These three cars account for 100% of the luxury cars in our data, suggesting that our model is misspecified. By the way, the point at the lower right of the graph, also cause for concern, is the Plymouth Arrow, our data-entry error. ◁

❑ Technical note

Stata's avplot command can be used with regressors already in the model, as we just did, or with potential regressors not yet in the model. In either case, avplot will produce the correct graph. The name "added-variable plot" is unfortunate in the case when the variable is already among the list of regressors but is, we think, still preferable to the name "partial-regression leverage plot" assigned by Belsley, Kuh, and Welsch (1980, 30) and more in the spirit of the original use of such plots by Mosteller and Tukey (1977, 271–279). Welsch (1986, 403), however, disagrees: "I am sorry to see that Chatterjee and Hadi [1986] endorse the term 'added-variable plot' when X_j is part of the original model" and goes on to suggest the name "adjusted partial residual plot". ❑

▷ Example 6: avplots

Added-variable plots are so useful that we should look at them for every regressor in the data. `avplots` makes this easy:

```
. avplots
```

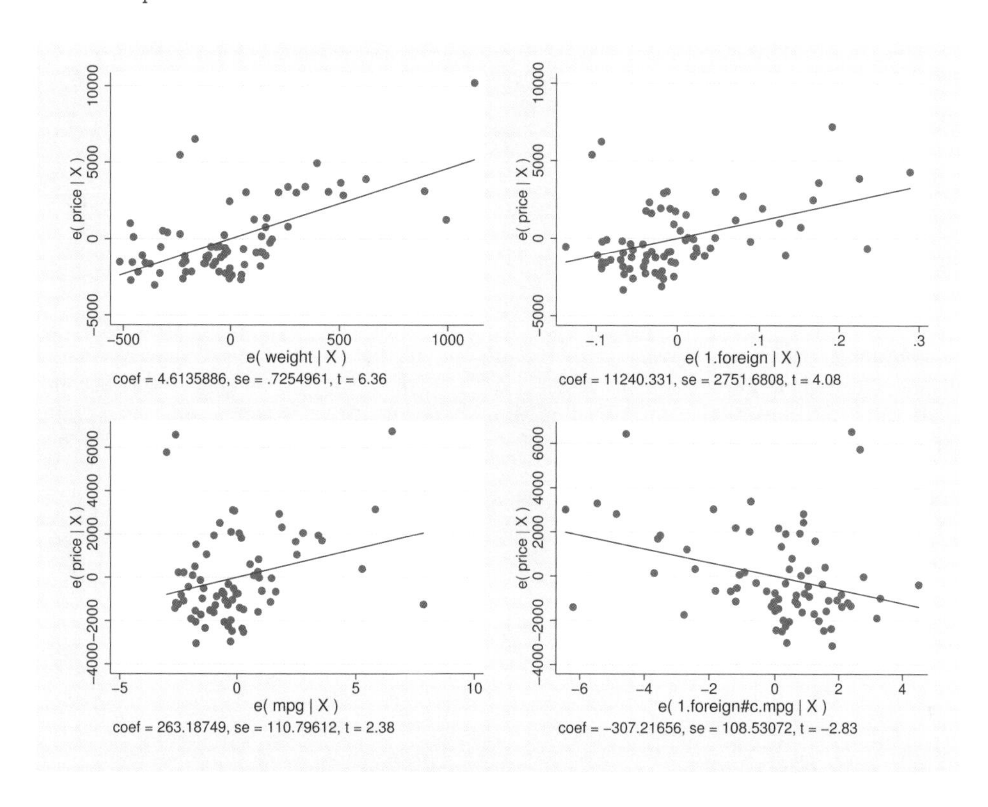

◁

Component-plus-residual plots

Added-variable plots are successful at identifying outliers, but they cannot be used to identify functional form. The component-plus-residual plot (Ezekiel 1924; Larsen and McCleary 1972) is another attempt at projecting multidimensional data into a two-dimensional form, but with different properties. Although the added-variable plot can identify outliers, the component-plus-residual plot cannot. It can, however, be used to examine the functional-form assumptions of the model. Both plots have the property that a regression line through the coordinates has a slope equal to the estimated coefficient in the regression model.

▷ Example 7: cprplot and acprplot

To illustrate these plots, we begin with a different model:

```
. use http://www.stata-press.com/data/r12/auto1, clear
(Automobile Models)

. regress price mpg weight

      Source |       SS       df       MS              Number of obs =      74
-------------+------------------------------           F(  2,    71) =   14.90
       Model |  187716578     2  93858289              Prob > F      =  0.0000
    Residual |  447348818    71  6300687.58            R-squared     =  0.2956
-------------+------------------------------           Adj R-squared =  0.2757
       Total |  635065396    73  8699525.97            Root MSE      =  2510.1

------------------------------------------------------------------------------
       price |      Coef.   Std. Err.      t    P>|t|     [95% Conf. Interval]
-------------+----------------------------------------------------------------
         mpg |   -55.9393   75.24136    -0.74   0.460    -205.9663    94.08771
      weight |   1.710992   .5861682     2.92   0.005     .5422063    2.879779
       _cons |     2197.9   3190.768     0.69   0.493    -4164.311     8560.11
------------------------------------------------------------------------------
```

In fact, we know that the effects of mpg in this model are nonlinear—if we added mpg squared to the model, its coefficient would have a t statistic of 2.38, the t statistic on mpg would become −2.48, and weight's effect would become about one-third of its current value and become statistically insignificant. Pretend that we do not know this.

The component-plus-residual plot for mpg is

```
. cprplot mpg, mspline msopts(bands(13))
```

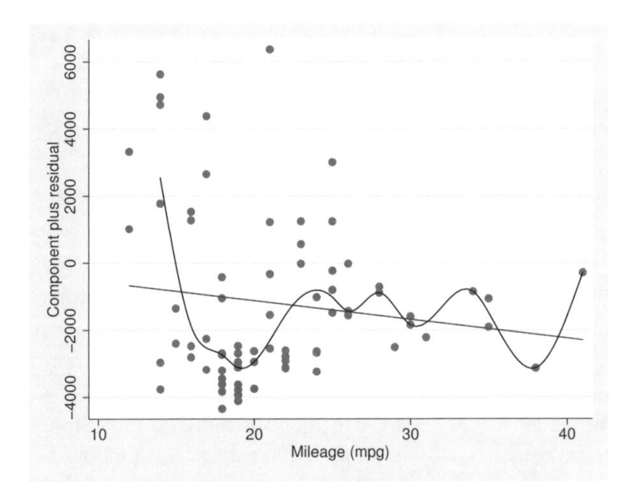

We are supposed to examine the above graph for nonlinearities or, equivalently, ask if the regression line, which has slope equal to the estimated effect of mpg in the original model, fits the data adequately. To assist our eyes, we added a median spline. Perhaps some people may detect nonlinearity from this graph, but we assert that if we had not previously revealed the nonlinearity of mpg and if we had not added the median spline, the graph would not overly bother us.

Mallows (1986) proposed an augmented component-plus-residual plot that is often more sensitive to detecting nonlinearity:

```
. acprplot mpg, mspline msopts(bands(13))
```

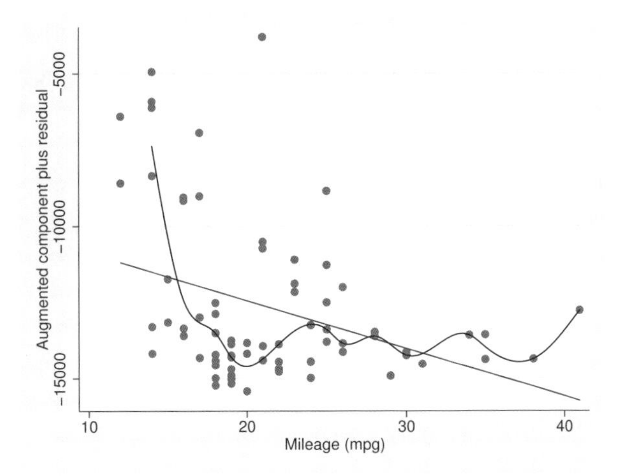

It does do somewhat better.

◁

Residual-versus-predictor plots

▷ Example 8: rvpplot

The residual-versus-predictor plot is a simple way to look for violations of the regression assumptions. If the assumptions are correct, there should be no pattern in the graph. Using our `price` on `mpg` and `weight` model, we type

```
. rvpplot mpg, yline(0)
```

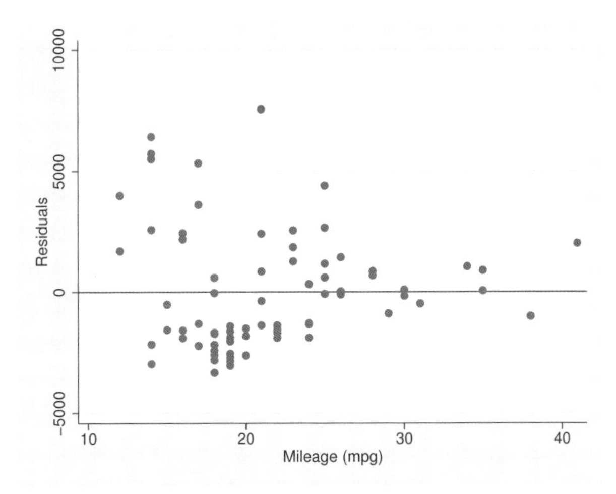

Remember, any pattern counts as a problem, and in this graph, we see that the variation in the residuals decreases as `mpg` increases.

◁

Leverage statistics

In addition to providing fitted values and the associated standard errors, the `predict` command can also be used to generate various statistics used to detect the influence of individual observations. This section provides a brief introduction to leverage (hat) statistics, and some of the following subsections discuss other influence statistics produced by `predict`.

▷ Example 9: diagonal elements of projection matrix

The diagonal elements of the projection matrix, obtained by the `hat` option, are a measure of distance in explanatory variable space. `leverage` is a synonym for `hat`.

```
. use http://www.stata-press.com/data/r12/auto
(1978 Automobile Data)

. regress mpg weight c.weight#c.weight foreign
  (output omitted)

. predict xdist, hat

. summarize xdist, detail
                           Leverage
-------------------------------------------------------------
      Percentiles      Smallest
 1%     .0251334       .0251334
 5%     .0255623       .0251334
10%     .0259213       .0253883       Obs                  74
25%     .0278442       .0255623       Sum of Wgt.          74

50%       .04103                      Mean           .0540541
                        Largest       Std. Dev.      .0459218
75%     .0631279       .1593606
90%     .0854584       .1593606       Variance       .0021088
95%     .1593606       .2326124       Skewness       3.440809
99%     .3075759       .3075759       Kurtosis       16.95135
```

Some 5% of our sample has an `xdist` measure in excess of 0.15. Let's force them to reveal their identities:

```
. list foreign make mpg if xdist>.15, divider

     +--------------------------------------+
     |  foreign   make                  mpg |
     |--------------------------------------|
 24. | Domestic   Ford Fiesta            28 |
 26. | Domestic   Linc. Continental      12 |
 27. | Domestic   Linc. Mark V           12 |
 43. | Domestic   Plym. Champ            34 |
     +--------------------------------------+
```

To understand why these cars are on this list, we must remember that the explanatory variables in our model are `weight` and `foreign` and that `xdist` measures distance in this metric. The Ford Fiesta and the Plymouth Champ are the two lightest domestic cars in our data. The Lincolns are the two heaviest domestic cars.

◁

L-R plots

▷ Example 10: lvr2plot

One of the most useful diagnostic graphs is provided by `lvr2plot` (leverage-versus-residual-squared plot), a graph of leverage against the (normalized) residuals squared.

```
. use http://www.stata-press.com/data/r12/auto, clear
(1978 Automobile Data)
. regress price weight foreign##c.mpg
  (output omitted)
. lvr2plot
```

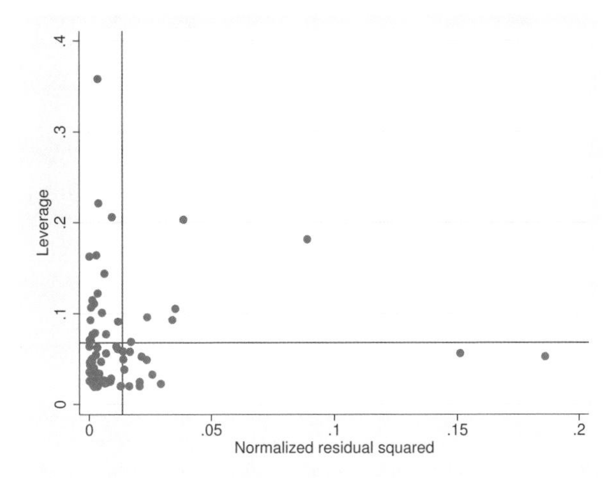

The lines on the chart show the average values of leverage and the (normalized) residuals squared. Points above the horizontal line have higher-than-average leverage; points to the right of the vertical line have larger-than-average residuals.

One point immediately catches our eye, and four more make us pause. The point at the top of the graph has high leverage and a smaller-than-average residual. The other points that bother us all have higher-than-average leverage, two with smaller-than-average residuals and two with larger-than-average residuals.

A less pretty but more useful version of the above graph specifies that `make` be used as the symbol (see [G-3] ***marker_label_options***):

```
. lvr2plot, mlabel(make) mlabp(0) m(none) mlabsize(small)
```

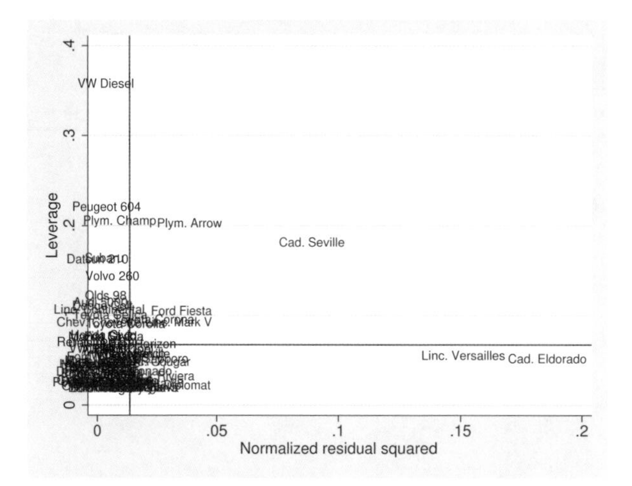

The VW Diesel, Plymouth Champ, Plymouth Arrow, and Peugeot 604 are the points that cause us the most concern. When we further examine our data, we discover that the VW Diesel is the only diesel in our data and that the data for the Plymouth Arrow were entered incorrectly into the computer. No such simple explanations were found for the Plymouth Champ and Peugeot 604.

◁

Standardized and Studentized residuals

The terms standardized and Studentized residuals have meant different things to different authors. In Stata, `predict` defines the standardized residual as $\widehat{e}_i = e_i/(s\sqrt{1-h_i})$ and the Studentized residual as $r_i = e_i/(s_{(i)}\sqrt{1-h_i})$, where $s_{(i)}$ is the root mean squared error of a regression with the ith observation removed. Stata's definition of the Studentized residual is the same as the one given in Bollen and Jackman (1990, 264) and is what Chatterjee and Hadi (1988, 74) call the "externally Studentized" residual. Stata's "standardized" residual is the same as what Chatterjee and Hadi (1988, 74) call the "internally Studentized" residual.

Standardized and Studentized residuals are attempts to adjust residuals for their standard errors. Although the ϵ_i theoretical residuals are homoskedastic by assumption (that is, they all have the same variance), the calculated e_i are not. In fact,

$$\mathrm{Var}(e_i) = \sigma^2(1 - h_i)$$

where h_i are the leverage measures obtained from the diagonal elements of hat matrix. Thus observations with the greatest leverage have corresponding residuals with the smallest variance.

Standardized residuals use the root mean squared error of the regression for σ. Studentized residuals use the root mean squared error of a regression omitting the observation in question for σ. In general, Studentized residuals are preferable to standardized residuals for purposes of outlier identification. Studentized residuals can be interpreted as the t statistic for testing the significance of a dummy variable equal to 1 in the observation in question and 0 elsewhere (Belsley, Kuh, and Welsch 1980). Such a dummy variable would effectively absorb the observation and so remove its influence in determining the other coefficients in the model. Caution must be exercised here, however, because of the simultaneous testing problem. You cannot simply list the residuals that would be individually significant at the 5% level—their joint significance would be far less (their joint significance *level* would be far greater).

▷ Example 11: standardized and Studentized residuals

In the opening remarks for this entry, we distinguished residuals from leverage and speculated on the impact of an observation with a small residual but large leverage. If we had adjusted the residuals for their standard errors, however, the adjusted residual would have been (relatively) larger and perhaps large enough so that we could simply examine the adjusted residuals. Taking our `price` on `weight` and `foreign##c.mpg` model, we can obtain the in-sample standardized and Studentized residuals by typing

```
. predict esta if e(sample), rstandard
. predict estu if e(sample), rstudent
```

In *L-R plots*, we discovered that the VW Diesel has the highest leverage in our data, but a corresponding small residual. The standardized and Studentized residuals for the VW Diesel are

```
. list make price esta estu if make=="VW Diesel"
```

	make	price	esta	estu
74.	VW Diesel	5,397	.6142691	.6114758

The Studentized residual of 0.611 can be interpreted as the t statistic for including a dummy variable for VW Diesel in our regression. Such a variable would not be significant.

◁

DFITS, Cook's Distance, and Welsch Distance

DFITS (Welsch and Kuh 1977), Cook's Distance (Cook 1977), and Welsch Distance (Welsch 1982) are three attempts to summarize the information in the leverage versus residual-squared plot into one statistic. That is, the goal is to create an index that is affected by the size of the residuals—outliers—and the size of h_i—leverage. Viewed mechanically, one way to write DFITS (Bollen and Jackman 1990, 265) is

$$\text{DFITS}_i = r_i\sqrt{\frac{h_i}{1-h_i}}$$

where r_i are the Studentized residuals. Thus large residuals increase the value of DFITS, as do large values of h_i. Viewed more traditionally, DFITS is a scaled difference between predicted values for the ith case when the regression is fit with and without the ith observation, hence the name.

The mechanical relationship between DFITS and Cook's Distance, D_i (Bollen and Jackman 1990, 266), is

$$D_i = \frac{1}{k}\frac{s_{(i)}^2}{s^2}\text{DFITS}_i^2$$

where k is the number of variables (including the constant) in the regression, s is the root mean squared error of the regression, and $s_{(i)}$ is the root mean squared error when the ith observation is omitted. Viewed more traditionally, D_i is a scaled measure of the distance between the coefficient vectors when the ith observation is omitted.

The mechanical relationship between DFITS and Welsch's Distance, W_i (Chatterjee and Hadi 1988, 123), is

$$W_i = \text{DFITS}_i\sqrt{\frac{n-1}{1-h_i}}$$

The interpretation of W_i is more difficult, as it is based on the empirical influence curve. Although DFITS and Cook's distance are similar, the Welsch distance measure includes another normalization by leverage.

Belsley, Kuh, and Welsch (1980, 28) suggest that DFITS values greater than $2\sqrt{k/n}$ deserve more investigation, and so values of Cook's distance greater than $4/n$ should also be examined (Bollen and Jackman 1990, 265–266). Through similar logic, the cutoff for Welsch distance is approximately $3\sqrt{k}$ (Chatterjee and Hadi 1988, 124).

▷ Example 12: DFITS influence measure

Using our model of `price` on `weight` and `foreign##c.mpg`, we can obtain the DFITS influence measure:

```
. use http://www.stata-press.com/data/r12/auto, clear
(1978 Automobile Data)
. regress price weight foreign##c.mpg
  (output omitted )
. predict e if e(sample), resid
. predict dfits, dfits
```

We did not specify `if e(sample)` in computing the DFITS statistic. DFITS is available only over the estimation sample, so specifying `if e(sample)` would have been redundant. It would have done no harm, but it would not have changed the results.

Our model has $k = 5$ independent variables (k includes the constant) and $n = 74$ observations; following the $2\sqrt{k/n}$ cutoff advice, we type

```
. list make price e dfits if abs(dfits) > 2*sqrt(5/74), divider
```

	make	price	e	dfits
12.	Cad. Eldorado	14,500	7271.96	.9564455
13.	Cad. Seville	15,906	5036.348	1.356619
24.	Ford Fiesta	4,389	3164.872	.5724172
27.	Linc. Mark V	13,594	3109.193	.5200413
28.	Linc. Versailles	13,466	6560.912	.8760136
42.	Plym. Arrow	4,647	-3312.968	-.9384231

We calculate Cook's distance and list the observations greater than the suggested $4/n$ cutoff:

```
. predict cooksd if e(sample), cooksd
. list make price e cooksd if cooksd > 4/74, divider
```

	make	price	e	cooksd
12.	Cad. Eldorado	14,500	7271.96	.1492676
13.	Cad. Seville	15,906	5036.348	.3328515
24.	Ford Fiesta	4,389	3164.872	.0638815
28.	Linc. Versailles	13,466	6560.912	.1308004
42.	Plym. Arrow	4,647	-3312.968	.1700736

Here we used `if e(sample)` because Cook's distance is not restricted to the estimation sample by default. It is worth comparing this list with the preceding one.

Finally, we use Welsch distance and the suggested $3\sqrt{k}$ cutoff:

```
. predict wd, welsch
. list make price e wd if abs(wd) > 3*sqrt(5), divider
```

	make	price	e	wd
12.	Cad. Eldorado	14,500	7271.96	8.394372
13.	Cad. Seville	15,906	5036.348	12.81125
28.	Linc. Versailles	13,466	6560.912	7.703005
42.	Plym. Arrow	4,647	-3312.968	-8.981481

Here we did not need to specify `if e(sample)` because `welsch` automatically restricts the prediction to the estimation sample.

◁

COVRATIO

COVRATIO (Belsley, Kuh, and Welsch 1980) measures the influence of the ith observation by considering the effect on the variance–covariance matrix of the estimates. The measure is the ratio of the determinants of the covariances matrix, with and without the ith observation. The resulting formula is

$$\text{COVRATIO}_i = \frac{1}{1-h_i}\left(\frac{n-k-\widehat{e}_i^2}{n-k-1}\right)^k$$

where $\widehat{e}_i$ is the standardized residual.

For noninfluential observations, the value of COVRATIO is approximately 1. Large values of the residuals or large values of leverage will cause deviations from 1, although if both are large, COVRATIO may tend back toward 1 and therefore not identify such observations (Chatterjee and Hadi 1988, 139).

Belsley, Kuh, and Welsch (1980) suggest that observations for which

$$|\text{COVRATIO}_i - 1| \geq \frac{3k}{n}$$

are worthy of further examination.

▷ Example 13: COVRATIO influence measure

Using our model of `price` on `weight` and `foreign##c.mpg`, we can obtain the COVRATIO measure and list the observations outside the suggested cutoff by typing

```
. predict covr, covratio
. list make price e covr if abs(covr-1) >= 3*5/74, divider
```

	make	price	e	covr
12.	Cad. Eldorado	14,500	7271.96	.3814242
13.	Cad. Seville	15,906	5036.348	.7386969
28.	Linc. Versailles	13,466	6560.912	.4761695
43.	Plym. Champ	4,425	1621.747	1.27782
53.	Audi 5000	9,690	591.2883	1.206842
57.	Datsun 210	4,589	19.81829	1.284801
64.	Peugeot 604	12,990	1037.184	1.348219
66.	Subaru	3,798	-909.5894	1.264677
71.	VW Diesel	5,397	999.7209	1.630653
74.	Volvo 260	11,995	1327.668	1.211888

The `covratio` option automatically restricts the prediction to the estimation sample.

◁

DFBETAs

DFBETAs are perhaps the most direct influence measure of interest to model builders. DFBETAs focus on one coefficient and measure the difference between the regression coefficient when the ith observation is included and excluded, the difference being scaled by the estimated standard error of the coefficient. Belsley, Kuh, and Welsch (1980, 28) suggest observations with $|\text{DFBETA}_i| > 2/\sqrt{n}$ as deserving special attention, but it is also common practice to use 1 (Bollen and Jackman 1990, 267), meaning that the observation shifted the estimate at least one standard error.

▷ Example 14: DFBETAs influence measure; the dfbeta() option

Using our model of `price` on `weight` and `foreign##c.mpg`, let's first ask which observations have the greatest impact on the determination of the coefficient on `1.foreign`. We will use the suggested $2/\sqrt{n}$ cutoff:

```
. sort foreign make
. predict dfor, dfbeta(1.foreign)
. list make price foreign dfor if abs(dfor) > 2/sqrt(74), divider
```

	make	price	foreign	dfor
12.	Cad. Eldorado	14,500	Domestic	-.5290519
13.	Cad. Seville	15,906	Domestic	.8243419
28.	Linc. Versailles	13,466	Domestic	-.5283729
42.	Plym. Arrow	4,647	Domestic	-.6622424
43.	Plym. Champ	4,425	Domestic	.2371104
64.	Peugeot 604	12,990	Foreign	.2552032
69.	Toyota Corona	5,719	Foreign	-.256431

The Cadillac Seville shifted the coefficient on `1.foreign` 0.82 standard deviations!

Now let us ask which observations have the greatest effect on the mpg coefficient:

```
. predict dmpg, dfbeta(mpg)
. list make price mpg dmpg if abs(dmpg) > 2/sqrt(74), divider
```

	make	price	mpg	dmpg
12.	Cad. Eldorado	14,500	14	-.5970351
13.	Cad. Seville	15,906	21	1.134269
28.	Linc. Versailles	13,466	14	-.6069287
42.	Plym. Arrow	4,647	28	-.8925859
43.	Plym. Champ	4,425	34	.3186909

Once again we see the Cadillac Seville heading the list, indicating that our regression results may be dominated by this one car.

◁

▷ Example 15: DFBETAs influence measure; the dfbeta command

We can use `predict, dfbeta()` or the `dfbeta` command to generate the DFBETAs. `dfbeta` makes up names for the new variables automatically and, without arguments, generates the DFBETAs for all the variables in the regression:

```
. dfbeta
                     _dfbeta_1: dfbeta(weight)
                     _dfbeta_2: dfbeta(1.foreign)
                     _dfbeta_3: dfbeta(mpg)
                     _dfbeta_4: dfbeta(1.foreign#c.mpg)
```

`dfbeta` created four new variables in our dataset: `_dfbeta_1`, containing the DFBETAs for `weight`; `_dfbeta_2`, containing the DFBETAs for `mpg`; and so on. Had we wanted only the DFBETAs for `mpg` and `weight`, we might have typed

```
. dfbeta mpg weight
                     _dfbeta_5: dfbeta(weight)
                     _dfbeta_6: dfbeta(mpg)
```

In the example above, we typed `dfbeta mpg weight` instead of `dfbeta`; if we had typed `dfbeta` followed by `dfbeta mpg weight`, here is what would have happened:

```
. dfbeta
                     _dfbeta_7: dfbeta(weight)
                     _dfbeta_8: dfbeta(1.foreign)
                     _dfbeta_9: dfbeta(mpg)
                    _dfbeta_10: dfbeta(1.foreign#c.mpg)
. dfbeta mpg weight
                    _dfbeta_11: dfbeta(weight)
                    _dfbeta_12: dfbeta(mpg)
```

`dfbeta` would have made up different names for the new variables. `dfbeta` never replaces existing variables—it instead makes up a different name, so we need to pay attention to `dfbeta`'s output.

◁

Formal tests for violations of assumptions

This section introduces some regression diagnostic commands that are designed to test for certain violations that `rvfplot` less formally attempts to detect. `estat ovtest` provides Ramsey's test for omitted variables—a pattern in the residuals. `estat hettest` provides a test for heteroskedasticity—the increasing or decreasing variation in the residuals with fitted values, with respect to the explanatory variables, or with respect to yet other variables. The score test implemented in `estat hettest` (Breusch and Pagan 1979; Cook and Weisberg 1983) performs a score test of the null hypothesis that $b = 0$ against the alternative hypothesis of multiplicative heteroskedasticity. `estat szroeter` provides a rank test for heteroskedasticity, which is an alternative to the score test computed by `estat hettest`. Finally, `estat imtest` computes an information matrix test, including an orthogonal decomposition into tests for heteroskedasticity, skewness, and kurtosis (Cameron and Trivedi 1990). The heteroskedasticity test computed by `estat imtest` is similar to the general test for heteroskedasticity that was proposed by White (1980). Cameron and Trivedi (2010, chap. 3) discuss most of these tests and provides more examples.

▷ Example 16: estat ovtest, estat hettest, estat szroeter, and estat imtest

We run these commands just mentioned on our model:

```
. estat ovtest

Ramsey RESET test using powers of the fitted values of price
       Ho:  model has no omitted variables
                  F(3, 66) =      7.77
                  Prob > F =      0.0002

. estat hettest

Breusch-Pagan / Cook-Weisberg tests for heteroskedasticity
         Ho: Constant variance
         variables: fitted values of price

         chi2(1)      =     6.50
         Prob > chi2  =   0.0108
```

Testing for heteroskedasticity in the right-hand-side variables is requested by specifying the `rhs` option. By specifying the `mtest(bonferroni)` option, we request that tests be conducted for each of the variables, with a Bonferroni adjustment for the p-values to accommodate our testing multiple hypotheses.

```
. estat hettest, rhs mtest(bonf)

Breusch-Pagan / Cook-Weisberg test for heteroskedasticity
         Ho: Constant variance
```

Variable	chi2	df	p
weight	15.24	1	0.0004 #
1.foreign	6.15	1	0.0525 #
mpg	9.04	1	0.0106 #
foreign# c.mpg 1	6.02	1	0.0566 #
simultaneous	15.60	4	0.0036

 # Bonferroni-adjusted p-values

```
. estat szroeter, rhs mtest(holm)
Szroeter's test for homoskedasticity
     Ho: variance constant
     Ha: variance monotonic in variable
```

Variable	chi2	df	p
weight	17.07	1	0.0001 #
1.foreign	6.15	1	0.0131 #
mpg	11.45	1	0.0021 #
foreign# c.mpg 1	6.17	1	0.0260 #

 # Holm adjusted p-values

Finally, we request the information matrix test, which is a conditional moments test with second-, third-, and fourth-order moment conditions.

```
. estat imtest
Cameron & Trivedi's decomposition of IM-test
```

Source	chi2	df	p
Heteroskedasticity	18.86	10	0.0420
Skewness	11.69	4	0.0198
Kurtosis	2.33	1	0.1273
Total	32.87	15	0.0049

We find evidence for omitted variables, heteroskedasticity, and nonnormal skewness.

So, why bother with the various graphical commands when the tests seem so much easier to interpret? In part, it is a matter of taste: both are designed to uncover the same problem, and both are, in fact, going about it in similar ways. One is based on a formal calculation, whereas the other is based on personal judgment in evaluating a graph. On the other hand, the tests are seeking evidence of specific problems, whereas judgment is more general. The careful analyst will use both.

We performed the omitted-variable test first. Omitted variables are a more serious problem than heteroskedasticity or the violations of higher moment conditions tested by `estat imtest`. If this were not a manual, having found evidence of omitted variables, we would never have run the `estat hettest`, `estat szroeter`, and `estat imtest` commands, at least not until we solved the omitted-variable problem.

◁

❑ Technical note

`estat ovtest` and `estat hettest` both perform two flavors of their respective tests. By default, `estat ovtest` looks for evidence of omitted variables by fitting the original model augmented by $\widehat{y}^2$, $\widehat{y}^3$, and $\widehat{y}^4$, which are the fitted values from the original model. Under the assumption of no misspecification, the coefficients on the powers of the fitted values will be zero. With the `rhs` option, `estat ovtest` instead augments the original model with powers (second through fourth) of the explanatory variables (except for dummy variables).

estat hettest, by default, looks for heteroskedasticity by modeling the variance as a function of the fitted values. If, however, we specify a variable or variables, the variance will be modeled as a function of the specified variables. In our example, if we had, a priori, some reason to suspect heteroskedasticity and that the heteroskedasticity is a function of a car's weight, then using a test that focuses on weight would be more powerful than the more general tests such as White's test or the first term in the Cameron–Trivedi decomposition test.

estat hettest, by default, computes the original Breusch–Pagan/Cook–Weisberg test, which includes the assumption of normally distributed errors. Koenker (1981) derived an $N * R^2$ version of this test that drops the normality assumption. Wooldridge (2009) gives an F-statistic version that does not require the normality assumption.

❑

Variance inflation factors

Problems arise in regression when the predictors are highly correlated. In this situation, there may be a significant change in the regression coefficients if you add or delete an independent variable. The estimated standard errors of the fitted coefficients are inflated, or the estimated coefficients may not be statistically significant even though a statistical relation exists between the dependent and independent variables.

Data analysts rely on these facts to check informally for the presence of multicollinearity. estat vif, another command for use after regress, calculates the variance inflation factors and tolerances for each of the independent variables.

The output shows the variance inflation factors together with their reciprocals. Some analysts compare the reciprocals with a predetermined tolerance. In the comparison, if the reciprocal of the VIF is smaller than the tolerance, the associated predictor variable is removed from the regression model. However, most analysts rely on informal rules of thumb applied to the VIF; see Chatterjee and Hadi (2006). According to these rules, there is evidence of multicollinearity if

1. The largest VIF is greater than 10 (some choose a more conservative threshold value of 30).
2. The mean of all the VIFs is considerably larger than 1.

▷ Example 17: estat vif

We examine a regression model fit using the ubiquitous automobile dataset:

```
. regress price mpg rep78 trunk headroom length turn displ gear_ratio
      Source |       SS       df       MS              Number of obs =      69
-------------+------------------------------           F(  8,    60) =    6.33
       Model |  264102049     8  33012756.2           Prob > F      =  0.0000
    Residual |  312694909    60  5211581.82           R-squared     =  0.4579
-------------+------------------------------           Adj R-squared =  0.3856
       Total |  576796959    68  8482308.22           Root MSE      =  2282.9

------------------------------------------------------------------------------
       price |      Coef.   Std. Err.      t    P>|t|     [95% Conf. Interval]
-------------+----------------------------------------------------------------
         mpg |    -144.84   82.12751    -1.76   0.083    -309.1195    19.43948
       rep78 |   727.5783   337.6107     2.16   0.035     52.25638      1402.9
       trunk |   44.02061    108.141     0.41   0.685    -172.2935    260.3347
    headroom |  -807.0996   435.5802    -1.85   0.069     -1678.39    64.19061
      length |  -8.688914   34.89848    -0.25   0.804    -78.49626    61.11843
        turn |  -177.9064   137.3455    -1.30   0.200    -452.6383    96.82551
displacement |   30.73146   7.576952     4.06   0.000      15.5753    45.88762
  gear_ratio |   1500.119   1110.959     1.35   0.182    -722.1303    3722.368
       _cons |   6691.976   7457.906     0.90   0.373    -8226.057    21610.01
------------------------------------------------------------------------------
```

```
. estat vif

    Variable |       VIF       1/VIF  
-------------+----------------------
      length |      8.22    0.121614
displacement |      6.50    0.153860
        turn |      4.85    0.205997
  gear_ratio |      3.45    0.290068
         mpg |      3.03    0.330171
       trunk |      2.88    0.347444
    headroom |      1.80    0.554917
       rep78 |      1.46    0.686147
-------------+----------------------
    Mean VIF |      4.02
```

The results are mixed. Although we have no VIFs greater than 10, the mean VIF is greater than 1, though not considerably so. We could continue the investigation of collinearity, but given that other authors advise that collinearity is a problem only when VIFs exist that are greater than 30 (contradicting our rule above), we will not do so here.

◁

▷ Example 18: estat vif, with strong evidence of multicollinearity

This example comes from a dataset described in Kutner, Nachtsheim, and Neter (2004, 257) that examines body fat as modeled by caliper measurements on the triceps, midarm, and thigh.

```
. use http://www.stata-press.com/data/r12/bodyfat, clear
(Body Fat)

. regress bodyfat tricep thigh midarm

      Source |       SS       df       MS              Number of obs =      20
-------------+------------------------------           F(  3,    16) =   21.52
       Model |  396.984607     3  132.328202           Prob > F      =  0.0000
    Residual |  98.4049068    16  6.15030667           R-squared     =  0.8014
-------------+------------------------------           Adj R-squared =  0.7641
       Total |  495.389513    19  26.0731323           Root MSE      =    2.48

------------------------------------------------------------------------------
     bodyfat |      Coef.   Std. Err.      t    P>|t|     [95% Conf. Interval]
-------------+----------------------------------------------------------------
     triceps |   4.334085   3.015511     1.44   0.170    -2.058512    10.72668
       thigh |  -2.856842   2.582015    -1.11   0.285    -8.330468    2.616785
      midarm |  -2.186056   1.595499    -1.37   0.190    -5.568362     1.19625
       _cons |   117.0844   99.78238     1.17   0.258    -94.44474    328.6136
------------------------------------------------------------------------------

. estat vif

    Variable |       VIF       1/VIF  
-------------+----------------------
     triceps |    708.84    0.001411
       thigh |    564.34    0.001772
      midarm |    104.61    0.009560
-------------+----------------------
    Mean VIF |    459.26
```

Here we see strong evidence of multicollinearity in our model. More investigation reveals that the measurements on the thigh and the triceps are highly correlated:

```
. corr triceps thigh midarm
(obs=20)

             |  triceps    thigh   midarm
-------------+---------------------------
     triceps |   1.0000
       thigh |   0.9238   1.0000
      midarm |   0.4578   0.0847   1.0000
```

If we remove the predictor `tricep` from the model (because it had the highest VIF), we get

```
. regress bodyfat thigh midarm

      Source |       SS       df       MS              Number of obs =      20
-------------+------------------------------           F(  2,    17) =   29.40
       Model |  384.279748     2  192.139874           Prob > F      =  0.0000
    Residual |  111.109765    17  6.53586854           R-squared     =  0.7757
-------------+------------------------------           Adj R-squared =  0.7493
       Total |  495.389513    19  26.0731323           Root MSE      =  2.5565

------------------------------------------------------------------------------
     bodyfat |      Coef.   Std. Err.      t    P>|t|     [95% Conf. Interval]
-------------+----------------------------------------------------------------
       thigh |   .8508818   .1124482     7.57   0.000     .6136367    1.088127
      midarm |   .0960295   .1613927     0.60   0.560    -.2444792    .4365383
       _cons |  -25.99696    6.99732    -3.72   0.002    -40.76001   -11.2339
------------------------------------------------------------------------------

. estat vif

    Variable |       VIF       1/VIF
-------------+----------------------
      midarm |      1.01    0.992831
       thigh |      1.01    0.992831
-------------+----------------------
    Mean VIF |      1.01
```

Note how the coefficients change and how the estimated standard errors for each of the regression coefficients become much smaller. The calculated value of R^2 for the overall regression for the subset model does not appreciably decline when we remove the correlated predictor. Removing an independent variable from the model is one way to deal with multicollinearity. Other methods include ridge regression, weighted least squares, and restricting the use of the fitted model to data that follow the same pattern of multicollinearity. In economic studies, it is sometimes possible to estimate the regression coefficients from different subsets of the data by using cross-section and time series.

◁

All examples above demonstrated the use of centered VIFs. As pointed out by Belsley (1991), the centered VIFs may fail to discover collinearity involving the constant term. One solution is to use the uncentered VIFs instead. According to the definition of the uncentered VIFs, the constant is viewed as a legitimate explanatory variable in a regression model, which allows one to obtain the VIF value for the constant term.

▷ Example 19: estat vif, with strong evidence of collinearity with the constant term

Consider the extreme example in which one of the regressors is highly correlated with the constant. We simulate the data and examine both centered and uncentered VIF diagnostics after fitted regression model as follows.

```
. use http://www.stata-press.com/data/r12/extreme_collin
. summarize
  (output omitted )
. regress y one x z

      Source |       SS       df       MS              Number of obs =     100
-------------+------------------------------           F(  3,    96) = 2710.27
       Model |  223801.985     3  74600.6617           Prob > F      =  0.0000
    Residual |  2642.42124    96  27.5252213           R-squared     =  0.9883
-------------+------------------------------           Adj R-squared =  0.9880
       Total |  226444.406    99  2287.31723           Root MSE      =  5.2464

------------------------------------------------------------------------------
           y |      Coef.   Std. Err.      t    P>|t|     [95% Conf. Interval]
-------------+----------------------------------------------------------------
         one |  -3.278582    10.5621    -0.31   0.757    -24.24419    17.68702
           x |   2.038696   .0242673    84.01   0.000     1.990526    2.086866
           z |   4.863137   .2681036    18.14   0.000     4.330956    5.395319
       _cons |   9.760075   10.50935     0.93   0.355    -11.10082    30.62097
------------------------------------------------------------------------------

. estat vif

    Variable |       VIF       1/VIF
-------------+----------------------
           z |      1.03    0.968488
           x |      1.03    0.971307
         one |      1.00    0.995425
-------------+----------------------
    Mean VIF |      1.02

. estat vif, uncentered

    Variable |       VIF       1/VIF
-------------+----------------------
         one |    402.94    0.002482
   intercept |    401.26    0.002492
           z |      2.93    0.341609
           x |      1.13    0.888705
-------------+----------------------
    Mean VIF |    202.06
```

According to the values of the centered VIFs (1.03, 1.03, 1.00), no harmful collinearity is detected in the model. However, by the construction of these simulated data, we know that `one` is highly collinear with the constant term. As such, the large values of uncentered VIFs for `one` (402.94) and `intercept` (401.26) reveal high collinearity of the variable `one` with the constant term.

◁

Saved results

`estat hettest` saves the following results for the (multivariate) score test in `r()`:

Scalars

`r(chi2)`	χ^2 test statistic
`r(df)`	#df for the asymptotic χ^2 distribution under H_0
`r(p)`	p-value

`estat hettest, fstat` saves results for the (multivariate) score test in `r()`:

Scalars

`r(F)`	test statistic
`r(df_m)`	#df of the test for the F distribution under H_0
`r(df_r)`	#df of the residuals for the F distribution under H_0
`r(p)`	p-value

`estat hettest` (if `mtest` is specified) and `estat szroeter` save the following in `r()`:

Matrices

`r(mtest)` a matrix of test results, with rows corresponding to the univariate tests

`mtest[.,1]`	χ^2 test statistic
`mtest[.,2]`	#df
`mtest[.,3]`	unadjusted p-value
`mtest[.,4]`	adjusted p-value (if an `mtest()` adjustment method is specified)

Macros

`r(mtmethod)`	adjustment method for p-values

`estat imtest` saves the following in `r()`:

Scalars

`r(chi2_t)`	IM-test statistic (= `r(chi2_h)` + `r(chi2_s)` + `r(chi2_k)`)
`r(df_t)`	df for limiting χ^2 distribution under H_0 (= `r(df_h)` + `r(df_s)` + `r(df_k)`)
`r(chi2_h)`	heteroskedasticity test statistic
`r(df_h)`	df for limiting χ^2 distribution under H_0
`r(chi2_s)`	skewness test statistic
`r(df_s)`	df for limiting χ^2 distribution under H_0
`r(chi2_k)`	kurtosis test statistic
`r(df_k)`	df for limiting χ^2 distribution under H_0
`r(chi2_w)`	White's heteroskedasticity test (if `white` specified)
`r(df_w)`	df for limiting χ^2 distribution under H_0

`estat ovtest` saves the following in `r()`:

Scalars

`r(p)`	two-sided p-value
`r(F)`	F statistic
`r(df)`	degrees of freedom
`r(df_r)`	residual degrees of freedom

Methods and formulas

All regression fit and diagnostic commands are implemented as ado-files.

See Hamilton (2009, chap. 7), Kohler and Kreuter (2009, sec. 8.3), or Baum (2006, chap. 5) for an overview of using Stata to perform regression diagnostics. See Peracchi (2001, chap. 8) for a mathematically rigorous discussion of diagnostics.

Methods and formulas are presented under the following headings:

predict
Special-interest postestimation commands

predict

Assume that you have already fit the regression model

$$\mathbf{y} = \mathbf{X}\mathbf{b} + \mathbf{e}$$

where $\mathbf{X}$ is $n \times k$.

Denote the previously estimated coefficient vector by $\mathbf{b}$ and its estimated variance matrix by $\mathbf{V}$. `predict` works by recalling various aspects of the model, such as $\mathbf{b}$, and combining that information with the data currently in memory. Let $\mathbf{x}_j$ be the jth observation currently in memory, and let s^2 be the mean squared error of the regression.

Let $\mathbf{V} = s^2(\mathbf{X}'\mathbf{X})^{-1}$. Let k be the number of independent variables including the intercept, if any, and let y_j be the observed value of the dependent variable.

The *predicted value* (`xb` option) is defined as $\widehat{y}_j = \mathbf{x}_j\mathbf{b}$.

Let ℓ_j represent a lower bound for an observation j and u_j represent an upper bound. The probability that $y_j|\mathbf{x}_j$ would be observed in the interval (ℓ_j, u_j)—the `pr(`ℓ`,` u`)` option—is

$$P(\ell_j, u_j) = \Pr(\ell_j < \mathbf{x}_j\mathbf{b} + e_j < u_j) = \Phi\left(\frac{u_j - \widehat{y}_j}{s}\right) - \Phi\left(\frac{\ell_j - \widehat{y}_j}{s}\right)$$

where for the `pr(`ℓ`,` u`)`, `e(`ℓ`,` u`)`, and `ystar(`ℓ`,` u`)` options, ℓ_j and u_j can be anywhere in the range $(-\infty, +\infty)$.

The option `e(`ℓ`,` u`)` computes the expected value of $y_j|\mathbf{x}_j$ conditional on $y_j|\mathbf{x}_j$ being in the interval (ℓ_j, u_j), that is, when $y_j|\mathbf{x}_j$ is truncated. It can be expressed as

$$E(\ell_j, u_j) = E(\mathbf{x}_j\mathbf{b} + e_j \mid \ell_j < \mathbf{x}_j\mathbf{b} + e_j < u_j) = \widehat{y}_j - s\frac{\phi\left(\frac{u_j - \widehat{y}_j}{s}\right) - \phi\left(\frac{\ell_j - \widehat{y}_j}{s}\right)}{\Phi\left(\frac{u_j - \widehat{y}_j}{s}\right) - \Phi\left(\frac{\ell_j - \widehat{y}_j}{s}\right)}$$

where ϕ is the normal density and Φ is the cumulative normal.

You can also compute `ystar(`ℓ, u`)`—the expected value of $y_j|\mathbf{x}_j$, where y_j is assumed censored at ℓ_j and u_j:

$$y_j^* = \begin{cases} \ell_j & \text{if } \mathbf{x}_j\mathbf{b} + e_j \le \ell_j \\ \mathbf{x}_j\mathbf{b} + u & \text{if } \ell_j < \mathbf{x}_j\mathbf{b} + e_j < u_j \\ u_j & \text{if } \mathbf{x}_j\mathbf{b} + e_j \ge u_j \end{cases}$$

This computation can be expressed in several ways, but the most intuitive formulation involves a combination of the two statistics just defined:

$$y_j^* = P(-\infty, \ell_j)\ell_j + P(\ell_j, u_j)E(\ell_j, u_j) + P(u_j, +\infty)u_j$$

A diagonal element of the projection matrix (`hat`) or (`leverage`) is given by

$$h_j = \mathbf{x}_j(\mathbf{X}'\mathbf{X})^{-1}\mathbf{x}_j'$$

The *standard error of the prediction* (the `stdp` option) is defined as $s_{p_j} = \sqrt{\mathbf{x}_j\mathbf{V}\mathbf{x}_j'}$ and can also be written as $s_{p_j} = s\sqrt{h_j}$.

The *standard error of the forecast* (stdf) is defined as $s_{f_j} = s\sqrt{1+h_j}$.

The *standard error of the residual* (stdr) is defined as $s_{r_j} = s\sqrt{1-h_j}$.

The *residuals* (residuals) are defined as $\widehat{e}_j = y_j - \widehat{y}_j$.

The *standardized residuals* (rstandard) are defined as $\widehat{e}_{s_j} = \widehat{e}_j / s_{r_j}$.

The *Studentized residuals* (rstudent) are defined as

$$r_j = \frac{\widehat{e}_j}{s_{(j)}\sqrt{1-h_j}}$$

where $s_{(j)}$ represents the root mean squared error with the jth observation removed, which is given by

$$s^2_{(j)} = \frac{s^2(T-k)}{T-k-1} - \frac{\widehat{e}_j^2}{(T-k-1)(1-h_j)}$$

Cook's D (cooksd) is given by

$$D_j = \frac{\widehat{e}^2_{s_j}(s_{p_j}/s_{r_j})^2}{k} = \frac{h_j \widehat{e}_j^2}{ks^2(1-h_j)^2}$$

DFITS (dfits) is given by

$$\text{DFITS}_j = r_j\sqrt{\frac{h_j}{1-h_j}}$$

Welsch distance (welsch) is given by

$$W_j = \frac{r_j\sqrt{h_j(n-1)}}{1-h_j}$$

COVRATIO (covratio) is given by

$$\text{COVRATIO}_j = \frac{1}{1-h_j}\left(\frac{n-k-\widehat{e}_j^2}{n-k-1}\right)^k$$

The DFBETAs (dfbeta) for a particular regressor x_i are given by

$$\text{DFBETA}_j = \frac{r_j u_j}{\sqrt{U^2(1-h_j)}}$$

where u_j are the residuals obtained from a regression of x_i on the remaining x's and $U^2 = \sum_j u_j^2$.

Special-interest postestimation commands

The lvr2plot command plots leverage against the squares of the normalized residuals. The normalized residuals are defined as $\widehat{e}_{n_j} = \widehat{e}_j/(\sum_i \widehat{e}_i^2)^{1/2}$.

The omitted-variable test (Ramsey 1969) reported by estat ovtest fits the regression $y_i = \mathbf{x}_i\mathbf{b} + \mathbf{z}_i\mathbf{t} + u_i$ and then performs a standard F test of $\mathbf{t} = \mathbf{0}$. The default test uses $\mathbf{z}_i = (\widehat{y}_i^2, \widehat{y}_i^3, \widehat{y}_i^4)$. If rhs is specified, $\mathbf{z}_i = (x_{1i}^2, x_{1i}^3, x_{1i}^4, x_{2i}^2, \ldots, x_{mi}^4)$. In either case, the variables are normalized to have minimum 0 and maximum 1 before powers are calculated.

The test for heteroskedasticity (Breusch and Pagan 1979; Cook and Weisberg 1983) models $\mathrm{Var}(e_i) = \sigma^2 \exp(\mathbf{zt})$, where $\mathbf{z}$ is a variable list specified by the user, the list of right-hand-side variables, or the fitted values $\mathbf{x}\widehat{\beta}$. The test is of $\mathbf{t} = \mathbf{0}$. Mechanically, estat hettest fits the augmented regression $\widehat{e}_i^2/\widehat{\sigma}^2 = a + \mathbf{z}_i\mathbf{t} + v_i$.

The original Breusch–Pagan/Cook–Weisberg version of the test assumes that the e_i are normally distributed under the null hypothesis which implies that the score test statistic S is equal to the model sum of squares from the augmented regression divided by 2. Under the null hypothesis, S has the χ^2 distribution with m degrees of freedom, where m is the number of columns of $\mathbf{z}$.

Koenker (1981) derived a score test of the null hypothesis that $\mathbf{t} = \mathbf{0}$ under the assumption that the e_i are independent and identically distributed (i.i.d.). Koenker showed that $S = N * R^2$ has a large-sample χ^2 distribution with m degrees of freedom, where N is the number of observations and R^2 is the R-squared in the augmented regression and m is the number of columns of $\mathbf{z}$. estat hettest, iid produces this version of the test.

Wooldridge (2009) showed that an F test of $\mathbf{t} = \mathbf{0}$ in the augmented regression can also be used under the assumption that the e_i are i.i.d. estat hettest, fstat produces this version of the test.

Szroeter's class of tests for homoskedasticity against the alternative that the residual variance increases in some variable x is defined in terms of

$$H = \frac{\sum_{i=1}^n h(x_i)e_i^2}{\sum_{i=1}^n e_i^2}$$

where $h(x)$ is some weight function that increases in x (Szroeter 1978). H is a weighted average of the $h(x)$, with the squared residuals serving as weights. Under homoskedasticity, H should be approximately equal to the unweighted average of $h(x)$. Large values of H suggest that e_i^2 tends to be large where $h(x)$ is large; that is, the variance indeed increases in x, whereas small values of H suggest that the variance actually decreases in x. estat szroeter uses $h(x_i) = \mathrm{rank}(x_i \text{ in } x_1 \ldots x_n)$; see Judge et al. [1985, 452] for details. estat szroeter displays a normalized version of H,

$$Q = \sqrt{\frac{6n}{n^2-1}}H$$

which is approximately $N(0,1)$ distributed under the null (homoskedasticity).

estat hettest and estat szroeter provide adjustments of p-values for multiple testing. The supported methods are described in [R] **test**.

estat imtest performs the information matrix test for the regression model, as well as an orthogonal decomposition into tests for heteroskedasticity δ_1, nonnormal skewness δ_2, and nonnormal kurtosis δ_3 (Cameron and Trivedi 1990; Long and Trivedi 1992). The decomposition is obtained via three auxiliary regressions. Let e be the regression residuals, $\widehat{\sigma}^2$ be the maximum likelihood estimate of σ^2 in the regression, n be the number of observations, X be the set of k variables specified with estat imtest, and R^2_{un} be the uncentered R^2 from a regression. δ_1 is obtained as nR^2_{un} from a regression of $e^2 - \widehat{\sigma}^2$ on the cross-products of the variables in X. δ_2 is computed as $nR^2_{\rm un}$ from a regression of $e^3 - 3\widehat{\sigma}^2 e$ on X. Finally, δ_3 is obtained as $nR^2_{\rm un}$ from a regression of $e^4 - 6\widehat{\sigma}^2 e^2 - 3\widehat{\sigma}^4$ on X. δ_1, δ_2, and δ_3 are asymptotically χ^2 distributed with $1/2k(k+1)$, K, and 1 degree of freedom. The information test statistic $\delta = \delta_1 + \delta_2 + \delta_3$ is asymptotically χ^2 distributed with $1/2k(k+3)$ degrees of freedom. White's test for heteroskedasticity is computed as nR^2 from a regression of $\widehat{u}^2$ on X and the cross-products of the variables in X. This test statistic is usually close to δ_1.

`estat vif` calculates the centered variance inflation factor (VIF$_c$) (Chatterjee and Hadi 2006, 235–239) for x_j, given by

$$\text{VIF}_c(x_j) = \frac{1}{1 - \widehat{R}^2_j}$$

where $\widehat{R}^2_j$ is the square of the centered multiple correlation coefficient that results when x_j is regressed with intercept against all the other explanatory variables.

The uncentered variance inflation factor (VIF$_{uc}$) (Belsley 1991, 28–29) for x_j is given by

$$\text{VIF}_{uc}(x_j) = \frac{1}{1 - \widetilde{R}^2_j}$$

where $\widetilde{R}^2_j$ is the square of the uncentered multiple correlation coefficient that results when x_j is regressed without intercept against all the other explanatory variables including the constant term.

Acknowledgments

`estat ovtest` and `estat hettest` are based on programs originally written by Richard Goldstein (1991, 1992). `estat imtest`, `estat szroeter`, and the current version of `estat hettest` were written by Jeroen Weesie, Department of Sociology, Utrecht University, The Netherlands; `estat imtest` is based in part on code written by J. Scott Long, Department of Sociology, Indiana University.

References

Adkins, L. C., and R. C. Hill. 2008. *Using Stata for Principles of Econometrics*. 3rd ed. Hoboken, NJ: Wiley.

Baum, C. F. 2006. *An Introduction to Modern Econometrics Using Stata*. College Station, TX: Stata Press.

Baum, C. F., N. J. Cox, and V. L. Wiggins. 2000. sg137: Tests for heteroskedasticity in regression error distribution. *Stata Technical Bulletin* 55: 15–17. Reprinted in *Stata Technical Bulletin Reprints*, vol. 10, pp. 147–149. College Station, TX: Stata Press.

Baum, C. F., and V. L. Wiggins. 2000a. sg135: Test for autoregressive conditional heteroskedasticity in regression error distribution. *Stata Technical Bulletin* 55: 13–14. Reprinted in *Stata Technical Bulletin Reprints*, vol. 10, pp. 143–144. College Station, TX: Stata Press.

——. 2000b. sg136: Tests for serial correlation in regression error distribution. *Stata Technical Bulletin* 55: 14–15. Reprinted in *Stata Technical Bulletin Reprints*, vol. 10, pp. 145–147. College Station, TX: Stata Press.

Belsley, D. A. 1991. *Conditional Diagnostics: Collinearity and Weak Data in Regression*. New York: Wiley.

Belsley, D. A., E. Kuh, and R. E. Welsch. 1980. *Regression Diagnostics: Identifying Influential Data and Sources of Collinearity*. New York: Wiley.

Bollen, K. A., and R. W. Jackman. 1990. Regression diagnostics: An expository treatment of outliers and influential cases. In *Modern Methods of Data Analysis*, ed. J. Fox and J. S. Long, 257–291. Newbury Park, CA: Sage.

Breusch, T. S., and A. R. Pagan. 1979. A simple test for heteroscedasticity and random coefficient variation. *Econometrica* 47: 1287–1294.

Cameron, A. C., and P. K. Trivedi. 1990. The information matrix test and its applied alternative hypotheses. Working paper 372, University of California–Davis, Institute of Governmental Affairs.

——. 2010. *Microeconometrics Using Stata*. Rev. ed. College Station, TX: Stata Press.

Chatterjee, S., and A. S. Hadi. 1986. Influential observations, high leverage points, and outliers in linear regression. *Statistical Science* 1: 379–393.

——. 1988. *Sensitivity Analysis in Linear Regression*. New York: Wiley.

——. 2006. *Regression Analysis by Example*. 4th ed. New York: Wiley.

Cook, R. D. 1977. Detection of influential observation in linear regression. *Technometrics* 19: 15–18.

Cook, R. D., and S. Weisberg. 1982. *Residuals and Influence in Regression*. New York: Chapman & Hall/CRC.

——. 1983. Diagnostics for heteroscedasticity in regression. *Biometrika* 70: 1–10.

Cox, N. J. 2004. Speaking Stata: Graphing model diagnostics. *Stata Journal* 4: 449–475.

DeMaris, A. 2004. *Regression with Social Data: Modeling Continuous and Limited Response Variables*. Hoboken, NJ: Wiley.

Ezekiel, M. 1924. A method of handling curvilinear correlation for any number of variables. *Journal of the American Statistical Association* 19: 431–453.

Garrett, J. M. 2000. sg157: Predicted values calculated from linear or logistic regression models. *Stata Technical Bulletin* 58: 27–30. Reprinted in *Stata Technical Bulletin Reprints*, vol. 10, pp. 258–261. College Station, TX: Stata Press.

Goldstein, R. 1991. srd5: Ramsey test for heteroscedasticity and omitted variables. *Stata Technical Bulletin* 2: 27. Reprinted in *Stata Technical Bulletin Reprints*, vol. 1, p. 177. College Station, TX: Stata Press.

——. 1992. srd14: Cook–Weisberg test of heteroscedasticity. *Stata Technical Bulletin* 10: 27–28. Reprinted in *Stata Technical Bulletin Reprints*, vol. 2, pp. 183–184. College Station, TX: Stata Press.

Hamilton, L. C. 1992. *Regression with Graphics: A Second Course in Applied Statistics*. Belmont, CA: Duxbury.

——. 2009. *Statistics with Stata (Updated for Version 10)*. Belmont, CA: Brooks/Cole.

Hardin, J. W. 1995. sg32: Variance inflation factors and variance-decomposition proportions. *Stata Technical Bulletin* 24: 17–22. Reprinted in *Stata Technical Bulletin Reprints*, vol. 4, pp. 154–160. College Station, TX: Stata Press.

Hill, R. C., W. E. Griffiths, and G. C. Lim. 2011. *Principles of Econometrics*. 4th ed. Hoboken, NJ: Wiley.

Hoaglin, D. C., and P. J. Kempthorne. 1986. Comment [on Chatterjee and Hadi 1986]. *Statistical Science* 1: 408–412.

Hoaglin, D. C., and R. E. Welsch. 1978. The hat matrix in regression and ANOVA. *American Statistician* 32: 17–22.

Judge, G. G., W. E. Griffiths, R. C. Hill, H. Lütkepohl, and T.-C. Lee. 1985. *The Theory and Practice of Econometrics*. 2nd ed. New York: Wiley.

Koenker, R. 1981. A note on studentizing a test for heteroskedasticity. *Journal of Econometrics* 17: 107–112.

Kohler, U., and F. Kreuter. 2009. *Data Analysis Using Stata*. 2nd ed. College Station, TX: Stata Press.

Kutner, M. H., C. J. Nachtsheim, and J. Neter. 2004. *Applied Linear Regression Models*. 4th ed. New York: McGraw–Hill/Irwin.

Larsen, W. A., and S. J. McCleary. 1972. The use of partial residual plots in regression analysis. *Technometrics* 14: 781–790.

Lindsey, C., and S. J. Sheather. 2010a. Optimal power transformation via inverse response plots. *Stata Journal* 10: 200–214.

——. 2010b. Model fit assessment via marginal model plots. *Stata Journal* 10: 215–225.

Long, J. S., and J. Freese. 2000. sg145: Scalar measures of fit for regression models. *Stata Technical Bulletin* 56: 34–40. Reprinted in *Stata Technical Bulletin Reprints*, vol. 10, pp. 197–205. College Station, TX: Stata Press.

Long, J. S., and P. K. Trivedi. 1992. Some specification tests for the linear regression model. *Sociological Methods and Research* 21: 161–204. Reprinted in *Testing Structural Equation Models*, ed. K. A. Bollen and J. S. Long, pp. 66–110. Newbury Park, CA: Sage.

Mallows, C. L. 1986. Augmented partial residuals. *Technometrics* 28: 313–319.

Mosteller, F., and J. W. Tukey. 1977. *Data Analysis and Regression: A Second Course in Statistics*. Reading, MA: Addison–Wesley.

Peracchi, F. 2001. *Econometrics*. Chichester, UK: Wiley.

Ramsey, J. B. 1969. Tests for specification errors in classical linear least-squares regression analysis. *Journal of the Royal Statistical Society, Series B* 31: 350–371.

Ramsey, J. B., and P. Schmidt. 1976. Some further results on the use of OLS and BLUS residuals in specification error tests. *Journal of the American Statistical Association* 71: 389–390.

Rousseeuw, P. J., and A. M. Leroy. 1987. *Robust Regression and Outlier Detection*. New York: Wiley.

Szroeter, J. 1978. A class of parametric tests for heteroscedasticity in linear econometric models. *Econometrica* 46: 1311–1327.

Velleman, P. F. 1986. Comment [on Chatterjee and Hadi 1986]. *Statistical Science* 1: 412–413.

Velleman, P. F., and R. E. Welsch. 1981. Efficient computing of regression diagnostics. *American Statistician* 35: 234–242.

Weesie, J. 2001. sg161: Analysis of the turning point of a quadratic specification. *Stata Technical Bulletin* 60: 18–20. Reprinted in *Stata Technical Bulletin Reprints*, vol. 10, pp. 273–277. College Station, TX: Stata Press.

Weisberg, S. 2005. *Applied Linear Regression*. 3rd ed. New York: Wiley.

Welsch, R. E. 1982. Influence functions and regression diagnostics. In *Modern Data Analysis*, ed. R. L. Launer and A. F. Siegel, 149–169. New York: Academic Press.

——. 1986. Comment [on Chatterjee and Hadi 1986]. *Statistical Science* 1: 403–405.

Welsch, R. E., and E. Kuh. 1977. Linear Regression Diagnostics. Technical Report 923-77, Massachusetts Institute of Technology, Cambridge, MA.

White, H. 1980. A heteroskedasticity-consistent covariance matrix estimator and a direct test for heteroskedasticity. *Econometrica* 48: 817–838.

Wooldridge, J. M. 2009. *Introductory Econometrics: A Modern Approach*. 4th ed. Cincinnati, OH: South-Western.

Also see

[R] **regress** — Linear regression

[R] **regress postestimation time series** — Postestimation tools for regress with time series

[U] **20 Estimation and postestimation commands**

Title

regress postestimation time series — Postestimation tools for regress with time series

Description

The following postestimation commands for time series are available for `regress`:

Command	Description
`estat archlm`	test for ARCH effects in the residuals
`estat bgodfrey`	Breusch–Godfrey test for higher-order serial correlation
`estat durbinalt`	Durbin's alternative test for serial correlation
`estat dwatson`	Durbin–Watson d statistic to test for first-order serial correlation

These commands provide regression diagnostic tools specific to time series. You must `tsset` your data before using these commands; see [TS] **tsset**.

`estat archlm` tests for time-dependent volatility. `estat bgodfrey`, `estat durbinalt`, and `estat dwatson` test for serial correlation in the residuals of a linear regression. For non–time-series regression diagnostic tools, see [R] **regress postestimation**.

`estat archlm` performs Engle's Lagrange multiplier (LM) test for the presence of autoregressive conditional heteroskedasticity.

`estat bgodfrey` performs the Breusch–Godfrey test for higher-order serial correlation in the disturbance. This test does not require that all the regressors be strictly exogenous.

`estat durbinalt` performs Durbin's alternative test for serial correlation in the disturbance. This test does not require that all the regressors be strictly exogenous.

`estat dwatson` computes the Durbin–Watson d statistic (Durbin and Watson 1950) to test for first-order serial correlation in the disturbance when all the regressors are strictly exogenous.

Syntax for estat archlm

`estat archlm` [, *archlm_options*]

archlm_options	Description
`lags(`*numlist*`)`	test *numlist* lag orders
`force`	allow test after `regress, vce(robust)`

Options for estat archlm

`lags(`*numlist*`)` specifies a list of numbers, indicating the lag orders to be tested. The test will be performed separately for each order. The default is order one.

`force` allows the test to be run after `regress, vce(robust)`. The command will not work if the `vce(cluster` *clustvar*`)` option is specified with `regress`; see [R] **regress**.

Syntax for estat bgodfrey

`estat bgodfrey [ , bgodfrey_options ]`

bgodfrey_options	Description
`lags(`*numlist*`)`	test *numlist* lag orders
`nomiss0`	do not use Davidson and MacKinnon's approach
`small`	obtain p-values using the F or t distribution

Options for estat bgodfrey

`lags(`*numlist*`)` specifies a list of numbers, indicating the lag orders to be tested. The test will be performed separately for each order. The default is order one.

`nomiss0` specifies that Davidson and MacKinnon's approach (1993, 358), which replaces the missing values in the initial observations on the lagged residuals in the auxiliary regression with zeros, not be used.

`small` specifies that the p-values of the test statistics be obtained using the F or t distribution instead of the default chi-squared or normal distribution.

Syntax for estat durbinalt

`estat durbinalt [ , durbinalt_options ]`

durbinalt_options	Description
`lags(`*numlist*`)`	test *numlist* lag orders
`nomiss0`	do not use Davidson and MacKinnon's approach
`robust`	compute standard errors using the robust/sandwich estimator
`small`	obtain p-values using the F or t distribution
`force`	allow test after `regress, vce(robust)` or after `newey`

Options for estat durbinalt

`lags(`*numlist*`)` specifies a list of numbers, indicating the lag orders to be tested. The test will be performed separately for each order. The default is order one.

`nomiss0` specifies that Davidson and MacKinnon's approach (1993, 358), which replaces the missing values in the initial observations on the lagged residuals in the auxiliary regression with zeros, not be used.

`robust` specifies that the Huber/White/sandwich robust estimator of the variance–covariance matrix be used in Durbin's alternative test.

`small` specifies that the p-values of the test statistics be obtained using the F or t distribution instead of the default chi-squared or normal distribution. This option may not be specified with `robust`, which always uses an F or t distribution.

`force` allows the test to be run after `regress, vce(robust)` and after `newey` (see [R] **regress** and [TS] **newey**). The command will not work if the `vce(cluster` *clustvar*`)` option is specified with `regress`.

Syntax for estat dwatson

```
estat dwatson
```

Remarks

The Durbin–Watson test is used to determine whether the error term in a linear regression model follows an AR(1) process. For the linear model

$$y_t = \mathbf{x}_t\boldsymbol{\beta} + u_t$$

the AR(1) process can be written as

$$u_t = \rho u_{t-1} + \epsilon_t$$

In general, an AR(1) process requires only that ϵ_t be independent and identically distributed (i.i.d.). The Durbin–Watson test, however, requires ϵ_t to be distributed $N(0, \sigma^2)$ for the statistic to have an exact distribution. Also, the Durbin–Watson test can be applied only when the regressors are strictly exogenous. A regressor x is strictly exogenous if Corr$(x_s, u_t) = 0$ for all s and t, which precludes the use of the Durbin–Watson statistic with models where lagged values of the dependent variable are included as regressors.

The null hypothesis of the test is that there is no first-order autocorrelation. The Durbin–Watson d statistic can take on values between 0 and 4 and under the null d is equal to 2. Values of d less than 2 suggest positive autocorrelation ($\rho > 0$), whereas values of d greater than 2 suggest negative autocorrelation ($\rho < 0$). Calculating the exact distribution of the d statistic is difficult, but empirical upper and lower bounds have been established based on the sample size and the number of regressors. Extended tables for the d statistic have been published by Savin and White (1977). For example, suppose you have a model with 30 observations and three regressors (including the constant term). For a test of the null hypothesis of no autocorrelation versus the alternative of positive autocorrelation, the lower bound of the d statistic is 1.284, and the upper bound is 1.567 at the 5% significance level. You would reject the null if $d < 1.284$, and you would fail to reject if $d > 1.567$. A value falling within the range (1.284, 1.567) leads to no conclusion about whether or not to reject the null hypothesis.

When lagged dependent variables are included among the regressors, the past values of the error term are correlated with those lagged variables at time t, implying that they are not strictly exogenous regressors. The inclusion of covariates that are not strictly exogenous causes the d statistic to be biased toward the acceptance of the null hypothesis. Durbin (1970) suggested an alternative test for models with lagged dependent variables and extended that test to the more general AR(p) serial correlation process

$$u_t = \rho_1 u_{t-1} + \cdots + \rho_p u_{t-p} + \epsilon_t$$

where ϵ_t is i.i.d. with variance σ^2 but is not assumed or required to be normal for the test.

The null hypothesis of Durbin's alternative test is

$$H_0 : \rho_1 = 0, \ldots, \rho_p = 0$$

and the alternative is that at least one of the ρ's is nonzero. Although the null hypothesis was originally derived for an AR(p) process, this test turns out to have power against MA(p) processes as well. Hence, the actual null of this test is that there is no serial correlation up to order p because the MA(p) and the AR(p) models are locally equivalent alternatives under the null. See Godfrey (1988, 113–115) for a discussion of this result.

Durbin's alternative test is in fact a LM test, but it is most easily computed with a Wald test on the coefficients of the lagged residuals in an auxiliary OLS regression of the residuals on their lags and all the covariates in the original regression. Consider the linear regression model

$$y_t = \beta_1 x_{1t} + \cdots + \beta_k x_{kt} + u_t \tag{1}$$

in which the covariates x_1 through x_k are not assumed to be strictly exogenous and u_t is assumed to be i.i.d. and to have finite variance. The process is also assumed to be stationary. (See Wooldridge [2009] for a discussion of stationarity.) Estimating the parameters in (1) by OLS obtains the residuals $\widehat{u}_t$. Next another OLS regression is performed of $\widehat{u}_t$ on $\widehat{u}_{t-1}, \ldots, \widehat{u}_{t-p}$ and the other regressors,

$$\widehat{u}_t = \gamma_1 \widehat{u}_{t-1} + \cdots + \gamma_p \widehat{u}_{t-p} + \beta_1 x_{1t} + \cdots + \beta_k x_{kt} + \epsilon_t \tag{2}$$

where ϵ_t stands for the random-error term in this auxiliary OLS regression. Durbin's alternative test is then obtained by performing a Wald test that $\gamma_1, \ldots, \gamma_p$ are jointly zero. The test can be made robust to an unknown form of heteroskedasticity by using a robust VCE estimator when estimating the regression in (2). When there are only strictly exogenous regressors and $p = 1$, this test is asymptotically equivalent to the Durbin–Watson test.

The Breusch–Godfrey test is also an LM test of the null hypothesis of no autocorrelation versus the alternative that u_t follows an AR(p) or MA(p) process. Like Durbin's alternative test, it is based on the auxiliary regression (2), and it is computed as NR^2, where N is the number of observations and R^2 is the simple R^2 from the regression. This test and Durbin's alternative test are asymptotically equivalent. The test statistic NR^2 has an asymptotic χ^2 distribution with p degrees of freedom. It is valid with or without the strict exogeneity assumption but is not robust to conditional heteroskedasticity, even if a robust VCE is used when fitting (2).

In fitting (2), the values of the lagged residuals will be missing in the initial periods. As noted by Davidson and MacKinnon (1993), the residuals will not be orthogonal to the other covariates in the model in this restricted sample, which implies that the R^2 from the auxiliary regression will not be zero when the lagged residuals are left out. Hence, Breusch and Godfrey's NR^2 version of the test may overreject in small samples. To correct this problem, Davidson and MacKinnon (1993) recommend setting the missing values of the lagged residuals to zero and running the auxiliary regression in (2) over the full sample used in (1). This small-sample correction has become conventional for both the Breusch–Godfrey and Durbin's alternative test, and it is the default for both commands. Specifying the `nomiss0` option overrides this default behavior and treats the initial missing values generated by regressing on the lagged residuals as missing. Hence, `nomiss0` causes these initial observations to be dropped from the sample of the auxiliary regression.

Durbin's alternative test and the Breusch–Godfrey test were originally derived for the case covered by `regress` without the `vce(robust)` option. However, after `regress, vce(robust)` and `newey`, Durbin's alternative test is still valid and can be invoked if the `robust` and `force` options are specified.

▷ Example 1: tests for serial correlation

Using data from Klein (1950), we first fit an OLS regression of consumption on the government wage bill:

```
. use http://www.stata-press.com/data/r12/klein
. tsset yr
        time variable:  yr, 1920 to 1941
                delta:  1 unit
. regress consump wagegovt
      Source |       SS       df       MS              Number of obs =      22
-------------+------------------------------           F(  1,    20) =   17.72
       Model |  532.567711     1  532.567711           Prob > F      =  0.0004
    Residual |  601.207167    20  30.0603584           R-squared     =  0.4697
-------------+------------------------------           Adj R-squared =  0.4432
       Total |  1133.77488    21  53.9892799           Root MSE      =  5.4827

------------------------------------------------------------------------------
     consump |      Coef.   Std. Err.      t    P>|t|     [95% Conf. Interval]
-------------+----------------------------------------------------------------
    wagegovt |    2.50744   .5957173     4.21   0.000     1.264796    3.750085
       _cons |   40.84699   3.192183    12.80   0.000     34.18821    47.50577
------------------------------------------------------------------------------
```

If we assume that wagegov is a strictly exogenous variable, we can use the Durbin–Watson test to check for first-order serial correlation in the errors.

```
. estat dwatson
Durbin-Watson d-statistic(  2,    22) =  .3217998
```

The Durbin–Watson d statistic, 0.32, is far from the center of its distribution ($d = 2.0$). Given 22 observations and two regressors (including the constant term) in the model, the lower 5% bound is about 0.997, much greater than the computed d statistic. Assuming that wagegov is strictly exogenous, we can reject the null of no first-order serial correlation. Rejecting the null hypothesis does not necessarily mean an AR process; other forms of misspecification may also lead to a significant test statistic. If we are willing to assume that the errors follow an AR(1) process and that wagegov is strictly exogenous, we could refit the model using arima or prais and model the error process explicitly; see [TS] **arima** and [TS] **prais**.

If we are not willing to assume that wagegov is strictly exogenous, we could instead use Durbin's alternative test or the Breusch–Godfrey to test for first-order serial correlation. Because we have only 22 observations, we will use the small option.

```
. estat durbinalt, small
Durbin's alternative test for autocorrelation
---------------------------------------------------------------------------
    lags(p)  |          F                 df                   Prob > F
-------------+-------------------------------------------------------------
       1     |       35.035          (  1,    19 )              0.0000
---------------------------------------------------------------------------
                        H0: no serial correlation
. estat bgodfrey, small
Breusch-Godfrey LM test for autocorrelation
---------------------------------------------------------------------------
    lags(p)  |          F                 df                   Prob > F
-------------+-------------------------------------------------------------
       1     |       14.264          (  1,    19 )              0.0013
---------------------------------------------------------------------------
                        H0: no serial correlation
```

Both tests strongly reject the null of no first-order serial correlation, so we decide to refit the model with two lags of `consump` included as regressors and then rerun `estat durbinalt` and `estat bgodfrey`. Because the revised model includes lagged values of the dependent variable, the Durbin–Watson test is not applicable.

```
. regress consump wagegovt L.consump L2.consump
      Source |       SS       df       MS              Number of obs =      20
-------------+------------------------------           F(  3,    16) =   44.01
       Model |  702.660311     3  234.220104           Prob > F      =  0.0000
    Residual |  85.1596011    16  5.32247507           R-squared     =  0.8919
-------------+------------------------------           Adj R-squared =  0.8716
       Total |  787.819912    19  41.4642059           Root MSE      =   2.307

------------------------------------------------------------------------------
     consump |      Coef.   Std. Err.      t    P>|t|     [95% Conf. Interval]
-------------+----------------------------------------------------------------
    wagegovt |   .6904282   .3295485     2.10   0.052    -.0081835     1.38904
             |
     consump |
         L1. |   1.420536    .197024     7.21   0.000     1.002864    1.838208
         L2. |   -.650888   .1933351    -3.37   0.004     -1.06074    -.241036
             |
       _cons |   9.209073   5.006701     1.84   0.084    -1.404659    19.82281
------------------------------------------------------------------------------

. estat durbinalt, small lags(1/2)
Durbin's alternative test for autocorrelation
---------------------------------------------------------------------------
    lags(p)  |          F                  df                 Prob > F
-------------+-------------------------------------------------------------
       1     |        0.080            (  1,    15 )           0.7805
       2     |        0.260            (  2,    14 )           0.7750
---------------------------------------------------------------------------
                        H0: no serial correlation

. estat bgodfrey, small lags(1/2)
Breusch-Godfrey LM test for autocorrelation
---------------------------------------------------------------------------
    lags(p)  |          F                  df                 Prob > F
-------------+-------------------------------------------------------------
       1     |        0.107            (  1,    15 )           0.7484
       2     |        0.358            (  2,    14 )           0.7056
---------------------------------------------------------------------------
                        H0: no serial correlation
```

Although `wagegov` and the constant term are no longer statistically different from zero at the 5% level, the output from `estat durbinalt` and `estat bgodfrey` indicates that including the two lags of `consump` has removed any serial correlation from the errors.

◁

Engle (1982) suggests an LM test for checking for autoregressive conditional heteroskedasticity (ARCH) in the errors. The pth-order ARCH model can be written as

$$\begin{aligned}\sigma_t^2 &= E(u_t^2|u_{t-1},\dots,u_{t-p}) \\ &= \gamma_0 + \gamma_1 u_{t-1}^2 + \dots + \gamma_p u_{t-p}^2\end{aligned}$$

To test the null hypothesis of no autoregressive conditional heteroskedasticity (that is, $\gamma_1 = \dots = \gamma_p = 0$), we first fit the OLS model (1), obtain the residuals $\widehat{u}_t$, and run another OLS regression on the lagged residuals:

$$\widehat{u}_t^2 = \gamma_0 + \gamma_1 \widehat{u}_{t-1}^2 + \dots + \gamma_p \widehat{u}_{t-p}^2 + \epsilon \tag{3}$$

The test statistic is NR^2, where N is the number of observations in the sample and R^2 is the R^2 from the regression in (3). Under the null hypothesis, the test statistic follows a χ^2_p distribution.

▷ Example 2: estat archlm

We refit the original model that does not include the two lags of `consump` and then use `estat archlm` to see if there is any evidence that the errors are autoregressive conditional heteroskedastic.

```
. regress consump wagegovt

      Source |       SS       df       MS              Number of obs =      22
-------------+------------------------------           F(  1,    20) =   17.72
       Model |  532.567711     1  532.567711           Prob > F      =  0.0004
    Residual |  601.207167    20  30.0603584           R-squared     =  0.4697
-------------+------------------------------           Adj R-squared =  0.4432
       Total |  1133.77488    21  53.9892799           Root MSE      =  5.4827

------------------------------------------------------------------------------
     consump |      Coef.   Std. Err.      t    P>|t|     [95% Conf. Interval]
-------------+----------------------------------------------------------------
    wagegovt |    2.50744   .5957173     4.21   0.000     1.264796    3.750085
       _cons |   40.84699   3.192183    12.80   0.000     34.18821    47.50577
------------------------------------------------------------------------------

. estat archlm, lags(1 2 3)
LM test for autoregressive conditional heteroskedasticity (ARCH)
---------------------------------------------------------------------------
    lags(p)  |          chi2               df                 Prob > chi2
-------------+-------------------------------------------------------------
       1     |          5.543               1                   0.0186
       2     |          9.431               2                   0.0090
       3     |          9.039               3                   0.0288
---------------------------------------------------------------------------
         H0: no ARCH effects        vs.  H1: ARCH(p) disturbance
```

`estat archlm` shows the results for tests of ARCH(1), ARCH(2), and ARCH(3) effects, respectively. At the 5% significance level, all three tests reject the null hypothesis that the errors are not autoregressive conditional heteroskedastic. See [TS] **arch** for information on fitting ARCH models.

◁

Saved results

`estat archlm` saves the following in `r()`:

Scalars

`r(N)`	number of observations	`r(N_gaps)`	number of gaps
`r(k)`	number of regressors		

Macros

`r(lags)`	lag order

Matrices

`r(arch)`	test statistic for each lag order	`r(p)`	two-sided p-values
`r(df)`	degrees of freedom		

`estat bgodfrey` saves the following in `r()`:

Scalars

`r(N)`	number of observations	`r(N_gaps)`	number of gaps
`r(k)`	number of regressors		

Macros

`r(lags)`	lag order

Matrices

`r(chi2)`	χ^2 statistic for each lag order	`r(p)`	two-sided p-values
`r(F)`	F statistic for each lag order (`small` only)	`r(df)`	degrees of freedom
`r(df_r)`	residual degrees of freedom (`small` only)		

`estat durbinalt` saves the following in `r()`:

Scalars

`r(N)`	number of observations	`r(N_gaps)`	number of gaps
`r(k)`	number of regressors		

Macros

`r(lags)`	lag order

Matrices

`r(chi2)`	χ^2 statistic for each lag order	`r(p)`	two-sided p-values
`r(F)`	F statistic for each lag order (`small` only)	`r(df)`	degrees of freedom
`r(df_r)`	residual degrees of freedom (`small` only)		

`estat dwatson` saves the following in `r()`:

Scalars

`r(N)`	number of observations	`r(N_gaps)`	number of gaps
`r(k)`	number of regressors	`r(dw)`	Durbin–Watson statistic

Methods and formulas

`estat archlm`, `estat bgodfrey`, `estat durbinalt`, and `estat dwatson` are implemented as ado-files.

Consider the regression

$$y_t = \beta_1 x_{1t} + \cdots + \beta_k x_{kt} + u_t \tag{4}$$

in which some of the covariates are not strictly exogenous. In particular, some of the x_{it} may be lags of the dependent variable. We are interested in whether the u_t are serially correlated.

The Durbin–Watson d statistic reported by `estat dwatson` is

$$d = \frac{\sum_{t=1}^{n-1} (\widehat{u}_{t+1} - \widehat{u}_t)^2}{\sum_{t=1}^{n} \widehat{u}_t^2}$$

where $\widehat{u}_t$ represents the residual of the tth observation.

To compute Durbin's alternative test and the Breusch–Godfrey test against the null hypothesis that there is no pth order serial correlation, we fit the regression in (4), compute the residuals, and then fit the following auxiliary regression of the residuals $\widehat{u}_t$ on p lags of $\widehat{u}_t$ and on all the covariates in the original regression in (4):

$$\widehat{u}_t = \gamma_1 \widehat{u}_{t-1} + \cdots + \gamma_p \widehat{u}_{t-p} + \beta_1 x_{1t} + \cdots + \beta_k x_{kt} + \epsilon \tag{5}$$

Durbin's alternative test is computed by performing a Wald test to determine whether the coefficients of $\widehat{u}_{t-1},\ldots,\widehat{u}_{t-p}$ are jointly different from zero. By default, the statistic is assumed to be distributed χ^2(p). When `small` is specified, the statistic is assumed to follow an $F(p,\ N-p-k)$ distribution. The reported p-value is a two-sided p-value. When `robust` is specified, the Wald test is performed using the Huber/White/sandwich estimator of the variance–covariance matrix, and the test is robust to an unspecified form of heteroskedasticity.

The Breusch–Godfrey test is computed as NR^2, where N is the number of observations in the auxiliary regression (5) and R^2 is the R^2 from the same regression (5). Like Durbin's alternative test, the Breusch–Godfrey test is asymptotically distributed $\chi^2(p)$, but specifying `small` causes the p-value to be computed using an $F(p,N-p-k)$.

By default, the initial missing values of the lagged residuals are replaced with zeros, and the auxiliary regression is run over the full sample used in the original regression of (4). Specifying the `nomiss0` option causes these missing values to be treated as missing values, and the observations are dropped from the sample.

Engle's LM test for ARCH(p) effects fits an OLS regression of $\widehat{u}_t^2$ on $\widehat{u}_{t-1}^2,\ldots,\widehat{u}_{t-p}^2$:

$$\widehat{u}_t^2=\gamma_0+\gamma_1\widehat{u}_{t-1}^2+\cdots+\gamma_p\widehat{u}_{t-p}^2+\epsilon$$

The test statistic is nR^2 and is asymptotically distributed $\chi^2(p)$.

Acknowledgment

The original versions of `estat archlm`, `estat bgodfrey`, and `estat durbinalt` were written by Christopher F. Baum, Boston College.

References

Baum, C. F. 2006. *An Introduction to Modern Econometrics Using Stata*. College Station, TX: Stata Press.

Baum, C. F., and V. L. Wiggins. 2000a. sg135: Test for autoregressive conditional heteroskedasticity in regression error distribution. *Stata Technical Bulletin* 55: 13–14. Reprinted in *Stata Technical Bulletin Reprints*, vol. 10, pp. 143–144. College Station, TX: Stata Press.

——. 2000b. sg136: Tests for serial correlation in regression error distribution. *Stata Technical Bulletin* 55: 14–15. Reprinted in *Stata Technical Bulletin Reprints*, vol. 10, pp. 145–147. College Station, TX: Stata Press.

Beran, R. J., and N. I. Fisher. 1998. A conversation with Geoff Watson. *Statistical Science* 13: 75–93.

Breusch, T. S. 1978. Testing for autocorrelation in dynamic linear models. *Australian Economic Papers* 17: 334–355.

Davidson, R., and J. G. MacKinnon. 1993. *Estimation and Inference in Econometrics*. New York: Oxford University Press.

Durbin, J. 1970. Testing for serial correlation in least-squares regressions when some of the regressors are lagged dependent variables. *Econometrica* 38: 410–421.

Durbin, J., and G. S. Watson. 1950. Testing for serial correlation in least squares regression. I. *Biometrika* 37: 409–428.

——. 1951. Testing for serial correlation in least squares regression. II. *Biometrika* 38: 159–177.

Engle, R. F. 1982. Autoregressive conditional heteroscedasticity with estimates of the variance of United Kingdom inflation. *Econometrica* 50: 987–1007.

Fisher, N. I., and P. Hall. 1998. Geoffrey Stuart Watson: Tributes and obituary (3 December 1921–3 January 1998). *Australian and New Zealand Journal of Statistics* 40: 257–267.

Godfrey, L. G. 1978. Testing against general autoregressive and moving average error models when the regressors include lagged dependent variables. *Econometrics* 46: 1293–1301.

——. 1988. *Misspecification Tests in Econometrics: The Lagrange Multiplier Principle and Other Approaches*. Econometric Society Monographs, No. 16. Cambridge: Cambridge University Press.

Greene, W. H. 2012. *Econometric Analysis*. 7th ed. Upper Saddle River, NJ: Prentice Hall.

Klein, L. R. 1950. *Economic Fluctuations in the United States 1921-1941*. New York: Wiley.

Phillips, P. C. B. 1988. The *ET* Interview: Professor James Durbin. *Econometric Theory* 4: 125–157.

Savin, N. E., and K. J. White. 1977. The Durbin–Watson test for serial correlation with extreme sample sizes or many regressors. *Econometrica* 45: 1989–1996.

Wooldridge, J. M. 2009. *Introductory Econometrics: A Modern Approach*. 4th ed. Cincinnati, OH: South-Western.

James Durbin (1923–) is a British statistician who was born in Wigan, near Manchester. He studied mathematics at Cambridge and after military service and various research posts joined the London School of Economics in 1950. His many contributions to statistics have centered on serial correlation, time series, sample survey methodology, goodness-of-fit tests, and sample distribution functions, with emphasis on applications in the social sciences.

Geoffrey Stuart Watson (1921–1998) was born in Victoria, Australia, and earned degrees at Melbourne University and North Carolina State University. After a visit to the University of Cambridge, he returned to Australia, working at Melbourne and then the Australian National University. Following periods at Toronto and Johns Hopkins, he settled at Princeton. Throughout his wide-ranging career, he made many notable accomplishments and important contributions, including the Durbin–Watson test for serial correlation, the Nadaraya–Watson estimator in nonparametric regression, and methods for analyzing directional data.

Leslie G. Godfrey (1946–) was born in London and earned degrees at the Universities of Exeter and London. He is now a professor of econometrics at the University of York. His interests center on implementation and interpretation of tests of econometric models, including nonnested models.

Trevor Stanley Breusch (1949–) was born in Queensland and earned degrees at the University of Queensland and Australian National University (ANU). After a post at the University of Southampton, he returned to work at ANU. His background is in econometric methods and his recent interests include political values and social attitudes, earnings and income, and measurement of underground economic activity.

Also see

[R] **regress** — Linear regression

[TS] **tsset** — Declare data to be time-series data

[R] **regress postestimation** — Postestimation tools for regress

Title

#review — Review previous commands

Syntax

#review [$\#_1$ [$\#_2$]]

Description

The #review command displays the last few lines typed at the terminal.

Remarks

#review (pronounced *pound-review*) is a Stata preprocessor command. #*commands* do not generate a return code or generate ordinary Stata errors. The only error message associated with #*commands* is "unrecognized #command".

The #review command displays the last few lines typed at the terminal. If no arguments follow #review, the last five lines typed at the terminal are displayed. The first argument specifies the number of lines to be reviewed, so #review 10 displays the last 10 lines typed. The second argument specifies the number of lines to be displayed, so #review 10 5 displays five lines, starting at the 10th previous line.

Stata reserves a buffer for #review lines and stores as many previous lines in the buffer as will fit, rolling out the oldest line to make room for the newest. Requests to #review lines no longer stored will be ignored. Only lines typed at the terminal are placed in the #review buffer. See [U] **10.5 Editing previous lines in Stata**.

▷ Example 1

Typing #review by itself will show the last five lines you typed at the terminal:

```
. #review
5 use mydata
4 * comments go into the #review buffer, too
3 describe
2 tabulate marriage educ [freq=number]
1 tabulate marriage educ [freq=number], chi2
. _
```

Typing #review 15 2 shows the 15th and 14th previous lines:

```
. #review 15 2
15 replace x=. if x<200
14 summarize x
. _
```

◁

Title

roc — Receiver operating characteristic (ROC) analysis

Description

ROC analysis quantifies the accuracy of diagnostic tests or other evaluation modalities used to discriminate between two states or conditions, which are here referred to as normal and abnormal or control and case. The discriminatory accuracy of a diagnostic test is measured by its ability to correctly classify known normal and abnormal subjects. For this reason, we often refer to the diagnostic test as a classifier. The analysis uses the ROC curve, a graph of the sensitivity versus 1 − specificity of the diagnostic test. The sensitivity is the fraction of positive cases that are correctly classified by the diagnostic test, whereas the specificity is the fraction of negative cases that are correctly classified. Thus the sensitivity is the true-positive rate, and the specificity is the true-negative rate.

There are six ROC commands:

Command	Entry	Description
`roccomp`	[R] **roccomp**	Tests of equality of ROC areas
`rocgold`	[R] **roccomp**	Tests of equality of ROC areas against a standard ROC curve
`rocfit`	[R] **rocfit**	Parametric ROC models
`rocreg`	[R] **rocreg**	Nonparametric and parametric ROC regression models
`rocregplot`	[R] **rocregplot**	Plot marginal and covariate-specific ROC curves
`roctab`	[R] **roctab**	Nonparametric ROC analysis

Postestimation commands are available after `rocfit` and `rocreg`; see [R] **rocfit postestimation** and [R] **rocreg postestimation**.

Both nonparametric and parametric (semiparametric) methods have been suggested for generating the ROC curve. The `roctab` command performs nonparametric ROC analysis for a single classifier. `roccomp` extends the nonparametric ROC analysis function of `roctab` to situations where we have multiple diagnostic tests of interest to be compared and tested. The `rocgold` command also provides ROC analysis for multiple classifiers. `rocgold` compares each classifier's ROC curve to a "gold standard" ROC curve and makes adjustments for multiple comparisons in the analysis. Both `rocgold` and `roccomp` also allow parametric estimation of the ROC curve through a binormal fit. In a binormal fit, both the control and the case populations are normal.

The `rocfit` command also estimates the ROC curve of a classifier through a binormal fit. Unlike `roctab`, `roccomp`, and `rocgold`, `rocfit` is an estimation command. In postestimation, graphs of the ROC curve and confidence bands can be produced. Additional tests on the parameters can also be conducted.

ROC analysis can be interpreted as a two-stage process. First, the control distribution of the classifier is estimated, assuming a normal model or using a distribution-free estimation technique. The classifier is standardized using the control distribution to 1 − percentile value, the false-positive rate. Second, the ROC curve is estimated as the case distribution of the standardized classifier values.

Covariates may affect both stages of ROC analysis. The first stage may be affected, yielding a covariate-adjusted ROC curve. The second stage may also be affected, producing multiple covariate-specific ROC curves.

The `rocreg` command performs ROC analysis under both types of covariate effects. Both parametric (semiparametric) and nonparametric methods may be used by `rocreg`. Like `rocfit`, `rocreg` is an estimation command and provides many postestimation capabilities.

The global performance of a diagnostic test is commonly summarized by the area under the ROC curve (AUC). This area can be interpreted as the probability that the result of a diagnostic test of a randomly selected abnormal subject will be greater than the result of the same diagnostic test from a randomly selected normal subject. The greater the AUC, the better the global performance of the diagnostic test. Each of the ROC commands provides computation of the AUC.

Citing a lack of clinical relevance for the AUC, other ROC summary measures have been suggested. These include the partial area under the ROC curve for a given false-positive rate t [pAUC(t)]. This is the area under the ROC curve from the false-positive rate of 0 to t. The ROC value at a particular false-positive rate and the false-positive rate for a particular ROC value are also useful summary measures for the ROC curve. These three measures are directly estimated by `rocreg` during the model fit or postestimation stages. Point estimates of ROC value are computed by the other ROC commands, but no standard errors are reported.

See Pepe (2003) for a discussion of ROC analysis. Pepe has posted Stata datasets and programs used to reproduce results presented in the book (http://www.stata.com/bookstore/pepe.html).

Reference

Pepe, M. S. 2003. *The Statistical Evaluation of Medical Tests for Classification and Prediction.* New York: Oxford University Press.

Title

roccomp — Tests of equality of ROC areas

Syntax

Test equality of ROC areas

roccomp *refvar classvar* [*classvars*] [*if*] [*in*] [*weight*] [, *roccomp_options*]

Test equality of ROC area against a standard ROC curve

rocgold *refvar goldvar classvar* [*classvars*] [*if*] [*in*] [*weight*] [, *rocgold_options*]

roccomp_options	Description
Main	
by(*varname*)	split into groups by variable
test(*matname*)	use contrast matrix for comparing ROC areas
graph	graph the ROC curve
norefline	suppress plotting the 45-degree reference line
separate	place each ROC curve on its own graph
summary	report the area under the ROC curve
binormal	estimate areas by using binormal distribution assumption
line#opts(*cline_options*)	affect rendition of the #th binormal fit line
level(#)	set confidence level; default is level(95)
Plot	
plot#opts(*plot_options*)	affect rendition of the #th ROC curve
Reference line	
rlopts(*cline_options*)	affect rendition of the reference line
Y axis, X axis, Titles, Legend, Overall	
twoway_options	any options other than by() documented in [G-3] ***twoway_options***

fweights are allowed; see [U] **11.1.6 weight**.

rocgold_options	Description
Main	
sidak	adjust the significance probability by using Šidák's method
test(*matname*)	use contrast matrix for comparing ROC areas
graph	graph the ROC curve
norefline	suppress plotting the 45-degree reference line
separate	place each ROC curve on its own graph
summary	report the area under the ROC curve
binormal	estimate areas by using binormal distribution assumption
line#opts(*cline_options*)	affect rendition of the #th binormal fit line
level(#)	set confidence level; default is level(95)
Plot	
plot#opts(*plot_options*)	affect rendition of the #th ROC curve; plot 1 is the "gold standard"
Reference line	
rlopts(*cline_options*)	affect rendition of the reference line
Y axis, X axis, Titles, Legend, Overall	
twoway_options	any options other than by() documented in [G-3] ***twoway_options***

fweights are allowed; see [U] **11.1.6 weight**.

plot_options	Description
marker_options	change look of markers (color, size, etc.)
marker_label_options	add marker labels; change look or position
cline_options	change the look of the line

Menu

roccomp

Statistics > Epidemiology and related > ROC analysis > Test equality of two or more ROC areas

rocgold

Statistics > Epidemiology and related > ROC analysis > Test equality of ROC area against gold standard

Description

The above commands are used to perform receiver operating characteristic (ROC) analyses with rating and discrete classification data.

The two variables *refvar* and *classvar* must be numeric. The reference variable indicates the true state of the observation, such as diseased and nondiseased or normal and abnormal, and must be coded as 0 and 1. The rating or outcome of the diagnostic test or test modality is recorded in *classvar*, which must be at least ordinal, with higher values indicating higher risk.

roccomp tests the equality of two or more ROC areas obtained from applying two or more test modalities to the same sample or to independent samples. roccomp expects the data to be in wide form when comparing areas estimated from the same sample and in long form for areas estimated from independent samples.

rocgold independently tests the equality of the ROC area of each of several test modalities, specified by *classvar*, against a "gold standard" ROC curve, *goldvar*. For each comparison, rocgold reports the raw and the Bonferroni-adjusted significance probability. Optionally, Šidák's adjustment for multiple comparisons can be obtained.

See [R] **rocfit** and [R] **rocreg** for commands that fit maximum-likelihood ROC models.

Options

Main

by(*varname*) (roccomp only) is required when comparing independent ROC areas. The by() variable identifies the groups to be compared.

sidak (rocgold only) requests that the significance probability be adjusted for the effect of multiple comparisons by using Šidák's method. Bonferroni's adjustment is reported by default.

test(*matname*) specifies the contrast matrix to be used when comparing ROC areas. By default, the null hypothesis that all areas are equal is tested.

graph produces graphical output of the ROC curve.

norefline suppresses plotting the 45-degree reference line from the graphical output of the ROC curve.

separate is meaningful only with roccomp and specifies that each ROC curve be placed on its own graph rather than one curve on top of the other.

summary reports the area under the ROC curve, its standard error, and its confidence interval. This option is needed only when also specifying graph.

binormal specifies that the areas under the ROC curves to be compared should be estimated using the binormal distribution assumption. By default, areas to be compared are computed using the trapezoidal rule.

line#opts(*cline_options*) affects the rendition of the line representing the #th ROC curve drawn using the binormal distribution assumption; see [G-3] ***cline_options***. These lines are drawn only if the binormal option is specified.

level(#) specifies the confidence level, as a percentage, for the confidence intervals. The default is level(95) or as set by set level; see [R] **level**.

Plot

plot#opts(*plot_options*) affects the rendition of the #th ROC curve—the curve's plotted points connected by lines. The *plot_options* can affect the size and color of markers, whether and how the markers are labeled, and whether and how the points are connected; see [G-3] ***marker_options***, [G-3] ***marker_label_options***, and [G-3] ***cline_options***.

For rocgold, plot1opts() are applied to the ROC for the gold standard.

Reference line

rlopts(*cline_options*) affects the rendition of the reference line; see [G-3] ***cline_options***.

Y axis, X axis, Titles, Legend, Overall

twoway_options are any of the options documented in [G-3] ***twoway_options***. These include options for titling the graph (see [G-3] ***title_options***), options for saving the graph to disk (see [G-3] ***saving_option***), and the `by()` option (see [G-3] ***by_option***).

Remarks

Remarks are presented under the following headings:

Introduction
Comparing areas under the ROC curve
Correlated data
Independent data
Comparing areas with a gold standard

Introduction

`roccomp` provides comparison of the ROC curves of multiple classifiers. `rocgold` compares the ROC curves of multiple classifiers with a single "gold standard" classifier. Adjustment of inference for multiple comparisons is also provided by `rocgold`.

See Pepe (2003) for a discussion of ROC analysis. Pepe has posted Stata datasets and programs used to reproduce results presented in the book (http://www.stata.com/bookstore/pepe.html).

Comparing areas under the ROC curve

The area under multiple ROC curves can be compared by using `roccomp`. The command syntax is slightly different if the ROC curves are correlated (that is, different diagnostic tests are applied to the same sample) or independent (that is, diagnostic tests are applied to different samples).

Correlated data

▷ Example 1

Hanley and McNeil (1983) presented data from an evaluation of two computer algorithms designed to reconstruct CT images from phantoms. We will call these two algorithms' modalities 1 and 2. A sample of 112 phantoms was selected; 58 phantoms were considered normal, and the remaining 54 were abnormal. Each of the two modalities was applied to each phantom, and the resulting images were rated by a reviewer using a six-point scale: 1 = definitely normal, 2 = probably normal, 3 = possibly normal, 4 = possibly abnormal, 5 = probably abnormal, and 6 = definitely abnormal. Because each modality was applied to the same sample of phantoms, the two sets of outcomes are correlated.

We list the first 7 observations:

```
. use http://www.stata-press.com/data/r12/ct
. list in 1/7, sep(0)
```

	mod1	mod2	status
1.	2	1	0
2.	5	5	1
3.	2	1	0
4.	2	3	0
5.	5	6	1
6.	2	2	0
7.	3	2	0

The data are in wide form, which is required when dealing with correlated data. Each observation corresponds to one phantom. The variable `mod1` identifies the rating assigned for the first modality, and `mod2` identifies the rating assigned for the second modality. The true status of the phantoms is given by `status=0` if they are normal and `status=1` if they are abnormal. The observations with at least one missing rating were dropped from the analysis.

We plot the two ROC curves and compare their areas.

```
. roccomp status mod1 mod2, graph summary
```

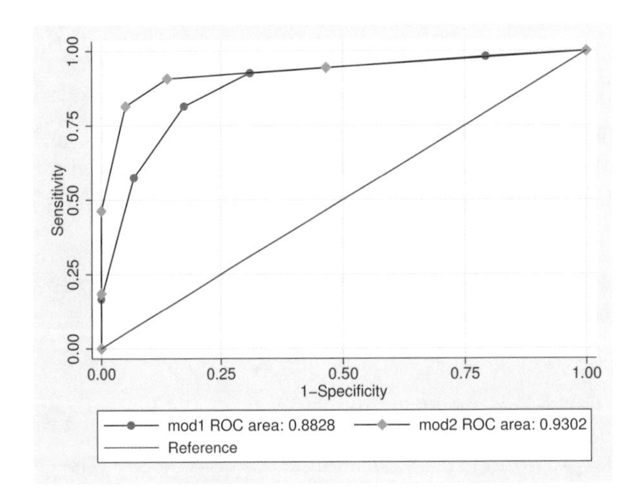

	Obs	ROC Area	Std. Err.	-Asymptotic Normal-- [95% Conf.	Interval]
mod1	112	0.8828	0.0317	0.82067	0.94498
mod2	112	0.9302	0.0256	0.88005	0.98042

```
Ho: area(mod1) = area(mod2)
    chi2(1) =     2.31       Prob>chi2 =    0.1282
```

By default, `roccomp`, with the `graph` option specified, plots the ROC curves on the same graph. Optionally the curves can be plotted side by side, each on its own graph, by also specifying `separate`.

For each curve, `roccomp` reports summary statistics and provides a test for the equality of the area under the curves, using an algorithm suggested by DeLong, DeLong, and Clarke-Pearson (1988).

Although the area under the ROC curve for modality 2 is larger than that of modality 1, the chi-squared test yielded a significance probability of 0.1282, suggesting that there is no significant difference between these two areas.

The `roccomp` command can also be used to compare more than two ROC areas. To illustrate this, we modified the previous dataset by including a fictitious third modality.

```
. use http://www.stata-press.com/data/r12/ct2
. roccomp status mod1 mod2 mod3, graph summary
```

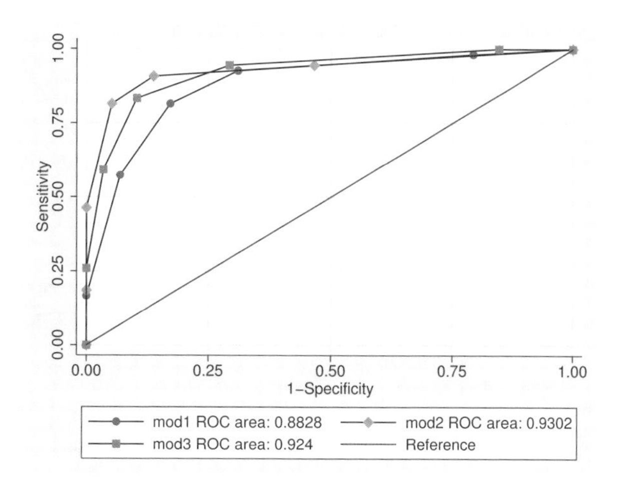


```
                               ROC                    -Asymptotic Normal--
                   Obs       Area     Std. Err.      [95% Conf. Interval]
-------------------------------------------------------------------------
mod1               112     0.8828       0.0317        0.82067     0.94498
mod2               112     0.9302       0.0256        0.88005     0.98042
mod3               112     0.9240       0.0241        0.87670     0.97132
-------------------------------------------------------------------------
Ho: area(mod1) = area(mod2) = area(mod3)
    chi2(2) =     6.54       Prob>chi2 =    0.0381
```

By default, `roccomp` tests whether the areas under the ROC curves are all equal. Other comparisons can be tested by creating a contrast matrix and specifying `test(`*matname*`)`, where *matname* is the name of the contrast matrix.

For example, assume that we are interested in testing whether the area under the ROC for `mod1` is equal to that of `mod3`. To do this, we can first create an appropriate contrast matrix and then specify its name with the `test()` option.

Of course, this is a trivial example because we could have just specified

```
. roccomp status mod1 mod3
```

without including `mod2` to obtain the same test results. However, for illustration, we will continue with this example.

The contrast matrix must have its number of columns equal to the number of *classvars* (that is, the total number of ROC curves) and a number of rows less than or equal to the number of *classvars*, and the elements of each row must add to zero.

```
. matrix C=(1,0,-1)
. roccomp status mod1 mod2 mod3, test(C)
                              ROC                    -Asymptotic Normal--
                  Obs        Area     Std. Err.      [95% Conf. Interval]
-------------------------------------------------------------------------
mod1              112      0.8828        0.0317       0.82067     0.94498
mod2              112      0.9302        0.0256       0.88005     0.98042
mod3              112      0.9240        0.0241       0.87670     0.97132
-------------------------------------------------------------------------
Ho: Comparison as defined by contrast matrix: C
    chi2(1) =      5.25       Prob>chi2 =    0.0220
```

Although all three areas are reported, the comparison is made using the specified contrast matrix.

Perhaps more interesting would be a comparison of the area from mod1 and the average area of mod2 and mod3.

```
. matrix C=(1,-.5,-.5)
. roccomp status mod1 mod2 mod3, test(C)
                              ROC                    -Asymptotic Normal--
                  Obs        Area     Std. Err.      [95% Conf. Interval]
-------------------------------------------------------------------------
mod1              112      0.8828        0.0317       0.82067     0.94498
mod2              112      0.9302        0.0256       0.88005     0.98042
mod3              112      0.9240        0.0241       0.87670     0.97132
-------------------------------------------------------------------------
Ho: Comparison as defined by contrast matrix: C
    chi2(1) =      3.43       Prob>chi2 =    0.0642
```

Other contrasts could be made. For example, we could test if mod3 is different from at least one of the other two by first creating the following contrast matrix:

```
. matrix C=(-1, 0, 1 \ 0, -1, 1)
. matrix list C
C[2,3]
    c1  c2  c3
r1  -1   0   1
r2   0  -1   1
```

◁

Independent data

▷ Example 2

In example 3, we noted that because each test modality was applied to the same sample of phantoms, the classification outcomes were correlated. Now assume that we have collected the same data presented by Hanley and McNeil (1983), except that we applied the first test modality to one sample of phantoms and the second test modality to a different sample of phantoms. The resulting measurements are now considered independent.

Here are a few of the observations.

```
. use http://www.stata-press.com/data/r12/ct3
. list in 1/7, sep(0)
```

	pop	status	rating	mod
1.	12	0	1	1
2.	31	0	1	2
3.	1	1	1	1
4.	3	1	1	2
5.	28	0	2	1
6.	19	0	2	2
7.	3	1	2	1

The data are in long form, which is required when dealing with independent data. The data consist of 24 observations: 6 observations corresponding to abnormal phantoms and 6 to normal phantoms evaluated using the first modality, and similarly 6 observations corresponding to abnormal phantoms and 6 to normal phantoms evaluated using the second modality. The number of phantoms corresponding to each observation is given by the `pop` variable. Once again we have frequency-weighted data. The variable `mod` identifies the modality, and `rating` is the assigned classification.

We can better view our data by using the `table` command.

```
. table status rating [fw=pop], by(mod) row col
```

mod and status	rating						
	1	2	3	4	5	6	Total
1							
0	12	28	8	6	4		58
1	1	3	6	13	22	9	54
Total	13	31	14	19	26	9	112
2							
0	31	19	5	3			58
1	3	2	5	19	15	10	54
Total	34	21	10	22	15	10	112

The `status` variable indicates the true status of the phantoms: `status` = 0 if they are normal and `status` = 1 if they are abnormal.

We now compare the areas under the two ROC curves.

```
. roccomp status rating [fw=pop], by(mod) graph summary
```

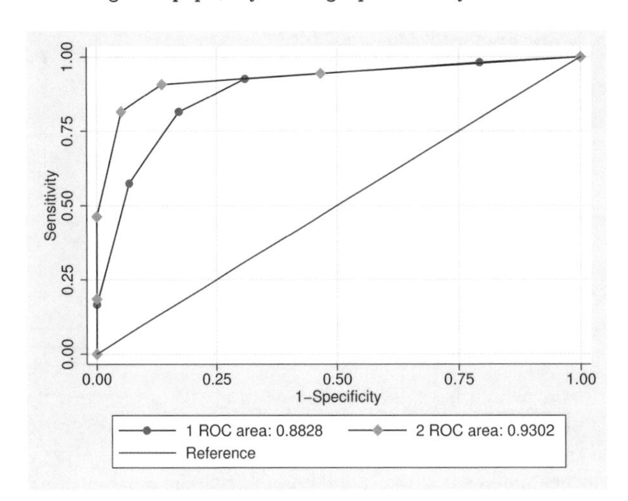


```
                              ROC                    -Asymptotic Normal--
mod                 Obs       Area     Std. Err.      [95% Conf. Interval]
-------------------------------------------------------------------------
1                   112     0.8828       0.0317        0.82067     0.94498
2                   112     0.9302       0.0256        0.88005     0.98042
-------------------------------------------------------------------------
Ho: area(1) = area(2)
    chi2(1) =     1.35       Prob>chi2 =    0.2447
```

◁

Comparing areas with a gold standard

The area under multiple ROC curves can be compared with a gold standard using `rocgold`. The command syntax is similar to that of `roccomp`. The tests are corrected for the effect of multiple comparisons.

▷ Example 3

We will use the same data (presented by Hanley and McNeil [1983]) as in the `roccomp` examples. Let's assume that the first modality is considered to be the standard against which both the second and third modalities are compared.

We want to plot and compare both the areas of the ROC curves of `mod2` and `mod3` with `mod1`. Because we consider `mod1` to be the gold standard, it is listed first after the reference variable in the `rocgold` command line.

```
. use http://www.stata-press.com/data/r12/ct2
. rocgold status mod1 mod2 mod3, graph summary
```

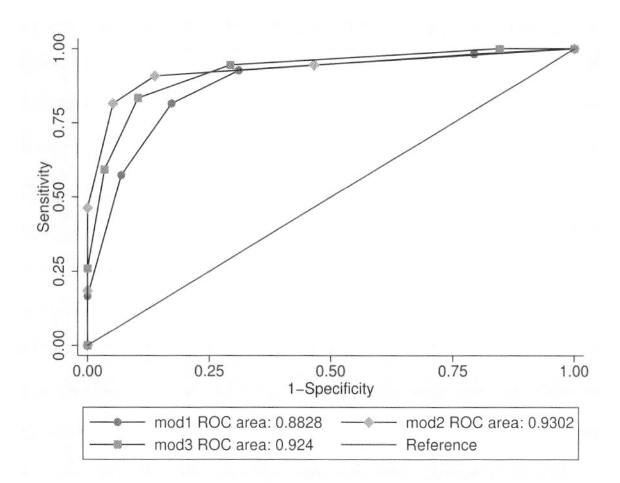

	ROC Area	Std. Err.	chi2	df	Pr>chi2	Bonferroni Pr>chi2
mod1 (standard)	0.8828	0.0317				
mod2	0.9302	0.0256	2.3146	1	0.1282	0.2563
mod3	0.9240	0.0241	5.2480	1	0.0220	0.0439

Equivalently, we could have done this in two steps by using the `roccomp` command.

```
. roccomp status mod1 mod2, graph summary
. roccomp status mod1 mod3, graph summary
```

◁

Saved results

`roccomp` saves the following in `r()`:

Scalars

`r(N_g)`	number of groups	`r(df)`	χ^2 degrees of freedom
`r(p)`	significance probability	`r(chi2)`	χ^2

Matrices

`r(V)`	variance–covariance matrix

`rocgold` saves the following in `r()`:

Scalars

`r(N_g)`	number of groups

Matrices

`r(V)`	variance–covariance matrix	`r(p)`	significance-probability vector
`r(chi2)`	χ^2 vector	`r(p_adj)`	adjusted significance-probability vector
`r(df)`	χ^2 degrees-of-freedom vector		

Methods and formulas

roccomp and rocgold are implemented as ado-files.

Assume that we applied a diagnostic test to each of N_n normal and N_a abnormal subjects. Further assume that the higher the outcome value of the diagnostic test, the higher the risk of the subject being abnormal. Let $\widehat{\theta}$ be the estimated area under the curve, and let $X_i, i = 1, 2, \dots, N_a$ and $Y_j, j = 1, 2, \dots, N_n$ be the values of the diagnostic test for the abnormal and normal subjects, respectively.

Areas under ROC curves are compared using an algorithm suggested by DeLong, DeLong, and Clarke-Pearson (1988). Let $\widehat{\theta} = (\widehat{\theta^1}, \widehat{\theta^2}, \dots, \widehat{\theta^k})$ be a vector representing the areas under k ROC curves. See *Methods and formulas* in [R] **roctab** for the definition of these area estimates.

For the rth area, define

$$V_{10}^r(X_i) = \frac{1}{N_n} \sum_{j=1}^{N_n} \psi(X_i^r, Y_j^r)$$

and for each normal subject, j, define

$$V_{01}^r(Y_j) = \frac{1}{N_a} \sum_{i=1}^{N_a} \psi(X_i^r, Y_j^r)$$

where

$$\psi(X^r, Y^r) = \begin{cases} 1 & Y^r < X^r \\ \frac{1}{2} & Y^r = X^r \\ 0 & Y^r > X^r \end{cases}$$

Define the $k \times k$ matrix $\mathbf{S_{10}}$ such that the (r, s)th element is

$$S_{10}^{r,s} = \frac{1}{N_a - 1} \sum_{i=1}^{N_a} \{V_{10}^r(X_i) - \widehat{\theta}^r\}\{V_{10}^s(X_i) - \widehat{\theta}^s\}$$

and $\mathbf{S_{01}}$ such that the (r, s)th element is

$$S_{01}^{r,s} = \frac{1}{N_n - 1} \sum_{j=1}^{N_n} \{V_{01}^r(Y_i) - \widehat{\theta}^r\}\{V_{01}^s(Y_i) - \widehat{\theta}^s\}$$

Then the covariance matrix is

$$S = \frac{1}{N_a} S_{10} + \frac{1}{N_n} S_{01}$$

Let $\mathbf{L}$ be a contrast matrix defining the comparison, so that

$$(\widehat{\theta} - \theta)'\mathbf{L}'\left(\mathbf{LSL}'\right)^{-1}\mathbf{L}(\widehat{\theta} - \theta)$$

has a chi-squared distribution with degrees of freedom equal to the rank of $\mathbf{LSL}'$.

References

Cleves, M. A. 1999. sg120: Receiver operating characteristic (ROC) analysis. *Stata Technical Bulletin* 52: 19–33. Reprinted in *Stata Technical Bulletin Reprints*, vol. 9, pp. 212–229. College Station, TX: Stata Press.

——. 2000. sg120.2: Correction to roccomp command. *Stata Technical Bulletin* 54: 26. Reprinted in *Stata Technical Bulletin Reprints*, vol. 9, p. 231. College Station, TX: Stata Press.

——. 2002a. Comparative assessment of three common algorithms for estimating the variance of the area under the nonparametric receiver operating characteristic curve. *Stata Journal* 2: 280–289.

——. 2002b. From the help desk: Comparing areas under receiver operating characteristic curves from two or more probit or logit models. *Stata Journal* 2: 301–313.

DeLong, E. R., D. M. DeLong, and D. L. Clarke-Pearson. 1988. Comparing the areas under two or more correlated receiver operating characteristic curves: A nonparametric approach. *Biometrics* 44: 837–845.

Erdreich, L. S., and E. T. Lee. 1981. Use of relative operating characteristic analysis in epidemiology: A method for dealing with subjective judgment. *American Journal of Epidemiology* 114: 649–662.

Hanley, J. A., and B. J. McNeil. 1983. A method of comparing the areas under receiver operating characteristic curves derived from the same cases. *Radiology* 148: 839–843.

Harbord, R. M., and P. Whiting. 2009. metandi: Meta-analysis of diagnostic accuracy using hierarchical logistic regression. *Stata Journal* 9: 211–229.

Juul, S., and M. Frydenberg. 2010. *An Introduction to Stata for Health Researchers*. 3rd ed. College Station, TX: Stata Press.

Ma, G., and W. J. Hall. 1993. Confidence bands for the receiver operating characteristic curves. *Medical Decision Making* 13: 191–197.

Pepe, M. S. 2003. *The Statistical Evaluation of Medical Tests for Classification and Prediction*. New York: Oxford University Press.

Reichenheim, M. E., and A. Ponce de Leon. 2002. Estimation of sensitivity and specificity arising from validity studies with incomplete design. *Stata Journal* 2: 267–279.

Seed, P. T., and A. Tobías. 2001. sbe36.1: Summary statistics for diagnostic tests. *Stata Technical Bulletin* 59: 25–27. Reprinted in *Stata Technical Bulletin Reprints*, vol. 10, pp. 90–93. College Station, TX: Stata Press.

Tobías, A. 2000. sbe36: Summary statistics report for diagnostic tests. *Stata Technical Bulletin* 56: 16–18. Reprinted in *Stata Technical Bulletin Reprints*, vol. 10, pp. 87–90. College Station, TX: Stata Press.

Working, H., and H. Hotelling. 1929. Application of the theory of error to the interpretation of trends. *Journal of the American Statistical Association* 24 (Suppl.): 73–85.

Also see

[R] **logistic postestimation** — Postestimation tools for logistic

[R] **roc** — Receiver operating characteristic (ROC) analysis

[R] **rocfit** — Parametric ROC models

[R] **rocreg** — Receiver operating characteristic (ROC) regression

[R] **roctab** — Nonparametric ROC analysis

Title

rocfit — Parametric ROC models

Syntax

`rocfit` *refvar classvar* [*if*] [*in*] [*weight*] [, *rocfit_options*]

rocfit_options	Description
Model	
`continuous(`*#*`)`	divide *classvar* into # groups of approximately equal length
`generate(`*newvar*`)`	create *newvar* containing classification groups
SE	
`vce(`*vcetype*`)`	*vcetype* may be `oim` or `opg`
Reporting	
`level(`*#*`)`	set confidence level; default is `level(95)`
Maximization	
maximize_options	control the maximization process; seldom used

`fweight`s are allowed; see [U] **11.1.6 weight**.

See [U] **20 Estimation and postestimation commands** for more capabilities of estimation commands.

Menu

Statistics > Epidemiology and related > ROC analysis > Parametric ROC analysis without covariates

Description

`rocfit` fits maximum-likelihood ROC models assuming a binormal distribution of the latent variable.

The two variables *refvar* and *classvar* must be numeric. The reference variable indicates the true state of the observation, such as diseased and nondiseased or normal and abnormal, and must be coded as 0 and 1. The rating or outcome of the diagnostic test or test modality is recorded in *classvar*, which must be at least ordinal, with higher values indicating higher risk.

See [R] **roc** for other commands designed to perform receiver operating characteristic (ROC) analyses with rating and discrete classification data.

Options

Model

`continuous(`*#*`)` specifies that the continuous *classvar* be divided into # groups of approximately equal length. This option is required when *classvar* takes on more than 20 distinct values.

continuous(.) may be specified to indicate that *classvar* be used as it is, even though it could have more than 20 distinct values.

generate(*newvar*) specifies the new variable that is to contain the values indicating the groups produced by continuous(#). generate() may be specified only with continuous().

SE

vce(*vcetype*) specifies the type of standard error reported. *vcetype* may be either oim or opg; see [R] ***vce_option***.

Reporting

level(#); see [R] **estimation options**.

Maximization

maximize_options: difficult, technique(*algorithm_spec*), iterate(#), [no]log, trace, gradient, showstep, hessian, showtolerance, tolerance(#), ltolerance(#), nrtolerance(#), nonrtolerance, and from(*init_specs*); see [R] **maximize**. These options are seldom used.

Setting the optimization type to technique(bhhh) resets the default *vcetype* to vce(opg).

Remarks

Dorfman and Alf (1969) developed a generalized approach for obtaining maximum likelihood estimates of the parameters for a smooth fitting ROC curve. The most commonly used method for ordinal data, and the one implemented here, is based upon the binormal model; see Pepe (2003), Pepe, Longton, and Janes (2009), and Janes, Longton, and Pepe (2009) for methods of ROC analysis for continuous data, including methods for adjusting for covariates.

The model assumes the existence of an unobserved, continuous, latent variable that is normally distributed (perhaps after a monotonic transformation) in both the normal and abnormal populations with means μ_n and μ_a and variances σ_n^2 and σ_a^2, respectively. The model further assumes that the K categories of the rating variable result from partitioning the unobserved latent variable by $K-1$ fixed boundaries. The method fits a straight line to the empirical ROC points plotted using normal probability scales on both axes. Maximum likelihood estimates of the line's slope and intercept and the $K-1$ boundaries are obtained simultaneously. See *Methods and formulas* for details.

The intercept from the fitted line is a measurement of $(\mu_a - \mu_n)/\sigma_a$, and the slope measures σ_n/σ_a.

Thus the intercept is the standardized difference between the two latent population means, and the slope is the ratio of the two standard deviations. The null hypothesis that there is no difference between the two population means is evaluated by testing that the intercept = 0, and the null hypothesis that the variances in the two populations are equal is evaluated by testing that the slope = 1.

▷ Example 1

We use Hanley and McNeil's (1982) dataset, described in example 1 of [R] **roctab**, to fit a smooth ROC curve assuming a binormal model.

```
. use http://www.stata-press.com/data/r12/hanley
. rocfit disease rating
Fitting binormal model:
Iteration 0:   log likelihood = -123.68069
Iteration 1:   log likelihood = -123.64867
Iteration 2:   log likelihood = -123.64855
Iteration 3:   log likelihood = -123.64855
Binormal model of disease on rating               Number of obs   =        109
Goodness-of-fit chi2(2) =        0.21
Prob > chi2             =      0.9006
Log likelihood          =  -123.64855

------------------------------------------------------------------------------
             |      Coef.   Std. Err.      z    P>|z|     [95% Conf. Interval]
-------------+----------------------------------------------------------------
   intercept |   1.656782   0.310456     5.34   0.000     1.048300    2.265265
   slope (*) |   0.713002   0.215882    -1.33   0.092     0.289881    1.136123
-------------+----------------------------------------------------------------
       /cut1 |   0.169768   0.165307     1.03   0.152    -0.154227    0.493764
       /cut2 |   0.463215   0.167235     2.77   0.003     0.135441    0.790990
       /cut3 |   0.766860   0.174808     4.39   0.000     0.424243    1.109477
       /cut4 |   1.797938   0.299581     6.00   0.000     1.210770    2.385106
------------------------------------------------------------------------------

------------------------------------------------------------------------------
             |           Indices from binormal fit
       Index |   Estimate    Std. Err.                [95% Conf. Interval]
-------------+----------------------------------------------------------------
    ROC area |   0.911331   0.029506                  0.853501    0.969161
    delta(m) |   2.323671   0.502370                  1.339044    3.308298
        d(e) |   1.934361   0.257187                  1.430284    2.438438
        d(a) |   1.907771   0.259822                  1.398530    2.417012
------------------------------------------------------------------------------
(*) z test for slope==1
```

`rocfit` outputs the MLE for the intercept and slope of the fitted regression line along with, here, four boundaries (because there are five ratings) labeled `/cut1` through `/cut4`. Also `rocfit` computes and reports four indices based on the fitted ROC curve: the area under the curve (labeled `ROC area`), $\delta(m)$ (labeled `delta(m)`), d_e (labeled `d(e)`), and d_a (labeled `d(a)`). More information about these indices can be found in *Methods and formulas* and in Erdreich and Lee (1981).

◁

Saved results

rocfit saves the following in e():

Scalars

e(N)	number of observations
e(k)	number of parameters
e(k_eq)	number of equations in e(b)
e(k_eq_model)	number of equations in overall model test
e(k_dv)	number of dependent variables
e(df_m)	model degrees of freedom
e(ll)	log likelihood
e(chi2_gf)	goodness-of-fit χ^2
e(df_gf)	goodness-of-fit degrees of freedom
e(p_gf)	χ^2 goodness-of-fit significance probability
e(area)	area under the ROC curve
e(se_area)	standard error for the area under the ROC curve
e(deltam)	delta(m)
e(se_delm)	standard area for delta(m)
e(de)	d(e) index
e(se_de)	standard error for d(e) index
e(da)	d(a) index
e(se_da)	standard error for d(a) index
e(rank)	rank of e(V)
e(ic)	number of iterations
e(rc)	return code
e(converged)	1 if converged, 0 otherwise

Macros

e(cmd)	rocfit
e(cmdline)	command as typed
e(depvar)	*refvar* and *classvar*
e(wtype)	weight type
e(wexp)	weight expression
e(title)	title in estimation output
e(chi2type)	GOF; type of model χ^2 test
e(vce)	*vcetype* specified in vce()
e(vcetype)	title used to label Std. Err.
e(opt)	type of optimization
e(which)	max or min; whether optimizer is to perform maximization or minimization
e(ml_method)	type of ml method
e(user)	name of likelihood-evaluator program
e(technique)	maximization technique
e(properties)	b V

Matrices

e(b)	coefficient vector
e(ilog)	iteration log (up to 20 iterations)
e(gradient)	gradient vector
e(V)	variance–covariance matrix of the estimators

Functions

e(sample)	marks estimation sample

Methods and formulas

rocfit is implemented as an ado-file.

Dorfman and Alf (1969) developed a general procedure for obtaining maximum likelihood estimates of the parameters of a smooth-fitting ROC curve. The most common method, and the one implemented in Stata, is based upon the binormal model.

The model assumes that there is an unobserved continuous latent variable that is normally distributed in both the normal and abnormal populations. The idea is better explained with the following illustration:

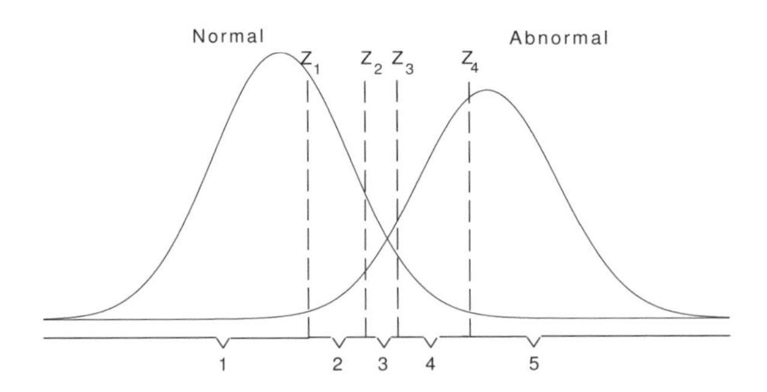

The latent variable is assumed to be normally distributed for both the normal and abnormal subjects, perhaps after a monotonic transformation, with means μ_n and μ_a and variances σ_n^2 and σ_a^2, respectively.

This latent variable is assumed to be partitioned into the k categories of the rating variable by $k-1$ fixed boundaries. In the above figure, the $k=5$ categories of the rating variable identified on the bottom result from the partition of the four boundaries Z_1 through Z_4.

Let R_j for $j=1,2,\ldots,k$ indicate the categories of the rating variable, let $i=1$ if the subject belongs to the normal group, and let $i=2$ if the subject belongs to the abnormal group.

Then

$$p(R_j|i=1) = F(Z_j) - F(Z_{j-1})$$

where $Z_k = (x_k - \mu_n)/\sigma_n$, F is the cumulative normal distribution, $F(Z_0)=0$, and $F(Z_k)=1$. Also,

$$p(R_j|i=2) = F(bZ_j - a) - F(bZ_{j-1} - a)$$

where $b = \sigma_n/\sigma_a$ and $a = (\mu_a - \mu_n)/\sigma_a$.

The parameters a, b and the $k-1$ fixed boundaries Z_j are simultaneously estimated by maximizing the log-likelihood function

$$\log L = \sum_{i=1}^{2}\sum_{j=1}^{k} r_{ij}\log\{p(R_j|i)\}$$

where r_{ij} is the number of R_js in group i.

The area under the fitted ROC curve is computed as

$$\Phi\left(\frac{a}{\sqrt{1+b^2}}\right)$$

where Φ is the standard normal cumulative distribution function.

Point estimates for the ROC curve indices are as follows:

$$\delta(m) = \frac{a}{b} \qquad d_e = \frac{2a}{b+1} \qquad d_a = \frac{a\sqrt{2}}{\sqrt{1+b^2}}$$

Variances for these indices are computed using the delta method.

The $\delta(m)$ estimates $(\mu_a - \mu_n)/\sigma_n$, d_e estimates $2(\mu_a - \mu_n)/(\sigma_a - \sigma_n)$, and d_a estimates $\sqrt{2}(\mu_a - \mu_n)/(\sigma_a^2 - \sigma_n^2)^2$.

Simultaneous confidence bands for the entire curve are obtained, as suggested by Ma and Hall (1993), by first obtaining Working–Hotelling (1929) confidence bands for the fitted straight line in normal probability coordinates and then transforming them back to ROC coordinates.

References

Bamber, D. 1975. The area above the ordinal dominance graph and the area below the receiver operating characteristic graph. *Journal of Mathematical Psychology* 12: 387–415.

Choi, B. C. K. 1998. Slopes of a receiver operating characteristic curve and likelihood ratios for a diagnostic test. *American Journal of Epidemiology* 148: 1127–1132.

Cleves, M. A. 1999. sg120: Receiver operating characteristic (ROC) analysis. *Stata Technical Bulletin* 52: 19–33. Reprinted in *Stata Technical Bulletin Reprints*, vol. 9, pp. 212–229. College Station, TX: Stata Press.

——. 2000. sg120.1: Two new options added to rocfit command. *Stata Technical Bulletin* 53: 18–19. Reprinted in *Stata Technical Bulletin Reprints*, vol. 9, pp. 230–231. College Station, TX: Stata Press.

Dorfman, D. D., and E. Alf, Jr. 1969. Maximum-likelihood estimation of parameters of signal-detection theory and determination of confidence intervals–rating-method data. *Journal of Mathematical Psychology* 6: 487–496.

Erdreich, L. S., and E. T. Lee. 1981. Use of relative operating characteristic analysis in epidemiology: A method for dealing with subjective judgment. *American Journal of Epidemiology* 114: 649–662.

Hanley, J. A., and B. J. McNeil. 1982. The meaning and use of the area under a receiver operating characteristic (ROC) curve. *Radiology* 143: 29–36.

Janes, H., G. Longton, and M. S. Pepe. 2009. Accommodating covariates in receiver operating characteristic analysis. *Stata Journal* 9: 17–39.

Ma, G., and W. J. Hall. 1993. Confidence bands for the receiver operating characteristic curves. *Medical Decision Making* 13: 191–197.

Pepe, M. S. 2003. *The Statistical Evaluation of Medical Tests for Classification and Prediction*. New York: Oxford University Press.

Pepe, M. S., G. Longton, and H. Janes. 2009. Estimation and comparison of receiver operating characteristic curves. *Stata Journal* 9: 1–16.

Working, H., and H. Hotelling. 1929. Application of the theory of error to the interpretation of trends. *Journal of the American Statistical Association* 24 (Suppl.): 73–85.

Also see

[R] **rocfit postestimation** — Postestimation tools for rocfit

[R] **roc** — Receiver operating characteristic (ROC) analysis

[R] **rocreg** — Receiver operating characteristic (ROC) regression

[U] **20 Estimation and postestimation commands**

Title

rocfit postestimation — Postestimation tools for rocfit

Description

The following command is of special interest after `rocfit`:

Command	Description
`rocplot`	plot the fitted ROC curve and simultaneous confidence bands

For information about `rocplot`, see below.

The following standard postestimation commands are also available:

Command	Description
`estat`	AIC, BIC, VCE, and estimation sample summary
`estimates`	cataloging estimation results
*`lincom`	point estimates, standard errors, testing, and inference for linear combinations of coefficients
*`test`	Wald tests of simple and composite linear hypotheses

*See *Using lincom and test* below.

See the corresponding entries in the *Base Reference Manual* for details.

Special-interest postestimation command

`rocplot` plots the fitted ROC curve and simultaneous confidence bands.

Syntax for rocplot

rocplot [, *rocplot_options*]

rocplot_options	Description
Main	
confband	display confidence bands
norefline	suppress plotting the reference line
level(#)	set confidence level; default is level(95)
Plot	
plotopts(*plot_options*)	affect rendition of the ROC points
Fit line	
lineopts(*cline_options*)	affect rendition of the fitted ROC line
CI plot	
ciopts(*area_options*)	affect rendition of the confidence bands
Reference line	
rlopts(*cline_options*)	affect rendition of the reference line
Add plots	
addplot(*plot*)	add other plots to the generated graph
Y axis, X axis, Titles, Legend, Overall	
twoway_options	any options other than by() documented in [G-3] ***twoway_options***

plot_options	Description
marker_options	change look of markers (color, size, etc.)
marker_label_options	add marker labels; change look or position
cline_options	change the look of the line

Menu

Statistics > Epidemiology and related > ROC analysis > ROC curves after rocfit

Options for rocplot

Main

confband specifies that simultaneous confidence bands be plotted around the ROC curve.

norefline suppresses plotting the 45-degree reference line from the graphical output of the ROC curve.

level(#) specifies the confidence level, as a percentage, for the confidence bands. The default is level(95) or as set by set level; see [R] **level**.

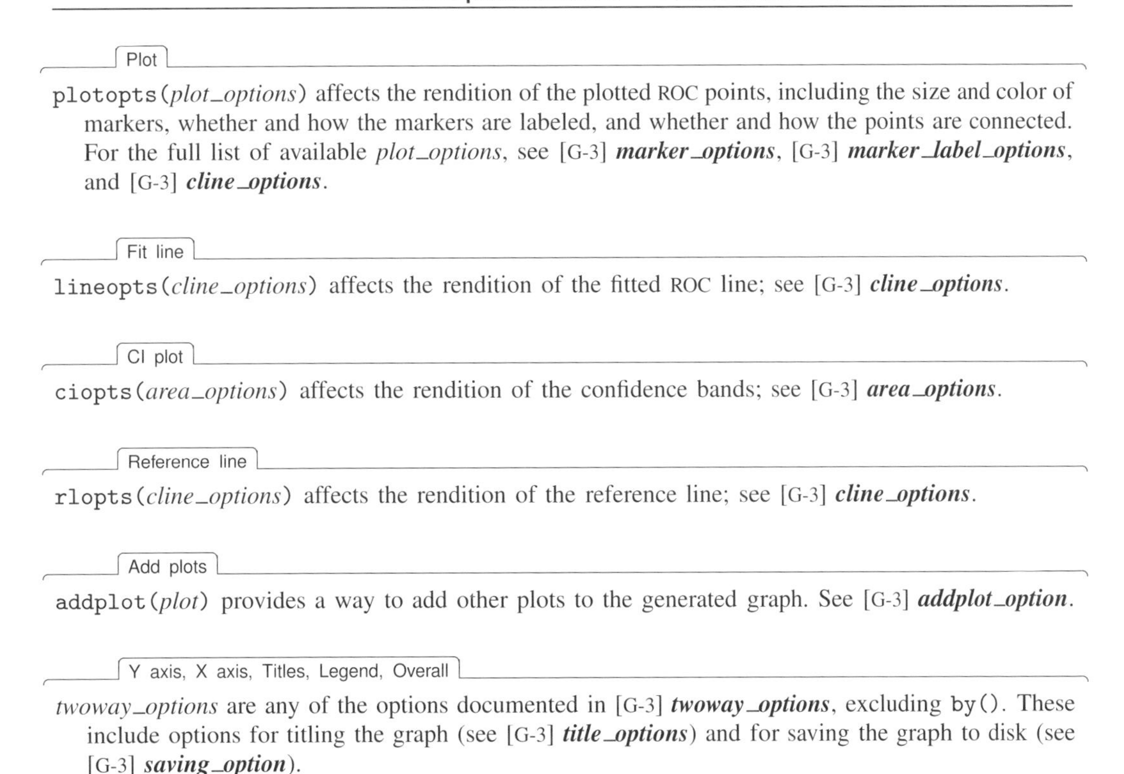

Plot

`plotopts(`*plot_options*`)` affects the rendition of the plotted ROC points, including the size and color of markers, whether and how the markers are labeled, and whether and how the points are connected. For the full list of available *plot_options*, see [G-3] ***marker_options***, [G-3] ***marker_label_options***, and [G-3] ***cline_options***.

Fit line

`lineopts(`*cline_options*`)` affects the rendition of the fitted ROC line; see [G-3] ***cline_options***.

CI plot

`ciopts(`*area_options*`)` affects the rendition of the confidence bands; see [G-3] ***area_options***.

Reference line

`rlopts(`*cline_options*`)` affects the rendition of the reference line; see [G-3] ***cline_options***.

Add plots

`addplot(`*plot*`)` provides a way to add other plots to the generated graph. See [G-3] ***addplot_option***.

Y axis, X axis, Titles, Legend, Overall

twoway_options are any of the options documented in [G-3] ***twoway_options***, excluding `by()`. These include options for titling the graph (see [G-3] ***title_options***) and for saving the graph to disk (see [G-3] ***saving_option***).

Remarks

Remarks are presented under the following headings:

Using lincom and test
Using rocplot

Using lincom and test

`intercept`, `slope`, and `cut#`, shown in example 1 of [R] **rocfit**, are equation names and not variable names, so they need to be referenced as described in *Special syntaxes after multiple-equation estimation* of [R] **test**. For example, instead of typing

```
. test intercept
intercept not found
r(111);
```

you should type

```
. test [intercept]_cons
 ( 1)  [intercept]_cons = 0
           chi2(  1) =   28.48
         Prob > chi2 =    0.0000
```

Using rocplot

▷ Example 1

In example 1 of [R] **rocfit**, we fit a ROC curve by typing `rocfit disease rating`.

In the output table for our model, we are testing whether the variances of the two latent populations are equal by testing that the slope = 1.

We plot the fitted ROC curve.

```
. rocplot, confband
```

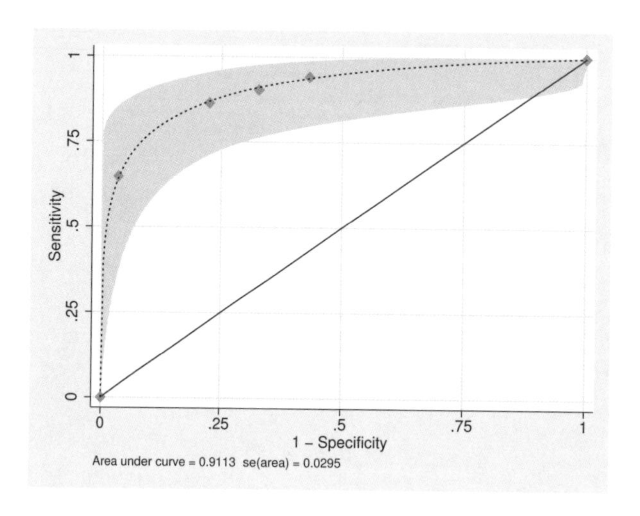

◁

Methods and formulas

All postestimation commands listed above are implemented as ado-files.

Also see

[R] **rocfit** — Parametric ROC models

[U] **20 Estimation and postestimation commands**

Title

rocreg — Receiver operating characteristic (ROC) regression

Syntax

Perform nonparametric analysis of ROC curve under covariates, using bootstrap

`rocreg` *refvar classvar* [*classvars*] [*if*] [*in*] [, *np_options boot_options*]

Perform parametric analysis of ROC curve under covariates, using bootstrap

`rocreg` *refvar classvar* [*classvars*] [*if*] [*in*], `probit`
[*probit_options boot_options*]

Perform parametric analysis of ROC curve under covariates, using maximum likelihood

`rocreg` *refvar classvar* [*classvars*] [*if*] [*in*] [*weight*], `probit ml`
[*probit_ml_options*]

np_options	Description
Model	
`auc`	estimate total area under the ROC curve; the default
`roc(`*numlist*`)`	estimate ROC for given false-positive rates
`invroc(`*numlist*`)`	estimate false-positive rates for given ROC values
`pauc(`*numlist*`)`	estimate partial area under the ROC curve (pAUC) up to each false-positive rate
`cluster(`*varname*`)`	variable identifying resampling clusters
`ctrlcov(`*varlist*`)`	adjust control distribution for covariates in *varlist*
`ctrlmodel(strata | linear)`	stratify or regress on covariates; default is `ctrlmodel(strata)`
`pvc(empirical | normal)`	use empirical or normal distribution percentile value estimates; default is `pvc(empirical)`
`tiecorrected`	adjust for tied observations; not allowed with `pvc(normal)`
Reporting	
`level(`*#*`)`	set confidence level; default is `level(95)`

probit_options	Description
Model	
*probit	fit the probit model
roccov(*varlist*)	covariates affecting ROC curve
fprpts(#)	number of false-positive rate points to use in fitting ROC curve; default is fprpts(10)
ctrlfprall	fit ROC curve at each false-positive rate in control population
cluster(*varname*)	variable identifying resampling clusters
ctrlcov(*varlist*)	adjust control distribution for covariates in *varlist*
ctrlmodel(strata | linear)	stratify or regress on covariates; default is ctrlmodel(strata)
pvc(empirical | normal)	use empirical or normal distribution percentile value estimates; default is pvc(empirical)
tiecorrected	adjust for tied observations; not allowed with pvc(normal)
Reporting	
level(#)	set confidence level; default is level(95)

*probit is required.

probit_ml_options	Description
Model	
*probit	fit the probit model
*ml	fit the probit model by maximum likelihood estimation
roccov(*varlist*)	covariates affecting ROC curve
cluster(*varname*)	variable identifying clusters
ctrlcov(*varlist*)	adjust control distribution for covariates in *varlist*
Reporting	
level(#)	set confidence level; default is level(95)
display_options	control column formats, line width, and display of omitted variables
Maximization	
maximize_options	control the maximization process; seldom used

*probit and ml are required.

fweights, iweights, and pweights are allowed with maximum likelihood estimation; see [U] **11.1.6 weight**.

boot_options	Description
Bootstrap	
nobootstrap	do not perform bootstrap, just output point estimates
bseed(#)	random-number seed for bootstrap
breps(#)	number of bootstrap replications; default is breps(1000)
bootcc	perform case–control (stratified on *refvar*) sampling rather than cohort sampling in bootstrap
nobstrata	ignore covariate stratification in bootstrap sampling
nodots	suppress bootstrap replication dots
*bsave(*filename*, ...)	save bootstrap replicates from parametric estimation
*bfile(*filename*)	use bootstrap replicates dataset for estimation replay

* bsave() and bfile() are allowed only when the probit option is also specified.

See [U] **20 Estimation and postestimation commands** for more capabilities of estimation commands.

Menu

Statistics > Epidemiology and related > ROC analysis > ROC regression models

Description

The rocreg command is used to perform receiver operating characteristic (ROC) analyses with rating and discrete classification data under the presence of covariates.

The two variables *refvar* and *classvar* must be numeric. The reference variable indicates the true state of the observation—such as diseased and nondiseased or normal and abnormal—and must be coded as 0 and 1. The *refvar* coded as 0 can also be called the control population, while the *refvar* coded as 1 comprises the case population. The rating or outcome of the diagnostic test or test modality is recorded in *classvar*, which must be ordinal, with higher values indicating higher risk.

rocreg can fit three models: a nonparametric model, a parametric probit model that uses the bootstrap for inference, and a parametric probit model fit using maximum likelihood.

Options for nonparametric ROC estimation, using bootstrap

Model

auc estimates the total area under the ROC curve. This is the default summary statistic.

roc(*numlist*) estimates the ROC corresponding to each of the false-positive rates in *numlist*. The values of *numlist* must be in the range (0,1).

invroc(*numlist*) estimates the false-positive rates corresponding to each of the ROC values in *numlist*. The values of *numlist* must be in the range (0,1).

pauc(*numlist*) estimates the partial area under the ROC curve up to each false-positive rate in *numlist*. The values of *numlist* must in the range (0,1].

cluster(*varname*) specifies the variable identifying resampling clusters.

ctrlcov(*varlist*) specifies the covariates to be used to adjust the control population.

`ctrlmodel(strata | linear)` specifies how to model the control population of classifiers on `ctrlcov()`. When `ctrlmodel(linear)` is specified, linear regression is used. The default is `ctrlmodel(strata)`; that is, the control population of classifiers is stratified on the control variables.

`pvc(empirical | normal)` determines how the percentile values of the control population will be calculated. When `pvc(normal)` is specified, the standard normal cumulative distribution function (CDF) is used for calculation. Specifying `pvc(empirical)` will use the empirical CDFs of the control population classifiers for calculation. The default is `pvc(empirical)`.

`tiecorrected` adjusts the percentile values for ties. For each value of the classifier, one half the probability that the classifier equals that value under the control population is added to the percentile value. `tiecorrected` is not allowed with `pvc(normal)`.

Reporting

`level(#)`; see [R] **estimation options**.

Also see *Options for rocreg, using bootstrap.*

Options for parametric ROC estimation, using bootstrap

Model

`probit` fits the probit model. This option is required and implies parametric estimation.

`roccov(`*varlist*`)` specifies the covariates that will affect the ROC curve.

`fprpts(#)` sets the number of false-positive rate points to use in modeling the ROC curve. These points form an equispaced grid on (0,1). The default is `fprpts(10)`.

`ctrlfprall` models the ROC curve at each false-positive rate in the control population.

`cluster(`*varname*`)` specifies the variable identifying resampling clusters.

`ctrlcov(`*varlist*`)` specifies the covariates to be used to adjust the control population.

`ctrlmodel(strata | linear)` specifies how to model the control population of classifiers on `ctrlcov()`. When `ctrlmodel(linear)` is specified, linear regression is used. The default is `ctrlmodel(strata)`; that is, the control population of classifiers is stratified on the control variables.

`pvc(empirical | normal)` determines how the percentile values of the control population will be calculated. When `pvc(normal)` is specified, the standard normal CDF is used for calculation. Specifying `pvc(empirical)` will use the empirical CDFs of the control population classifiers for calculation. The default is `pvc(empirical)`.

`tiecorrected` adjusts the percentile values for ties. For each value of the classifier, one half the probability that the classifier equals that value under the control population is added to the percentile value. `tiecorrected` is not allowed with `pvc(normal)`.

Reporting

`level(#)`; see [R] **estimation options**.

Also see *Options for rocreg, using bootstrap.*

Options for parametric ROC estimation, using maximum likelihood

Model

probit fits the probit model. This option is required and implies parametric estimation.

ml fits the probit model by maximum likelihood estimation. This option is required and must be specified with probit.

roccov(*varlist*) specifies the covariates that will affect the ROC curve.

cluster(*varname*) specifies the variable used for clustering.

ctrlcov(*varlist*) specifies the covariates to be used to adjust the control population.

Reporting

level(#); see [R] **estimation options**.

display_options: noomitted, cformat(%*fmt*), pformat(%*fmt*), sformat(%*fmt*), and nolstretch; see [R] **estimation options**.

Maximization

maximize_options: difficult, technique(*algorithm_spec*), iterate(#), [no]log, trace, gradient, showstep, hessian, showtolerance, tolerance(#), ltolerance(#), nrtolerance(#), nonrtolerance, and from(*init_specs*); see [R] **maximize**. These options are seldom used. The technique(bhhh) option is not allowed.

Options for rocreg, using bootstrap

Bootstrap

nobootstrap specifies that bootstrap standard errors not be calculated.

bseed(#) specifies the random-number seed to be used in the bootstrap.

breps(#) sets the number of bootstrap replications. The default is breps(1000).

bootcc performs case–control (stratified on *refvar*) sampling rather than cohort bootstrap sampling.

nobstrata ignores covariate stratification in bootstrap sampling.

nodots suppresses bootstrap replicate dots.

bsave(*filename*, ...) saves bootstrap replicates from parametric estimation in the given *filename* with specified options (that is, replace). bsave() is only allowed with parametric analysis using bootstrap.

bfile(*filename*) specifies to use the bootstrap replicates dataset for estimation replay. bfile() is only allowed with parametric analysis using bootstrap.

Remarks

Remarks are presented under the following headings:

Introduction

Receiver operating characteristic (ROC) analysis provides a quantitative measure of the accuracy of diagnostic tests to discriminate between two states or conditions. These conditions may be referred to as normal and abnormal, nondiseased and diseased, or control and case. We will use these terms interchangeably. The discriminatory accuracy of a diagnostic test is measured by its ability to correctly classify known control and case subjects.

The analysis uses the ROC curve, a graph of the sensitivity versus 1 − specificity of the diagnostic test. The sensitivity is the fraction of positive cases that are correctly classified by the diagnostic test, whereas the specificity is the fraction of negative cases that are correctly classified. Thus the sensitivity is the true-positive rate, and the specificity is the true-negative rate. We also call 1 − specificity the false-positive rate.

These rates are functions of the possible outcomes of the diagnostic test. At each outcome, a decision will be made by the user of the diagnostic test to classify the tested subject as either normal or abnormal. The true-positive and false-positive rates measure the probability of correct classification or incorrect classification of the subject as abnormal. Given the classification role of the diagnostic test, we will refer to it as the classifier.

Using this basic definition of the ROC curve, Pepe (2000) and Pepe (2003) describe how ROC analysis can be performed as a two-stage process. In the first stage, the control distribution of the classifier is estimated. The specificity is then determined as the percentiles of the classifier values calculated based on the control population. The false-positive rates are calculated as 1 − specificity. In the second stage, the ROC curve is estimated as the cumulative distribution of the case population's "false-positive" rates, also known as the survival function under the case population of the previously calculated percentiles. We use the terms ROC value and true-positive value interchangeably.

This formulation of ROC curve analysis provides simple, nonparametric estimates of several ROC curve summary parameters: area under the ROC curve, partial area under the ROC curve, ROC value for a given false-positive rate, and false-positive rate (also known as invROC) for a given ROC value. In the next section, we will show how to use `rocreg` to compute these estimates with bootstrap inference. There we will also show how `rocreg` complements the other nonparametric Stata ROC commands `roctab` and `roccomp`.

Other factors beyond condition status and the diagnostic test may affect both stages of ROC analysis. For example, a test center may affect the control distribution of the diagnostic test. Disease severity may affect the distribution of the standardized diagnostic test under the case population. Our analysis of the ROC curve in these situations will be more accurate if we take these covariates into account.

In a nonparametric ROC analysis, covariates may only affect the first stage of estimation; that is, they may be used to adjust the control distribution of the classifier. In a parametric ROC analysis, it is assumed that ROC follows a normal distribution, and thus covariates may enter the model at both stages; they may be used to adjust the control distribution and to model ROC as a function of these covariates and the false-positive rate. In parametric models, both sets of covariates need not be distinct but, in fact, they are often the same.

To model covariate effects on the first stage of ROC analysis, Janes and Pepe (2009) propose a covariate-adjusted ROC curve. We will demonstrate the covariate adjustment capabilities of `rocreg` in *Covariate-adjusted ROC curves*.

To account for covariate effects at the second stage, we assume a parametric model. Particularly, the ROC curve is a generalized linear model of the covariates. We will thus have a separate ROC curve for each combination of the relevant covariates. In *Parametric ROC curves: Estimating equations*, we show how to fit the model with estimating equations and bootstrap inference using `rocreg`.

This method, documented as the "pdf" approach in Alonzo and Pepe (2002), works well with weak assumptions about the control distribution.

Also in *Parametric ROC curves: Estimating equations*, we show how to fit a constant-only parametric model (involving no covariates) of the ROC curve with weak assumptions about the control distribution. The constant-only model capabilities of `rocreg` in this context will be compared with those of `rocfit`. `roccomp` has the `binormal` option, which will allow it to compute area under the ROC curve according to a normal ROC curve, equivalent to that obtained by `rocfit`. We will compare this functionality with that of `rocreg`.

In *Parametric ROC curves: Maximum likelihood*, we demonstrate maximum likelihood estimation of the ROC curve model with `rocreg`. There we assume a normal linear model for the classifier on the covariates and case–control status. This method is documented in Pepe (2003). We will also demonstrate how to use this method with no covariates, and we will compare `rocreg` under the constant-only model with `rocfit` and `roccomp`.

The `rocregplot` command is used repeatedly in this entry. This command provides graphical output for `rocreg` and is documented in [R] **rocregplot**.

ROC statistics

`roctab` computes the ROC curve by calculating the false-positive rate and true-positive rate empirically at every value of the input classifier. It makes no distributional assumptions about the case or control distributions. We can get identical behavior from `rocreg` by using the default option settings.

▷ Example 1: Nonparametric ROC, AUC

Hanley and McNeil (1982) presented data from a study in which a reviewer was asked to classify, using a five-point scale, a random sample of 109 tomographic images from patients with neurological problems. The rating scale was as follows: 1 is definitely normal, 2 is probably normal, 3 is questionable, 4 is probably abnormal, and 5 is definitely abnormal. The true disease status was normal for 58 of the patients and abnormal for the remaining 51 patients.

Here we list 9 of the 109 observations:

```
. use http://www.stata-press.com/data/r12/hanley
. list disease rating in 1/9
```

	disease	rating
1.	1	5
2.	0	1
3.	1	5
4.	0	4
5.	0	1
6.	0	3
7.	1	5
8.	0	5
9.	0	1

For each observation, `disease` identifies the true disease status of the subject (`0` is normal, `1` is abnormal), and `rating` contains the classification value assigned by the reviewer.

We run `roctab` on these data, specifying the `graph` option so that the ROC curve is rendered. We then calculate the false-positive and true-positive rates of the ROC curve by using `rocreg`. We graph the rates with `rocregplot`. Because we focus on `rocreg` output later, for now we use the `quietly` prefix to omit the output of `rocreg`. Both graphs are combined using `graph combine` (see [G-2] **graph combine**) for comparison. To ease the comparison, we specify the `aspectratio(1)` option in `roctab`; this is the default aspect ratio in `rocregplot`.

```
. roctab disease rating, graph aspectratio(1) name(a) nodraw title("roctab")
. quietly rocreg disease rating
. rocregplot, name(b) nodraw legend(off) title("rocreg")
. graph combine a b
```

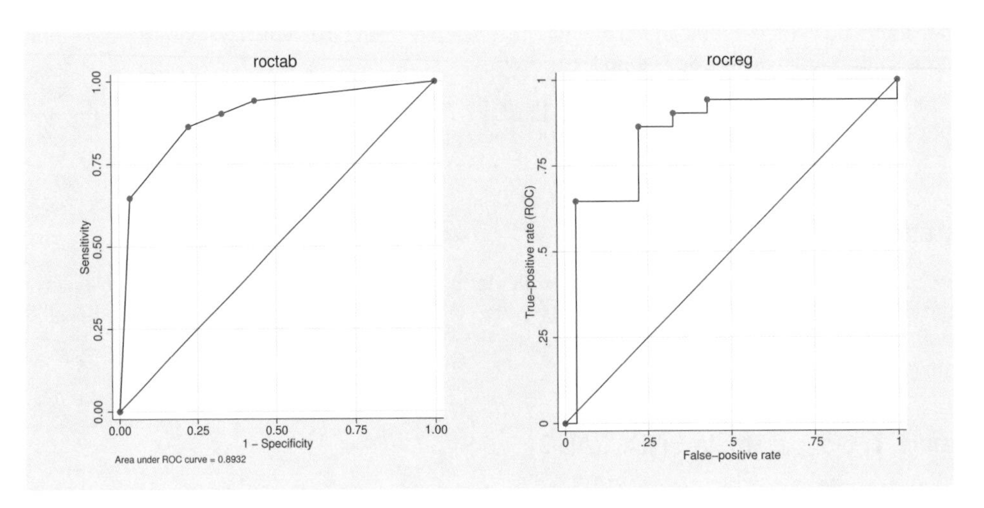

Both `roctab` and `rocreg` compute the same false-positive rate and ROC values. The stairstep line connection style of the graph on the right emphasizes the empirical nature of its estimates. The control distribution of the classifier is estimated using the empirical CDF estimate. Similarly, the ROC curve, the distribution of the resulting case observation false-positive rate values, is estimated using the empirical CDF. Note the footnote in the `roctab` plot. By default, `roctab` will estimate the area under the ROC curve (AUC) using a trapezoidal approximation to the estimated false-positive rate and true-positive rate points.

The AUC can be interpreted as the probability that a randomly selected member of the case population will have a larger classifier value than a randomly selected member of the control population. It can also be viewed as the average ROC value, averaged uniformly over the (0,1) false-positive rate domain (Pepe 2003).

The nonparametric estimator of the AUC (DeLong, DeLong, and Clarke-Pearson 1988; Hanley and Hajian-Tilaki 1997) used by `rocreg` is equivalent to the sample mean of the percentile values of the case observations. Thus to calculate the nonparametric AUC estimate, we only need to calculate the percentile values of the case observations with respect to the control distribution.

This estimate can differ from the trapezoidal approximation estimate. Under discrete classification data, like we have here, there may be ties between classifier values from case to control. The trapezoidal approximation uses linear interpolation between the classifier values to correct for ties. Correcting the nonparametric estimator involves adding a correction term to each observation's percentile value, which measures the probability that the classifier is equal to (instead of less than) the observation's classifier value.

The tie-corrected nonparametric estimate (trapezoidal approximation) is used when we think the true ROC curve is smooth. This means that the classifier we measure is a discretized approximation of a true latent and a continuous classifier.

We now recompute the ROC curve of `rating` for classifying `disease` and calculate the AUC. Specifying the `tiecorrected` option allows tie correction to be used in the `rocreg` calculation. Under nonparametric estimation, `rocreg` bootstraps to obtain standard errors and confidence intervals for requested statistics. We use the default 1,000 bootstrap replications to obtain confidence intervals for our parameters. This is a reasonable lower bound to the number of replications (Mooney and Duval 1993) required for estimating percentile confidence intervals. By specifying the `summary` option in `roctab`, we will obtain output showing the trapezoidal approximation of the AUC estimate, along with standard error and confidence-interval estimates for the trapezoidal approximation suggested by DeLong, DeLong, and Clarke-Pearson (1988).

```
. roctab disease rating, summary
                      ROC                    —Asymptotic Normal——
           Obs       Area     Std. Err.      [95% Conf. Interval]
          ————————————————————————————————————————————————————————
           109     0.8932       0.0307        0.83295     0.95339
. rocreg disease rating, auc tiecorrected bseed(29092)
(running rocregstat on estimation sample)
Bootstrap replications (1000)
———+——— 1 ———+——— 2 ———+——— 3 ———+——— 4 ———+——— 5
..................................................    50
..................................................   100
  (output omitted )
..................................................   950
..................................................  1000
Bootstrap results                                Number of obs      =        109
                                                 Replications       =       1000
Nonparametric ROC estimation
Control standardization: empirical, corrected for ties
ROC method             : empirical
Area under the ROC curve
   Status    : disease
   Classifier: rating
```

AUC	Observed Coef.	Bias	Bootstrap Std. Err.	[95% Conf. Interval]		
	.8931711	.000108	.0292028	.8359347	.9504075	(N)
				.8290958	.9457951	(P)
				.8280714	.9450642	(BC)

The estimates of AUC match well. The standard error from `roctab` is close to the bootstrap standard error calculated by `rocreg`. The bootstrap standard error generalizes to the more complex models that we consider later, whereas the `roctab` standard-error calculation does not.

◁

The AUC can be used to compare different classifiers. It is the most popular summary statistic for comparisons (Pepe, Longton, and Janes 2009). `roccomp` will compute the trapezoidal approximation of the AUC and graph the ROC curves of multiple classifiers. Using the DeLong, DeLong, and Clarke-Pearson (1988) covariance estimates for the AUC estimate, `roccomp` performs a Wald test of the null hypothesis that all classifier AUC values are equal. `rocreg` has similar capabilities.

▷ Example 2: Nonparametric ROC, AUC, multiple classifiers

Hanley and McNeil (1983) presented data from an evaluation of two computer algorithms designed to reconstruct CT images from phantoms. We will call these two algorithms modalities 1 and 2. A sample of 112 phantoms was selected; 58 phantoms were considered normal, and the remaining 54 were abnormal. Each of the two modalities was applied to each phantom, and the resulting images were rated by a reviewer using a six-point scale: 1 is definitely normal, 2 is probably normal, 3 is possibly normal, 4 is possibly abnormal, 5 is probably abnormal, and 6 is definitely abnormal. Because each modality was applied to the same sample of phantoms, the two sets of outcomes are correlated.

We list the first seven observations:

```
. use http://www.stata-press.com/data/r12/ct, clear
. list in 1/7, sep(0)

     ┌───────────────────────┐
     │ mod1   mod2   status  │
     ├───────────────────────┤
  1. │    2      1        0  │
  2. │    5      5        1  │
  3. │    2      1        0  │
  4. │    2      3        0  │
  5. │    5      6        1  │
  6. │    2      2        0  │
  7. │    3      2        0  │
     └───────────────────────┘
```

Each observation corresponds to one phantom. The mod1 variable identifies the rating assigned for the first modality, and the mod2 variable identifies the rating assigned for the second modality. The true status of the phantoms is given by status==0 if they are normal and status==1 if they are abnormal. The observations with at least one missing rating were dropped from the analysis.

A fictitious dataset was created from this true dataset, adding a third test modality. We will use roccomp to compute the AUC statistic for each modality in these data and compare the AUC of the three modalities. We obtain the same behavior from rocreg. As before, the tiecorrected option is specified so that the AUC is calculated with the trapezoidal approximation.

```
. use http://www.stata-press.com/data/r12/ct2
. roccomp status mod1 mod2 mod3, summary

                              ROC                    -Asymptotic Normal--
                  Obs       Area     Std. Err.      [95% Conf. Interval]
-------------------------------------------------------------------------
mod1              112     0.8828       0.0317        0.82067     0.94498
mod2              112     0.9302       0.0256        0.88005     0.98042
mod3              112     0.9240       0.0241        0.87670     0.97132
-------------------------------------------------------------------------
Ho: area(mod1) = area(mod2) = area(mod3)
    chi2(2) =      6.54       Prob>chi2 =    0.0381
```

```
. rocreg status mod1 mod2 mod3, tiecorrected bseed(38038) nodots
Bootstrap results                                 Number of obs      =       112
                                                  Replications       =      1000

Nonparametric ROC estimation

Control standardization: empirical, corrected for ties
ROC method             : empirical

Area under the ROC curve

   Status    : status
   Classifier: mod1

             Observed                 Bootstrap
       AUC       Coef.       Bias     Std. Err.     [95% Conf. Interval]

              .8828225  -.0006367     .0322291     .8196546    .9459903  (N)
                                                   .8147518    .9421572  (P)
                                                   .8124397    .9394085 (BC)

   Status    : status
   Classifier: mod2

             Observed                 Bootstrap
       AUC       Coef.       Bias     Std. Err.     [95% Conf. Interval]

              .9302363  -.0015402     .0259593     .8793569    .9811156  (N)
                                                   .8737522    .9737432  (P)
                                                   .8739467    .9737768 (BC)

   Status    : status
   Classifier: mod3

             Observed                 Bootstrap
       AUC       Coef.       Bias     Std. Err.     [95% Conf. Interval]

              .9240102  -.0003528     .0247037     .8755919    .9724286  (N)
                                                   .8720036    .9674485  (P)
                                                   .8693548        .965 (BC)

Ho: All classifiers have equal AUC values.
Ha: At least one classifier has a different AUC value.
P-value:      .0389797                 Test based on bootstrap (N) assumptions.
```

We see that the AUC estimates are equivalent, and the standard errors are quite close as well. The p-value for the tests of equal AUC under `rocreg` leads to similar inference as the p-value from `roccomp`. The Wald test performed by `rocreg` uses the joint bootstrap estimate variance matrix of the three AUC estimators rather than the DeLong, DeLong, and Clarke-Pearson (1988) variance estimate used by `roccomp`.

`roccomp` is used here on potentially correlated classifiers that are recorded in wide-format data. It can also be used on long-format data to compare independent classifiers. Further details can be found in [R] **roccomp**.

◁

Citing the AUC's lack of clinical relevance, there is argument against using it as a key summary statistic of the ROC curve (Pepe 2003; Cook 2007). Pepe, Longton, and Janes (2009) suggest using the estimate of the ROC curve itself at a particular point, or the estimate of the false-positive rate at a given ROC value, also known as invROC.

Recall from example 1 how nonparametric rocreg graphs look, with the stairstep pattern in the ROC curve. In an ideal world, the graph would be a smooth one-to-one function, and it would be trivial to map a false-positive rate to its corresponding true-positive rate and vice versa.

However, smooth ROC curves can only be obtained by assuming a parametric model that uses linear interpolation between observed false-positive rates and between observed true-positive rates, and rocreg is certainly capable of that; see example 1 of [R] **rocregplot**. However, under nonparametric estimation, the mapping between false-positive rates and true-positive rates is not one to one, and estimates tend to be less reliable the further you are from an observed data point. This is somewhat mitigated by using tie-corrected rates (the tiecorrected option).

When we examine continuous data, the difference between the tie-corrected estimates and the standard estimates becomes negligible, and the empirical estimate of the ROC curve becomes close to the smooth ROC curve obtained by linear interpolation. So the nonparametric ROC and invROC estimates work well.

Fixing one rate value of interest can be difficult and subjective (Pepe 2003). A compromise measure is the partial area under the ROC curve (pAUC) (McClish 1989; Thompson and Zucchini 1989). This is the integral of the ROC curve from 0 and above to a given false-positive rate (perhaps the largest clinically acceptable value). Like the AUC estimate, the nonparametric estimate of the pAUC can be written as a sample average of the case observation percentiles, but with an adjustment based on the prescribed maximum false-positive rate (Dodd and Pepe 2003). A tie correction may also be applied so that it reflects the trapezoidal approximation.

We cannot compare rocreg with roctab or roccomp on the estimation of pAUC, because pAUC is not computed by the latter two.

▷ Example 3: Nonparametric ROC, other statistics

To see how rocreg estimates ROC, invROC, and pAUC, we will examine a new study. Wieand et al. (1989) examined a pancreatic cancer study with two continuous classifiers, here called y1 (CA 19-9) and y2 (CA 125). This study was also examined in Pepe, Longton, and Janes (2009). The indicator of cancer in a subject is recorded as d. The study was a case–control study, stratifying participants on disease status.

We list the first five observations:

```
. use http://labs.fhcrc.org/pepe/book/data/wiedat2b.dta, clear
(S. Wieand - Pancreatic cancer diagnostic marker data)
. list in 1/5
```

	y1	y2	d
1.	28	13.3	no
2.	15.5	11.1	no
3.	8.2	16.7	no
4.	3.4	12.6	no
5.	17.3	7.4	no

We will estimate the ROC curves at a large value (0.7) and a small value (0.2) of the false-positive rate. These values are specified in roc(). The false-positive rate for ROC or sensitivity value of 0.6 will also be estimated by specifying invroc(). Percentile confidence intervals for these parameters are displayed in the graph obtained by rocregplot after rocreg. The pAUC statistic will be calculated for the false-positive rate of 0.5, which is specified as an argument to the pauc() option. Following Pepe, Longton, and Janes (2009), we use a stratified bootstrap, sampling separately from the case and control populations by specifying the bootcc option. This reflects the case–control nature of the study.

All four statistics can be estimated simultaneously by rocreg. For clarity, however, we will estimate each statistic with a separate call to rocreg. rocregplot is used after estimation to graph the ROC and false-positive rate estimates. The display of the individual, observation-specific false-positive rate and ROC values will be omitted in the plot. This is accomplished by specifying msymbol(i) in our plot1opts() and plot2opts() options to rocregplot.

```
. rocreg d y1 y2, roc(.7) bseed(8378923) bootcc nodots
Bootstrap results
Number of strata   =         2                Number of obs      =        141
                                              Replications       =       1000
Nonparametric ROC estimation
Control standardization: empirical
ROC method             : empirical
ROC curve
   Status    : d
   Classifier: y1

              Observed                 Bootstrap
        ROC      Coef.        Bias     Std. Err.     [95% Conf. Interval]

         .7   .9222222   -.0021889     .0323879      .8587432   .9857013  (N)
                                                     .8444445   .9777778  (P)
                                                     .8555555   .9777778 (BC)

   Status    : d
   Classifier: y2

              Observed                 Bootstrap
        ROC      Coef.        Bias     Std. Err.     [95% Conf. Interval]

         .7   .8888889   -.0035556     .0414215      .8077043   .9700735  (N)
                                                           .8   .9611111  (P)
                                                     .7888889   .9555556 (BC)

Ho: All classifiers have equal ROC values.
Ha: At least one classifier has a different ROC value.
Test based on bootstrap (N) assumptions.
ROC         P-value

     .7    .5423044
```

```
. rocregplot, plot1opts(msymbol(i)) plot2opts(msymbol(i))
```

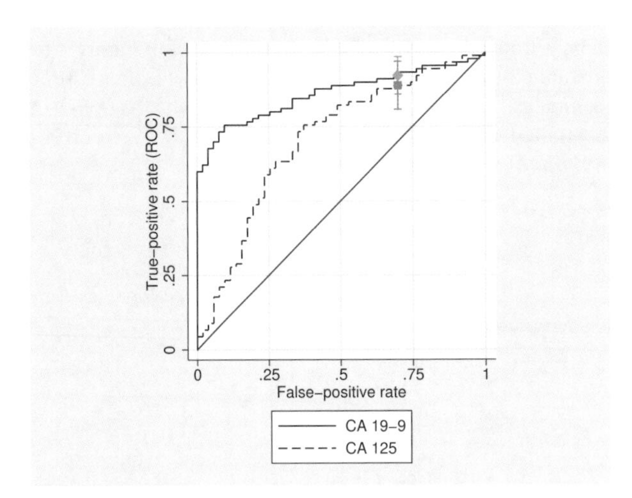

In this study, we see that classifier `y1` (CA 19-9) is a uniformly better test than is classifier `y2` (CA 125) until high levels of false-positive rate and sensitivity or ROC value are reached. At the high level of false-positive rate, 0.7, the ROC value does not significantly differ between the two classifiers. This can be seen in the plot by the overlapping confidence intervals.

```
. rocreg d y1 y2, roc(.2) bseed(8378923) bootcc nodots
Bootstrap results
Number of strata   =          2              Number of obs      =        141
                                             Replications       =       1000
Nonparametric ROC estimation
Control standardization: empirical
ROC method             : empirical
ROC curve
   Status    : d
   Classifier: y1

              Observed                Bootstrap
         ROC     Coef.        Bias    Std. Err.     [95% Conf. Interval]

          .2   .7777778   .0011778    .0483655     .6829831    .8725725  (N)
                                                   .6888889    .8777778  (P)
                                                   .6777778    .8666667 (BC)

   Status    : d
   Classifier: y2

              Observed                Bootstrap
         ROC     Coef.        Bias    Std. Err.     [95% Conf. Interval]

          .2   .4888889  -.0091667    .1339863     .2262806    .7514971  (N)
                                                   .2222222          .7  (P)
                                                   .2111111          .7 (BC)

Ho: All classifiers have equal ROC values.
Ha: At least one classifier has a different ROC value.
Test based on bootstrap (N) assumptions.
ROC           P-value

      .2     .043234
. rocregplot, plot1opts(msymbol(i)) plot2opts(msymbol(i))
```

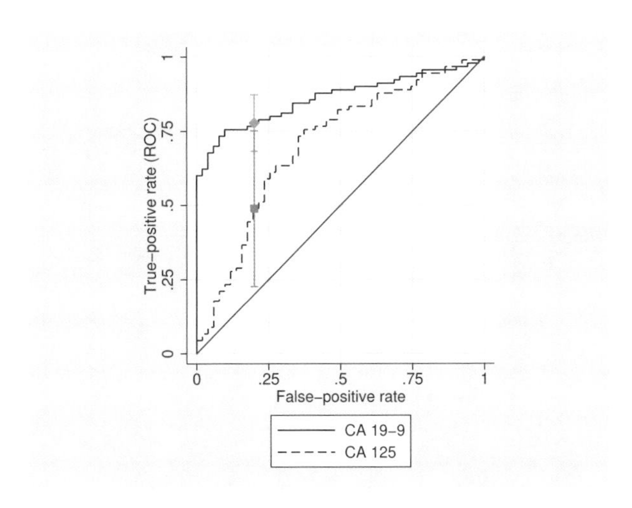

The sensitivity for the false-positive rate of 0.2 is found to be higher under y1 than under y2, and this difference is significant at the 0.05 level. In the plot, this is shown by the vertical confidence intervals.

```
. rocreg d y1 y2, invroc(.6) bseed(8378923) bootcc nodots
Bootstrap results
Number of strata   =          2                Number of obs      =       141
                                               Replications       =      1000
Nonparametric ROC estimation
Control standardization: empirical
ROC method             : empirical
False-positive rate
   Status    : d
   Classifier: y1

             Observed                  Bootstrap
      invROC     Coef.        Bias     Std. Err.     [95% Conf. Interval]

          .6          0   .0158039     .0267288    -.0523874    .0523874  (N)
                                                           0    .0784314  (P)
                                                           0    .1372549 (BC)

   Status    : d
   Classifier: y2

             Observed                  Bootstrap
      invROC     Coef.        Bias     Std. Err.     [95% Conf. Interval]

          .6    .254902   .0101961     .0757902     .1063559     .403448  (N)
                                                    .1372549    .4313726  (P)
                                                    .1176471    .3921569 (BC)

Ho: All classifiers have equal invROC values.
Ha: At least one classifier has a different invROC value.
Test based on bootstrap (N) assumptions.
invROC       P-value

      .6    .0016562
. rocregplot, plot1opts(msymbol(i)) plot2opts(msymbol(i))
```

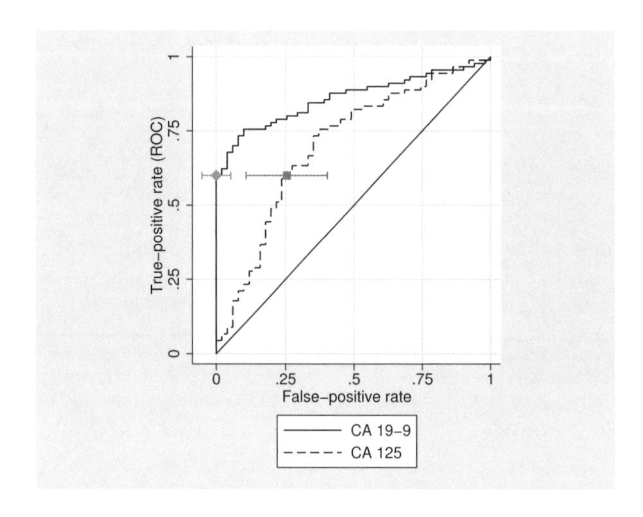

We find significant evidence that false-positive rates corresponding to a sensitivity of 0.6 are different from y1 to y2. This is visually indicated by the horizontal confidence intervals, which are separated from each other.

```
. rocreg d y1 y2, pauc(.5) bseed(8378923) bootcc nodots
Bootstrap results
Number of strata    =          2                Number of obs      =        141
                                                Replications       =       1000

Nonparametric ROC estimation
Control standardization: empirical
ROC method             : empirical
Partial area under the ROC curve
   Status    : d
   Classifier: y1

------------------------------------------------------------------------------
             |   Observed                  Bootstrap
        pAUC |      Coef.        Bias      Std. Err.     [95% Conf. Interval]
-------------+----------------------------------------------------------------
          .5 |   .3932462   -.0000769      .021332      .3514362   .4350562  (N)
             |                                          .3492375    .435512  (P)
             |                                          .3492375    .435403 (BC)
------------------------------------------------------------------------------
   Status    : d
   Classifier: y2

------------------------------------------------------------------------------
             |   Observed                  Bootstrap
        pAUC |      Coef.        Bias      Std. Err.     [95% Conf. Interval]
-------------+----------------------------------------------------------------
          .5 |   .2496732    .0019168     .0374973      .1761798   .3231666  (N)
             |                                           .177451   .3253268  (P)
             |                                          .1738562   .3233115 (BC)
------------------------------------------------------------------------------
Ho: All classifiers have equal pAUC values.
Ha: At least one classifier has a different pAUC value.
Test based on bootstrap (N) assumptions.
pAUC        |  P-value
------------+-----------
         .5 |  .0011201
```

We also find significant evidence supporting the hypothesis that the pAUC for y1 up to a false-positive rate of 0.5 differs from the area of the same region under the ROC curve of y2.

◁

Covariate-adjusted ROC curves

When covariates affect the control distribution of the diagnostic test, thresholds for the test being classified as abnormal may be chosen that vary with the covariate values. These conditional thresholds will be more accurate than the marginal thresholds that would normally be used, because they take into account the specific distribution of the diagnostic test under the given covariate values as opposed to the marginal distribution over all covariate values.

By using these covariate-specific thresholds, we are essentially creating new classifiers for each covariate-value combination, and thus we are creating multiple ROC curves. As explained in Pepe (2003), when the case and control distributions of the covariates are the same, the marginal ROC curve will always be bound above by these covariate-specific ROC curves. So using conditional thresholds will never provide a less powerful test diagnostic in this case.

In the marginal ROC curve calculation, the classifiers are standardized to percentiles according to the control distribution, marginalized over the covariates. Thus the ROC curve is the CDF of the standardized case observations. The covariate-adjusted ROC curve is the CDF of one minus the conditional control percentiles for the case observations, and the marginal ROC curve is the CDF of one minus the marginal control percentiles for the case observations (Pepe and Cai 2004). Thus the standardization of classifier to false-positive rate value is conditioned on the specific covariate values under the covariate-adjusted ROC curve.

The covariate-adjusted ROC curve (Janes and Pepe 2009) at a given false-positive rate t is equivalent to the expected value of the covariate-specific ROC at t over all covariate combinations. When the covariates in question do not affect the case distribution of the classifier, the covariate-specific ROC will have the same value at each covariate combination. So here the covariate-adjusted ROC is equivalent to the covariate-specific ROC, regardless of covariate values.

When covariates do affect the case distribution of the classifier, users of the diagnostic test would likely want to model the covariate-specific ROC curves separately. Tools to do this can be found in the parametric modeling discussion in the following two sections. Regardless, the covariate-adjusted ROC curve can serve as a meaningful summary of covariate-adjusted accuracy.

Also note that the ROC summary statistics defined in the previous section have covariate-adjusted analogs. These analogs are estimated in a similar manner as under the marginal ROC curve (Janes, Longton, and Pepe 2009). The options for their calculation in `rocreg` are identical to those given in the previous section. Further details can be found in *Methods and formulas*.

▷ Example 4: Nonparametric ROC, linear covariate adjustment

Norton et al. (2000) studied data from a neonatal audiology study on three tests to identify hearing impairment in newborns. These data were also studied in Janes, Longton, and Pepe (2009). Here we list 5 of the 5,058 observations.

```
. use http://www.stata-press.com/data/r12/nnhs, clear
(Norton - neonatal audiology data)

. list in 1/5
```

	id	ear	male	currage	d	y1	y2	y3
1.	B0157	R	M	42.42	0	-3.1	-9	-1.5
2.	B0157	L	M	42.42	0	-4.5	-8.7	-2.71
3.	B0158	R	M	40.14	1	-3.2	-13.2	-2.64
4.	B0161	L	F	38.14	0	-22.1	-7.8	-2.59
5.	B0167	R	F	37	0	-10.9	-6.6	-1.42

The classifiers `y1` (DPOAE 65 at 2 kHz), `y2` (TEOAE 80 at 2 kHz), and `y3` (ABR) and the hearing impairment indicator `d` are recorded along with some relevant covariates. The infant's age is recorded in months as `currage`, and the infant's gender is indicated by `male`. Over 90% of the newborns were tested in each ear (`ear`), so we will cluster on infant ID (`id`).

Following the strategy of Janes, Longton, and Pepe (2009), we will first perform ROC analysis for the classifiers while adjusting for the covariate effects of the infant's gender and age. This is done by specifying these variables in the `ctrlcov()` option. We adjust using a linear regression rule, by specifying `ctrlmodel(linear)`. This means that when a user of the diagnostic test chooses a threshold conditional on the age and gender covariates, they assume that the diagnostic test classifier has some linear dependence on age and gender and equal variance as their levels vary. Our cluster adjustment is made by specifying the `cluster()` option.

We will focus on the first classifier. The percentile, or specificity, values are calculated empirically by default, and thus so are the false-positive rates, (1 − specificity). Also by default, the ROC curve values are empirically defined by the false-positive rates. To draw the ROC curve, we again use `rocregplot`.

The AUC is calculated by default. For brevity, we specify the `nobootstrap` option so that bootstrap sampling is not performed. The AUC point estimate will be sufficient for our purposes.

```
. rocreg d y1, ctrlcov(male currage) ctrlmodel(linear) cluster(id) nobootstrap
Nonparametric ROC estimation

Covariate control      : linear regression
Control variables      : male currage
Control standardization: empirical
ROC method             : empirical

Status    : d
Classifier: y1
Covariate control adjustment model:

Linear regression                                      Number of obs =     4907
                                                       F(  2,  2685) =    13.80
                                                       Prob > F      =   0.0000
                                                       R-squared     =   0.0081
                                                       Root MSE      =   7.7515

                                  (Std. Err. adjusted for 2686 clusters in id)

                             Robust
          y1       Coef.   Std. Err.      t    P>|t|     [95% Conf. Interval]

        male    .2471744   .2603598     0.95   0.343    -.2633516    .7577005
     currage   -.2032456   .0389032    -5.22   0.000    -.2795288   -.1269624
       _cons   -1.239484   1.487855    -0.83   0.405    -4.156942    1.677973

Area under the ROC curve

   Status    : d
   Classifier: y1

              Observed                Bootstrap
         AUC      Coef.        Bias    Std. Err.     [95% Conf. Interval]

               .6293994          .           .            .          .  (N)
                                                          .          .  (P)
                                                          .          .  (BC)
```

```
. rocregplot
```

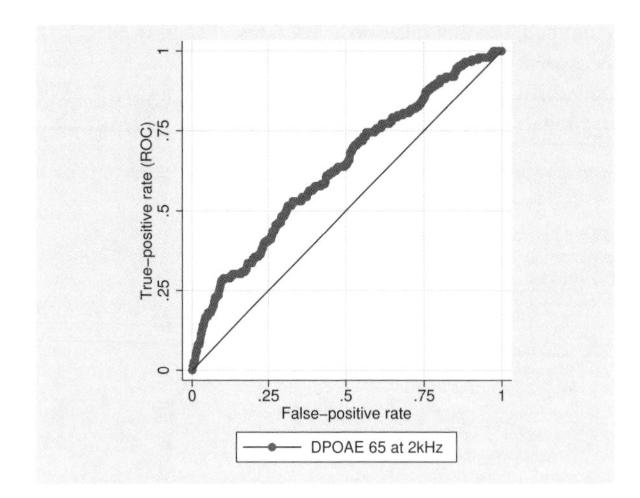

Our covariate control adjustment model shows that `currage` has a negative effect on `y1` (DPOAE 65 at 2 kHz) under the control population. At the 0.001 significance level, we reject that its contribution to `y1` is zero, and the point estimate has a negative sign. This result does not directly tell us about the effect of `currage` on the ROC curve of `y1` as a classifier of `d`. None of the case observations are used in the linear regression, so information on `currage` for abnormal cases is not used in the model. This result does show us how to calculate false-positive rates for tests that use thresholds conditional on a child's sex and current age. We will see how `currage` affects the ROC curve when `y1` is used as a classifier and conditional thresholds are used based on `male` and `currage` in the following section, *Parametric ROC curves: Estimating equations.*

❑ Technical note

Under this nonparametric estimation, `rocreg` saved the false-positive rate for each observation's `y1` values in the utility variable `_fpr_y1`. The true-positive rates are stored in the utility variable `_roc_y1`. For other models, say with classifier *yname*, these variables would be named `_fpr_`*yname* and `_roc_`*yname*. They will also be overwritten with each call of `rocreg`. The variables `_roc_*` and `_fpr_*` are usually for internal `rocreg` use only and are overwritten with each call of `rocreg`. They are only created for nonparametric models or parametric models that do not involve ROC covariates. In these models, covariates may only affect the first stage of estimation, the control distribution, and not the ROC curve itself. In parametric models that allow ROC covariates, different covariate values would lead to different ROC curves.

❑

To see how the covariate-adjusted ROC curve estimate differs from the standard marginal estimate, we will reestimate the ROC curve for classifier `y1` without covariate adjustment. We rename these variables before the new estimation and then draw an overlaid `twoway line` (see [G-2] **graph twoway line**) plot to compare the two.

```
. rename _fpr_y1 o_fpr_y1
. rename _roc_y1 o_roc_y1
. label variable o_roc_y1 "covariate_adjusted"
. rocreg d y1, cluster(id) nobootstrap
Nonparametric ROC estimation
Control standardization: empirical
ROC method             : empirical
Area under the ROC curve
   Status    : d
   Classifier: y1
```

AUC	Observed Coef.	Bias	Bootstrap Std. Err.	[95% Conf. Interval]		
	.6279645	.	.	.	.	(N)
				.	.	(P)
				.	.	(BC)

```
. label variable _roc_y1 "marginal"
. twoway line _roc_y1 _fpr_y1, sort(_fpr_y1 _roc_y1) connect(J) ||
     line o_roc_y1 o_fpr_y1, sort(o_fpr_y1 o_roc_y1)
     connect(J) lpattern(dash) aspectratio(1) legend(cols(1))
```

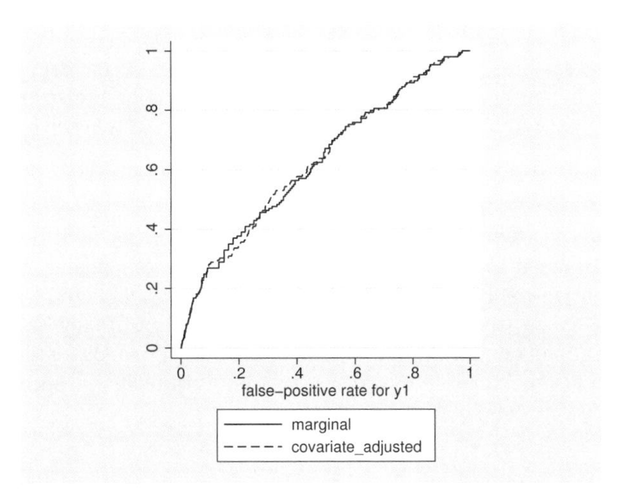

Though they are close, particularly in AUC, there are clearly some points of difference between the estimates. So the covariate-adjusted ROC curve may be useful here.

◁

In our examples thus far, we have used the empirical CDF estimator to estimate the control distribution. `rocreg` allows some flexibility here. The `pvc(normal)` option may be specified to calculate the percentile values according to a Gaussian distribution of the control.

Covariate adjustment in `rocreg` may also be performed with stratification instead of linear regression. Under the stratification method, the unique values of the stratified covariates each define separate parameters for the control distribution of the classifier. A user of the diagnostic test chooses a threshold based on the control distribution conditioned on the unique covariate value parameters.

We will demonstrate the use of normal percentile values and covariate stratification in our next example.

▷ Example 5: Nonparametric ROC, covariate stratification

The hearing test study of Stover et al. (1996) examined the effectiveness of negative signal-to-noise ratio, nsnr, as a classifier of hearing loss. The test was administered under nine different settings, corresponding to different frequency, xf, and intensity, xl, combinations. Here we list 10 of the 1,848 observations.

```
. use http://www.stata-press.com/data/r12/dp, clear
(Stover - DPOAE test data)

. list in 1/10

     +---------------------------------------------+
     |  id   d   nsnr      xf    xl    xd |
     |---------------------------------------------|
  1. | 101   1     18   10.01   5.5   3.5 |
  2. | 101   1     19   20.02   5.5     3 |
  3. | 101   1    7.6   10.01     6   3.5 |
  4. | 101   1     15   20.02     6     3 |
  5. | 101   1     16   10.01   6.5   3.5 |
     |---------------------------------------------|
  6. | 101   1    5.8   20.02   6.5     3 |
  7. | 102   0   -2.6   10.01   5.5     . |
  8. | 102   0     -3   14.16   5.5     . |
  9. | 102   1     10   20.02   5.5     1 |
 10. | 102   0   -5.8   10.01     6     . |
     +---------------------------------------------+
```

Hearing loss is represented by d. The covariate xd is a measure of the degree of hearing loss. We will use this covariate in later analysis, because it only affects the case distribution of the classifier. Multiple measurements are taken for each individual, id, so we will cluster by individual.

We evaluate the effectiveness of nsnr using xf and xl as stratification covariates with rocreg; the default method of covariate adjustment.

As mentioned before, the default false-positive rate calculation method in rocreg estimates the conditional control distribution of the classifiers empirically. For comparison, we will also estimate a separate ROC curve using false-positive rates assuming the conditional control distribution is normal. This behavior is requested by specifying the pvc(normal) option. Using the rocregplot option name() to store the ROC plots and using the graph combine command, we are able to compare the Gaussian and empirical ROC curves side by side. As before, for brevity we specify the nobootstrap option to suppress bootstrap sampling.

```
. rocreg d nsnr, ctrlcov(xf xl) cluster(id) nobootstrap
Nonparametric ROC estimation

Covariate control       : stratification
Control variables       : xf xl
Control standardization: empirical
ROC method              : empirical

Area under the ROC curve

   Status    : d
   Classifier: nsnr

------------------------------------------------------------------------------
             |    Observed                Bootstrap
         AUC |       Coef.       Bias     Std. Err.     [95% Conf. Interval]
-------------+----------------------------------------------------------------
             |    .9264192          .             .            .          .  (N)
             |                                                 .          .  (P)
             |                                                 .          . (BC)
------------------------------------------------------------------------------

. rocregplot, title(Empirical FPR) name(a) nodraw
```

```
. rocreg d nsnr, pvc(normal) ctrlcov(xf xl) cluster(id) nobootstrap
Nonparametric ROC estimation

Covariate control       : stratification
Control variables       : xf xl
Control standardization: normal
ROC method              : empirical

Area under the ROC curve

   Status    : d
   Classifier: nsnr

------------------------------------------------------------------------------
             |  Observed               Bootstrap
         AUC |     Coef.       Bias    Std. Err.     [95% Conf. Interval]
-------------+----------------------------------------------------------------
             |  .9309901          .            .             .          .  (N)
             |                                               .          .  (P)
             |                                               .          . (BC)
------------------------------------------------------------------------------

. rocregplot, title(Normal FPR) name(b) nodraw
. graph combine a b, xsize(5)
```

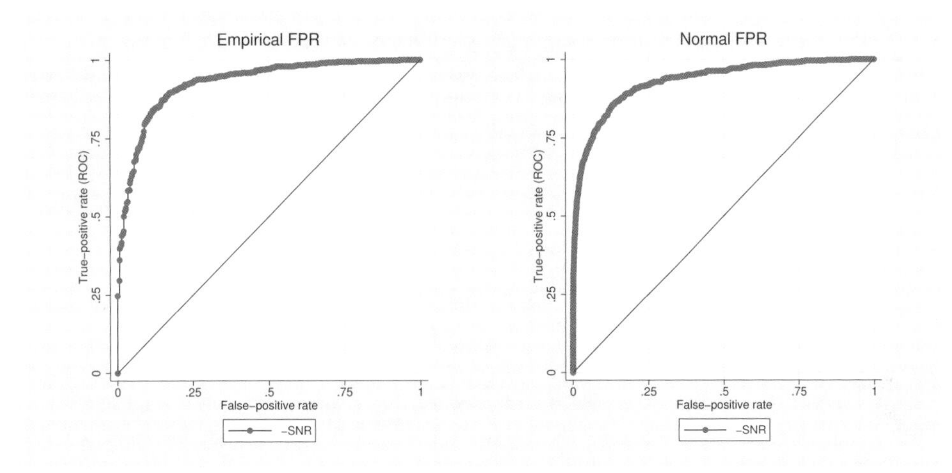

On cursory visual inspection, we see little difference between the two curves. The AUC values are close as well. So it is sensible to assume that we have Gaussian percentile values for control standardization. ◁

Parametric ROC curves: Estimating equations

We now assume a parametric model for covariate effects on the second stage of ROC analysis. Particularly, the ROC curve is a probit model of the covariates. We will thus have a separate ROC curve for each combination of the relevant covariates.

Under weak assumptions about the control distribution of the classifier, we can fit this model by using estimating equations as described in Alonzo and Pepe (2002). This method can be also be used without covariate effects in the second stage, assuming a parametric model for the single (constant only) ROC curve. Covariates may still affect the first stage of estimation, so we parametrically model the single covariate-adjusted ROC curve (from the previous section). The marginal ROC curve, involving no covariates in either stage of estimation, can be fit parametrically as well.

In addition to the Alonzo and Pepe (2002) explanation, further details are given in Pepe, Longton, and Janes (2009); Janes, Longton, and Pepe (2009); Pepe (2003); and Janes and Pepe (2009).

The parametric models that we consider assume that the ROC curve is a cumulative distribution function g invoked with input of a linear polynomial in the corresponding quantile function invoked on the false-positive rate u. In this context, we assume that g corresponds to a standard normal cumulative distribution function, Φ. So the corresponding quantile function is Φ^{-1}. The constant intercept of the polynomial may depend on covariates, but the slope term α (the quantile coefficient) may not.

$$\text{ROC}\,(u) = g\{\mathbf{x}'\boldsymbol{\beta} + \alpha g^{-1}\,(u)\}$$

The first step of the algorithm involves the choice of false-positive rates to use in the parametric fit. These are typically a set of equispaced points spanning the interval (0,1). Alonzo and Pepe (2002) examined the effect of fitting large and small sets of points, finding that relatively small sets could be used with little loss of efficiency. Alternatively, the set can be formed by using the observed false-positive rates in the data (Pepe 2003). Further details on the algorithm are provided in *Methods and formulas*.

Under parametric estimation, all the summary measures we defined earlier, except the AUC, are not calculated until postestimation. In models with covariates, each covariate combination would yield a different ROC curve and thus different summary parameters, so no summary parameters are initially estimated. In marginal parametric models (where there are no ROC covariates, but there are potentially control covariates), we will calculate the AUC and leave the other measures for postestimation; see [R] **rocreg postestimation**. As with the other parameters, we bootstrap for standard errors and inference.

We will now demonstrate how `rocreg` performs the Alonzo and Pepe (2002) algorithm using the previous section's examples and others.

▷ Example 6: Parametric ROC, linear covariate adjustment

We return to the neonatal audiology study with gender and age covariates (Norton et al. 2000), which we discussed in example 4. Janes, Longton, and Pepe (2009) suspected the current age of the infant would play a role in the case distribution of the classifier `y1` (DPOAE 65 at 2 kHz). They postulated a probit link between the ROC curve and the covariate-adjusted false-positive rates. We follow their investigation and reach similar results.

In example 4, we saw the results of adjusting for the `currage` and `male` variables in the control population for classifier `y1`. Now we see how `currage` affects the ROC curve when `y1` is used with thresholds conditioned on `male` and `currage`.

We specify the covariates that should affect the ROC curve in the `roccov()` option. By default, `rocreg` will choose 10 equally spaced false-positive rates in the (0,1) interval as fitting points. The `fprpts()` option allows the user to specify more or fewer points. We specify the `bsave()` option with the `nnhs2y1` dataset so that we can use the bootstrap resamples in postestimation.

```
. use http://www.stata-press.com/data/r12/nnhs, clear
(Norton - neonatal audiology data)

. rocreg d y1, probit ctrlcov(currage male) ctrlmodel(linear) roccov(currage)
> cluster(id) bseed(56930) bsave(nnhs2y1) nodots

Bootstrap results                               Number of obs      =      5056
                                                Replications       =      1000

Parametric ROC estimation

Covariate control       : linear regression
Control variables       : currage male
Control standardization: empirical
ROC method              : parametric            Link: probit

Status    : d
Classifier: y1
Covariate control adjustment model:

Linear regression                                      Number of obs =    4907
                                                       F(  2,  2685) =   13.80
                                                       Prob > F      =  0.0000
                                                       R-squared     =  0.0081
                                                       Root MSE      =  7.7515

                                   (Std. Err. adjusted for 2686 clusters in id)

------------------------------------------------------------------------------
             |               Robust
          y1 |      Coef.   Std. Err.      t    P>|t|     [95% Conf. Interval]
-------------+----------------------------------------------------------------
     currage |  -.2032456   .0389032    -5.22   0.000    -.2795288   -.1269624
        male |   .2471744   .2603598     0.95   0.343    -.2633516    .7577005
       _cons |  -1.239484   1.487855    -0.83   0.405    -4.156942    1.677973
------------------------------------------------------------------------------

   Status    : d
   Classifier: y1
   ROC Model :
                                  (Replications based on 2741 clusters in id)

------------------------------------------------------------------------------
             |    Observed               Bootstrap
          y1 |       Coef.       Bias    Std. Err.   [95% Conf. Interval]
-------------+----------------------------------------------------------------
       _cons |   -1.272505  -.0566737    1.076706     -3.38281   .8377993  (N)
             |                                       -3.509356   .7178385  (P)
             |                                       -3.487457   .7813575 (BC)
     currage |    .0448228   .0015878    .0280384    -.0101316   .0997771  (N)
             |                                        -.007932   .1033131  (P)
             |                                       -.0102905    .101021 (BC)
-------------+----------------------------------------------------------------
      probit |
-------------+----------------------------------------------------------------
       _cons |    .9372393   .0128376    .0747228     .7907853   1.083693  (N)
             |                                        .8079087   1.101941  (P)
             |                                        .7928988   1.083399 (BC)
------------------------------------------------------------------------------
```

Note how the number of clusters—here infants—changes from the covariate control adjustment model fit to the ROC model. The control fit is limited to control cases and thus fewer infants. The ROC is fit on all the data, so the variance is adjusted for all clustering on all infants.

With a 0.05 level of statistical significance, we cannot reject the null hypothesis that `currage` has no effect on the ROC curve at a given false-positive rate. This is because each of our 95% bootstrap confidence intervals contains 0. This corresponds with the finding in Janes, Longton, and Pepe (2009) where the reported 95% intervals each contained 0. We cannot reject that the intercept parameter β_0, reported as `_cons` in the main table, is 0 at the 0.05 level either. The slope parameter α, reported as _cons in the probit table, is close to 1 and cannot be rejected as being 1 at the 0.05 level. Under the assumption that the ROC coefficients except α are 0 and that $\alpha = 1$, the ROC curve at false-positive rate u is equal to u. In other words, we cannot reject that the false-positive rate is equal to the true-positive rate, and so the test is noninformative. Further investigation of the results requires postestimation; see [R] **rocreg postestimation**.

◁

The fitting point set can be formed by using the observed false-positive rates (Pepe 2003). Our next example will illustrate this.

▷ Example 7: Parametric ROC, covariate stratification

We return to the hearing test study of Stover et al. (1996), which we discussed in example 5. Pepe (2003) suspected that intensity, xd, would play a role in the case distribution of the negative signal-to-noise ratio (nsnr) classifier. A ROC regression was fit with covariate adjustment for xf and xl with stratification, and for ROC covariates xf, xl, and xd. There is no prohibition against the same covariate being used in the first and second stages of ROC calculation. The false-positive rate fitting point set was composed of all observed false-positive rates in the control data.

We fit the model with rocreg here. Using observed false-positive rates as the fitting point set can make the dataset very large, so fitting the model is computationally intensive. We demonstrate the fitting algorithm without precise confidence intervals, focusing instead on the coefficient estimates and standard errors. We will thus perform only 50 bootstrap replications, a reasonable number to obtain accurate standard error estimates (Mooney and Duval 1993). The number of replications is specified in the breps() option.

The ROC covariates are specified in roccov(). We specify that all observed false-positive rates in the control observations be used as fitting points with the ctrlfprall option. The nobstrata option specifies that the bootstrap is not stratified. The covariate stratification in the first stage of estimation does not affect the resampling. We will return to this example in postestimation, so we save the bootstrap results in the nsnrf dataset with the bsave() option.

```
. use http://www.stata-press.com/data/r12/dp
(Stover - DPOAE test data)

. rocreg d nsnr, probit ctrlcov(xf xl) roccov(xf xl xd) ctrlfprall cluster(id)
> nobstrata bseed(156385) breps(50) bsave(nsnrf)
(running rocregstat on estimation sample)

Bootstrap replications (50)
----+--- 1 ---+--- 2 ---+--- 3 ---+--- 4 ---+--- 5
..................................................    50

Bootstrap results                            Number of obs      =      1848
                                             Replications       =        50

Parametric ROC estimation

Covariate control       : stratification
Control variables       : xf xl
Control standardization: empirical
ROC method              : parametric         Link: probit

   Status    : d
   Classifier: nsnr
   ROC Model :
                                   (Replications based on 208 clusters in id)

------------------------------------------------------------------------------
             |   Observed               Bootstrap
        nsnr |      Coef.       Bias    Std. Err.     [95% Conf. Interval]
-------------+----------------------------------------------------------------
       _cons |   3.247872  -.0846178    .8490006     1.583862   4.911883  (N)
             |                                       1.598022   4.690076  (P)
             |                                       1.346904   4.690076 (BC)
          xf |   .0502557    .014478    .0329044    -.0142357    .1147471  (N)
             |                                      -.0031814    .1186107  (P)
             |                                      -.0053095    .1132185 (BC)
          xl |  -.4327223  -.0194846    .1116309    -.6515149  -.2139298  (N)
             |                                      -.6570321  -.2499706  (P)
             |                                      -.6570321   -.231854 (BC)
          xd |   .4431764   .0086147    .0936319     .2596612   .6266916  (N)
             |                                        .330258   .6672749  (P)
             |                                       .3487118   .7674865 (BC)
-------------+----------------------------------------------------------------
      probit |
-------------+----------------------------------------------------------------
       _cons |   1.032657  -.0188887    .1224993     .7925628   1.272751  (N)
             |                                       .7815666   1.236179  (P)
             |                                       .7815666   1.237131 (BC)
------------------------------------------------------------------------------
```

We obtain results similar to those reported in Pepe (2003, 159). Unlike in our previous example, we find that the coefficients for `xl` and `xd` differ from 0 at the 0.05 level of significance. So over certain covariate combinations, we can have a variety of informative tests using `nsnr` as a classifier. ◁

As mentioned before, when there are no covariates, `rocreg` can still fit a parametric model for the ROC curve of a classifier by using the Alonzo and Pepe (2002) method. `roccomp` and `rocfit` can fit marginal probit models as well. We will compare the behavior of `rocreg` with that of `roccomp` and `rocfit` for probit models without covariates.

When the `binormal` option is specified, `roccomp` calculates the AUC for input classifiers according to the maximum likelihood algorithm of `rocfit`. The `rocfit` algorithm expects discrete classifiers but can slice continuous classifiers into discrete partitions. Further, the case and control distributions are both assumed normal. Actually, the observed classification values are taken as discrete indicators of the latent normally distributed classification values. This method is documented in Dorfman and Alf (1969).

Alonzo and Pepe (2002) compared their estimating equations probability density function method (with empirical estimation of the false-positive rates) to the maximum likelihood approach of Dorfman and Alf (1969) and found that they had similar efficiency and mean squared error. So we should expect `rocfit` and `rocreg` to give similar results when fitting a simple probit model.

▷ Example 8: Parametric ROC, marginal model

We return to the Hanley and McNeil (1982) data. We will fit a probit model to the ROC curve, assuming that the `rating` variable is a discrete indicator of an underlying latent normal random variable in both the case and control populations of `disease`. We invoke `rocfit` with the default options. `rocreg` is invoked with the `probit` option. The percentile values are calculated empirically. Because there are fewer categories than 10, there will be fewer than 10 false-positive rates that trigger a different true-positive rate value. So for efficiency, we invoke `rocreg` with the `ctrlfprall` option.

```
. use http://www.stata-press.com/data/r12/hanley
. rocfit disease rating, nolog
Binormal model of disease on rating                 Number of obs   =        109
Goodness-of-fit chi2(2) =       0.21
Prob > chi2             =     0.9006
Log likelihood          = -123.64855

------------------------------------------------------------------------------
             |      Coef.   Std. Err.      z    P>|z|     [95% Conf. Interval]
-------------+----------------------------------------------------------------
   intercept |   1.656782   0.310456     5.34   0.000     1.048300    2.265265
   slope (*) |   0.713002   0.215882    -1.33   0.184     0.289881    1.136123
-------------+----------------------------------------------------------------
       /cut1 |   0.169768   0.165307     1.03   0.304    -0.154227    0.493764
       /cut2 |   0.463215   0.167235     2.77   0.006     0.135441    0.790990
       /cut3 |   0.766860   0.174808     4.39   0.000     0.424243    1.109477
       /cut4 |   1.797938   0.299581     6.00   0.000     1.210770    2.385106
------------------------------------------------------------------------------

------------------------------------------------------------------------------
             |              Indices from binormal fit
       Index |   Estimate   Std. Err.                     [95% Conf. Interval]
-------------+----------------------------------------------------------------
    ROC area |   0.911331   0.029506                      0.853501    0.969161
    delta(m) |   2.323671   0.502370                      1.339044    3.308298
        d(e) |   1.934361   0.257187                      1.430284    2.438438
        d(a) |   1.907771   0.259822                      1.398530    2.417012
------------------------------------------------------------------------------
(*) z test for slope==1
```

```
. rocreg disease rating, probit ctrlfprall bseed(8574309) nodots
Bootstrap results                               Number of obs      =       109
                                                Replications       =      1000

Parametric ROC estimation

Control standardization: empirical
ROC method              : parametric            Link: probit

   Status    : disease
   Classifier: rating
   ROC Model :

------------------------------------------------------------------------------
             |   Observed               Bootstrap
      rating |      Coef.       Bias    Std. Err.     [95% Conf. Interval]
-------------+----------------------------------------------------------------
       _cons |   1.635041   .0588548    .3609651     .9275621    2.342519  (N)
             |                                       1.162363    2.556508  (P)
             |                                       1.164204    2.566174 (BC)
-------------+----------------------------------------------------------------
      probit |
-------------+----------------------------------------------------------------
       _cons |   .6951252   .0572146    .3241451     .0598125    1.330438  (N)
             |                                       .3500569    1.430441  (P)
             |                                       .3372983    1.411953 (BC)
------------------------------------------------------------------------------

------------------------------------------------------------------------------
             |   Observed               Bootstrap
         AUC |      Coef.       Bias    Std. Err.     [95% Conf. Interval]
-------------+----------------------------------------------------------------
             |   .9102903  -.0051749    .0314546     .8486405    .9719402  (N)
             |                                        .837113    .9605498  (P)
             |                                       .8468336    .9630486 (BC)
------------------------------------------------------------------------------
```

We see that the intercept and slope parameter estimates are close. The intercept (_cons in the main table) is clearly nonzero. Under rocreg, the slope (_cons in the probit table) and its percentile and bias-corrected confidence intervals are close to those of rocfit. The area under the ROC curve for each of the rocreg and rocfit estimators also matches closely.

◁

Now we will compare the parametric fit of rocreg under the constant probit model with roccomp.

▷ Example 9: Parametric ROC, marginal model, multiple classifiers

We now use the fictitious dataset generated from Hanley and McNeil (1983). To fit a probit model using roccomp, we specify the binormal option. Our specification of rocreg remains the same as before.

rocregplot is used to render the model produced by rocreg. We specify several graph options to both roccomp and rocregplot to ease comparison. When the binormal option is specified along with graph, roccomp will draw the binormal fitted lines in addition to connected line plots of the empirical false-positive and true-positive rates.

In this plot, we overlay scatterplots of the empirical false-positive rates (because percentile value calculation defaulted to pvc(empirical)) and the parametric true-positive rates.

```
. use http://www.stata-press.com/data/r12/ct2, clear
. roccomp status mod1 mod2 mod3, summary binormal graph aspectratio(1)
>         plot1opts(connect(i) msymbol(o))
>         plot2opts(connect(i) msymbol(s))
>         plot3opts(connect(i) msymbol(t))
>         legend(label(1 "mod1") label(3 "mod2") label(5 "mod3")
>              label(2 "mod1 fit") label(4 "mod2 fit")
>              label(6 "mod3 fit")  order(1 3 5 2 4 6) cols(1))
>         title(roccomp) name(a) nodraw
Fitting binormal model for: mod1
Fitting binormal model for: mod2
Fitting binormal model for: mod3
                                 ROC
                       Obs       Area     Std. Err.      [95% Conf. Interval]
--------------------------------------------------------------------------------
mod1                   112     0.8945       0.0305        0.83482     0.95422
mod2                   112     0.9382       0.0264        0.88647     0.99001
mod3                   112     0.9376       0.0223        0.89382     0.98139
--------------------------------------------------------------------------------
Ho: area(mod1) = area(mod2) = area(mod3)
    chi2(2) =      8.27       Prob>chi2 =    0.0160
. rocreg status mod1 mod2 mod3, probit ctrlfprall bseed(867340912) nodots
Bootstrap results                                Number of obs      =       112
                                                 Replications       =      1000
Parametric ROC estimation
Control standardization: empirical
ROC method             : parametric              Link: probit
   Status    : status
   Classifier: mod1
   ROC Model :
------------------------------------------------------------------------------
             |   Observed               Bootstrap
        mod1 |      Coef.       Bias    Std. Err.     [95% Conf. Interval]
-------------+----------------------------------------------------------------
       _cons |   1.726034   .1363112    .5636358     .6213277    2.83074   (N)
             |                                       1.162477    3.277376   (P)
             |                                       1.152112    3.187595 (BC)
-------------+----------------------------------------------------------------
      probit |
-------------+----------------------------------------------------------------
       _cons |   .9666323   .0872018    .4469166     .0906919    1.842573   (N)
             |                                        .518082    2.219548   (P)
             |                                       .5568404    2.394036 (BC)
------------------------------------------------------------------------------

------------------------------------------------------------------------------
             |   Observed               Bootstrap
         AUC |      Coef.       Bias    Std. Err.     [95% Conf. Interval]
-------------+----------------------------------------------------------------
             |   .8927007  -.0011794    .0313951     .8311675     .954234   (N)
             |                                       .8245637    .9466904   (P)
             |                                       .8210562    .9432855 (BC)
------------------------------------------------------------------------------
```

```
Status    : status
Classifier: mod2
ROC Model :

------------------------------------------------------------------------------
             |   Observed               Bootstrap
        mod2 |      Coef.       Bias    Std. Err.     [95% Conf. Interval]
-------------+----------------------------------------------------------------
       _cons |   1.696811   .0918364    .5133386     .6906858    2.702936  (N)
             |                                       1.21812    2.973929  (P)
             |                                       1.22064    3.068454 (BC)
-------------+----------------------------------------------------------------
      probit |
-------------+----------------------------------------------------------------
       _cons |   .4553828    .047228    .3345303    -.2002845    1.11105  (N)
             |                                      .1054933     1.18013  (P)
             |                                      .1267796    1.272523 (BC)
------------------------------------------------------------------------------

------------------------------------------------------------------------------
             |   Observed               Bootstrap
         AUC |      Coef.       Bias    Std. Err.     [95% Conf. Interval]
-------------+----------------------------------------------------------------
             |    .938734  -.0037989    .0261066     .8875659    .9899021  (N)
             |                                       .8777664    .9778214  (P)
             |                                       .8823555    .9792451 (BC)
------------------------------------------------------------------------------

Status    : status
Classifier: mod3
ROC Model :

------------------------------------------------------------------------------
             |   Observed               Bootstrap
        mod3 |      Coef.       Bias    Std. Err.     [95% Conf. Interval]
-------------+----------------------------------------------------------------
       _cons |   2.281359   .1062846    .6615031     .9848363    3.577881  (N)
             |                                      1.637764    4.157873  (P)
             |                                      1.666076    4.474779 (BC)
-------------+----------------------------------------------------------------
      probit |
-------------+----------------------------------------------------------------
       _cons |   1.107736   .0514693    .4554427     .2150843    2.000387  (N)
             |                                        .58586     2.28547  (P)
             |                                      .6385949    2.671192 (BC)
------------------------------------------------------------------------------

------------------------------------------------------------------------------
             |   Observed               Bootstrap
         AUC |      Coef.       Bias    Std. Err.     [95% Conf. Interval]
-------------+----------------------------------------------------------------
             |   .9368321  -.0023853    .0231363     .8914859    .9821784  (N)
             |                                      .8844096    .9722485  (P)
             |                                      .8836259    .9718463 (BC)
------------------------------------------------------------------------------

Ho: All classifiers have equal AUC values.
Ha: At least one classifier has a different AUC value.

P-value:     .0778556             Test based on bootstrap (N) assumptions.

. rocregplot, title(rocreg) nodraw name(b) plot1opts(msymbol(o))
> plot2opts(msymbol(s)) plot3opts(msymbol(t))
```

```
. graph combine a b
```

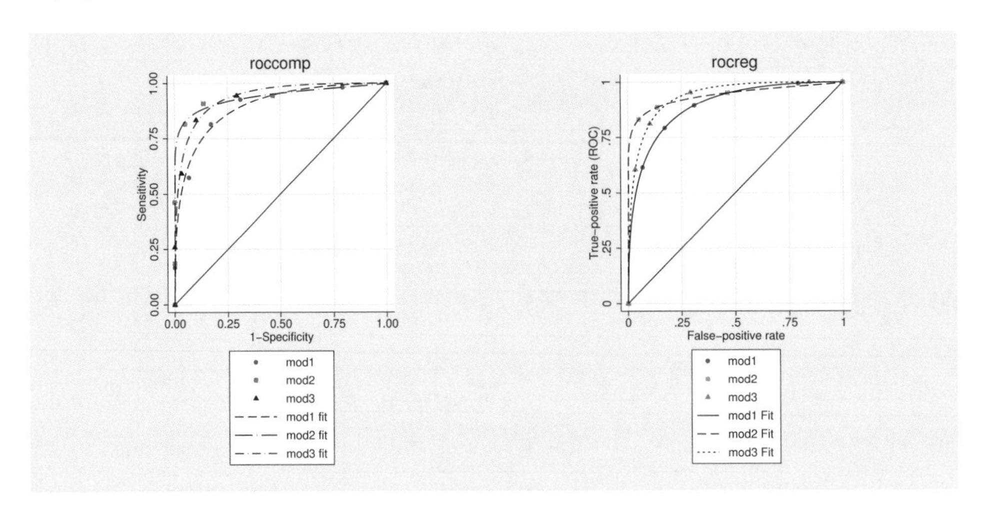

We see differing true-positive rate values in the scattered points, which is expected because `roccomp` gives the empirical estimate and `rocreg` gives the parametric estimate. However, the estimated curves and areas under the ROC curve look similar. Using the Wald test based on the bootstrap covariance, `rocreg` rejects the null hypothesis that each test has the same AUC at the 0.1 significance level. `roccomp` formulates the asymptotic covariance using the `rocfit` estimates of AUC. Examination of its output leads to rejection of the null hypothesis that the AUCs are equal across each test at the 0.05 significance level.

◁

Parametric ROC curves: Maximum likelihood

The Alonzo and Pepe (2002) method of fitting a parametric model to the ROC curve is powerful because it can be generally applied, but that can be a limitation as well. Whenever we invoke the method and want anything other than point estimates of the parameters, we must perform bootstrap resampling.

An alternative is to use maximum likelihood inference to fit the ROC curve. This method can save computational time by avoiding the bootstrap.

`rocreg` implements maximum likelihood estimation for ROC curve analysis when both the case and control populations are normal. Particularly, the classifier is a normal linear model on certain covariates, and the covariate effect and variance of the classifier may change between the case and control populations. This model is defined in Pepe (2003, 145).

$$y = \mathbf{z}'\boldsymbol{\beta}_0 + D\mathbf{x}'\boldsymbol{\beta}_1 + \sigma\left(D\right)\epsilon$$

Our error term, ϵ, is a standard normal random variable. The variable D is our true status variable, being 1 for the case population observations and 0 for the control population observations. The variance function σ is defined as

$$\sigma\left(D\right) = \sigma_0\left(D=0\right) + \sigma_1\left(D=1\right)$$

This provides two variance parameters in the model and does not depend on covariate values.

Suppose a covariate x_i is present in $\mathbf{z}$ and $\mathbf{x}$. The coefficient β_{1i} represents the interaction effect of the x_i and D. It is the extra effect that x_i has on classifier y under the case population, $D = 1$, beyond the main effect β_{0i}. These $\boldsymbol{\beta_1}$ coefficients are directly related to the ROC curve of y.

Under this model, the ROC curve is derived to be

$$\text{ROC}\,(u) = \Phi\left[\frac{1}{\sigma_1}\{\mathbf{x}'\boldsymbol{\beta_1} + \sigma_0\Phi^{-1}\,(u)\}\right]$$

For convenience, we reparameterize the model at this point, creating the parameters $\beta_i = \sigma_1^{-1}\beta_{1i}$ and $\alpha = \sigma_1^{-1}\sigma_0$. We refer to β_0 as the constant intercept, `i_cons`. The parameter α is referred to as the constant slope, `s_cons`.

$$\text{ROC}\,(u) = \Phi\{\mathbf{x}'\boldsymbol{\beta} + \alpha\Phi^{-1}\,(u)\}$$

We may interpret the final coefficients as the standardized linear effect of the ROC covariate on the classifier under the case population. The marginal effect of the covariate on the classifier in the control population is removed, and it is rescaled by the case population standard deviation of the classifier when all ROC covariate effects are removed. An appreciable effect on the classifier by a ROC covariate in this measure leads to an appreciable effect on the classifier's ROC curve by the ROC covariate.

The advantage of estimating the control coefficients $\boldsymbol{\beta_0}$ is similar to the gains of estimating the covariate control models in the estimating equations ROC method and nonparametric ROC estimation. This model would similarly apply when evaluating a test that is conditioned on control covariates.

Again we note that under parametric estimation, all the summary measures we defined earlier except the AUC are not calculated until postestimation. In models with covariates, each covariate combination would yield a different ROC curve and thus different summary parameters, so no summary parameters are estimated initially. In marginal parametric models, we will calculate the AUC and leave the other measures for postestimation. There is a simple closed-form formula for the AUC under the probit model. Using this formula, the delta method can be invoked for inference on the AUC. Details on AUC estimation for probit marginal models are found in *Methods and formulas*.

We will demonstrate the maximum likelihood method of `rocreg` by revisiting the models of the previous section.

▷ Example 10: Maximum likelihood ROC, single classifier

Returning to the hearing test study of Stover et al. (1996), we use a similar covariate grouping as before. The frequency `xf` and intensity `xl` are control covariates ($\mathbf{z}$), while all three covariates `xf`, `xl`, and hearing loss degree `xd` are case covariates ($\mathbf{x}$). In example 7, we fit this model using the Alonzo and Pepe (2002) method. Earlier we stratified on the control covariates and estimated the conditioned control distribution of `nsnr` empirically. Now we assume a normal linear model for `nsnr` on `xf` and `xl` under the control population.

We fit the model by specifying the control covariates in the `ctrlcov()` option and the case covariates in the `roccov()` option. The `ml` option tells `rocreg` to perform maximum likelihood estimation.

```
. use http://www.stata-press.com/data/r12/dp, clear
(Stover - DPOAE test data)
. rocreg d nsnr, ctrlcov(xf xl) roccov(xf xl xd) probit ml cluster(id) nolog
Parametric ROC estimation
Covariate control        : linear regression
Control variables        : xf xl
Control standardization: normal
ROC method               : parametric                Link: probit
  Status     : d
  Classifiers: nsnr
  Classifier : nsnr
  Covariate control adjustment model:
                                  (Std. Err. adjusted for 208 clusters in id)
```

	Coef.	Robust Std. Err.	z	P>|z|	[95% Conf.	Interval]
casecov						
xf	.4690907	.1408683	3.33	0.001	.192994	.7451874
xl	-3.187785	.8976521	-3.55	0.000	-4.947151	-1.42842
xd	3.042998	.3569756	8.52	0.000	2.343339	3.742657
_cons	23.48064	5.692069	4.13	0.000	12.32439	34.63689
casesd						
_cons	7.979708	.354936	22.48	0.000	7.284047	8.67537
ctrlcov						
xf	-.1447499	.0615286	-2.35	0.019	-.2653438	-.0241561
xl	-.8631348	.2871976	-3.01	0.003	-1.426032	-.3002378
_cons	1.109477	1.964004	0.56	0.572	-2.7399	4.958854
ctrlsd						
_cons	7.731203	.3406654	22.69	0.000	7.063511	8.398894

```
   Status    : d
   ROC Model :
                                  (Std. Err. adjusted for 208 clusters in id)
```

	Coef.	Robust Std. Err.	z	P>|z|	[95% Conf.	Interval]
nsnr						
i_cons	2.942543	.7569821	3.89	0.000	1.458885	4.426201
xf	.0587854	.0175654	3.35	0.001	.024358	.0932129
xl	-.3994865	.1171914	-3.41	0.001	-.6291775	-.1697955
xd	.381342	.0449319	8.49	0.000	.2932771	.4694068
s_cons	.9688578	.0623476	15.54	0.000	.8466587	1.091057

We find the results are similar to those of example 7. Frequency (xf) and intensity (xl) have a negative effect on the classifier nsnr in the control population.

The negative control effect is mitigated for xf in the case population, but the effect for xl is even more negative there. Hearing loss severity, xd, has a positive effect on nsnr in the case population, and it is undefined in the control population.

The ROC coefficients are shown in the ROC Model table. Each are different from 0 at the 0.05 level. At this level, we also cannot conclude that the variances differ from case to control populations, because 1 is in the 95% confidence interval for s_cons, the ratio of the case to control standard deviation parameters.

Both frequency (`xf`) and hearing loss severity (`xd`) make a positive contribution to the ROC curve and thus make the test more powerful. Intensity (`xl`) has a negative effect on the ROC curve and weakens the test. We previously saw in example 5 that the control distribution appears to be normal, so using maximum likelihood to fit this model is a reasonable approach.

This model was also fit in Pepe (2003, 147). Pepe used separate least-squares estimates for the case and control samples. We obtain similar results for the coefficients, but the maximum likelihood fitting yields slightly different standard deviations by considering both case and control observations concurrently. In addition, a misprint in Pepe (2003, 147) reports a coefficient of −4.91 for `xl` in the case population instead of −3.19 as reported by Stata.

◁

Inference on multiple classifiers using the Alonzo and Pepe (2002) estimating equation method is performed by fitting each model separately and bootstrapping to determine the dependence of the estimates. Using the maximum likelihood method, we also fit each model separately. We use `suest` (see [R] **suest**) to estimate the joint variance–covariance of our parameter estimates.

For our models, we can view the score equation for each model as an estimating equation. The estimate that solves the estimating equation (that makes the score 0) is asymptotically normal with a variance matrix that can be estimated using the inverse of the squared scores. By stacking the score equations of the separate models, we can estimate the variance matrix for all the parameter estimates by using this rule. This is an informal explanation; further details can be found in [R] **suest** and in the references Rogers (1993); White (1982 and 1996).

Now we will examine a case with multiple classification variables.

▷ Example 11: Maximum likelihood ROC, multiple classifiers

We return to the neonatal audiology study with gender and age covariates (Norton et al. 2000). In example 6, we fit a model with `male` and `currage` as control covariates, and `currage` as a ROC covariate for the classifier `y1` (DPOAE 65 at 2 kHz). We will refit this model, extending it to include the classifier `y2` (TEOAE 80 at 2 kHz).

```
. use http://www.stata-press.com/data/r12/nnhs
(Norton - neonatal audiology data)

. rocreg d y1 y2, probit ml ctrlcov(currage male) roccov(currage) cluster(id) nolog
Parametric ROC estimation

Covariate control        : linear regression
Control variables        : currage male
Control standardization: normal
ROC method               : parametric                Link: probit

  Status     : d
  Classifiers: y1 y2

  Classifier : y1
  Covariate control adjustment model:

------------------------------------------------------------------------------
             |      Coef.   Std. Err.      z    P>|z|     [95% Conf. Interval]
-------------+----------------------------------------------------------------
casecov      |
     currage |    .494211    .2126672     2.32   0.020      .077391    .9110311
       _cons |  -15.00403    8.238094    -1.82   0.069     -31.1504    1.142338
-------------+----------------------------------------------------------------
casesd       |
       _cons |    8.49794    .4922792    17.26   0.000     7.533091     9.46279
-------------+----------------------------------------------------------------
ctrlcov      |
     currage |  -.2032048    .0323803    -6.28   0.000     -.266669   -.1397406
        male |   .2369359    .2201391     1.08   0.282    -.1945288    .6684006
       _cons |   -1.23534    1.252775    -0.99   0.324    -3.690734    1.220055
-------------+----------------------------------------------------------------
ctrlsd       |
       _cons |   7.749156    .0782225    99.07   0.000     7.595843    7.902469
------------------------------------------------------------------------------

  Classifier : y2
  Covariate control adjustment model:

------------------------------------------------------------------------------
             |      Coef.   Std. Err.      z    P>|z|     [95% Conf. Interval]
-------------+----------------------------------------------------------------
casecov      |
     currage |   .5729861    .2422662     2.37   0.018     .0981532    1.047819
       _cons |   -18.2597    9.384968    -1.95   0.052     -36.6539    .1344949
-------------+----------------------------------------------------------------
casesd       |
       _cons |   9.723858    .5632985    17.26   0.000     8.619813     10.8279
-------------+----------------------------------------------------------------
ctrlcov      |
     currage |  -.1694575    .0291922    -5.80   0.000    -.2266732   -.1122419
        male |   .7122587    .1993805     3.57   0.000     .3214802    1.103037
       _cons |  -5.651728    1.129452    -5.00   0.000    -7.865415   -3.438042
-------------+----------------------------------------------------------------
ctrlsd       |
       _cons |   6.986167    .0705206    99.07   0.000      6.84795    7.124385
------------------------------------------------------------------------------
```

```
Status    : d
ROC Model :
                              (Std. Err. adjusted for 2741 clusters in id)

                             Robust
                  Coef.    Std. Err.      z    P>|z|     [95% Conf. Interval]

y1
      i_cons  -1.765608   1.105393    -1.60   0.110    -3.932138    .4009225
     currage   .0581566   .0290177     2.00   0.045     .0012828    .1150303
      s_cons   .9118864   .0586884    15.54   0.000     .7968593    1.026913

y2
      i_cons  -1.877825    .905174    -2.07   0.038    -3.651933   -.1037167
     currage   .0589258   .0235849     2.50   0.012     .0127002    .1051514
      s_cons   .7184563   .0565517    12.70   0.000      .607617    .8292957
```

Both classifiers have similar results. The results for y1 show the same direction as the estimating equation results in example 6. However, we can now reject the null hypothesis that the ROC currage coefficient is 0 at the 0.05 level.

In example 6, we could not reject that the slope parameter s_cons was 1 and that the constant intercept or ROC coefficient for current age was 0. The resulting ROC curve implied a noninformative test using y1 as a classifier. This is not the case with our current results. As currage increases, we expect a steeper ROC curve and thus a more powerful test, for both classifiers y1 (DPOAE 65 at 2 kHz) and y2 (TEOAE 80 at 2 kHz).

In example 10, the clustering of observations within infant id was adjusted in the individual fit of nsnr. In our current example, the adjustment for the clustering of observations within id is performed during concurrent estimation, as opposed to during the individual classifier fits (as in example 10). This adjustment, performed by suest, is still accurate.

◁

Now we will fit constant probit models and compare rocreg with rocfit and roccomp with the binormal option. Our first applications of rocfit and roccomp are taken directly from examples 8 and 9. The Dorfman and Alf (1969) algorithm that rocfit works with uses discrete classifiers or uses slicing to make a classifier discrete. So we are applying the maximum likelihood method of rocreg on discrete classification data here, where it expects continuous data. We expect to see some discrepancies, but we do not find great divergence in the estimates. After revisiting examples 8 and 9, we will fit a probit model with a continuous classifier and no covariates using rocreg, and we will compare the results with those from rocfit.

▷ Example 12: Maximum likelihood ROC, marginal model

Using the Hanley and McNeil (1982) data, discussed in example 1 and in example 8, we fit a constant probit model of the classifier rating with true status disease. rocreg is invoked with the ml option and compared with rocfit.

```
. use http://www.stata-press.com/data/r12/hanley, clear

. rocfit disease rating, nolog

Binormal model of disease on rating                Number of obs    =        109
Goodness-of-fit chi2(2) =         0.21
Prob > chi2             =       0.9006
Log likelihood          =   -123.64855

------------------------------------------------------------------------------
             |     Coef.   Std. Err.       z    P>|z|      [95% Conf. Interval]
-------------+----------------------------------------------------------------
   intercept |  1.656782   0.310456     5.34   0.000      1.048300    2.265265
   slope (*) |  0.713002   0.215882    -1.33   0.184      0.289881    1.136123
-------------+----------------------------------------------------------------
       /cut1 |  0.169768   0.165307     1.03   0.304     -0.154227    0.493764
       /cut2 |  0.463215   0.167235     2.77   0.006      0.135441    0.790990
       /cut3 |  0.766860   0.174808     4.39   0.000      0.424243    1.109477
       /cut4 |  1.797938   0.299581     6.00   0.000      1.210770    2.385106
------------------------------------------------------------------------------

------------------------------------------------------------------------------
             |               Indices from binormal fit
       Index |  Estimate   Std. Err.                   [95% Conf. Interval]
-------------+----------------------------------------------------------------
    ROC area |  0.911331   0.029506                    0.853501    0.969161
    delta(m) |  2.323671   0.502370                    1.339044    3.308298
        d(e) |  1.934361   0.257187                    1.430284    2.438438
        d(a) |  1.907771   0.259822                    1.398530    2.417012
------------------------------------------------------------------------------
(*) z test for slope==1

. rocreg disease rating, probit ml nolog

Parametric ROC estimation

Control standardization: normal
ROC method             : parametric                Link: probit

  Status     : disease
  Classifiers: rating

  Classifier : rating
  Covariate control adjustment model:

------------------------------------------------------------------------------
             |      Coef.   Std. Err.      z    P>|z|     [95% Conf. Interval]
-------------+----------------------------------------------------------------
casecov      |
       _cons |     2.3357   .2334285    10.01   0.000     1.878188    2.793211
-------------+----------------------------------------------------------------
casesd       |
       _cons |   1.117131   .1106124    10.10   0.000     .9003344    1.333927
-------------+----------------------------------------------------------------
ctrlcov      |
       _cons |   2.017241   .1732589    11.64   0.000      1.67766    2.356823
-------------+----------------------------------------------------------------
ctrlsd       |
       _cons |   1.319501   .1225125    10.77   0.000      1.07938    1.559621
------------------------------------------------------------------------------

   Status    : disease
   ROC Model :

------------------------------------------------------------------------------
             |      Coef.   Std. Err.      z    P>|z|     [95% Conf. Interval]
-------------+----------------------------------------------------------------
rating       |
      i_cons |   2.090802   .2941411     7.11   0.000     1.514297    2.667308
      s_cons |   1.181151   .1603263     7.37   0.000     .8669177    1.495385
         auc |   .9116494   .0261658    34.84   0.000     .8603654    .9629333
------------------------------------------------------------------------------
```

We compare the estimates for these models:

	rocfit	rocreg, ml
slope	0.7130	1.1812
SE of slope	0.2159	0.1603
intercept	1.6568	2.0908
SE of intercept	0.3105	0.2941
AUC	0.9113	0.9116
SE of AUC	0.0295	0.0262

We find that both the intercept and the slope are estimated as higher with the maximum likelihood method under `rocreg` than with `rocfit`. The AUC (ROC area in `rocfit`) is close for both commands. We find that the standard errors of each of these estimates is slightly lower under `rocreg` than `rocfit` as well.

Both `rocfit` and `rocreg` suggest that the slope parameter of the ROC curve (`slope` in `rocfit` and `s_cons` in `rocreg`) is not significantly different from 1. Thus we cannot reject that the classifier has the same variance in both case and control populations. There is, however, significant evidence that the intercepts (`i_cons` in `rocreg` and `intercept` in `rocfit`) differ from 0. Because of the positive direction of the intercept estimates, the ROC curve for `rating` as a classifier of `disease` suggests that `rating` provides an informative test. This is also suggested by the high AUC, which is significantly different from 0.5, that is, a flip of a coin.

◁

▷ Example 13: Maximum likelihood ROC, marginal model, multiple classifiers

We use the fictitious dataset generated from Hanley and McNeil (1983), which we previously used in example 2 and in example 9. To fit a probit model using `roccomp`, we specify the `binormal` option. We perform parametric, maximum likelihood ROC analysis using `rocreg`. We use `rocregplot` to plot the ROC curves created by `rocreg`.

```
. use http://www.stata-press.com/data/r12/ct2, clear
. roccomp status mod1 mod2 mod3, summary binormal graph aspectratio(1)
>         plot1opts(connect(i) msymbol(o))
>         plot2opts(connect(i) msymbol(s))
>         plot3opts(connect(i) msymbol(t))
>         legend(label(1 "mod1") label(3 "mod2") label(5 "mod3")
>                label(2 "mod1 fit") label(4 "mod2 fit") label(6 "mod3 fit")
>         order(1 3 5 2 4 6) cols(1)) title(roccomp) name(a) nodraw
Fitting binormal model for: mod1
Fitting binormal model for: mod2
Fitting binormal model for: mod3
                                  ROC
                      Obs        Area      Std. Err.      [95% Conf. Interval]
-------------------------------------------------------------------------------
mod1                  112      0.8945       0.0305        0.83482     0.95422
mod2                  112      0.9382       0.0264        0.88647     0.99001
mod3                  112      0.9376       0.0223        0.89382     0.98139
-------------------------------------------------------------------------------
Ho: area(mod1) = area(mod2) = area(mod3)
    chi2(2) =     8.27       Prob>chi2 =    0.0160
```

```
. rocreg status mod1 mod2 mod3, probit ml nolog
Parametric ROC estimation

Control standardization: normal
ROC method             : parametric               Link: probit

  Status     : status
  Classifiers: mod1 mod2 mod3

  Classifier : mod1
  Covariate control adjustment model:
```

	Coef.	Std. Err.	z	P>|z|	[95% Conf.	Interval]
casecov						
_cons	2.118135	.2165905	9.78	0.000	1.693626	2.542645
casesd						
_cons	1.166078	.1122059	10.39	0.000	.9461589	1.385998
ctrlcov						
_cons	2.344828	.1474147	15.91	0.000	2.0559	2.633755
ctrlsd						
_cons	1.122677	.1042379	10.77	0.000	.9183746	1.32698

```
  Classifier : mod2
  Covariate control adjustment model:
```

	Coef.	Std. Err.	z	P>|z|	[95% Conf.	Interval]
casecov						
_cons	2.659642	.2072731	12.83	0.000	2.253395	3.06589
casesd						
_cons	1.288468	.1239829	10.39	0.000	1.045466	1.53147
ctrlcov						
_cons	1.655172	.1105379	14.97	0.000	1.438522	1.871823
ctrlsd						
_cons	.8418313	.0781621	10.77	0.000	.6886365	.9950262

```
  Classifier : mod3
  Covariate control adjustment model:
```

	Coef.	Std. Err.	z	P>|z|	[95% Conf.	Interval]
casecov						
_cons	2.353768	.1973549	11.93	0.000	1.966959	2.740576
casesd						
_cons	1.143359	.1100198	10.39	0.000	.9277243	1.358994
ctrlcov						
_cons	2.275862	.1214094	18.75	0.000	2.037904	2.51382
ctrlsd						
_cons	.9246267	.0858494	10.77	0.000	.7563649	1.092888

```
   Status     : status
   ROC Model  :

              |               Robust
              |      Coef.   Std. Err.      z    P>|z|     [95% Conf. Interval]
--------------+----------------------------------------------------------------
mod1          |
       i_cons |    1.81646    .3144804     5.78   0.000      1.20009    2.432831
       s_cons |   .9627801    .1364084     7.06   0.000     .6954245    1.230136
          auc |    .904657    .0343518    26.34   0.000     .8373287    .9719853
--------------+----------------------------------------------------------------
mod2          |
       i_cons |   2.064189    .3267274     6.32   0.000     1.423815    2.704563
       s_cons |   .6533582    .1015043     6.44   0.000     .4544135    .8523029
          auc |   .9580104    .0219713    43.60   0.000     .9149473    1.001073
--------------+----------------------------------------------------------------
mod3          |
       i_cons |   2.058643    .2890211     7.12   0.000     1.492172    2.625113
       s_cons |   .8086932    .1163628     6.95   0.000     .5806262     1.03676
          auc |   .9452805    .0236266    40.01   0.000     .8989732    .9915877
-------------------------------------------------------------------------------
Ho: All classifiers have equal AUC values.
Ha: At least one classifier has a different AUC value.
P-value:     .0808808
. rocregplot, title(rocreg) nodraw name(b)
> plot1opts(msymbol(o)) plot2opts(msymbol(s)) plot3opts(msymbol(t))
. graph combine a b, xsize(5)
```

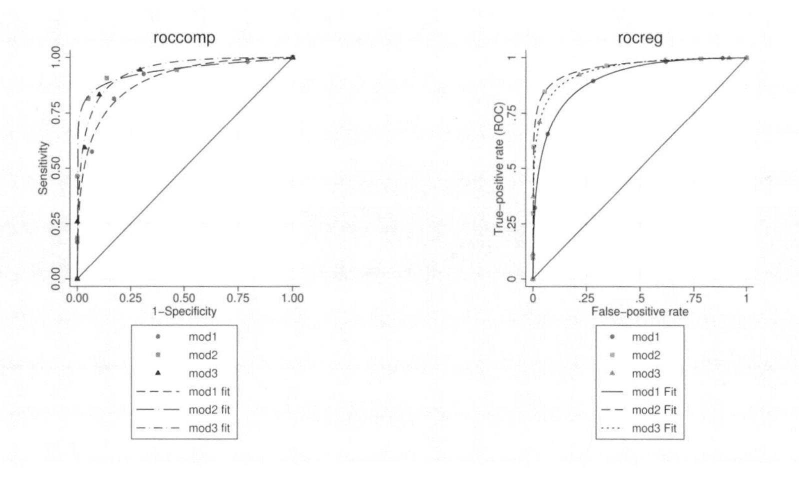

We compare the AUC estimates for these models:

	roccomp	rocreg, ml
mod1	0.8945	0.9047
mod2	0.9382	0.9580
mod3	0.9376	0.9453

Each classifier has a higher estimated AUC under `rocreg` than `roccomp`. Each curve appears to be raised and smoothed in the `rocreg` fit as compared with `roccomp`. They are different, but not drastically different. The inference on whether the curve areas are the same is similar to example 9. We reject equality at the 0.10 level under `rocreg` and at the 0.05 level under `roccomp`.

Each intercept is significantly different from 0 at the 0.05 level and is estimated in a positive direction. Though all but classifier `mod2` has 1 in their slope confidence intervals, the high intercepts suggest steep ROC curves and powerful tests.

Also note that the false-positive and true-positive rate points are calculated empirically in the `roccomp` graph and parametrically in `rocreg`. In example 9, the false-positive rates calculated by `rocreg` were calculated empirically, similar to `roccomp`. But in this example, the rates are calculated based on normal percentiles.

◁

Now we will generate an example to compare `rocfit` and `rocreg` under maximum likelihood estimation of a continuous classifier.

▷ Example 14: Maximum likelihood ROC, graphical comparison with rocfit

We generate 500 realizations of a population under threat of disease. One quarter of the population has the disease. A classifier `x` is measured, which has a control distribution of $N(1,3)$ and a case distribution of $N(1+5,2)$. We will invoke `rocreg` with the `ml` option on this generated data. We specify the `continuous()` option for `rocfit` and invoke it on the data as well. The `continuous()` option tells `rocfit` how many discrete slices to partition the data into before fitting.

For comparison of the two curves, we will use the `rocfit` postestimation command, `rocplot`; see [R] **rocfit postestimation**. This command graphs the empirical false-positive and true-positive rates with an overlaid fit of the binormal curve estimated by `rocfit`. `rocplot` also supports an `addplot()` option. We use the saved variables from `rocreg` in this option to overlay a line plot of the `rocreg` fit.

```
. clear
. set seed 8675309
. set obs 500
obs was 0, now 500
. generate d = runiform() < .25
. quietly generate double epsilon = 3*invnormal(runiform()) if d == 0
. quietly replace epsilon = 2*invnormal(runiform()) if d == 1
. quietly generate double x = 1 + d*5 + epsilon
```

```
. rocreg d x, probit ml nolog
Parametric ROC estimation

Control standardization: normal
ROC method             : parametric                Link: probit

  Status     : d
  Classifiers: x

  Classifier : x
  Covariate control adjustment model:

------------------------------------------------------------------------------
             |      Coef.   Std. Err.      z    P>|z|     [95% Conf. Interval]
-------------+----------------------------------------------------------------
casecov      |
       _cons |   4.905612   .2411624    20.34   0.000     4.432943    5.378282
-------------+----------------------------------------------------------------
casesd       |
       _cons |   2.038278   .1299559    15.68   0.000     1.783569    2.292987
-------------+----------------------------------------------------------------
ctrlcov      |
       _cons |   1.010382   .1561482     6.47   0.000     .7043377    1.316427
-------------+----------------------------------------------------------------
ctrlsd       |
       _cons |   3.031849   .1104134    27.46   0.000     2.815443    3.248255
------------------------------------------------------------------------------

   Status    : d
   ROC Model :

------------------------------------------------------------------------------
             |      Coef.   Std. Err.      z    P>|z|     [95% Conf. Interval]
-------------+----------------------------------------------------------------
x            |
      i_cons |   2.406743    .193766    12.42   0.000     2.026969    2.786518
      s_cons |   1.487456   .1092172    13.62   0.000     1.273394    1.701518
         auc |   .9103292    .012754    71.38   0.000     .8853318    .9353266
------------------------------------------------------------------------------

. rocfit d x, continuous(10) nolog
Binormal model of d on x                          Number of obs   =        500
Goodness-of-fit chi2(7) =         1.69
Prob > chi2             =       0.9751
Log likelihood          =   -911.91338

------------------------------------------------------------------------------
             |      Coef.   Std. Err.      z    P>|z|     [95% Conf. Interval]
-------------+----------------------------------------------------------------
   intercept |   2.207250   0.232983     9.47   0.000     1.750611    2.663888
   slope (*) |   1.281443   0.158767     1.77   0.076     0.970265    1.592620
-------------+----------------------------------------------------------------
       /cut1 |  -1.895707   0.130255   -14.55   0.000    -2.151001   -1.640412
       /cut2 |  -1.326900   0.089856   -14.77   0.000    -1.503015   -1.150784
       /cut3 |  -0.723677   0.070929   -10.20   0.000    -0.862695   -0.584660
       /cut4 |  -0.116960   0.064666    -1.81   0.070    -0.243702    0.009782
       /cut5 |   0.442769   0.066505     6.66   0.000     0.312422    0.573116
       /cut6 |   1.065183   0.075744    14.06   0.000     0.916728    1.213637
       /cut7 |   1.689570   0.102495    16.48   0.000     1.488683    1.890457
       /cut8 |   2.495841   0.185197    13.48   0.000     2.132861    2.858821
       /cut9 |   3.417994   0.348485     9.81   0.000     2.734976    4.101012
------------------------------------------------------------------------------
```

Index	Estimate	Std. Err.		[95% Conf. Interval]	
	Indices from binormal fit				
ROC area	0.912757	0.013666		0.885972	0.939542
delta(m)	1.722473	0.127716		1.472153	1.972792
d(e)	1.934960	0.125285		1.689405	2.180515
d(a)	1.920402	0.121804		1.681670	2.159135

```
(*) z test for slope==1
. rocplot, plotopts(msymbol(i)) lineopts(lpattern(dash))
>         norefline addplot(line _roc_x _fpr_x, sort(_fpr_x _roc_x)
>         lpattern(solid)) aspectratio(1) legend(off)
```

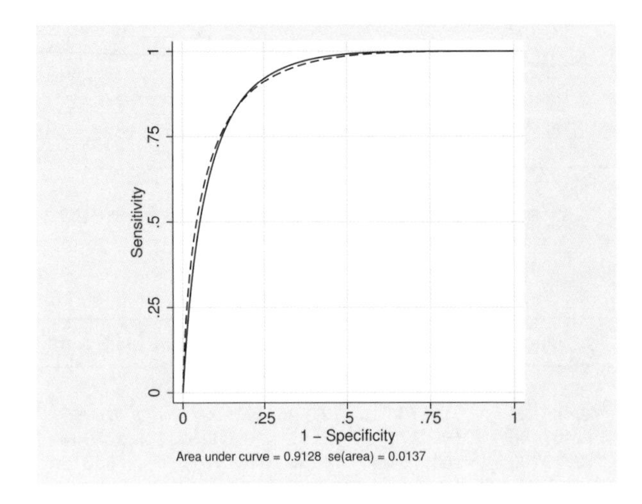

We find that the curves are close. As before, the `rocfit` estimates are lower for the slope and intercept than under `rocreg`. The AUC estimates are close. Though the slope confidence interval contains 1, a high ROC intercept suggests a steep ROC curve and thus a powerful test.

◁

Saved results

Nonparametric `rocreg` saves the following in `e()`:

Scalars

`e(N)`	number of observations
`e(N_strata)`	number of covariate strata
`e(N_clust)`	number of clusters
`e(rank)`	rank of `e(V)`

Macros

`e(cmd)`	`rocreg`
`e(cmdline)`	command as typed
`e(classvars)`	classification variable list
`e(refvar)`	status variable, reference variable
`e(ctrlmodel)`	covariate-adjustment specification
`e(ctrlcov)`	covariate-adjustment variables
`e(pvc)`	percentile value calculation method
`e(title)`	title in estimation output
`e(tiecorrected)`	indicates whether tie correction was used
`e(nobootstrap)`	indicates that bootstrap was performed
`e(bseed)`	seed used in bootstrap, if bootstrap performed
`e(breps)`	number of bootstrap resamples, if bootstrap performed
`e(cc)`	indicates whether case–control groups were used as resampling strata
`e(nobstrata)`	indicates whether resampling should stratify based on control covariates
`e(clustvar)`	name of cluster variable
`e(roc)`	false-positive rates where ROC was estimated
`e(invroc)`	ROC values where false-positive rates were estimated
`e(pauc)`	false-positive rates where pAUC was estimated
`e(auc)`	indicates that AUC was calculated
`e(vce)`	`bootstrap`
`e(properties)`	`b V` (or `b` if bootstrap not performed)

Matrices

`e(b)`	coefficient vector
`e(V)`	variance–covariance matrix of the estimators
`e(b_bs)`	bootstrap estimates
`e(bias)`	estimated biases
`e(se)`	estimated standard errors
`e(z0)`	median biases
`e(ci_normal)`	normal-approximation confidence intervals
`e(ci_percentile)`	percentile confidence intervals
`e(ci_bc)`	bias-corrected confidence intervals

Functions

`e(sample)`	marks estimation sample

Parametric, bootstrap rocreg saves the following in e():

Scalars

e(N)	number of observations
e(N_strata)	number of covariate strata
e(N_clust)	number of clusters
e(rank)	rank of e(V)

Macros

e(cmd)	rocreg
e(cmdline)	command as typed
e(title)	title in estimation output
e(classvars)	classification variable list
e(refvar)	status variable, reference variable
e(ctrlmodel)	covariate-adjustment specification
e(ctrlcov)	covariate-adjustment variables
e(pvc)	percentile value calculation method
e(title)	title in estimation output
e(tiecorrected)	indicates whether tie correction was used
e(probit)	probit
e(roccov)	ROC covariates
e(fprpts)	number of points used as false-positive rate fit points
e(ctrlfprall)	indicates whether all observed false-positive rates were used as fit points
e(nobootstrap)	indicates that bootstrap was performed
e(bseed)	seed used in bootstrap
e(breps)	number of bootstrap resamples
e(cc)	indicates whether case–control groups were used as resampling strata
e(nobstrata)	indicates whether resampling should stratify based on control covariates
e(clustvar)	name of cluster variable
e(vce)	bootstrap
e(properties)	b V (or b if nobootstrap is specified)
e(predict)	program used to implement predict

Matrices

e(b)	coefficient vector
e(V)	variance–covariance matrix of the estimators
e(b_bs)	bootstrap estimates
e(reps)	number of nonmissing results
e(bias)	estimated biases
e(se)	estimated standard errors
e(z0)	median biases
e(ci_normal)	normal-approximation confidence intervals
e(ci_percentile)	percentile confidence intervals
e(ci_bc)	bias-corrected confidence intervals

Functions

e(sample)	marks estimation sample

Parametric, maximum likelihood `rocreg` saves the following in `e()`:

Scalars

`e(N)`	number of observations
`e(N_clust)`	number of clusters
`e(rank)`	rank of `e(V)`

Macros

`e(cmd)`	`rocreg`
`e(cmdline)`	command as typed
`e(classvars)`	classification variable list
`e(refvar)`	status variable
`e(ctrlmodel)`	`linear`
`e(ctrlcov)`	control population covariates
`e(roccov)`	ROC covariates
`e(probit)`	`probit`
`e(pvc)`	`normal`
`e(wtype)`	weight type
`e(wexp)`	weight expression
`e(title)`	title in estimation output
`e(clustvar)`	name of cluster variable
`e(vce)`	`cluster` if clustering used
`e(vcetype)`	`robust` if multiple classifiers or clustering used
`e(ml)`	indicates that maximum likelihood estimation was used
`e(predict)`	program used to implement `predict`

Matrices

`e(b)`	coefficient vector
`e(V)`	variance–covariance matrix of the estimators

Functions

`e(sample)`	marks estimation sample

Methods and formulas

`rocreg` is implemented as an ado-file.

Assume that we applied a diagnostic test to each of N_0 control and N_1 case subjects. Further assume that the higher the outcome value of the diagnostic test, the higher the risk of the subject being abnormal. Let $y_{1i}, i = 1, 2, \ldots, N_1$, and $y_{0j}, j = 1, 2, \ldots, N_0$, be the values of the diagnostic test for the case and control subjects, respectively. The true status variable D identifies an observation as case $D = 1$ or control $D = 0$. The CDF of the classifier Y is F. Conditional on D, we write the CDF as F_D.

Methods and formulas are presented under the following headings:

ROC statistics
Covariate-adjusted ROC curves
Parametric ROC curves: Estimating equations
Parametric ROC curves: Maximum likelihood

ROC statistics

We obtain these definitions and their estimates from Pepe (2003) and Pepe, Longton, and Janes (2009). The false-positive and true-positive rates at cutoff c are defined as

$$\text{FPR}\,(y) = P\left(Y \geq y \middle| D = 0\right)$$

$$\text{TPR}\,(y) = P\left(Y \geq y \middle| D = 1\right)$$

The true-positive rate, or ROC value at false-positive rate u, is given by

$$\text{ROC}\,(u) = P\left(1 - F_0\,(Y) \leq u \middle| D = 1\right)$$

When Y is continuous, the false-positive rate can be written as

$$\text{FPR}\,(y) = 1 - F_0\,(y)$$

The empirical CDF for the sample $z_1, \ldots, z_n$ is given by

$$\widehat{F}(z) = \sum_{i=1}^{n} \frac{I\,(z < z_i)}{n}$$

The empirical estimates $\widehat{\text{FPR}}$ and $\widehat{\text{ROC}}$ both use this empirical CDF estimator.

The area under the ROC curve is defined as

$$\text{AUC} = \int_0^1 \text{ROC}\,(u)\, d_u$$

The partial area under the ROC curve for false-positive rate a is defined as

$$\text{pAUC}\,(a) = \int_0^a \text{ROC}\,(u)\, d_u$$

The nonparametric estimate for the AUC is given by

$$\widehat{\text{AUC}} = \sum_{i=1}^{N_1} \frac{1 - \widehat{\text{FPR}}\,(y_{1i})}{N_1}$$

The nonparametric estimate of pAUC is given by

$$\widehat{\text{pAUC}}\,(a) = \sum_{i=1}^{N_1} \frac{\max\left\{1 - \widehat{\text{FPR}}\,(y_{1i}) - (1-a), 0\right\}}{N_1}$$

For discrete classifiers, a correction term is subtracted from the false-positive rate estimate so that the $\widehat{\text{AUC}}$ and $\widehat{\text{pAUC}}$ estimates correspond with a trapezoidal approximation to the area of the ROC curve.

$$\text{FPR}^c\,(y) = 1 - \widehat{F}_0\,(y) - \frac{1}{2}\sum_{j=1}^{N_0} \frac{I\,(y = y_{0j})}{N_0}$$

In the nonparametric estimation of the ROC curve, all inference is performed using the `bootstrap` command (see [R] **bootstrap**). `rocreg` also allows users to calculate the ROC curve and related statistics by assuming a normal control distribution. So these formulas are updated by replacing F_0 by Φ (with adjustment of the marginal mean and variance of the control distribution).

Covariate-adjusted ROC curves

Suppose we observe covariate vector Z in addition to the classifier Y. Let $Z_{1i}, i = 1, 2, \ldots, N_1$, and $Z_{0j}, j = 1, 2, \ldots, N_0$, be the values of the covariates for the case and control subjects, respectively.

The covariate-adjusted ROC curve is defined by Janes and Pepe (2009) as

$$\text{AROC}\,(t) = E\left\{\text{ROC}\left(t \middle| Z_0\right)\right\}$$

It is calculated by replacing the marginal control CDF estimate, $\widehat{F}_0$, with the conditional control CDF estimate, $\widehat{F}_{0Z}$. If we used a normal control CDF, then we would replace the marginal control mean and variance with the conditional control mean and variance. The formulas of the previous section can be updated for covariate-adjustment by making this substitution of the conditional CDF for the marginal CDF in the false-positive rate calculation.

Because the calculation of the ROC value is now performed based on the conditionally calculated false-positive rate, no further conditioning is made in its calculation under nonparametric estimation.

`rocreg` supports covariate adjustment with stratification and linear regression. Under stratification, separate parameters are estimated for the control distribution at each level of the covariates. Under linear regression, the classifier is regressed on the covariates over the control distribution, and the resulting coefficients serve as parameters for $\widehat{F}_{0Z}$.

Parametric ROC curves: Estimating equations

Under nonparametric estimation of the ROC curve with covariate adjustment, no further conditioning occurs in the ROC curve calculation beyond the use of covariate-adjusted false-positive rates as inputs.

Under parametric estimation of the ROC curve, we can relax this restriction. We model the ROC curve as a cumulative distribution function g (standard normal Φ) invoked with input of a linear polynomial in the corresponding quantile function (here Φ^{-1}) invoked on the false-positive rate u. The constant intercept of the polynomial may depend on covariates; the slope term α (quantile coefficient) may not.

$$\text{ROC}\,(u) = g\{\mathbf{x}'\boldsymbol{\beta} + \alpha g^{-1}\,(u)\}$$

Pepe (2003) notes that having a binormal ROC ($g = \Phi$) is equivalent to specifying that some monotone transformation of the data exists to make the case and control classifiers normally distributed. This specification applies to the marginal case and control.

Under weak assumptions about the control distribution of the classifier, we can fit this model by using estimating equations (Alonzo and Pepe 2002). The method can be used without covariate effects in the second stage, assuming a parametric model for the single ROC curve. Using the Alonzo and Pepe (2002) method, the covariate-adjusted ROC curve may be fit parametrically. The marginal ROC curve, involving no covariates in either stage of estimation, can be fit parametrically as well. In addition to the Alonzo and Pepe (2002) explanation, further details are given in Pepe, Longton, and Janes (2009); Janes, Longton, and Pepe (2009); Pepe (2003); and Janes and Pepe (2009).

The algorithm can be described as follows:

1. Estimate the false-positive rates of the classifier `fpr`. These may be computed in any fashion outlined so far: covariate-adjusted, empirically, etc.
2. Determine a set of n_p false-positive rates to use as fitting points $f_1, \ldots, f_{n_p}$. These may be an equispaced grid on $(0, 1)$ or the set of observed false-positive rates from part 1.

3. Expand the case observation portion of the data to include a subobservation for each fitting point. So there are now $N_1(n_p - 1)$ additional observations in the data.
4. Generate a new dummy variable u. For subobservation j, u $= I\,(\texttt{fpr} \leq f_j)$.
5. Generate a new variable quant containing the quantiles of the false-positive rate fitting points. For subobservation j, quant $= g^{-1}\,(f_j)$.
6. Perform a binary regression (probit, $g = \Phi$) of fpr on the covariates x and quantile variable quant.

The coefficients of part 6 are the coefficients of the ROC model. The coefficients of the covariates coincide naturally with estimates of $\boldsymbol{\beta}$, and the α parameter is estimated by the coefficient on quant. Because the method is so general and makes few distributional assumptions, bootstrapping must be performed for inference. If multiple classifiers are to be fit, the algorithm is performed separately for each in each bootstrap, and the bootstrap is used to estimate covariances.

We mentioned earlier that in parametric estimation, the AUC was the only summary parameter that could be estimated initially. This is true when we fit the marginal probit model because there are no covariates in part 6 of the algorithm.

To calculate the AUC statistic under a marginal probit model, we use the formula

$$\text{AUC} = \Phi\left(\frac{\beta_0}{\sqrt{1+\alpha^2}}\right)$$

Alternatively, the AUC for the probit model can be calculated as pAUC(1) in postestimation. Under both models, bootstrapping is performed for inference on the AUC.

Parametric ROC curves: Maximum likelihood

rocreg supports another form of parametric ROC estimation: maximum likelihood with a normally distributed classifier. This method assumes that the classifier is a normal linear model on certain covariates, and the covariate effect and variance of the classifier may change between the case and control populations. The model is defined in Pepe (2003, 145).

$$y = \mathbf{z}'\boldsymbol{\beta_0} + D\mathbf{x}'\boldsymbol{\beta_1} + \sigma\,(D)\,\epsilon$$

Our error term, ϵ, is a standard normal random variable. The variable D is our true status variable, being 1 for the case population observations and 0 for the control population observations. The variance function σ is defined as

$$\sigma\,(D) = \sigma_0\,(D=0) + \sigma_1\,(D=1)$$

This provides two variance parameters in the model and does not depend on covariate values.

Under this model, the ROC curve is easily derived to be

$$\text{ROC}\,(u) = \Phi\left[\frac{1}{\sigma_1}\left\{\mathbf{x}'\boldsymbol{\beta_1} + \sigma_0\Phi^{-1}\,(u)\right\}\right]$$

We reparameterize the model, creating the parameters $\beta_i = \sigma_1^{-1}\beta_{1i}$ and $\alpha = \sigma_1^{-1}\sigma_0$. We refer to β_0 as the constant intercept, i_cons. The parameter α is referred to as the constant slope, s_cons.

$$\text{ROC}\,(u) = \Phi\{\mathbf{x}'\boldsymbol{\beta} + \alpha\Phi^{-1}\,(u)\}$$

The original model defining the classifier y leads to the following single observation likelihoods for $D = 0$ and $D = 1$:

$$L(\boldsymbol{\beta_0}, \boldsymbol{\beta_1}, \sigma_1, \sigma_0, \big| D = 0, y, \mathbf{z}, \mathbf{x}) = \frac{1}{\sqrt{2\pi}\sigma_0} \exp \frac{-(y - \mathbf{z}'\boldsymbol{\beta_0})^2}{2\sigma_0^2}$$

$$L(\boldsymbol{\beta_0}, \boldsymbol{\beta_1}, \sigma_1, \sigma_0, \big| D = 1, y, \mathbf{z}, \mathbf{x}) = \frac{1}{\sqrt{2\pi}\sigma_1} \exp \frac{-(y - \mathbf{z}'\boldsymbol{\beta_0} - \mathbf{x}'\boldsymbol{\beta_1})^2}{2\sigma_1^2}$$

These can be combined to yield the observation-level log likelihood:

$$\begin{aligned} \ln L(\boldsymbol{\beta_0}, \boldsymbol{\beta_1}, \sigma_1, \sigma_0, \big| D, y, \mathbf{z}, \mathbf{x}) = & -\frac{\ln 2\pi}{2} \\ & - I\,(D = 0) \left\{ \ln\sigma_0 + \frac{(y - \mathbf{z}'\boldsymbol{\beta_0})^2}{2\sigma_0^2} \right\} \\ & - I\,(D = 1) \left\{ \ln\sigma_1 + \frac{(y - \mathbf{z}'\boldsymbol{\beta_0} - \mathbf{x}'\boldsymbol{\beta_1})^2}{2\sigma_1^2} \right\} \end{aligned}$$

When there are multiple classifiers, each classifier is fit separately with maximum likelihood. Then the results are combined by stacking the scores and using the sandwich variance estimator. For more information, see [R] **suest** and the references White (1982); Rogers (1993); and White (1996).

Acknowledgments

We thank Margaret S. Pepe, Holly Janes, and Gary Longton of the Fred Hutchinson Cancer Research Center for providing the inspiration for the `rocreg` command and for illuminating many useful datasets for its documentation.

References

Alonzo, T. A., and M. S. Pepe. 2002. Distribution-free ROC analysis using binary regression techniques. *Biostatistics* 3: 421–432.

Cleves, M. A. 1999. sg120: Receiver operating characteristic (ROC) analysis. *Stata Technical Bulletin* 52: 19–33. Reprinted in *Stata Technical Bulletin Reprints*, vol. 9, pp. 212–229. College Station, TX: Stata Press.

——. 2000. sg120.2: Correction to roccomp command. *Stata Technical Bulletin* 54: 26. Reprinted in *Stata Technical Bulletin Reprints*, vol. 9, p. 231. College Station, TX: Stata Press.

——. 2002a. Comparative assessment of three common algorithms for estimating the variance of the area under the nonparametric receiver operating characteristic curve. *Stata Journal* 2: 280–289.

——. 2002b. From the help desk: Comparing areas under receiver operating characteristic curves from two or more probit or logit models. *Stata Journal* 2: 301–313.

Cook, N. R. 2007. Use and misuse of the receiver operating characteristic curve in risk prediction. *Circulation* 115: 928–935.

DeLong, E. R., D. M. DeLong, and D. L. Clarke-Pearson. 1988. Comparing the areas under two or more correlated receiver operating characteristic curves: A nonparametric approach. *Biometrics* 44: 837–845.

Dodd, L. E., and M. S. Pepe. 2003. Partial AUC estimation and regression. *Biometrics* 59: 614–623.

Dorfman, D. D., and E. Alf, Jr. 1969. Maximum-likelihood estimation of parameters of signal-detection theory and determination of confidence intervals–rating-method data. *Journal of Mathematical Psychology* 6: 487–496.

Hanley, J. A., and K. O. Hajian-Tilaki. 1997. Sampling variability of nonparametric estimates of the areas under receiver operating characteristic curves: An update. *Academic Radiology* 4: 49–58.

Hanley, J. A., and B. J. McNeil. 1982. The meaning and use of the area under a receiver operating characteristic (ROC) curve. *Radiology* 143: 29–36.

——. 1983. A method of comparing the areas under receiver operating characteristic curves derived from the same cases. *Radiology* 148: 839–843.

Janes, H., G. Longton, and M. S. Pepe. 2009. Accommodating covariates in receiver operating characteristic analysis. *Stata Journal* 9: 17–39.

Janes, H., and M. S. Pepe. 2009. Adjusting for covariate effects on classification accuracy using the covariate-adjusted receiver operating characteristic curve. *Biometrika* 96: 371–382.

McClish, D. K. 1989. Analyzing a portion of the ROC curve. *Medical Decision Making* 9: 190–195.

Mooney, C. Z., and R. D. Duval. 1993. *Bootstrapping: A Nonparametric Approach to Statistical Inference*. Newbury Park, CA: Sage.

Norton, S. J., M. P. Gorga, J. E. Widen, R. C. Folsom, Y. Sininger, B. Cone-Wesson, B. R. Vohr, K. Mascher, and K. Fletcher. 2000. Identification of neonatal hearing impairment: Evaluation of transient evoked otoacoustic emission, distortion product otoacoustic emission, and auditory brain stem response test performance. *Ear and Hearing* 21: 508–528.

Pepe, M. S. 1998. Three approaches to regression analysis of receiver operating characteristic curves for continuous test results. *Biometrics* 54: 124–135.

——. 2000. Receiver operating characteristic methodology. *Journal of the American Statistical Association* 95: 308–311.

——. 2003. *The Statistical Evaluation of Medical Tests for Classification and Prediction*. New York: Oxford University Press.

Pepe, M. S., and T. Cai. 2004. The analysis of placement values for evaluating discriminatory measures. *Biometrics* 60: 528–535.

Pepe, M. S., G. Longton, and H. Janes. 2009. Estimation and comparison of receiver operating characteristic curves. *Stata Journal* 9: 1–16.

Rogers, W. H. 1993. sg16.4: Comparison of nbreg and glm for negative binomial. *Stata Technical Bulletin* 16: 7. Reprinted in *Stata Technical Bulletin Reprints*, vol. 3, pp. 82–84. College Station, TX: Stata Press.

Stover, L., M. P. Gorga, S. T. Neely, and D. Montoya. 1996. Toward optimizing the clinical utility of distortion product otoacoustic emission measurements. *Journal of the Acoustical Society of America* 100: 956–967.

Thompson, M. L., and W. Zucchini. 1989. On the statistical analysis of ROC curves. *Statistics in Medicine* 8: 1277–1290.

White, H. 1982. Maximum likelihood estimation of misspecified models. *Econometrica* 50: 1–25.

——. 1996. *Estimation, Inference and Specification Analysis*. Cambridge: Cambridge University Press.

Wieand, S., M. H. Gail, B. R. James, and K. L. James. 1989. A family of nonparametric statistics for comparing diagnostic markers with paired or unpaired data. *Biometrika* 76: 585–592.

Also see

[R] **rocreg postestimation** — Postestimation tools for rocreg

[R] **rocregplot** — Plot marginal and covariate-specific ROC curves after rocreg

[R] **rocfit** — Parametric ROC models

[R] **roc** — Receiver operating characteristic (ROC) analysis

Title

rocreg postestimation — Postestimation tools for rocreg

Description

The following commands are of special interest after `rocreg`:

Command	Description
`estat nproc`	nonparametric ROC curve estimation, keeping fit information from `rocreg`
`rocregplot`	plot marginal and covariate-specific ROC curves

For information about `estat nproc`, see below.
For information about `rocregplot`, see [R] **rocregplot**.

The following standard postestimation commands are also available:

Command	Description
`estimates`	cataloging estimation results
`lincom`	point estimates, standard errors, testing, and inference for linear combinations
`nlcom`	point estimates, standard errors, testing, and inference for nonlinear combinations of coefficients
`predict`	predictions for parametric ROC curve estimation
`test`	Wald tests of simple and composite linear hypotheses
`testnl`	Wald tests of nonlinear hypotheses

See the corresponding entries in the *Base Reference Manual* for details.

Special-interest postestimation command

The `estat nproc` command allows calculation of all the ROC curve summary statistics for covariate-specific ROC curves, as well as for a nonparametric ROC estimation. Under nonparametric estimation, a single ROC curve is estimated by `rocreg`. Covariates can affect this estimation, but there are no separate covariate-specific ROC curves. Thus the input arguments for `estat nproc` are taken in the command line rather than from the data as variable values.

Syntax for predict

predict [*type*] *newvar* [*if*] [*in*] [, *statistic options*]

statistic	Description
Main	
at(*varname*)	input variable for statistic
auc	total area under the ROC curve; the default
roc	ROC values for given false-positive rates in at()
invroc	false-positive rate for given ROC values in at()
pauc	partial area under the ROC curve up to each false-positive rate in at()
classvar(*varname*)	statistic for given classifier

options	Description
Options	
intpts(*#*)	points in numeric integration of pAUC calculation
se(*newvar*)	predict standard errors
ci(*stubname*)	produce confidence intervals, stored as variables with prefix *stubname* and suffixes _l and _u
level(*#*)	set confidence level; default is level(95)
*bfile(*filename*, ...)	load dataset containing bootstrap replicates from rocreg
*btype(n|p|bc)	produce normal-based (n), percentile (p), or bias-corrected (bc) confidence intervals; default is btype(n)

* bfile() and btype() are only allowed with parametric analysis using bootstrap inference.

Menu

Statistics > Postestimation > Predictions, residuals, etc.

Options for predict

Main

at(*varname*) records the variable to be used as input for the above predictions.

auc predicts the total area under the ROC curve defined by the covariate values in the data. This is the default statistic.

roc predicts the ROC values for false-positive rates stored in *varname* specified in at().

invroc predicts the false-positive rates for given ROC values stored in *varname* specified in at().

pauc predicts the partial area under the ROC curve up to each false-positive rate stored in *varname* specified in at().

classvar(*varname*) performs the prediction for the specified classifier.

Options

intpts(#) specifies that # points be used in the pAUC calculation.

se(*newvar*) specifies that standard errors be produced and stored in *newvar*.

ci(*stubname*) requests that confidence intervals be produced and the lower and upper bounds be stored in *stubname*_l and *stubname*_u, respectively.

level(#) specifies the confidence level, as a percentage, for confidence intervals. The default is level(95) or as set by set level; see [U] **20.7 Specifying the width of confidence intervals**.

bfile(*filename*, ...) uses bootstrap replicates of parameters from rocreg stored in *filename* to estimate standard errors and confidence intervals of predictions.

btype(n | p | bc) specifies whether to produce normal-based (n), percentile (p), or bias-corrected (bc) confidence intervals. The default is btype(n).

Syntax for estat nproc

estat nproc [, *estat_nproc_options*]

estat_nproc_options	Description
Main	
auc	estimate total area under the ROC curve
roc(*numlist*)	estimate ROC values for given false-positive rates
invroc(*numlist*)	estimate false-positive rate for given ROC values
pauc(*numlist*)	estimate partial area under the ROC curve up to each false-positive rate

At least one option must be specified.

Menu

Statistics > Postestimation > Reports and statistics

Options for estat nproc

Main

auc estimates the total area under the ROC curve.

roc(*numlist*) estimates the ROC for each of the false-positive rates in *numlist*. The values in *numlist* must be in the range (0,1).

invroc(*numlist*) estimates the false-positive rate for each of the ROC values in *numlist*. The values in *numlist* must be in the range (0,1).

pauc(*numlist*) estimates the partial area under the ROC curve up to each false-positive rate in *numlist*. The values in *numlist* must in the range (0,1].

Remarks

Remarks are presented under the following headings:

Using predict after rocreg
Using estat nproc

Using predict after rocreg

predict, after parametric rocreg, predicts the AUC, the ROC value, the false-positive rate (invROC), or the pAUC value. The default is auc.

We begin by estimating the area under the ROC curve for each of the three age-specific ROC curves in example 1 of [R] **rocregplot**: 30, 40, and 50 months.

▷ Example 1

In example 6 of [R] **rocreg**, a probit ROC model was fit to audiology test data from Norton et al. (2000). The estimating equations method of Alonzo and Pepe (2002) was used to fit the model. Gender and age were covariates that affected the control distribution of the classifier y1 (DPOAE 65 at 2 kHz). Age was a ROC covariate for the model, so we fit separate ROC curves at each age.

Following Janes, Longton, and Pepe (2009), we drew the ROC curves for ages 30, 40, and 50 months in example 1 of [R] **rocregplot**. Now we use predict to estimate the AUC for the ROC curve at each of those ages.

The bootstrap dataset saved by rocreg in example 6 of [R] **rocreg**, nnhs2y1.dta, is used in the bfile() option.

We will store the AUC prediction in the new variable predAUC. We specify the se() option with the new variable name seAUC to produce an estimate of the prediction's standard error. By specifying the stubname cin in ci(), we tell predict to create normal-based confidence intervals (the default) as new variables cin_l and cin_u.

```
. use http://www.stata-press.com/data/r12/nnhs
(Norton - neonatal audiology data)

. rocreg d y1, probit ctrlcov(currage male) ctrlmodel(linear) roccov(currage)
> cluster(id) bseed(56930) bsave(nnhs2y1)
  (output omitted )

. set obs 5061
obs was 5058, now 5061

. quietly replace currage = 30 in 5059

. quietly replace currage = 40 in 5060

. quietly replace currage = 50 in 5061

. predict predAUC in 5059/5061, auc se(seAUC) ci(cin) bfile(nnhs2y1)

. list currage predAUC seAUC cin* in 5059/5061
```

	currage	predAUC	seAUC	cin_l	cin_u
5059.	30	.5209999	.0712928	.3812686	.6607312
5060.	40	.6479176	.0286078	.5918474	.7039879
5061.	50	.7601378	.0746157	.6138937	.9063819

As expected, we find the AUC to increase with age.

Essentially, we have a stored bootstrap sample of ROC covariate coefficient estimates in nnhs2y1.dta. We calculate the AUC using each set of coefficient estimates, resulting in a sample of AUC estimates. Then the bootstrap standard error and confidence intervals are calculated based on this AUC sample. Further details of the computation of the standard error and percentile confidence intervals can be found in *Methods and formulas* and in [R] **bootstrap**.

We can also produce percentile or bias-corrected confidence intervals by specifying btype(p) or btype(bc), which we now demonstrate.

```
. drop *AUC*
. predict predAUC in 5059/5061, auc se(seAUC) ci(cip) bfile(nnhs2y1) btype(p)
. list currage predAUC cip* in 5059/5061
```

	currage	predAUC	cip_l	cip_u
5059.	30	.5209999	.3760555	.6513149
5060.	40	.6479176	.5893397	.7032645
5061.	50	.7601378	.5881404	.8836223

```
. drop *AUC*
. predict predAUC in 5059/5061, auc se(seAUC) ci(cibc) bfile(nnhs2y1) btype(bc)
. list currage predAUC cibc* in 5059/5061
```

	currage	predAUC	cibc_l	cibc_u
5059.	30	.5209999	.3736968	.6500064
5060.	40	.6479176	.588947	.7010052
5061.	50	.7601378	.5812373	.8807758

◁

predict can also estimate the ROC value and the false-positive rate (invROC).

▷ Example 2

In example 7 of [R] **rocreg**, we fit the ROC curve for status variable hearing loss (d) and classifier negative signal-to-noise ratio nsnr with ROC covariates frequency (xf), intensity (xl), and hearing loss severity (xd). The data were obtained from Stover et al. (1996). The model fit was probit with bootstrap resampling. We saved 50 bootstrap replications in the dataset nsnrf.dta.

The covariate value combinations xf = 10.01, xl = 5.5, and xd = .5, and xf = 10.01, xl = 6.5, and xd = 4 are of interest. In example 3 of [R] **rocregplot**, we estimated the ROC values for false-positive rates 0.2 and 0.7 and the false-positive rate for a ROC value of 0.5 by using rocregplot. We will use predict to replicate the estimation.

We begin by appending observations with our desired covariate combinations to the data. We also create two new variables: rocinp, which contains the ROC values for which we wish to predict the corresponding invROC values, and invrocinp, which contains the invROC values corresponding to the ROC values we wish to predict.

```
. clear
. input xf xl xd rocinp invrocinp
            xf          xl          xd      rocinp   invrocinp
  1. 10.01 5.5 .5 .2 .
  2. 10.01 6.5 4  .2 .
  3. 10.01 5.5 .5 .7 .5
  4. 10.01 6.5 4  .7 .5
  5. end
. save newdata
file newdata.dta saved
. use http://www.stata-press.com/data/r12/dp
(Stover - DPOAE test data)
. quietly rocreg d nsnr, ctrlcov(xf xl) roccov(xf xl xd) probit cluster(id)
> nobstrata ctrlfprall bseed(156385) breps(50) ctrlmodel(strata) bsave(nsnrf)
. append using newdata
. list xf xl xd invrocinp rocinp in 1849/1852
```

	xf	xl	xd	invroc~p	rocinp
1849.	10.01	5.5	.5	.	.2
1850.	10.01	6.5	4	.	.2
1851.	10.01	5.5	.5	.5	.7
1852.	10.01	6.5	4	.5	.7

Now we will use `predict` to estimate the ROC value for the false-positive rates stored in `rocinp`. We specify the `roc` option, and we specify `rocinp` in the `at()` option. The other options, `se()` and `ci()`, are used to obtain standard errors and confidence intervals, respectively. The dataset of bootstrap samples, `nsnrf.dta`, is specified in `bfile()`. After prediction, we list the point estimates and standard errors.

```
. predict rocit in 1849/1852, roc at(rocinp) se(seroc) ci(cin) bfile(nsnrf)
. list xf xl xd rocinp rocit seroc if !missing(rocit)
```

	xf	xl	xd	rocinp	rocit	seroc
1849.	10.01	5.5	.5	.2	.7652956	.0735506
1850.	10.01	6.5	4	.2	.9672505	.0227977
1851.	10.01	5.5	.5	.7	.9835816	.0204353
1852.	10.01	6.5	4	.7	.999428	.0011309

These results match example 3 of [R] **rocregplot**. We list the confidence intervals next. These also conform to the `rocregplot` results from example 3 in [R] **rocregplot**. We begin with the confidence intervals for ROC under the covariate values `xf=10.01`, `xl=5.5`, and `xd=.5`.

```
. list xf xl xd rocinp rocit cin* if inlist(_n, 1849, 1851)
```

	xf	xl	xd	rocinp	rocit	cin_l	cin_u
1849.	10.01	5.5	.5	.2	.7652956	.6211391	.9094521
1851.	10.01	5.5	.5	.7	.9835816	.9435292	1.023634

Now we list the ROC confidence intervals under the covariate values xf=10.01, xl=6.5, and xd=4.

```
. list xf xl xd rocinp rocit cin* if inlist(_n, 1850, 1852)
```

	xf	xl	xd	rocinp	rocit	cin_l	cin_u
1850.	10.01	6.5	4	.2	.9672505	.9225678	1.011933
1852.	10.01	6.5	4	.7	.999428	.9972115	1.001644

Now we will predict the false-positive rate for a ROC value by specifying the invroc option. We pass the invrocinp variable as an argument to the at() option. Again we list the point estimates and standard errors first.

```
. drop ci*
. predict invrocit in 1849/1852, invroc at(invrocinp) se(serocinv) ci(cin)
> bfile(nsnrf)
. list xf xl xd invrocinp invrocit serocinv if !missing(invrocit)
```

	xf	xl	xd	invroc~p	invrocit	serocinv
1851.	10.01	5.5	.5	.5	.0615144	.0254042
1852.	10.01	6.5	4	.5	.0043298	.0045938

These also match those of example 3 of [R] **rocregplot**. Listing the confidence intervals shows identical results as well. First we list the confidence intervals under the covariate values xf=10.01, xl=5.5, and xd=.5.

```
. list xf xl xd invrocinp invrocit cin* in 1851
```

	xf	xl	xd	invroc~p	invrocit	cin_l	cin_u
1851.	10.01	5.5	.5	.5	.0615144	.0117231	.1113057

Now we list the confidence intervals for false-positive rate under the covariate values xf=10.01, xl=6.5, and xd=4.

```
. list xf xl xd invrocinp invrocit cin* in 1852
```

	xf	xl	xd	invroc~p	invrocit	cin_l	cin_u
1852.	10.01	6.5	4	.5	.0043298	-.004674	.0133335

◁

The predict command can also be used after a maximum-likelihood ROC model is fit.

▷ Example 3

In the previous example, we revisited the estimating equations fit of a probit model with ROC covariates frequency (xf), intensity (xl), and hearing loss severity (xd) to the Stover et al. (1996) audiology study data. A maximum likelihood fit of the same model was performed in example 10 of [R] **rocreg**. In example 2 of [R] **rocregplot**, we used rocregplot to estimate ROC values and false-positive rates for this model under two covariate configurations. We will use predict to obtain the same estimates. We will also estimate the partial area under the ROC curve.

We append the data as in the previous example. This leads to the following four final observations in the data.

```
. use http://www.stata-press.com/data/r12/dp, clear
(Stover - DPOAE test data)
. rocreg d nsnr, probit ctrlcov(xf xl) roccov(xf xl xd) ml cluster(id)
  (output omitted)
. append using newdata
. list xf xl xd invrocinp rocinp in 1849/1852
```

	xf	xl	xd	invroc~p	rocinp
1849.	10.01	5.5	.5	.	.2
1850.	10.01	6.5	4	.	.2
1851.	10.01	5.5	.5	.5	.7
1852.	10.01	6.5	4	.5	.7

Now we predict the ROC value for false-positive rates of 0.2 and 0.7. Under maximum likelihood prediction, only Wald-type confidence intervals are produced. We specify a new variable name for the standard error in the se() option and a stubname for the confidence-interval variables in the ci() option.

```
. predict rocit in 1849/1852, roc at(rocinp) se(seroc) ci(ci)
. list xf xl xd rocinp rocit seroc ci_l ci_u if !missing(rocit), noobs
```

xf	xl	xd	rocinp	rocit	seroc	ci_l	ci_u
10.01	5.5	.5	.2	.7608593	.0510501	.660803	.8609157
10.01	6.5	4	.2	.9499408	.0179824	.914696	.9851856
10.01	5.5	.5	.7	.978951	.0097382	.9598644	.9980376
10.01	6.5	4	.7	.9985001	.0009657	.9966073	1.000393

These results match our estimates in example 2 of [R] **rocregplot**. We also match example 2 of [R] **rocregplot** when we estimate the false-positive rate for a ROC value of 0.5.

```
. drop ci*
. predict invrocit in 1851/1852, invroc at(invrocinp) se(serocinv) ci(ci)
. list xf xl xd invrocinp invrocit serocinv ci_l ci_u if !missing(invrocit), noobs
```

xf	xl	xd	invroc~p	invrocit	serocinv	ci_l	ci_u
10.01	5.5	.5	.5	.0578036	.0198626	.0188736	.0967336
10.01	6.5	4	.5	.0055624	.0032645	-.0008359	.0119607

◁

▷ Example 4

In example 13 of [R] **rocreg**, we fit a maximum-likelihood marginal probit model to each classifier of the fictitious dataset generated from Hanley and McNeil (1983). In example 5 of [R] **rocregplot**, `rocregplot` was used to draw the ROC for the `mod1` and `mod3` classifiers. Estimates of the ROC value and false-positive rate were also obtained with Wald-type confidence intervals.

We return to this example, this time using `predict` to estimate the ROC value and false-positive rate. We will also estimate the pAUC for the false-positive rates of 0.3 and 0.8.

First, we add the input variables to the data. The variable `paucinp` will hold the 0.3 and 0.8 false-positive rates that we will input to pAUC. The variable `invrocinp` holds the ROC value of 0.8 for which we will estimate the false-positive rate. Finally, the variable `rocinp` holds the false-positive rates of 0.15 and 0.75 for which we will estimate the ROC value.

```
. use http://www.stata-press.com/data/r12/ct2, clear
. rocreg status mod1 mod2 mod3, probit ml
  (output omitted)
. quietly generate paucinp = .3 in 111
. quietly replace paucinp = .8 in 112
. quietly generate invrocinp = .8 in 112
. quietly generate rocinp = .15 in 111
. quietly replace rocinp = .75 in 112
```

Then, we estimate the ROC value for false-positive rates 0.15 and 0.75 under classifier `mod1`. The point estimate is stored in `roc1`. Wald confidence intervals and standard errors are also estimated. We find that these results match those of example 5 of [R] **rocregplot**.

```
. predict roc1 in 111/112, classvar(mod1) roc at(rocinp) se(sr1) ci(cir1)
. list rocinp roc1 sr1 cir1* in 111/112
```

	rocinp	roc1	sr1	cir1_l	cir1_u
111.	.15	.7934935	.0801363	.6364293	.9505578
112.	.75	.9931655	.0069689	.9795067	1.006824

Now we perform the same estimation under the classifier `mod3`.

```
. predict roc3 in 111/112, classvar(mod3) roc at(roci) se(sr3) ci(cir3)
. list rocinp roc3 sr3 cir3* in 111/112
```

	rocinp	roc3	sr3	cir3_l	cir3_u
111.	.15	.8888596	.0520118	.7869184	.9908009
112.	.75	.9953942	.0043435	.9868811	1.003907

Next we estimate the false-positive rate for the ROC value of 0.8. These results also match example 5 of [R] **rocregplot**.

```
. predict invroc1 in 112, classvar(mod1) invroc at(invrocinp) se(sir1) ci(ciir1)
. list invrocinp invroc1 sir1 ciir1* in 112
```

	invroc~p	invroc1	sir1	ciir1_l	ciir1_u
112.	.8	.1556435	.069699	.0190361	.292251

```
. predict invroc3 in 112, classvar(mod3) invroc at(invrocinp) se(sir3) ci(ciir3)
. list invrocinp invroc3 sir3 ciir3* in 112
```

	invroc~p	invroc3	sir3	ciir3_l	ciir3_u
112.	.8	.0661719	.045316	-.0226458	.1549896

Finally, we estimate the pAUC for false-positive rates of 0.3 and 0.8. The point estimate is calculated by numeric integration. Wald confidence intervals are obtained with the delta method. Further details are presented in *Methods and formulas*.

```
. predict pauc1 in 111/112, classvar(mod1) pauc at(paucinp) se(sp1) ci(cip1)
. list paucinp pauc1 sp1 cip1* in 111/112
```

	paucinp	pauc1	sp1	cip1_l	cip1_u
111.	.3	.221409	.0240351	.174301	.268517
112.	.8	.7033338	.0334766	.6377209	.7689466

```
. predict pauc3 in 111/112, classvar(mod3) pauc at(paucinp) se(sp3) ci(cip3)
. list paucinp pauc3 sp3 cip3* in 111/112
```

	paucinp	pauc3	sp3	cip3_l	cip3_u
111.	.3	.2540215	.0173474	.2200213	.2880217
112.	.8	.7420408	.0225192	.6979041	.7861776

◁

Using estat nproc

When you initially use `rocreg` to fit a nonparametric ROC curve, you can obtain bootstrap estimates of a ROC value, false-positive rate, area under the ROC curve, and partial area under the ROC curve. The `estat nproc` command allows the user to estimate these parameters after `rocreg` has originally been used.

The seed and resampling settings used by `rocreg` are used by `estat nproc`. So the results for these new statistics are identical to what they would be if they had been initially estimated in the `rocreg` command. These new statistics, together with those previously estimated in `rocreg`, are returned in `r()`.

We demonstrate with an example.

▷ Example 5

In example 3 of [R] **rocreg**, we examined data from a pancreatic cancer study (Wieand et al. 1989). Two continuous classifiers, `y1` (CA 19-9) and `y2` (CA 125), were used for the true status variable `d`. In that example, we estimated various quantities including the false-positive rate for a ROC value of 0.6 and the pAUC for a false-positive rate of 0.5. Here we replicate that estimation with a call to `rocreg` to estimate the former and follow that with a call to `estat nproc` to estimate the latter. For simplicity, we restrict estimation to classifier `y1` (CA 19-9).

We start by executing `rocreg`, estimating the false-positive rate for a ROC value of 0.6. This value is specified in `invroc()`. Case–control resampling is used by specifying the `bootcc` option.

```
. use http://labs.fhcrc.org/pepe/book/data/wiedat2b, clear
(S. Wieand - Pancreatic cancer diagnostic marker data)
. rocreg d y1, invroc(.6) bseed(8378923) bootcc nodots
Bootstrap results
Number of strata    =          2                Number of obs      =        141
                                                Replications       =       1000
Nonparametric ROC estimation
Control standardization: empirical
ROC method             : empirical
False-positive rate
   Status    : d
   Classifier: y1
```

invROC	Observed Coef.	Bias	Bootstrap Std. Err.	[95% Conf.	Interval]	
.6	0	.0158039	.0267288	-.0523874	.0523874	(N)
				0	.0784314	(P)
				0	.1372549	(BC)

Now we will estimate the pAUC for the false-positive rate of 0.5 using `estat nproc` and the `pauc()` option.

```
. matrix list e(b)
symmetric e(b)[1,1]
           y1:
    invroc_1
y1         0
. estat nproc, pauc(.5)
Bootstrap results
Number of strata    =          2                Number of obs      =        141
                                                Replications       =       1000
Nonparametric ROC estimation
Control standardization: empirical
ROC method             : empirical
False-positive rate
   Status    : d
   Classifier: y1
```

invROC	Observed Coef.	Bias	Bootstrap Std. Err.	[95% Conf.	Interval]	
.6	0	.0158039	.0267288	-.0523874	.0523874	(N)
				0	.0784314	(P)
				0	.1372549	(BC)

```
Partial area under the ROC curve
   Status    : d
   Classifier: y1
```

pAUC	Observed Coef.	Bias	Bootstrap Std. Err.	[95% Conf.	Interval]	
.5	.3932462	-.0000769	.021332	.3514362	.4350562	(N)
				.3492375	.435512	(P)
				.3492375	.435403	(BC)

```
. matrix list r(b)

r(b)[1,2]
             y1:         y1:
       invroc_1      pauc_1
y1            0   .39324619

. matrix list e(b)

symmetric e(b)[1,1]
            y1:
      invroc_1
y1           0

. matrix list r(V)

symmetric r(V)[2,2]
                      y1:         y1:
                invroc_1      pauc_1
y1:invroc_1    .00071443
  y1:pauc_1    -.000326   .00045506

. matrix list e(V)

symmetric e(V)[1,1]
                      y1:
                invroc_1
y1:invroc_1    .00071443
```

◁

The advantages of using `estat nproc` are twofold. First, you can estimate additional parameters of interest without having to respecify the bootstrap settings you did with `rocreg`; instead `estat nproc` uses the bootstrap settings that were stored by `rocreg`. Second, parameters estimated with `estat nproc` are added to those parameters estimated by `rocreg` and returned in the matrices `r(b)` (parameter estimates) and `r(V)` (variance–covariance matrix). Thus you can also obtain correlations between any quantities you wish to estimate.

Saved results

`estat nproc` saves the following in `r()`:

`r(b)`	coefficient vector
`r(V)`	variance–covariance matrix of the estimators
`r(ci_normal)`	normal-approximation confidence intervals
`r(ci_percentile)`	percentile confidence intervals
`r(ci_bc)`	bias-corrected confidence intervals

Methods and formulas

All postestimation commands listed above are implemented as ado-files.

Details on computation of the nonparametric ROC curve and the estimation of the parametric ROC curve model coefficients can be found in [R] **rocreg**. Here we describe how to estimate the ROC curve summary statistics for a parametric model. The cumulative distribution function, g, can be the standard normal cumulative distribution function, Φ.

Methods and formulas are presented under the following headings:

Parametric model: Summary parameter definition
Maximum likelihood estimation
Estimating equations estimation

Parametric model: Summary parameter definition

Conditioning on covariates $\mathbf{x}$, we have the following ROC curve model:

$$\text{ROC}\,(u) = g\{\mathbf{x}'\boldsymbol{\beta} + \alpha g^{-1}\,(u)\}$$

$\mathbf{x}$ can be constant, and $\boldsymbol{\beta} = \beta_0$, the constant intercept.

We can solve this equation to obtain the false-positive rate value u for a ROC value of r:

$$u = g\left[\{g^{-1}\,(r) - \mathbf{x}'\boldsymbol{\beta}\}\alpha^{-1}\right]$$

The partial area under the ROC curve for the false-positive rate u is defined by

$$\text{pAUC}\,(u) = \int_o^u g\{\mathbf{x}'\boldsymbol{\beta} + \alpha g^{-1}\,(t)\}dt$$

The area under the ROC curve is defined by

$$\text{AUC} = \int_o^1 g\{\mathbf{x}'\boldsymbol{\beta} + \alpha g^{-1}\,(t)\}dt$$

When g is the standard normal cumulative distribution function Φ, we can express the AUC as

$$\text{AUC} = \Phi\left(\frac{\mathbf{x}'\boldsymbol{\beta}}{\sqrt{1+\alpha^2}}\right)$$

Maximum likelihood estimation

We allow maximum likelihood estimation under probit parametric models, so $g = \Phi$. The ROC value, false-positive rate, and AUC parameters all have closed-form expressions in terms of the covariate values $\mathbf{x}$, coefficient vector $\boldsymbol{\beta}$, and slope parameter α. So to estimate these three types of summary parameters, we use the delta method (Oehlert 1992; Phillips and Park 1988). Particularly, we use the `nlcom` command (see [R] **nlcom**) to implement the delta method.

To estimate the partial area under the ROC curve for false-positive rate u, we use numeric integration. A trapezoidal approximation is used in calculating the integrals. A numeric integral of the ROC(t) function conditioned on the covariate values $\mathbf{x}$, coefficient vector estimate $\widehat{\boldsymbol{\beta}}$, and slope parameter estimate $\widehat{\alpha}$ is computed over the range $t = [0, u]$. This gives us the point estimate of pAUC(u).

To calculate the standard error and confidence intervals for the point estimate of pAUC(u), we again use the delta method. Details on the delta method algorithm can be found in *Methods and formulas* of [R] **nlcom** and the earlier mentioned references.

Under maximum likelihood estimation, the coefficient estimates $\widehat{\boldsymbol{\beta}}$ and slope estimate $\widehat{\alpha}$ are asymptotically normal with variance matrix $\mathbf{V}$. For convenience, we rename the parameter vector $[\boldsymbol{\beta}', \alpha]$ to the k-parameter vector $\boldsymbol{\theta} = [\theta_1, \ldots, \theta_k]$. We will also explicitly refer to the conditioning of the ROC curve by $\boldsymbol{\theta}$ in its mention as ROC$(t, \boldsymbol{\theta})$.

Under the delta method, the continuous scalar function of the estimate $\widehat{\boldsymbol{\theta}}$, $f(\widehat{\boldsymbol{\theta}})$ has asymptotic mean $f(\boldsymbol{\theta})$ and asymptotic covariance

$$\widehat{\text{Var}}\left\{f(\widehat{\boldsymbol{\theta}})\right\} = \mathbf{fVf}'$$

where $\mathbf{f}$ is the $1 \times k$ matrix of derivatives for which

$$\mathbf{f}_{1j} = \frac{\partial f(\boldsymbol{\theta})}{\partial \theta_j} \qquad j = 1, \ldots, k$$

The asymptotic covariance of $f(\widehat{\boldsymbol{\theta}})$ is estimated and then used in conjunction with $f(\widehat{\boldsymbol{\theta}})$ for further inference, including Wald confidence intervals, standard errors, and hypothesis testing.

In the case of pAUC(u) estimation, our $f(\widehat{\boldsymbol{\theta}})$ is the aforementioned numeric integral of the ROC curve. It estimates $f(\boldsymbol{\theta})$, the true integral of the ROC curve on the $[0, u]$ range. The $\mathbf{V}$ variance matrix is estimated using the likelihood information that `rocreg` calculated, and the estimation is performed by `rocreg` itself.

The partial derivatives of $f(\boldsymbol{\theta})$ can be determined by using Leibnitz's rule (Weisstein 2011):

$$\mathbf{f}_{1j} = \frac{\partial}{\partial \theta_j} \int_0^u \text{ROC}(t, \boldsymbol{\theta}) dt = \int_0^u \frac{\partial}{\partial \theta_j} \text{ROC}(t, \boldsymbol{\theta}) dt \qquad j = 1, \ldots, k$$

When θ_j corresponds with the slope parameter α, we obtain the following partial derivative:

$$\frac{\partial}{\partial \alpha} \text{pAUC}(u) = \int_0^u \phi\{\mathbf{x}'\boldsymbol{\beta} + \alpha \Phi^{-1}(t)\} \Phi^{-1}(t)\, dt$$

The partial derivative of $f(\boldsymbol{\theta})$ [pAUC(u)] for β_0 is the following:

$$\frac{\partial}{\partial \beta_0} \text{pAUC}(u) = \int_0^u \phi\{\mathbf{x}'\boldsymbol{\beta} + \alpha \Phi^{-1}(t)\} dt$$

For a nonintercept coefficient, we obtain the following:

$$\frac{\partial}{\partial \beta_i} \text{pAUC}(u) = \int_0^u x_i \phi\{\mathbf{x}'\boldsymbol{\beta} + \alpha \Phi^{-1}(t)\} dt$$

We can estimate each of these integrals by numeric integration, plugging in the estimates $\widehat{\boldsymbol{\beta}}$ and $\widehat{\alpha}$ for the parameters. This, together with the previously calculated estimate $\widehat{\mathbf{V}}$, provides an estimate of the asymptotic covariance of $f(\widehat{\boldsymbol{\theta}}) = \widehat{\text{pAUC}}(u)$, which allows us to perform further statistical inference on pAUC(u).

Estimating equations estimation

When we fit a model using the Alonzo and Pepe (2002) estimating equations method, we use the bootstrap to perform inference on the ROC curve summary parameters. Each bootstrap sample provides a sample of the coefficient estimates $\boldsymbol{\beta}$ and the slope estimates α. Using the formulas in *Parametric model: Summary parameter definition* under *Methods and formulas*, we can obtain an estimate of the ROC, false-positive rate, or AUC for each resample. Using numeric integration (with the trapezoidal approximation), we can also estimate the pAUC of the resample.

By making these calculations, we obtain a bootstrap sample of our summary parameter estimate. We then obtain bootstrap standard errors, normal approximation confidence intervals, percentile confidence intervals, and bias-corrected confidence intervals using this bootstrap sample. Further details can be found in [R] **bootstrap**.

References

Alonzo, T. A., and M. S. Pepe. 2002. Distribution-free ROC analysis using binary regression techniques. *Biostatistics* 3: 421–432.

Choi, B. C. K. 1998. Slopes of a receiver operating characteristic curve and likelihood ratios for a diagnostic test. *American Journal of Epidemiology* 148: 1127–1132.

Cleves, M. A. 1999. sg120: Receiver operating characteristic (ROC) analysis. *Stata Technical Bulletin* 52: 19–33. Reprinted in *Stata Technical Bulletin Reprints*, vol. 9, pp. 212–229. College Station, TX: Stata Press.

——. 2000. sg120.1: Two new options added to rocfit command. *Stata Technical Bulletin* 53: 18–19. Reprinted in *Stata Technical Bulletin Reprints*, vol. 9, pp. 230–231. College Station, TX: Stata Press.

Hanley, J. A., and B. J. McNeil. 1983. A method of comparing the areas under receiver operating characteristic curves derived from the same cases. *Radiology* 148: 839–843.

Janes, H., G. Longton, and M. S. Pepe. 2009. Accommodating covariates in receiver operating characteristic analysis. *Stata Journal* 9: 17–39.

Norton, S. J., M. P. Gorga, J. E. Widen, R. C. Folsom, Y. Sininger, B. Cone-Wesson, B. R. Vohr, K. Mascher, and K. Fletcher. 2000. Identification of neonatal hearing impairment: Evaluation of transient evoked otoacoustic emission, distortion product otoacoustic emission, and auditory brain stem response test performance. *Ear and Hearing* 21: 508–528.

Oehlert, G. W. 1992. A note on the delta method. *American Statistician* 46: 27–29.

Phillips, P. C. B., and J. Y. Park. 1988. On the formulation of Wald tests of nonlinear restrictions. *Econometrica* 56: 1065–1083.

Stover, L., M. P. Gorga, S. T. Neely, and D. Montoya. 1996. Toward optimizing the clinical utility of distortion product otoacoustic emission measurements. *Journal of the Acoustical Society of America* 100: 956–967.

Weisstein, E. W. 2011. Leibniz integral rule. From *Mathworld*—A Wolfram Web Resource. http://mathworld.wolfram.com/LeibnizIntegralRule.html.

Wieand, S., M. H. Gail, B. R. James, and K. L. James. 1989. A family of nonparametric statistics for comparing diagnostic markers with paired or unpaired data. *Biometrika* 76: 585–592.

Also see

[R] **rocreg** — Receiver operating characteristic (ROC) regression

[R] **rocregplot** — Plot marginal and covariate-specific ROC curves after rocreg

[U] **20 Estimation and postestimation commands**

Title

rocregplot — Plot marginal and covariate-specific ROC curves after rocreg

Syntax

Plot ROC curve after nonparametric analysis

`rocregplot` [, *boot_options common_options*]

Plot ROC curve after parametric analysis using bootstrap

`rocregplot` [, *probit_options boot_options common_options*]

Plot ROC curve after parametric analysis using maximum likelihood

`rocregplot` [, *probit_options common_options*]

probit_options	Description
Main	
`at(`*varname=#* [*varname=#*...]`)` [`at1(`*varname=#* [*varname=#*...]`)` [`at2(`*varname=#* [*varname=#*...]`)` [...]]]	value of specified covariates/mean of unspecified covariates
* `roc(`*numlist*`)`	show estimated ROC values for given false-positive rates
* `invroc(`*numlist*`)`	show estimated false-positive rates for given ROC values
`level(`*#*`)`	set confidence level; default is `level(95)`
Curve	
`line#opts(`*cline_options*`)`	affect rendition of ROC curve #

* Only one of `roc()` or `invroc()` may be specified.

common_options	Description
Main	
`classvars(`*varlist*`)`	restrict plotting of ROC curves to specified classifiers
`norefline`	suppress plotting the reference line
Scatter	
`plot#opts(`*scatter_options*`)`	affect rendition of classifier #s false-positive rate and ROC scatter points; not allowed with `at()`
Reference line	
`rlopts(`*cline_options*`)`	affect rendition of the reference line
Y axis, X axis, Titles, Legend, Overall	
twoway_options	any options other than `by()` documented in [G-3] ***twoway_options***

boot_options	Description
Bootstrap	
† bfile(*filename*)	load dataset containing bootstrap replicates from rocreg
btype(n | p | bc)	plot normal-based (n), percentile (p), or bias-corrected (bc) confidence intervals; default is btype(n)

† bfile() is only allowed with parametric analysis using bootstrap inference; in which case this option is required with roc() or invroc().

Menu

Statistics > Epidemiology and related > ROC analysis > ROC curves after rocreg

Description

Under parametric estimation, rocregplot plots the fitted ROC curves for specified covariate values and classifiers. If rocreg, probit or rocreg, probit ml were previously used, the false-positive rates (for specified ROC values) and ROC values (for specified false-positive rates) for each curve may also be plotted, along with confidence intervals.

Under nonparametric estimation, rocregplot will plot the fitted ROC curves using the _fpr_* and _roc_* variables produced by rocreg. Point estimates and confidence intervals for false-positive rates and ROC values that were computed in rocreg may be plotted as well.

probit_options

Main

at(*varname=#* ...) requests that the covariates specified by *varname* be set to #. By default, rocreg evaluates the function by setting each covariate to its mean value. This option causes the ROC curve to be evaluated at the value of the covariates listed in at() and at the mean of all unlisted covariates.

at1(*varname=#* ...), at2(*varname=#* ...), ..., at10(*varname=#* ...) specify that ROC curves (up to 10) be plotted on the same graph. at1(), at2(), ..., at10() work like the at() option. They request that the function be evaluated at the value of the covariates specified and at the mean of all unlisted covariates. at1() specifies the values of the covariates for the first curve, at2() specifies the values of the covariates for the second curve, and so on.

roc(*numlist*) specifies that estimated ROC values for given false-positive rates be graphed.

invroc(*numlist*) specifies that estimated false-positive rates for given ROC values be graphed.

level(#) specifies the confidence level, as a percentage, for confidence intervals. The default is level(95) or as set by set level; see [U] **20.7 Specifying the width of confidence intervals**. level() may be specified with either roc() or invroc().

Curve

line#opts(*cline_options*) affects the rendition of ROC curve #. See [G-3] ***cline_options***.

common_options

Main

classvars(*varlist*) restricts plotting ROC curves to specified classification variables.

norefline suppresses plotting the reference line.

Scatter

plot#opts(*scatter_options*) affects the rendition of classifier #'s false-positive rate and ROC scatter points. This option applies only to non-ROC covariate estimation graphing. See [G-2] **graph twoway scatter**.

Reference line

rlopts(*cline_options*) affects rendition of the reference line. See [G-3] ***cline_options***.

Y axis, X axis, Titles, Legend, Overall

twoway_options are any of the options documented in [G-3] ***twoway_options***, excluding by(). These include options for titling the graph (see [G-3] ***title_options***) and options for saving the graph to disk (see [G-3] ***saving_option***).

boot_options

Bootstrap

bfile(*filename*) uses bootstrap replicates of parameters from rocreg stored in *filename* to estimate standard errors and confidence intervals of predictions. bfile() must be specified with either roc() or invroc() if parametric estimation with bootstrapping was used.

btype(n | p | bc) indicates the desired type of confidence-interval rendering. n draws normal-based, p draws percentile, and bc draws bias-corrected confidence intervals for specified false-positive rates and ROC values in roc() and invroc(). The default is btype(n).

Remarks

Remarks are presented under the following headings:

Plotting covariate-specific ROC curves
Plotting marginal ROC curves

Plotting covariate-specific ROC curves

The rocregplot command is also demonstrated in [R] **rocreg**. We will further demonstrate its use with several examples. Particularly, we will show how rocregplot can draw the ROC curves of covariate models that have been fit using rocreg.

▷ Example 1

In example 6 of [R] **rocreg**, we fit a probit ROC model to audiology test data from Norton et al. (2000). The estimating equation method of Alonzo and Pepe (2002) was used to the fit the model. Gender and age were covariates that affected the control distribution of the classifier `y1` (DPOAE 65 at 2 kHz). Age was a ROC covariate for the model, so we fit separate ROC curves at each age.

Following Janes, Longton, and Pepe (2009), we draw the ROC curves for ages 30, 40, and 50 months. The `at1()`, `at2()`, and `at3()` options are used to specify the age covariates.

```
. use http://www.stata-press.com/data/r12/nnhs
(Norton - neonatal audiology data)
. rocreg d y1, probit ctrlcov(currage male) ctrlmodel(linear) roccov(currage)
    cluster(id) bseed(56930) bsave(nnhs2y1, replace)
  (output omitted)
. rocregplot, at1(currage=30) at2(currage=40) at3(currage=50)
```

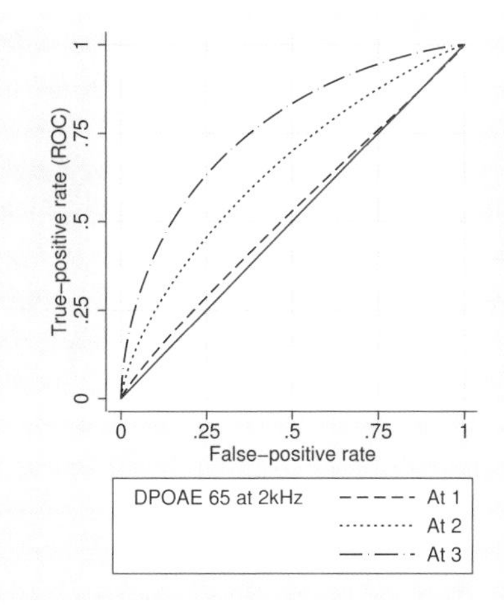

Here we use the default entries of the legend, which indicate the "at #" within the specified `at*` options and the classifier to which the curve corresponds. ROC curve one corresponds with `currage=30`, two with `currage=40`, and three with `currage=50`. The positive effect of age on the ROC curve is evident. At an age of 30 months (`currage=30`), the ROC curve of `y1` (DPOAE 65 at 2 kHz) is nearly equivalent to that of a noninformative test that gives equal probability to hearing loss. At age 50 months (`currage=50`), corresponding to some of the oldest children in the study, the ROC curve shows that test `y1` (DPOAE 65 at 2 kHz) is considerably more powerful than the noninformative test.

You may create your own legend by specifying the `legend()` option. The default legend is designed for the possibility of multiple covariates. Here we could change the legend entries to `currage` values and gain some extra clarity. However, this may not be feasible when there are many covariates present.

◁

We can also use `rocregplot` after maximum likelihood estimation.

▷ Example 2

We return to the audiology study with frequency (`xf`), intensity (`xl`), and hearing loss severity (`xd`) covariates from Stover et al. (1996) that we examined in example 10 of [R] **rocreg**. Negative signal-to-noise ratio is again used as a classifier. Using maximum likelihood, we fit a probit model to these data with the indicated ROC covariates.

After fitting the model, we wish to compare the ROC curves of two covariate combinations. The first has an intensity value of 5.5 (the lowest intensity, corresponding to 55 decibels) and a frequency of 10.01 (the lowest frequency, corresponding to 1001 hertz). We give the first combination a hearing loss severity value of 0.5 (the lowest). The second covariate combination has the same frequency, but the highest intensity value of 6.5 (65 decibels). We give this second covariate set a higher severity value of 4. We will visually compare the two ROC curves resulting from these two covariate value combinations.

We specify false-positive rates of 0.7 first followed by 0.2 in the `roc()` option to visually compare the size of the ROC curve at large and small false-positive rates. Because maximum likelihood estimation was used to fit the model, a Wald confidence interval is produced for the estimated ROC value and false-positive rate parameters. Further details are found in *Methods and formulas*.

```
. use http://www.stata-press.com/data/r12/dp
(Stover - DPOAE test data)

. rocreg d nsnr, probit ctrlcov(xf xl) roccov(xf xl xd) ml cluster(id)
  (output omitted)

. rocregplot, at1(xf=10.01, xl=5.5, xd=.5) at2(xf=10.01, xl=6.5, xd=4) roc(.7)
ROC curve

   Status    : d
   Classifier: nsnr

Under covariates:

             |    at1
-------------+-----------
          xf |    10.01
          xl |      5.5
          xd |       .5

--------------------------------------------------------------------
         ROC |     Coef.   Std. Err.       [95% Conf. Interval]
-------------+------------------------------------------------------
          .7 |   .978951   .0097382        .9598645    .9980376
--------------------------------------------------------------------

Under covariates:

             |    at2
-------------+-----------
          xf |    10.01
          xl |      6.5
          xd |        4

--------------------------------------------------------------------
         ROC |     Coef.   Std. Err.       [95% Conf. Interval]
-------------+------------------------------------------------------
          .7 |  .9985001   .0009657        .9966073    1.000393
--------------------------------------------------------------------
```

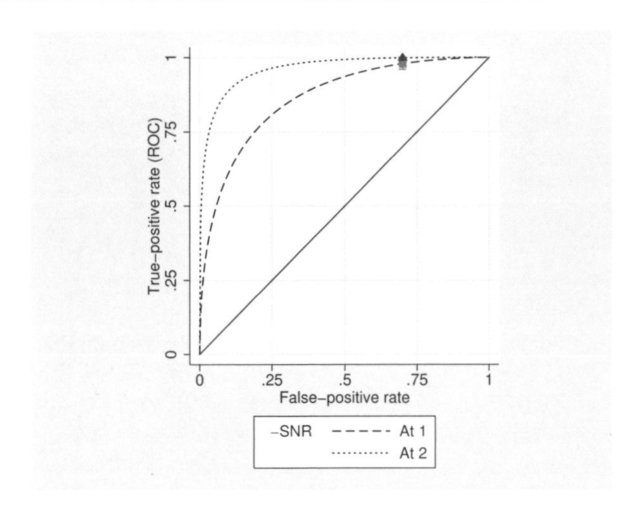

At the higher false-positive rate value of 0.7, we see little difference in the ROC values and note that the confidence intervals nearly overlap. Now we view the same curves with the lower false-positive rate compared.

```
. rocregplot, at1(xf=10.01, xl=5.5, xd=.5) at2(xf=10.01, xl=6.5, xd=4) roc(.2)
ROC curve

   Status    : d
   Classifier: nsnr

Under covariates:

             |      at1
-------------+------------
          xf |    10.01
          xl |      5.5
          xd |       .5

-----------------------------------------------------------------------
         ROC |      Coef.   Std. Err.      [95% Conf. Interval]
-------------+---------------------------------------------------------
          .2 |   .7608593   .0510501       .660803    .8609157
-----------------------------------------------------------------------

Under covariates:

             |      at2
-------------+------------
          xf |    10.01
          xl |      6.5
          xd |        4

-----------------------------------------------------------------------
         ROC |      Coef.   Std. Err.      [95% Conf. Interval]
-------------+---------------------------------------------------------
          .2 |   .9499408   .0179824       .914696    .9851856
-----------------------------------------------------------------------
```

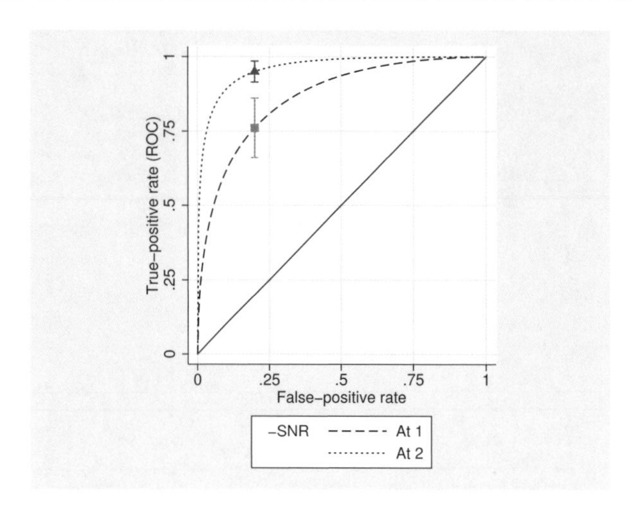

The lower false-positive rate of 0.2 shows clearly distinguishable ROC values. Now we specify option `invroc(.5)` to view how the false-positive rates vary at a ROC value of 0.5.

```
. rocregplot, at1(xf=10.01, xl=5.5, xd=.5) at2(xf=10.01, xl=6.5, xd=4) invroc(.5)
False-positive rate

   Status    : d
   Classifier: nsnr

Under covariates:

             |    at1
-------------+-----------
          xf |  10.01
          xl |    5.5
          xd |     .5

------------------------------------------------------------
      invROC |     Coef.   Std. Err.     [95% Conf. Interval]
-------------+----------------------------------------------
          .5 |  .0578036   .0198626      .0188736   .0967336
------------------------------------------------------------

Under covariates:

             |    at2
-------------+-----------
          xf |  10.01
          xl |    6.5
          xd |      4

------------------------------------------------------------
      invROC |     Coef.   Std. Err.     [95% Conf. Interval]
-------------+----------------------------------------------
          .5 |  .0055624   .0032645     -.0008359   .0119607
------------------------------------------------------------
```

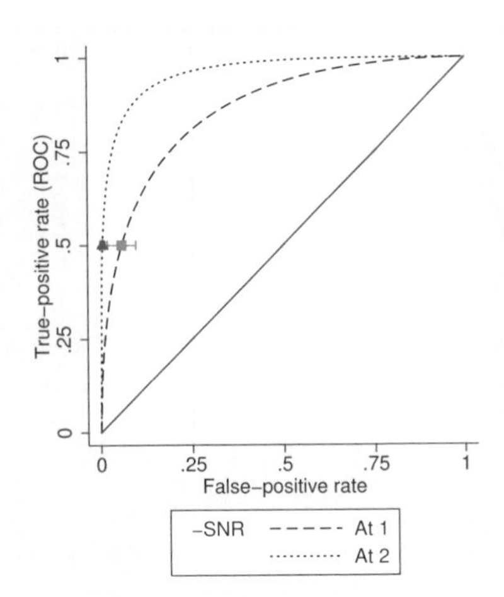

At a ROC value of 0.5, the false-positive rates for both curves are small and close to one another. ◁

❑ Technical note

We can use the `testnl` command to support our visual observations with statistical inference. We use it to perform a Wald test of the null hypothesis that the two ROC curves just rendered are equal at a false-positive rate of 0.7.

```
. testnl normal(_b[i_cons]+10.01*_b[xf]+5.5*_b[xl]
>                + .5*_b[xd]+_b[s_cons]*invnormal(.7)) =
>       normal(_b[i_cons]+10.01*_b[xf]+6.5*_b[xl]
>                + 4*_b[xd]+_b[s_cons]*invnormal(.7))

  (1)  normal(_b[i_cons]+10.01*_b[xf]+5.5*_b[xl]  +.5*_b[xd]+_b[s_cons]*invnormal(.7))=
       normal(_b[i_cons]+10.01*_b[xf]+6.5*_b[xl]  + 4*_b[xd]+_b[s_cons]*invnormal(.7))

               chi2(1) =        4.53
           Prob > chi2 =        0.0332
```

The test is significant at the 0.05 level, and thus we find that the two curves are significantly different. Now we will use `testnl` again to test equality of the false-positive rates for each curve with a ROC value of 0.5. The inverse ROC formula used is derived in *Methods and formulas*.

```
. testnl normal((invnormal(.5)-(_b[i_cons]+10.01*_b[xf]+5.5*_b[xl]+.5*_b[xd]))
>                /_b[s_cons]) =
>       normal((invnormal(.5)-(_b[i_cons]+10.01*_b[xf]+6.5*_b[xl]+4*_b[xd]))
>                /_b[s_cons])

  (1)  normal((invnormal(.5)-(_b[i_cons]+10.01*_b[xf]+5.5*_b[xl]+.5*_b[xd]))
              /_b[s_cons]) =
       normal((invnormal(.5)-(_b[i_cons]+10.01*_b[xf]+6.5*_b[xl]+4*_b[xd]))
              /_b[s_cons])

chi2(1) =        8.01
           Prob > chi2 =        0.0046
```

We again reject the null hypothesis that the two curves are equal at the 0.05 level. ❑

The model of our last example was also fit using the estimating equations method in example 7 of [R] **rocreg**. We will demonstrate `rocregplot` after that model fit as well.

▷ Example 3

In example 2, we used `rocregplot` after a maximum likelihood model fit of the ROC curve for classifier `nsnr` and covariates frequency (`xf`), intensity (`xl`), and hearing loss severity (`xd`). The data were obtained from the audiology study described in Stover et al. (1996). In example 7 of [R] **rocreg**, we fit the model using the estimating equations method of Alonzo and Pepe (2002). Under this method, bootstrap resampling is used to make inferences. We saved 50 bootstrap replications in `nsnrf.dta`, which we re-create below.

We use `rocregplot` to draw the ROC curves for `nsnr` under the covariate values `xf = 10.01`, `xl = 5.5`, and `xd = .5`, and `xf = 10.01`, `xl = 6.5`, and `xd = 4`. The `at#()` options are used to specify the covariate values. The previous bootstrap results are made available to `rocregplot` with the `bfile()` option. As before, we will specify 0.2 and 0.7 as false-positive rates in the `roc()` option and 0.5 as a ROC value in the `invroc()` option. We do not specify `btype()` and thus our graph will contain normal-based bootstrap confidence bands, the default.

```
. use http://www.stata-press.com/data/r12/dp
(Stover - DPOAE test data)

. rocreg d nsnr, probit ctrlcov(xf xl) roccov(xf xl xd) cluster(id)
> nobstrata ctrlfprall bseed(156385) breps(50) bsave(nsnrf, replace)
  (output omitted)

. rocregplot, at1(xf=10.01, xl=5.5, xd=.5) at2(xf=10.01, xl=6.5, xd=4)
> roc(.7) bfile(nsnrf)

ROC curve

   Status    : d
   Classifier: nsnr

Under covariates:

             |       at1
-------------+-----------
          xf |     10.01
          xl |       5.5
          xd |        .5

                                  (Replications based on 208 clusters in id)
-----------------------------------------------------------------------------
             |   Observed               Bootstrap
         ROC |      Coef.       Bias    Std. Err.    [95% Conf. Interval]
-------------+---------------------------------------------------------------
          .7 |   .9835816   .0087339     .0204353    .9435292    1.023634  (N)
             |                                       .9155462    .9974037  (P)
             |                                       .9392258    .9976629 (BC)
-----------------------------------------------------------------------------

Under covariates:

             |       at2
-------------+-----------
          xf |     10.01
          xl |       6.5
          xd |         4

                                  (Replications based on 208 clusters in id)
-----------------------------------------------------------------------------
             |   Observed               Bootstrap
         ROC |      Coef.       Bias    Std. Err.    [95% Conf. Interval]
-------------+---------------------------------------------------------------
          .7 |    .999428   .0006059     .0011309    .9972115    1.001644  (N)
             |                                       .9958003    .9999675  (P)
             |                                       .9968304    .9999901 (BC)
-----------------------------------------------------------------------------
```

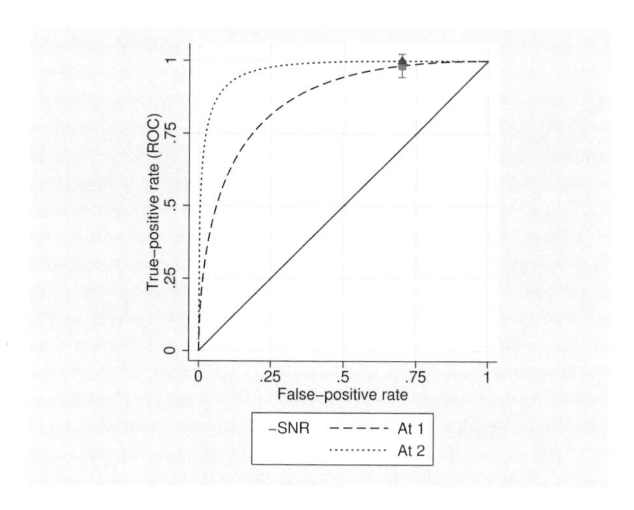

As shown in the graph, we find that the ROC values at a false-positive rate of 0.7 are close together, as they were in the maximum likelihood estimation in example 2. We now repeat this process for the lower false-positive rate of 0.2 by using the `roc(.2)` option.

```
. rocregplot, at1(xf=10.01, xl=5.5, xd=.5) at2(xf=10.01, xl=6.5, xd=4)
> roc(.2) bfile(nsnrf)
ROC curve

   Status    : d
   Classifier: nsnr

Under covariates:
```

	at1
xf	10.01
xl	5.5
xd	.5

(Replications based on 208 clusters in id)

ROC	Observed Coef.	Bias	Bootstrap Std. Err.	[95% Conf.	Interval]	
.2	.7652956	.0145111	.0735506	.6211391	.9094522	(N)
				.6054495	.878052	(P)
				.6394838	.9033081	(BC)

```
Under covariates:
```

	at2
xf	10.01
xl	6.5
xd	4

(Replications based on 208 clusters in id)

ROC	Observed Coef.	Bias	Bootstrap Std. Err.	[95% Conf.	Interval]	
.2	.9672505	.0072429	.0227977	.9225679	1.011933	(N)
				.9025254	.9931714	(P)
				.9235289	.9979637	(BC)

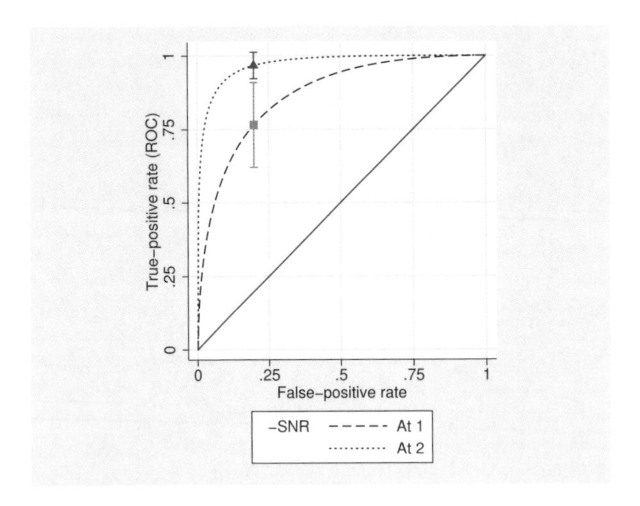

The ROC values are slightly higher at the false-positive rate of 0.2 than they were in the maximum likelihood estimation in example 2. To see if the false-positive rates differ at a ROC value of 0.5, we specify the `invroc(.5)` option.

```
. rocregplot, at1(xf=10.01, xl=5.5, xd=.5) at2(xf=10.01, xl=6.5, xd=4)
> invroc(.5) bfile(nsnrf)
False-positive rate
   Status    : d
   Classifier: nsnr
Under covariates:

             |      at1
-------------+-----------
          xf |    10.01
          xl |      5.5
          xd |       .5

                                 (Replications based on 208 clusters in id)
------------------------------------------------------------------------------
             |  Observed                Bootstrap
      invROC |     Coef.       Bias     Std. Err.     [95% Conf. Interval]
-------------+----------------------------------------------------------------
          .5 |  .0615144  -.0063531    .0254042     .0117231    .1113057  (N)
             |                                      .0225159    .1265046  (P)
             |                                      .0224352    .1265046 (BC)
------------------------------------------------------------------------------

Under covariates:

             |      at2
-------------+-----------
          xf |    10.01
          xl |      6.5
          xd |        4

                                 (Replications based on 208 clusters in id)
------------------------------------------------------------------------------
             |  Observed                Bootstrap
      invROC |     Coef.       Bias     Std. Err.     [95% Conf. Interval]
-------------+----------------------------------------------------------------
          .5 |  .0043298  -.0012579    .0045938     -.004674    .0133335  (N)
             |                                      .0002773    .0189199  (P)
             |                                      .0001292    .0134801 (BC)
------------------------------------------------------------------------------
```

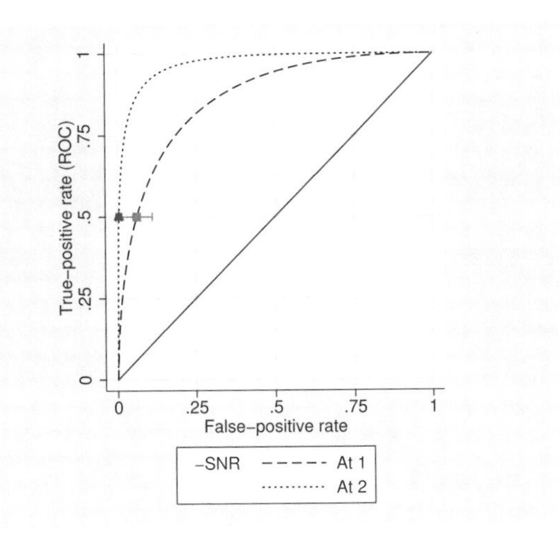

The point estimates of the ROC value and false-positive rate are both computed directly using the point estimates of the ROC coefficients. Calculation of the standard errors and confidence intervals is slightly more complicated. Essentially, we have stored a sample of our ROC covariate coefficient estimates in `nsnrf.dta`. We then calculate the ROC value or false-positive rate estimates using each set of coefficient estimates, resulting in a sample of point estimates. Then the bootstrap standard error and confidence intervals are calculated based on these bootstrap samples. Details of the computation of the standard error and percentile confidence intervals can be found in *Methods and formulas* and in [R] **bootstrap**.

As mentioned in [R] **rocreg**, 50 resamples is a reasonable lower bound for obtaining bootstrap standard errors (Mooney and Duval 1993). However, it may be too low for obtaining percentile and bias-corrected confidence intervals. Normal-based confidence intervals are valid when the bootstrap distribution exhibits normality. See [R] **bootstrap postestimation** for more details.

We can assess the normality of the bootstrap distribution by using a normal probability plot. Stata provides this in the `pnorm` command (see [R] **diagnostic plots**). We will use `nsnrf.dta` to draw a normal probability plot for the ROC estimate corresponding to a false-positive rate of 0.2. We use the covariate values `xf = 10.01`, `xl = 6.5`, and `xd = 4`.

```
. use nsnrf
(bootstrap: rocregnewstat)
. generate double rocp2 = nsnr_b_i_cons + 10.01*nsnr_b_xf + 6.5*nsnr_b_xl +
> 4*nsnr_b_xd+nsnr_b_s_cons*invnormal(.2)
. replace rocp2 = normal(rocp2)
(50 real changes made)
```

```
. pnorm rocp2
```

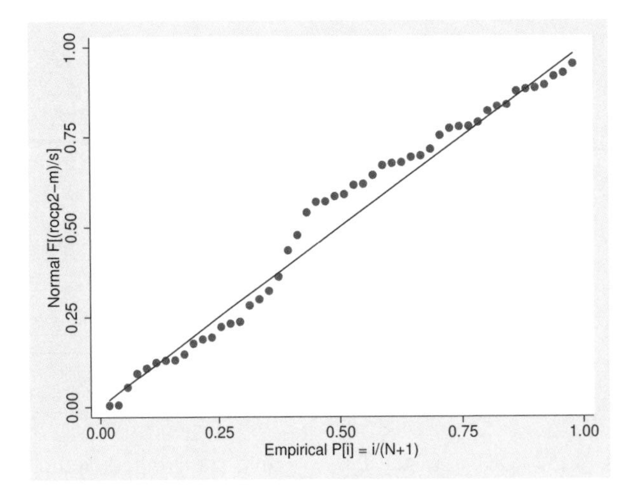

The closeness of the points to the horizontal line on the normal probability plot shows us that the bootstrap distribution is approximately normal. So it is reasonable to use the normal-based confidence intervals for ROC at a false-positive rate of 0.2 under covariate values `xf = 10.01`, `xl = 6.5`, and `xd = 4`.

◁

Plotting marginal ROC curves

The `rocregplot` command can also be used after fitting models with no covariates. We will demonstrate this with an empirical ROC model fit in [R] **rocreg**.

▷ Example 4

We run `rocregplot` after fitting the single-classifier, empirical ROC model shown in example 1 of [R] **rocreg**. There we empirically predicted the ROC curve of the classifier `rating` for the true status variable `disease` from the Hanley and McNeil (1982) data. The `rocreg` command saves variables `_roc_rating` and `_fpr_rating`, which give the ROC values and false-positive rates, respectively, for every value of `rating`. These variables are used by `rocregplot` to render the ROC curve.

```
. use http://www.stata-press.com/data/r12/hanley, clear
. rocreg disease rating, noboot
Nonparametric ROC estimation
Control standardization: empirical
ROC method             : empirical
Area under the ROC curve
   Status    : disease
   Classifier: rating
```

AUC	Observed Coef.	Bias	Bootstrap Std. Err.	[95% Conf.	Interval]	
	.8407708	.	.	.	.	(N)
				.	.	(P)
				.	.	(BC)

```
. rocregplot
```

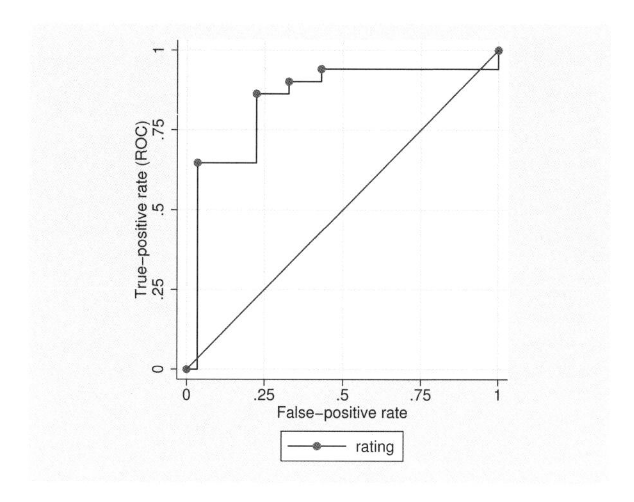

◁

We end our discussion of `rocregplot` by showing its use after a marginal probit model.

▷ Example 5

In example 13 of [R] **rocreg**, we fit a maximum-likelihood probit model to each classifier of the fictitious dataset generated from Hanley and McNeil (1983).

We use `rocregplot` after the original `rocreg` command to draw the ROC curves for classifiers `mod1` and `mod3`. This is accomplished by specifying the two variables in the `classvars()` option. We will use the `roc()` option to obtain confidence intervals for ROC values at false-positive rates of 0.15 and 0.75. We will specify the `invroc()` option to obtain false-positive rate confidence intervals for a ROC value of 0.8. As mentioned previously, these are Wald confidence intervals.

First, we will view results for a false-positive rate of 0.75.

```
. use http://www.stata-press.com/data/r12/ct2, clear
. rocreg status mod1 mod2 mod3, probit ml
  (output omitted)
. rocregplot, classvars(mod1 mod3) roc(.75)
ROC curve

   Status    : status
   Classifier: mod1
```

ROC	Coef.	Std. Err.	[95% Conf.	Interval]
.75	.9931655	.0069689	.9795067	1.006824

```
   Status    : status
   Classifier: mod3
```

ROC	Coef.	Std. Err.	[95% Conf.	Interval]
.75	.9953942	.0043435	.9868811	1.003907

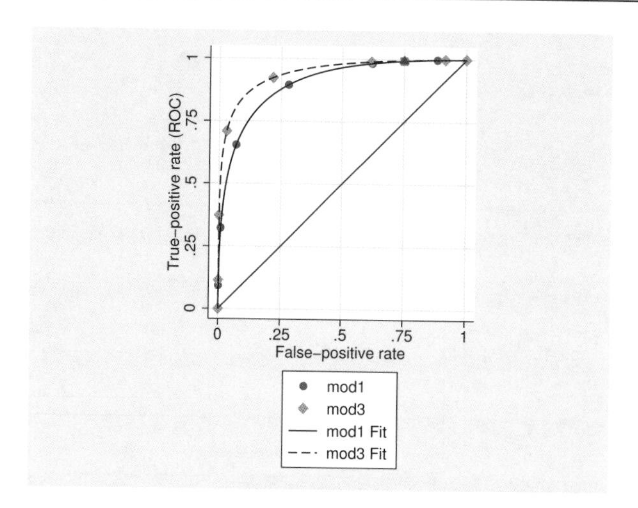

We see that the estimates for each of the two ROC curves are close. Because this is a marginal model, the actual false-positive rate and the true-positive rate for each observation are plotted in the graph. The added point estimates of the ROC value at false-positive rate 0.75 are shown as diamond (`mod3`) and circle (`mod1`) symbols in the upper-right-hand corner of the graph at FPR = 0.75. Confidence bands are also plotted at FPR = 0.75 but are so narrow that they are barely noticeable. Under both classifiers, the ROC value at 0.75 is very high. Now we will compare these results to those with a lower false-positive rate of 0.15.

```
. rocregplot, classvars(mod1 mod3) roc(.15)
ROC curve

   Status    : status
   Classifier: mod1

          ROC |      Coef.   Std. Err.     [95% Conf. Interval]
--------------+------------------------------------------------
          .15 |   .7934935    .0801363      .6364292    .9505578

   Status    : status
   Classifier: mod3

          ROC |      Coef.   Std. Err.     [95% Conf. Interval]
--------------+------------------------------------------------
          .15 |   .8888596    .0520118      .7869184    .9908008
```

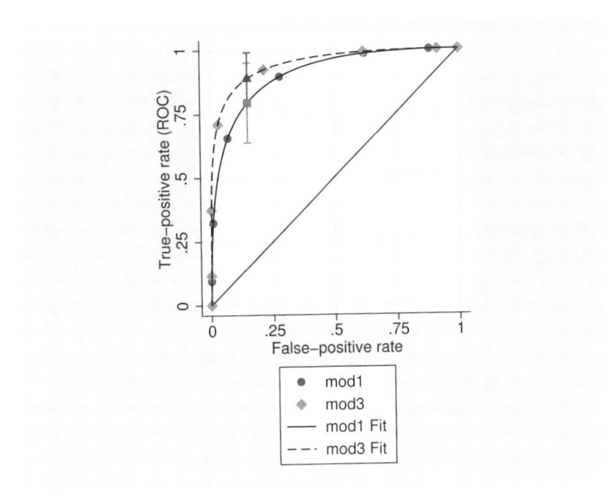

The ROC value for the false-positive rate of 0.15 is more separated in the two classifiers. Here we see that `mod3` has a larger ROC value than `mod1` for this false-positive rate, but the confidence intervals of the estimates overlap.

By specifying `invroc(.8)`, we obtain invROC confidence intervals corresponding to a ROC value of 0.8.

```
. rocregplot, classvars(mod1 mod3) invroc(.8)
False-positive rate
   Status    : status
   Classifier: mod1
```

invROC	Coef.	Std. Err.	[95% Conf. Interval]	
.8	.1556435	.069699	.019036	.2922509

```
   Status    : status
   Classifier: mod3
```

invROC	Coef.	Std. Err.	[95% Conf. Interval]	
.8	.0661719	.045316	-.0226458	.1549896

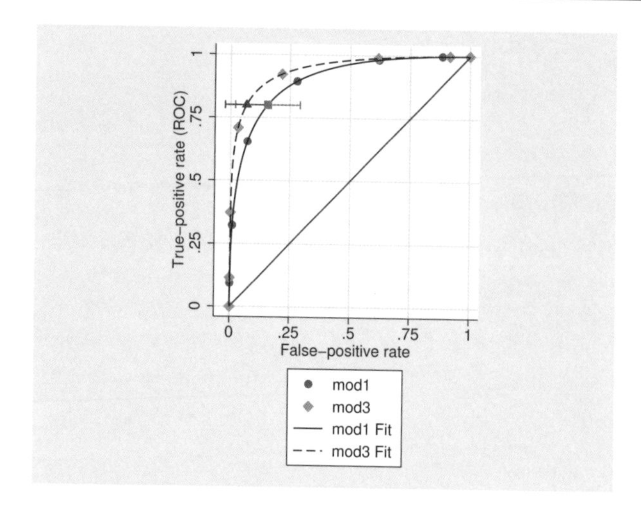

For estimation of the false-positive rate at a ROC value of 0.8, the confidence intervals overlap. Both classifiers require only a small false-positive rate to achieve a ROC value of 0.8.

◁

Methods and formulas

`rocregplot` is implemented as an ado-file.

Details on computation of the nonparametric ROC curve and the estimation of the parametric ROC curve model coefficients can be found in [R] **rocreg**. Here we describe how to estimate the ROC values and false-positive rates of a parametric model. The cumulative distribution function g can be the standard normal cumulative distribution function.

Methods and formulas are presented under the following headings:

Parametric model: Summary parameter definition
Maximum likelihood estimation
Estimating equations estimation

Parametric model: Summary parameter definition

Conditioning on covariates $\mathbf{x}$, we have the following ROC curve model:

$$\mathrm{ROC}\,(u) = g\{\mathbf{x}'\boldsymbol{\beta} + \alpha g^{-1}\,(u)\}$$

$\mathbf{x}$ can be constant, and $\boldsymbol{\beta} = \beta_0$, the constant intercept.

With simple algebra, we can solve this equation to obtain the false-positive rate value u for a ROC value of r:

$$u = g\left[\{g^{-1}\,(r) - \mathbf{x}'\boldsymbol{\beta}\}\alpha^{-1}\right]$$

Maximum likelihood estimation

We allow maximum likelihood estimation under probit parametric models, so $g = \Phi$. The ROC value and false-positive rate parameters all have closed-form expressions in terms of the covariate values $\mathbf{x}$, coefficient vector $\boldsymbol{\beta}$, and slope parameter α. Thus to estimate these two types of summary parameters, we use the delta method (Oehlert 1992; Phillips and Park 1988). Particularly, we use the `nlcom` command (see [R] **nlcom**) to implement the delta method.

Under maximum likelihood estimation, the coefficient estimates $\widehat{\boldsymbol{\beta}}$ and slope estimate $\widehat{\alpha}$ are asymptotically normal with variance matrix $\mathbf{V}$. For convenience, we rename the parameter vector $[\boldsymbol{\beta}', \alpha]$ to the k-parameter vector $\boldsymbol{\theta} = [\theta_1, \ldots, \theta_k]$. We will also explicitly refer to the conditioning of the ROC curve by $\boldsymbol{\theta}$ in its mention as $\text{ROC}(t, \boldsymbol{\theta})$.

Under the delta method, the continuous scalar function of the estimate $\widehat{\boldsymbol{\theta}}$, $f(\widehat{\boldsymbol{\theta}})$ has asymptotic mean $f(\boldsymbol{\theta})$ and asymptotic covariance

$$\widehat{\text{Var}}\left\{f(\widehat{\boldsymbol{\theta}})\right\} = \mathbf{f}\mathbf{V}\mathbf{f}'$$

where $\mathbf{f}$ is the $1 \times k$ matrix of derivatives for which

$$\mathbf{f}_{1j} = \frac{\partial f(\boldsymbol{\theta})}{\partial \theta_j} \qquad j = 1, \ldots, k$$

The asymptotic covariance of $f(\widehat{\boldsymbol{\theta}})$ is estimated and then used in conjunction with $f(\widehat{\boldsymbol{\theta}})$ for further inference, including Wald confidence intervals, standard errors, and hypothesis testing.

Estimating equations estimation

When we fit a model using the Alonzo and Pepe (2002) estimating equations method, we use the bootstrap to perform inference on the ROC curve summary parameters. Each bootstrap sample provides a sample of the coefficient estimates $\boldsymbol{\beta}$ and the slope estimates α. Using the formulas above, we can obtain an estimate of the ROC value or false-positive rate for each resample.

By making these calculations, we obtain a bootstrap sample of our summary parameter estimate. We then obtain bootstrap standard errors, normal approximation confidence intervals, percentile confidence intervals, and bias-corrected confidence intervals using this bootstrap sample. Further details can be found in [R] **bootstrap**.

References

Alonzo, T. A., and M. S. Pepe. 2002. Distribution-free ROC analysis using binary regression techniques. *Biostatistics* 3: 421–432.

Bamber, D. 1975. The area above the ordinal dominance graph and the area below the receiver operating characteristic graph. *Journal of Mathematical Psychology* 12: 387–415.

Choi, B. C. K. 1998. Slopes of a receiver operating characteristic curve and likelihood ratios for a diagnostic test. *American Journal of Epidemiology* 148: 1127–1132.

Cleves, M. A. 1999. sg120: Receiver operating characteristic (ROC) analysis. *Stata Technical Bulletin* 52: 19–33. Reprinted in *Stata Technical Bulletin Reprints*, vol. 9, pp. 212–229. College Station, TX: Stata Press.

——. 2000. sg120.1: Two new options added to rocfit command. *Stata Technical Bulletin* 53: 18–19. Reprinted in *Stata Technical Bulletin Reprints*, vol. 9, pp. 230–231. College Station, TX: Stata Press.

Hanley, J. A., and B. J. McNeil. 1982. The meaning and use of the area under a receiver operating characteristic (ROC) curve. *Radiology* 143: 29–36.

——. 1983. A method of comparing the areas under receiver operating characteristic curves derived from the same cases. *Radiology* 148: 839–843.

Janes, H., G. Longton, and M. S. Pepe. 2009. Accommodating covariates in receiver operating characteristic analysis. *Stata Journal* 9: 17–39.

Mooney, C. Z., and R. D. Duval. 1993. *Bootstrapping: A Nonparametric Approach to Statistical Inference*. Newbury Park, CA: Sage.

Norton, S. J., M. P. Gorga, J. E. Widen, R. C. Folsom, Y. Sininger, B. Cone-Wesson, B. R. Vohr, K. Mascher, and K. Fletcher. 2000. Identification of neonatal hearing impairment: Evaluation of transient evoked otoacoustic emission, distortion product otoacoustic emission, and auditory brain stem response test performance. *Ear and Hearing* 21: 508–528.

Oehlert, G. W. 1992. A note on the delta method. *American Statistician* 46: 27–29.

Phillips, P. C. B., and J. Y. Park. 1988. On the formulation of Wald tests of nonlinear restrictions. *Econometrica* 56: 1065–1083.

Stover, L., M. P. Gorga, S. T. Neely, and D. Montoya. 1996. Toward optimizing the clinical utility of distortion product otoacoustic emission measurements. *Journal of the Acoustical Society of America* 100: 956–967.

Also see

[R] **rocreg** — Receiver operating characteristic (ROC) regression

[R] **rocreg postestimation** — Postestimation tools for rocreg

[U] **20 Estimation and postestimation commands**

Title

roctab — Nonparametric ROC analysis

Syntax

`roctab` *refvar classvar* [*if*] [*in*] [*weight*] [, *options*]

roctab_options	Description
Main	
`lorenz`	report Gini and Pietra indices
`binomial`	calculate exact binomial confidence intervals
`detail`	show details on sensitivity/specificity for each cutpoint
`table`	display the raw data in a $2 \times k$ contingency table
`bamber`	calculate standard errors by using the Bamber method
`hanley`	calculate standard errors by using the Hanley method
`graph`	graph the ROC curve
`norefline`	suppress plotting the 45-degree reference line
`summary`	report the area under the ROC curve
`specificity`	graph sensitivity versus specificity
`level(#)`	set confidence level; default is `level(95)`
Plot	
`plotopts(`*plot_options*`)`	affect rendition of the ROC curve
Reference line	
`rlopts(`*cline_options*`)`	affect rendition of the reference line
Add plots	
`addplot(`*plot*`)`	add other plots to the generated graph
Y axis, X axis, Titles, Legend, Overall	
twoway_options	any options other than `by()` documented in [G-3] ***twoway_options***

`fweight`s are allowed; see [U] **11.1.6 weight**.

plot_options	Description
marker_options	change look of markers (color, size, etc.)
marker_label_options	add marker labels; change look or position
cline_options	change the look of the line

Menu

Statistics > Epidemiology and related > ROC analysis > Nonparametric ROC analysis without covariates

Description

The above command is used to perform receiver operating characteristic (ROC) analyses with rating and discrete classification data.

The two variables *refvar* and *classvar* must be numeric. The reference variable indicates the true state of the observation, such as diseased and nondiseased or normal and abnormal, and must be coded as 0 and 1. The rating or outcome of the diagnostic test or test modality is recorded in *classvar*, which must be at least ordinal, with higher values indicating higher risk.

`roctab` performs nonparametric ROC analyses. By default, `roctab` calculates the area under the ROC curve. Optionally, `roctab` can plot the ROC curve, display the data in tabular form, and produce Lorenz-like plots.

See [R] **rocfit** for a command that fits maximum-likelihood ROC models.

Options

Main

`lorenz` specifies that Gini and Pietra indices be reported. Optionally, `graph` will plot the Lorenz-like curve.

`binomial` specifies that exact binomial confidence intervals be calculated.

`detail` outputs a table displaying the sensitivity, specificity, the percentage of subjects correctly classified, and two likelihood ratios for each possible cutpoint of *classvar*.

`table` outputs a $2 \times k$ contingency table displaying the raw data.

`bamber` specifies that the standard error for the area under the ROC curve be calculated using the method suggested by Bamber (1975). Otherwise, standard errors are obtained as suggested by DeLong, DeLong, and Clarke-Pearson (1988).

`hanley` specifies that the standard error for the area under the ROC curve be calculated using the method suggested by Hanley and McNeil (1982). Otherwise, standard errors are obtained as suggested by DeLong, DeLong, and Clarke-Pearson (1988).

`graph` produces graphical output of the ROC curve. If `lorenz` is specified, graphical output of a Lorenz-like curve will be produced.

`norefline` suppresses plotting the 45-degree reference line from the graphical output of the ROC curve.

`summary` reports the area under the ROC curve, its standard error, and its confidence interval. If `lorenz` is specified, Lorenz indices are reported. This option is needed only when also specifying `graph`.

`specificity` produces a graph of sensitivity versus specificity instead of sensitivity versus (1 − specificity). `specificity` implies `graph`.

`level(#)` specifies the confidence level, as a percentage, for the confidence intervals. The default is `level(95)` or as set by `set level`; see [R] **level**.

Plot

`plotopts(`*plot_options*`)` affects the rendition of the plotted ROC curve—the curve's plotted points connected by lines. The *plot_options* can affect the size and color of markers, whether and how the markers are labeled, and whether and how the points are connected; see [G-3] ***marker_options***, [G-3] ***marker_label_options***, and [G-3] ***cline_options***.

Reference line

rlopts(*cline_options*) affects the rendition of the reference line; see [G-3] ***cline_options***.

Add plots

addplot(*plot*) provides a way to add other plots to the generated graph; see [G-3] ***addplot_option***.

Y axis, X axis, Titles, Legend, Overall

twoway_options are any of the options documented in [G-3] ***twoway_options***, excluding by(). These include options for titling the graph (see [G-3] ***title_options***) and for saving the graph to disk (see [G-3] ***saving_option***).

Remarks

Remarks are presented under the following headings:

Introduction
Nonparametric ROC curves
Lorenz-like curves

Introduction

The roctab command provides nonparametric estimation of the ROC for a given classifier and true-status reference variable. The Lorenz curve functionality of roctab, which provides an alternative to standard ROC analysis, is discussed in *Lorenz-like curves*.

See Pepe (2003) for a discussion of ROC analysis. Pepe has posted Stata datasets and programs used to reproduce results presented in the book (http://www.stata.com/bookstore/pepe.html).

Nonparametric ROC curves

The points on the nonparametric ROC curve are generated using each possible outcome of the diagnostic test as a classification cutpoint and computing the corresponding sensitivity and $1-$specificity. These points are then connected by straight lines, and the area under the resulting ROC curve is computed using the trapezoidal rule.

▷ Example 1

Hanley and McNeil (1982) presented data from a study in which a reviewer was asked to classify, using a five-point scale, a random sample of 109 tomographic images from patients with neurological problems. The rating scale was as follows: 1 = definitely normal, 2 = probably normal, 3 = questionable, 4 = probably abnormal, and 5 = definitely abnormal. The true disease status was normal for 58 of the patients and abnormal for the remaining 51 patients.

Here we list 9 of the 109 observations:

```
. use http://www.stata-press.com/data/r12/hanley
. list disease rating in 1/9
```

	disease	rating
1.	1	5
2.	0	1
3.	1	5
4.	0	4
5.	0	1
6.	0	3
7.	1	5
8.	0	5
9.	0	1

For each observation, disease identifies the true disease status of the subject (0 = normal, 1 = abnormal), and rating contains the classification value assigned by the reviewer.

We can use roctab to calculate and plot the nonparametric ROC curve by specifying both the summary and graph options. By also specifying the table option, we obtain a contingency table summarizing our dataset.

```
. roctab disease rating, table graph summary
```

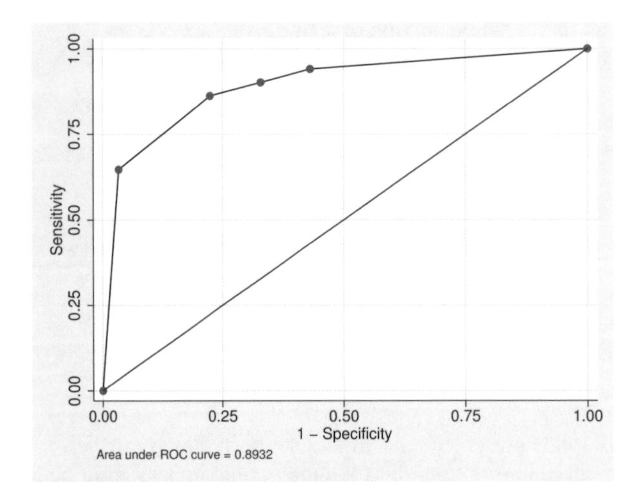

disease	rating 1	2	3	4	5	Total
0	33	6	6	11	2	58
1	3	2	2	11	33	51
Total	36	8	8	22	35	109

Obs	ROC Area	Std. Err.	–Asymptotic Normal– [95% Conf.	Interval]
109	0.8932	0.0307	0.83295	0.95339

By default, `roctab` reports the area under the curve, its standard error, and its confidence interval. The `graph` option can be used to plot the ROC curve.

The ROC curve is plotted by computing the sensitivity and specificity using each value of the rating variable as a possible cutpoint. A point is plotted on the graph for each of the cutpoints. These plotted points are joined by straight lines to form the ROC curve, and the area under the ROC curve is computed using the trapezoidal rule.

We can tabulate the computed sensitivities and specificities for each of the possible cutpoints by specifying `detail`.

```
. roctab disease rating, detail
Detailed report of Sensitivity and Specificity
```

Cutpoint	Sensitivity	Specificity	Correctly Classified	LR+	LR-
(>= 1)	100.00%	0.00%	46.79%	1.0000	
(>= 2)	94.12%	56.90%	74.31%	2.1835	0.1034
(>= 3)	90.20%	67.24%	77.98%	2.7534	0.1458
(>= 4)	86.27%	77.59%	81.65%	3.8492	0.1769
(>= 5)	64.71%	96.55%	81.65%	18.7647	0.3655
(> 5)	0.00%	100.00%	53.21%		1.0000

Obs	ROC Area	Std. Err.	–Asymptotic Normal– [95% Conf. Interval]	
109	0.8932	0.0307	0.83295	0.95339

Each cutpoint in the table indicates the ratings used to classify tomographs as being from an abnormal subject. For example, the first cutpoint (`>= 1`) indicates that all tomographs rated as 1 or greater are classified as coming from abnormal subjects. Because all tomographs have a rating of 1 or greater, all are considered abnormal. Consequently, all abnormal cases are correctly classified (sensitivity = 100%), but none of the normal patients is classified correctly (specificity = 0%). For the second cutpoint (`>=2`), tomographs with ratings of 1 are classified as normal, and those with ratings of 2 or greater are classified as abnormal. The resulting sensitivity and specificity are 94.12% and 56.90%, respectively. Using this cutpoint, we correctly classified 74.31% of the 109 tomographs. Similar interpretations can be used on the remaining cutpoints. As mentioned, each cutpoint corresponds to a point on the nonparametric ROC curve. The first cutpoint (`>=1`) corresponds to the point at (1,1), and the last cutpoint (> 5) corresponds to the point at (0,0).

`detail` also reports two likelihood ratios suggested by Choi (1998): the likelihood ratio for a positive test result (LR+) and the likelihood ratio for a negative test result (LR–). The LR+ is the ratio of the probability of a positive test among the truly positive subjects to the probability of a positive test among the truly negative subjects. The LR– is the ratio of the probability of a negative test among the truly positive subjects to the probability of a negative test among the truly negative subjects. Choi points out that LR+ corresponds to the slope of the line from the origin to the point on the ROC curve determined by the cutpoint. Similarly, LR– corresponds to the slope from the point (1,1) to the point on the ROC curve determined by the cutpoint.

By default, `roctab` calculates the standard error for the area under the curve by using an algorithm suggested by DeLong, DeLong, and Clarke-Pearson (1988) and asymptotic normal confidence intervals. Optionally, standard errors based on methods suggested by Bamber (1975) or Hanley and McNeil (1982) can be computed by specifying `bamber` or `hanley`, respectively, and an exact binomial confidence interval can be obtained by specifying `binomial`.

```
. roctab disease rating, bamber
                      ROC         Bamber          -Asymptotic Normal--
         Obs          Area        Std. Err.       [95% Conf. Interval]

         109          0.8932      0.0306           0.83317     0.95317
. roctab disease rating, hanley binomial
                      ROC         Hanley          -- Binomial Exact --
         Obs          Area        Std. Err.       [95% Conf. Interval]

         109          0.8932      0.0320           0.81559     0.94180
```

◁

Lorenz-like curves

For applications where it is known that the risk status increases or decreases monotonically with increasing values of the diagnostic test, the ROC curve and associated indices are useful in assessing the overall performance of a diagnostic test. When the risk status does not vary monotonically with increasing values of the diagnostic test, however, the resulting ROC curve can be nonconvex and its indices can be unreliable. For these situations, Lee (1999) proposed an alternative to the ROC analysis based on Lorenz-like curves and the associated Pietra and Gini indices.

Lee (1999) mentions at least three specific situations where results from Lorenz curves are superior to those obtained from ROC curves: 1) a diagnostic test with similar means but very different standard deviations in the abnormal and normal populations, 2) a diagnostic test with bimodal distributions in either the normal or abnormal population, and 3) a diagnostic test distributed symmetrically in the normal population and skewed in the abnormal.

When the risk status increases or decreases monotonically with increasing values of the diagnostic test, the ROC and Lorenz curves yield interchangeable results.

▷ Example 2

To illustrate the use of the `lorenz` option, we constructed a fictitious dataset that yields results similar to those presented in Table III of Lee (1999). The data assume that a 12-point rating scale was used to classify 442 diseased and 442 healthy subjects. We list a few of the observations.

```
. use http://www.stata-press.com/data/r12/lorenz, clear
. list in 1/7, noobs sep(0)
```

disease	class	pop
0	5	66
1	11	17
0	6	85
0	3	19
0	10	19
0	2	7
1	4	16

The data consist of 24 observations: 12 observations from diseased individuals and 12 from nondiseased individuals. Each observation corresponds to one of the 12 classification values of the rating-scale variable, `class`. The number of subjects represented by each observation is given by the `pop` variable, making this a frequency-weighted dataset. The data were generated assuming a binormal distribution of the latent variable with similar means for the normal and abnormal populations but with the standard deviation for the abnormal population five times greater than that of the normal population.

```
. roctab disease class [fweight=pop], graph summary
```

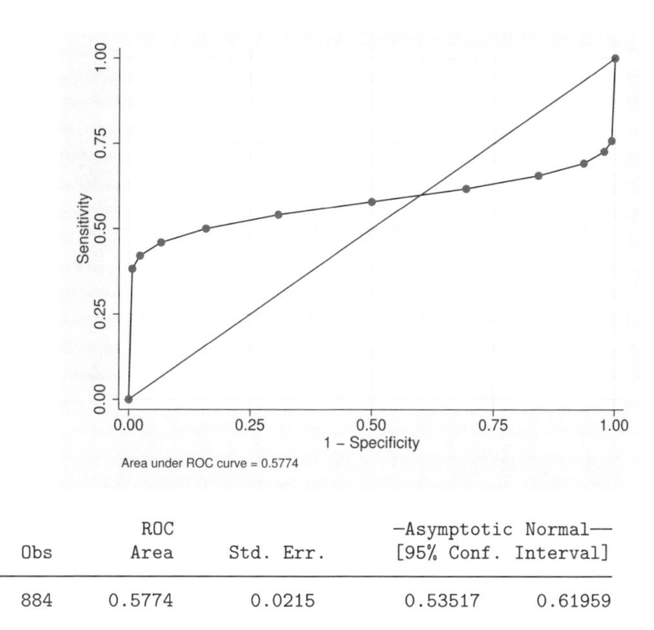


```
                      ROC                    -Asymptotic Normal--
           Obs       Area     Std. Err.     [95% Conf. Interval]
         ------------------------------------------------------------
           884     0.5774       0.0215        0.53517     0.61959
```

The resulting ROC curve is nonconvex or, as termed by Lee, "wiggly". Lee argues that for this and similar situations, the Lorenz curve and indices are preferred.

```
. roctab disease class [fweight=pop], lorenz graph summary
```

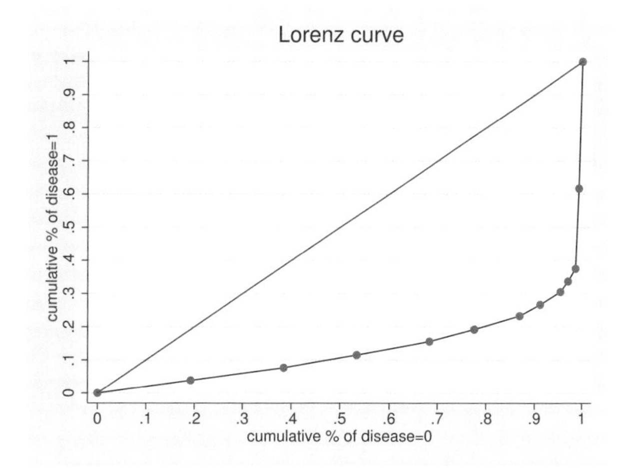


```
 Lorenz curve
---------------------------
  Pietra index =    0.6493
  Gini index   =    0.7441
```

Like ROC curves, a more bowed Lorenz curve suggests a better diagnostic test. This bowedness is quantified by the Pietra index, which is geometrically equivalent to twice the largest triangle that can be inscribed in the area between the curve and the diagonal line, and the Gini index, which is equivalent to twice the area between the Lorenz curve and the diagonal. Lee (1999) provides several additional interpretations for the Pietra and Gini indices.

◁

Saved results

`roctab` saves the following in `r()`:

Scalars

`r(N)`	number of observations
`r(se)`	standard error for the area under the ROC curve
`r(lb)`	lower bound of CI for the area under the ROC curve
`r(ub)`	upper bound of CI for the area under the ROC curve
`r(area)`	area under the ROC curve
`r(pietra)`	Pietra index
`r(gini)`	Gini index

Methods and formulas

`roctab` is implemented as ado-files.

Assume that we applied a diagnostic test to each of N_n normal and N_a abnormal subjects. Further assume that the higher the outcome value of the diagnostic test, the higher the risk of the subject being abnormal. Let $\widehat{\theta}$ be the estimated area under the curve, and let $X_i, i = 1, 2, \dots, N_a$ and $Y_j, j = 1, 2, \dots, N_n$ be the values of the diagnostic test for the abnormal and normal subjects, respectively.

The points on the nonparametric ROC curve are generated using each possible outcome of the diagnostic test as a classification cutpoint and computing the corresponding sensitivity and $1-$specificity. These points are then connected by straight lines, and the area under the resulting ROC curve is computed using the trapezoidal rule.

The default standard error for the area under the ROC curve is computed using the algorithm described by DeLong, DeLong, and Clarke-Pearson (1988). For each abnormal subject, i, define

$$V_{10}(X_i) = \frac{1}{N_n} \sum_{j=1}^{N_n} \psi(X_i, Y_j)$$

and for each normal subject, j, define

$$V_{01}(Y_j) = \frac{1}{N_a} \sum_{i=1}^{N_a} \psi(X_i, Y_j)$$

where

$$\psi(X, Y) = \begin{cases} 1 & Y < X \\ \frac{1}{2} & Y = X \\ 0 & Y > X \end{cases}$$

Define

$$S_{10} = \frac{1}{N_a - 1} \sum_{i=1}^{N_a} \{V_{10}(X_i) - \widehat{\theta}\}^2$$

and

$$S_{01} = \frac{1}{N_n - 1} \sum_{j=1}^{N_n} \{V_{01}(Y_j) - \widehat{\theta}\}^2$$

The variance of the estimated area under the ROC curve is given by

$$\text{var}(\widehat{\theta}) = \frac{1}{N_a} S_{10} + \frac{1}{N_n} S_{01}$$

The `hanley` standard error for the area under the ROC curve is computed using the algorithm described by Hanley and McNeil (1982). It requires the calculation of two quantities: Q_1 is Pr(two randomly selected abnormal subjects will both have a higher score than a randomly selected normal subject), and Q_2 is Pr(one randomly selected abnormal subject will have a higher score than any two randomly selected normal subjects). The Hanley and McNeil variance of the estimated area under the ROC curve is

$$\text{var}(\widehat{\theta}) = \frac{\widehat{\theta}(1-\widehat{\theta}) + (N_a - 1)(Q_1 - \widehat{\theta}^2) + (N_n - 1)(Q_2 - \widehat{\theta}^2)}{N_a N_n}$$

The `bamber` standard error for the area under the ROC curve is computed using the algorithm described by Bamber (1975). For any two Y values, Y_j and Y_k, and any X_i value, define

$$b_{yyx} = p(Y_j, Y_k < X_i) + p(X_i < Y_j, Y_k) - 2p(Y_j < X_i < Y_k)$$

and similarly, for any two X values, X_i and X_l, and any Y_j value, define

$$b_{xxy} = p(X_i, X_l < Y_j) + p(Y_j < X_i, X_l) - 2p(X_i < Y_j < X_l)$$

Bamber's unbiased estimate of the variance for the area under the ROC curve is

$$\text{var}(\widehat{\theta}) = \frac{1}{4}(N_a - 1)(N_n - 1)\{p(X \neq Y) + (N_a - 1)b_{xxy} + (N_n - 1)b_{yyx} - 4(N_a + N_n - 1)(\widehat{\theta} - 0.5)^2\}$$

Asymptotic confidence intervals are constructed and reported by default, assuming a normal distribution for the area under the ROC curve.

Exact binomial confidence intervals are calculated as described in [R] **ci**, with p equal to the area under the ROC curve.

References

Bamber, D. 1975. The area above the ordinal dominance graph and the area below the receiver operating characteristic graph. *Journal of Mathematical Psychology* 12: 387–415.

Choi, B. C. K. 1998. Slopes of a receiver operating characteristic curve and likelihood ratios for a diagnostic test. *American Journal of Epidemiology* 148: 1127–1132.

Cleves, M. A. 1999. sg120: Receiver operating characteristic (ROC) analysis. *Stata Technical Bulletin* 52: 19–33. Reprinted in *Stata Technical Bulletin Reprints*, vol. 9, pp. 212–229. College Station, TX: Stata Press.

——. 2000. sg120.2: Correction to roccomp command. *Stata Technical Bulletin* 54: 26. Reprinted in *Stata Technical Bulletin Reprints*, vol. 9, p. 231. College Station, TX: Stata Press.

——. 2002a. Comparative assessment of three common algorithms for estimating the variance of the area under the nonparametric receiver operating characteristic curve. *Stata Journal* 2: 280–289.

——. 2002b. From the help desk: Comparing areas under receiver operating characteristic curves from two or more probit or logit models. *Stata Journal* 2: 301–313.

DeLong, E. R., D. M. DeLong, and D. L. Clarke-Pearson. 1988. Comparing the areas under two or more correlated receiver operating characteristic curves: A nonparametric approach. *Biometrics* 44: 837–845.

Erdreich, L. S., and E. T. Lee. 1981. Use of relative operating characteristic analysis in epidemiology: A method for dealing with subjective judgment. *American Journal of Epidemiology* 114: 649–662.

Hanley, J. A., and B. J. McNeil. 1982. The meaning and use of the area under a receiver operating characteristic (ROC) curve. *Radiology* 143: 29–36.

Harbord, R. M., and P. Whiting. 2009. metandi: Meta-analysis of diagnostic accuracy using hierarchical logistic regression. *Stata Journal* 9: 211–229.

Juul, S., and M. Frydenberg. 2010. *An Introduction to Stata for Health Researchers*. 3rd ed. College Station, TX: Stata Press.

Lee, W. C. 1999. Probabilistic analysis of global performances of diagnostic tests: Interpreting the Lorenz curve-based summary measures. *Statistics in Medicine* 18: 455–471.

Ma, G., and W. J. Hall. 1993. Confidence bands for the receiver operating characteristic curves. *Medical Decision Making* 13: 191–197.

Pepe, M. S. 2003. *The Statistical Evaluation of Medical Tests for Classification and Prediction*. New York: Oxford University Press.

Reichenheim, M. E., and A. Ponce de Leon. 2002. Estimation of sensitivity and specificity arising from validity studies with incomplete design. *Stata Journal* 2: 267–279.

Seed, P. T., and A. Tobías. 2001. sbe36.1: Summary statistics for diagnostic tests. *Stata Technical Bulletin* 59: 25–27. Reprinted in *Stata Technical Bulletin Reprints*, vol. 10, pp. 90–93. College Station, TX: Stata Press.

Tobías, A. 2000. sbe36: Summary statistics report for diagnostic tests. *Stata Technical Bulletin* 56: 16–18. Reprinted in *Stata Technical Bulletin Reprints*, vol. 10, pp. 87–90. College Station, TX: Stata Press.

Working, H., and H. Hotelling. 1929. Application of the theory of error to the interpretation of trends. *Journal of the American Statistical Association* 24 (Suppl.): 73–85.

Also see

[R] **logistic postestimation** — Postestimation tools for logistic

[R] **roc** — Receiver operating characteristic (ROC) analysis

[R] **roccomp** — Tests of equality of ROC areas

[R] **rocfit** — Parametric ROC models

[R] **rocreg** — Receiver operating characteristic (ROC) regression

Title

rologit — Rank-ordered logistic regression

Syntax

`rologit` *depvar indepvars* [*if*] [*in*] [*weight*] , `group(`*varname*`)` [*options*]

options	Description
Model	
* `group(`*varname*`)`	identifier variable that links the alternatives
`offset(`*varname*`)`	include *varname* in model with coefficient constrained to 1
`incomplete(`*#*`)`	use # to code unranked alternatives; default is `incomplete(0)`
`reverse`	reverse the preference order
`notestrhs`	keep right-hand-side variables that do not vary within group
`ties(`*spec*`)`	method to handle ties: `exactm`, `breslow`, `efron`, or `none`
SE/Robust	
`vce(`*vcetype*`)`	*vcetype* may be `oim`, `robust`, `cluster` *clustvar*, `bootstrap`, or `jackknife`
Reporting	
`level(`*#*`)`	set confidence level; default is `level(95)`
display_options	control column formats, row spacing, line width, and display of omitted variables and base and empty cells
Maximization	
maximize_options	control the maximization process; seldom used
`coeflegend`	display legend instead of statistics

*`group(`*varname*`)` is required.
indepvars may contain factor variables; see [U] **11.4.3 Factor variables**.
`bootstrap`, `by`, `jackknife`, `rolling`, and `statsby` are allowed; see [U] **11.1.10 Prefix commands**.
Weights are not allowed with the `bootstrap` prefix; see [R] **bootstrap**.
`fweight`s, `iweight`s, and `pweight`s are allowed, except with `ties(efron)`; see [U] **11.1.6 weight**.
`coeflegend` does not appear in the dialog box.
See [U] **20 Estimation and postestimation commands** for more capabilities of estimation commands.

Menu

Statistics > Ordinal outcomes > Rank-ordered logistic regression

Description

`rologit` fits the rank-ordered logistic regression model by maximum likelihood (Beggs, Cardell, and Hausman 1981). This model is also known as the Plackett–Luce model (Marden 1995), as the exploded logit model (Punj and Staelin 1978), and as the choice-based method of conjoint analysis (Hair et al. 2010).

rologit expects the data to be in long form, similar to clogit (see [R] **clogit**), in which each of the ranked alternatives forms an observation; all observations related to an individual are linked together by the variable that you specify in the group() option. The distinction from clogit is that *depvar* in rologit records the rankings of the alternatives, whereas for clogit, *depvar* marks only the best alternative by a value not equal to zero. rologit interprets equal scores of *depvar* as ties. The ranking information may be incomplete "at the bottom" (least preferred alternatives). That is, unranked alternatives may be coded as 0 or as a common value that may be specified with the incomplete() option.

If your data record only the unique best alternative, rologit fits the same model as clogit.

Options

Model

group(*varname*) is required, and it specifies the identifier variable (numeric or string) that links the alternatives for an individual, which have been compared and rank ordered with respect to one another.

offset(*varname*); see [R] **estimation options**.

incomplete(*#*) specifies the numeric value used to code alternatives that are not ranked. It is assumed that unranked alternatives are less preferred than the ranked alternatives (that is, the data record the ranking of the most preferred alternatives). It is not assumed that subjects are indifferent between the unranked alternatives. # defaults to 0.

reverse specifies that in the preference order, a higher number means a less attractive alternative. The default is that higher values indicate more attractive alternatives. The rank-ordered logit model is not symmetric in the sense that reversing the ordering simply leads to a change in the signs of the coefficients.

notestrhs suppresses the test that the independent variables vary within (at least some of) the groups. Effects of variables that are always constant are not identified. For instance, a rater's gender cannot directly affect his or her rankings; it could affect the rankings only via an interaction with a variable that does vary over alternatives.

ties(*spec*) specifies the method for handling ties (indifference between alternatives) (see [ST] **stcox** for details):

exactm	exact marginal likelihood (default)
breslow	Breslow's method (default if pweights specified)
efron	Efron's method (default if robust VCE)
none	no ties allowed

SE/Robust

vce(*vcetype*) specifies the type of standard error reported, which includes types that are derived from asymptotic theory, that are robust to some kinds of misspecification, that allow for intragroup correlation, and that use bootstrap or jackknife methods; see [R] ***vce_option***.

If ties(exactm) is specified, *vcetype* may be only oim, bootstrap, or jackknife.

Reporting

level(*#*); see [R] **estimation options**.

display_options: `noomitted`, `vsquish`, `noemptycells`, `baselevels`, `allbaselevels`, `cformat(%fmt)`, `pformat(%fmt)`, `sformat(%fmt)`, and `nolstretch`; see [R] **estimation options**.

Maximization

maximize_options: `iterate(#)`, `trace`, `[no]log`, `tolerance(#)`, `ltolerance(#)`, `nrtolerance(#)`, and `nonrtolerance`; see [R] **maximize**. These options are seldom used.

The following option is available with `rologit` but is not shown in the dialog box:

`coeflegend`; see [R] **estimation options**.

Remarks

The rank-ordered logit model can be applied to analyze how decision makers combine attributes of alternatives into overall evaluations of the attractiveness of these alternatives. The model generalizes a version of McFadden's choice model without alternative-specific covariates, as fit by the `clogit` command. It uses richer information about the comparison of alternatives, namely, how decision-makers rank the alternatives rather than just specifying the alternative that they like best.

Remarks are presented under the following headings:

Examples
Comparing respondents
Incomplete rankings and ties
Clustered choice data
Comparison of rologit and clogit
On reversals of rankings

Examples

A popular way to study employer preferences for characteristics of employees is the quasi-experimental "vignette method". As an example, we consider the research by de Wolf on the labor market position of social science graduates (de Wolf 2000). This study addresses how the educational portfolio (for example, general skills versus specific knowledge) affects short-term and long-term labor-market opportunities. De Wolf asked 22 human resource managers (the respondents) to rank order the six most suitable candidates of 20 fictitious applicants and to rank order these six candidates for three jobs, namely, 1) researcher, 2) management trainee, and 3) policy adviser. Applicants were described by 10 attributes, including their age, gender, details of their portfolio, and work experience. In this example, we analyze a subset of the data. Also, to simplify the output, we drop, at random, 10 nonselected applicants per case. The resulting dataset includes 29 cases, consisting of 10 applicants each. The data are in long form: observations correspond to alternatives (the applications), and alternatives that figured in one decision task are identified by the variable `caseid`. We list the observations for `caseid==7`, in which the respondent considered applicants for a social-science research position.

```
. use http://www.stata-press.com/data/r12/evignet
(Vignet study employer prefs (Inge de Wolf 2000))
. list pref female age grades edufit workexp boardexp if caseid==7, noobs
```

pref	female	age	grades	edufit	workexp	boardexp
0	yes	28	A/B	no	none	no
0	no	25	C/D	yes	one year	no
0	no	25	C/D	yes	none	yes
0	yes	25	C/D	no	internship	yes
1	no	25	C/D	yes	one year	yes
2	no	25	A/B	yes	none	no
3	yes	25	A/B	yes	one year	no
4	yes	25	A/B	yes	none	yes
5	no	25	A/B	yes	internship	no
6	yes	28	A/B	yes	one year	yes

Here six applicants were selected. The rankings are stored in the variable `pref`, where a value of 6 corresponds to "best among the candidates", a value of 5 corresponds to "second-best among the candidates", etc. The applicants with a ranking of 0 were not among the best six candidates for the job. The respondent was not asked to express his preferences among these four applicants, but by the elicitation procedure, it is known that he ranks these four applicants below the six selected applicants. The best candidate was a female, 28 years old, with education fitting the job, with good grades (A/B), with 1 year of work experience, and with experience being a board member of a fraternity, a sports club, etc. The profiles of the other candidates read similarly. Here the respondent completed the task; that is, he selected and rank ordered the six most suitable applicants. Sometimes the respondent performed only part of the task.

```
. list pref female age grades edufit workexp boardexp if caseid==18, noobs
```

pref	female	age	grades	edufit	workexp	boardexp
0	no	25	C/D	yes	none	yes
0	no	25	C/D	no	internship	yes
0	no	28	C/D	no	internship	yes
0	yes	25	A/B	no	one year	no
2	yes	25	A/B	no	none	yes
2	no	25	A/B	no	none	yes
2	no	25	A/B	no	one year	yes
5	no	25	A/B	no	none	yes
5	no	25	A/B	no	none	yes
5	yes	25	A/B	no	none	no

The respondent selected the six best candidates and segmented these six candidates into two groups: one group with the three best candidates, and a second group of three candidates that were "still acceptable". The numbers 2 and 5, indicating these two groups, are arbitrary apart from the implied ranking of the groups. The ties between the candidates in a group indicate that the respondent was not able to rank the candidates within the group.

The purpose of the vignette experiment was to explore and test hypotheses about which of the employees' attributes are valued by employers, how these attributes are weighted depending on the type of job (described by variable `job` in these data), etc. In the psychometric tradition of Thurstone (1927), *value* is assumed to be linear in the attributes, with the coefficients expressing the direction and weight of the attributes. In addition, it is assumed that *valuation* is to some extent a random procedure, captured by an additive random term. For instance, if value depends only on an applicant's age and gender, we would have

$$\text{value}(\texttt{female}_i, \texttt{age}_i) = \beta_1 \texttt{female}_i + \beta_2 \texttt{age}_i + \epsilon_i$$

where the random residual, ϵ_i, captures all omitted attributes. Thus $\beta_1 > 0$ means that the employer assigns higher value to a woman than to a man. Given this conceptualization of value, it is straightforward to model the decision (selection) among alternatives or the ranking of alternatives: the alternative with the highest value is selected (chosen), or the alternatives are ranked according to their value. To complete the specification of a model of choice and of ranking, we assume that the random residual ϵ_i follows an "extreme value distribution of type I", introduced in this context by Luce (1959). This specific assumption is made mostly for computational convenience.

This model is known by many names. Among others, it is known as the rank-ordered logit model in economics (Beggs, Cardell, and Hausman 1981), as the exploded logit model in marketing research (Punj and Staelin 1978), as the choice-based conjoint analysis model (Hair et al. 2010), and as the Plackett–Luce model (Marden 1995). The model coefficients are estimated using the method of maximum likelihood. The implementation in `rologit` uses an analogy between the rank-ordered logit model and the Cox regression model observed by Allison and Christakis (1994); see *Methods and formulas*. The `rologit` command implements this method for rankings, whereas `clogit` deals with the variant of choices, that is, only the most highly valued alternative is recorded. In the latter case, the model is also known as the Luce–McFadden choice model. In fact, when the data record the most preferred (unique) alternative and no additional ranking information about preferences is available, `rologit` and `clogit` return the same information, though formatted somewhat differently.

```
. rologit pref female age grades edufit workexp boardexp if job==1, group(caseid)
Iteration 0:   log likelihood =  -95.41087
Iteration 1:   log likelihood = -71.180903
Iteration 2:   log likelihood =  -68.47734
Iteration 3:   log likelihood = -68.345918
Iteration 4:   log likelihood = -68.345389
Refining estimates:
Iteration 0:   log likelihood = -68.345389
Rank-ordered logistic regression                Number of obs      =        80
Group variable: caseid                          Number of groups   =         8
No ties in data                                 Obs per group: min =        10
                                                               avg =     10.00
                                                               max =        10
                                                LR chi2(6)         =     54.13
Log likelihood = -68.34539                      Prob > chi2        =    0.0000

------------------------------------------------------------------------------
        pref |      Coef.   Std. Err.      z    P>|z|     [95% Conf. Interval]
-------------+----------------------------------------------------------------
      female |  -.4487287   .3671307    -1.22   0.222    -1.168292    .2708343
         age |  -.0984926   .0820473    -1.20   0.230    -.2593024    .0623172
      grades |   3.064534   .6148245     4.98   0.000       1.8595    4.269568
      edufit |   .7658064   .3602366     2.13   0.034     .0597556    1.471857
     workexp |   1.386427    .292553     4.74   0.000     .8130341    1.959821
    boardexp |   .6944377   .3762596     1.85   0.065    -.0430176    1.431893
------------------------------------------------------------------------------
```

Focusing only on the variables whose coefficients are significant at the 10% level (we are analyzing 8 respondents only!), the estimated value of an applicant for a job of type 1 (research positions) can be written as

```
value = 3.06*grades + 0.77*edufit + 1.39*workexp + 0.69*boardexp
```

Thus employers prefer applicants for a research position (`job==1`) whose educational portfolio fits the job, who have better grades, who have more relevant work experience, and who have (extracurricular) board experience. They do not seem to care much about the sex and age of applicants, which is comforting.

Given these estimates of the valuation by employers, we consider the probabilities that each of the applications is ranked first. Under the assumption that the ϵ_i are independent and follow an extreme value type I distribution, Luce (1959) showed that the probability, π_i, that alternative i is valued higher than alternatives $2, \ldots, k$ can be written in the multinomial logit form

$$\pi_i = \Pr\left\{\text{value}_1 > \max(\text{value}_2, \ldots, \text{value}_m)\right\} = \frac{\exp(\text{value}_i)}{\sum_{j=1}^{k} \exp(\text{value}_i)}$$

The probability of observing a specific ranking can be written as the *product* of such terms, representing a sequential decision interpretation in which the rater first chooses the most preferred alternative, and then the most preferred alternative among the rest, etc.

The probabilities for alternatives to be ranked first are conveniently computed by `predict`.

```
. predict p if e(sample)
(option pr assumed; conditional probability that alternative is ranked first)
(210 missing values generated)

. sort caseid pref p

. list pref p grades edufit workexp boardexp if caseid==7, noobs
```

pref	p	grades	edufit	workexp	boardexp
0	.0027178	C/D	yes	none	yes
0	.0032275	C/D	no	internship	yes
0	.0064231	A/B	no	none	no
0	.0217202	C/D	yes	one year	no
1	.0434964	C/D	yes	one year	yes
2	.0290762	A/B	yes	none	no
3	.2970933	A/B	yes	one year	no
4	.0371747	A/B	yes	none	yes
5	.1163203	A/B	yes	internship	no
6	.4427504	A/B	yes	one year	yes

There clearly is a positive relation between the stated ranking and the predicted probabilities for alternatives to be ranked first, but the association is not perfect. In fact, we would not have expected a perfect association, as the model specifies a (nondegenerate) probability distribution over the possible rankings of the alternatives. These predictions for sets of 10 candidates can also be used to make predictions for subsets of the alternatives. For instance, suppose that only the last three candidates listed in this table would be available. According to parameter estimates of the rank-ordered logit model, the probability that the last of these candidates is selected equals $0.443/(0.037+0.116+0.443) = 0.743$.

Comparing respondents

The `rologit` model assumes that all respondents, HR managers in large public-sector organizations in The Netherlands, use the *same* valuation function; that is, they apply the same decision weights. This is the substantive interpretation of the assumption that the β's are constant between the respondents. To probe this assumption, we could test whether the coefficients vary between different groups of respondents. For a metric characteristic of the HR manager, such as `firmsize`, we can consider a trend-model in the valuation weights,

$$\beta_{ij} = \alpha_{i0} + \alpha_{i1}\texttt{firmsize}_j$$

and we can test that the slopes α_{i1} of `firmsize` are zero.

```
. generate firmsize = employer
. rologit pref edufit grades workexp c.firmsize#c.(edufit grades workexp boardexp)
> if job==1, group(caseid) nolog
Rank-ordered logistic regression                Number of obs      =        80
Group variable: caseid                          Number of groups   =         8
No ties in data                                 Obs per group: min =        10
                                                               avg =     10.00
                                                               max =        10
                                                LR chi2(7)         =     57.17
Log likelihood = -66.82346                      Prob > chi2        =    0.0000

------------------------------------------------------------------------------
        pref |      Coef.   Std. Err.      z    P>|z|     [95% Conf. Interval]
-------------+----------------------------------------------------------------
      edufit |    1.29122    1.13764     1.13   0.256    -.9385127    3.520953
      grades |   6.439776   2.288056     2.81   0.005     1.955267    10.92428
     workexp |    1.23342   .8065067     1.53   0.126     -.347304    2.814144
             |
  c.firmsize#|
    c.edufit |  -.0173333   .0711942    -0.24   0.808    -.1568714    .1222048
             |
  c.firmsize#|
    c.grades |  -.2099279   .1218251    -1.72   0.085    -.4487008     .028845
             |
  c.firmsize#|
   c.workexp |   .0097508   .0525081     0.19   0.853    -.0931632    .1126649
             |
  c.firmsize#|
  c.boardexp |   .0382304   .0227545     1.68   0.093    -.0063676    .0828284
------------------------------------------------------------------------------

. testparm c.firmsize#c.(edufit grades workexp boardexp)
 ( 1)  c.firmsize#c.edufit = 0
 ( 2)  c.firmsize#c.grades = 0
 ( 3)  c.firmsize#c.workexp = 0
 ( 4)  c.firmsize#c.boardexp = 0
           chi2(  4) =    7.14
         Prob > chi2 =    0.1288
```

The Wald test that the slopes of the interacted `firmsize` variables are jointly zero provides no evidence upon which we would reject the null hypothesis; that is, we do not find evidence against the assumption of constant valuation weights of the attributes by firms of different size. We did not enter `firmsize` as a predictor variable. Characteristics of the decision-making agent do not vary between alternatives. Thus an additive effect of these characteristics on the valuation of alternatives does *not* affect the agent's ranking of alternatives and his choice. Consequently the coefficient of `firmsize` is not identified. `rologit` would in fact have diagnosed the problem and dropped `firmsize` from the analysis. Diagnosing this problem can slow the estimation considerably; the test may be suppressed by specifying the `notestrhs` option.

Incomplete rankings and ties

rologit allows incomplete rankings and ties in the rankings as proposed by Allison and Christakis (1994). rologit permits rankings to be incomplete only "at the bottom"; namely, that the ranking of the least attractive alternatives for subjects may not be known—do not confuse this with the situation that a subject is indifferent between these alternatives. This form of incompleteness occurred in the example discussed here, because the respondents were instructed to select and rank only the top six alternatives. It may also be that respondents refused to rank the alternatives that are very unattractive. rologit does not allow other forms of incompleteness, for instance, data in which respondents indicate which of four cars they like best, and which one they like least, but not how they rank the two intermediate cars. Another example of incompleteness that cannot be analyzed with rologit is data in which respondents select the three alternatives they like best but are not requested to express their preferences among the three selected alternatives.

rologit also permits ties in rankings. rologit assumes that if a subject expresses a tie between two or more alternatives, he or she actually holds one particular strict preference ordering, but with all possibilities of a strict ordering consistent with the expressed weak ordering being equally probable. For instance, suppose that a respondent ranks alternative 1 highest. He prefers alternatives 2 and 3 over alternative 4, and he is indifferent between alternatives 2 and 3. We assume that this respondent either has the strict preference ordering $1 > 2 > 3 > 4$ or $1 > 3 > 2 > 4$, with both possibilities being equally likely. From a psychometric perspective, it may actually be more appropriate to also assume that the alternatives 2 and 3 are close; for instance, the difference between the associated valuations (utilities) is less than some threshold or minimally discernible difference. Computationally, however, this is a more demanding model.

Clustered choice data

We have seen that applicants with work experience are in a relatively favorable position. To test whether the effects of work experience vary between the jobs, we can include interactions between the type of job and the attributes of applicants. Such interactions can be obtained using factor variables.

Because some HR managers contributed data for more than one job, we cannot assume that their selection decisions for different jobs are independent. We can account for this by specifying the vce(cluster *clustvar*) option. By treating choice data as incomplete ranking data with only the most preferred alternative marked, rologit may be used to estimate the model parameters for clustered choice data.

```
. rologit pref job##c.(female grades edufit workexp), group(caseid)
> vce(cluster employer) nolog
2.job 3.job omitted because of no within-caseid variance
Rank-ordered logistic regression                Number of obs      =       290
Group variable: caseid                          Number of groups   =        29
Ties handled via the Efron method               Obs per group: min =        10
                                                               avg =     10.00
                                                               max =        10
                                                Wald chi2(12)      =     79.57
Log pseudolikelihood = -296.3855                Prob > chi2        =    0.0000
                                  (Std. Err. adjusted for 22 clusters in employer)

------------------------------------------------------------------------------
             |               Robust
        pref |      Coef.   Std. Err.      z    P>|z|     [95% Conf. Interval]
-------------+----------------------------------------------------------------
         job |
          2  |          0  (omitted)
          3  |          0  (omitted)
             |
      female |  -.2286609   .2519883    -0.91   0.364    -.7225489    .2652272
      grades |   2.812555   .8517878     3.30   0.001     1.143081    4.482028
      edufit |   .7027757   .2398396     2.93   0.003     .2326987    1.172853
     workexp |   1.224453   .3396773     3.60   0.000     .5586978    1.890208
             |
job#c.female |
          2  |   .0293815   .4829166     0.06   0.951    -.9171177    .9758808
          3  |   .1195538   .3688844     0.32   0.746    -.6034463    .8425538
             |
job#c.grades |
          2  |  -2.364247   1.005963    -2.35   0.019    -4.335898   -.3925961
          3  |   -1.88232   .8995277    -2.09   0.036    -3.645362   -.1192782
             |
job#c.edufit |
          2  |   -.267475   .4244964    -0.63   0.529    -1.099473    .5645226
          3  |  -.3182995   .3689972    -0.86   0.388    -1.041521    .4049217
             |
job#c.workexp|
          2  |  -.6870077   .3692946    -1.86   0.063    -1.410812    .0367964
          3  |  -.4656993   .4515712    -1.03   0.302    -1.350763    .4193639
------------------------------------------------------------------------------
```

The parameter estimates for the first job type are very similar to those that would have been obtained from an analysis isolated to these data. Differences are due only to an implied change in the method of handling ties. With clustered observations, `rologit` uses Efron's method. If we had specified the `ties(efron)` option with the separate analyses, then the parameter estimates would have been identical to the simultaneous results. Another difference is that `rologit` now reports robust standard errors, adjusted for clustering within respondents. These could have been obtained for the separate analyses, as well by specifying the `vce(robust)` option. In fact, this option would also have forced `rologit` to switch to Efron's method as well.

Given the combined results for the three types of jobs, we can test easily whether the weights for the attributes of applicants vary between the jobs, in other words, whether employers are looking for different qualifications in applicants for different jobs. A Wald test for the equality hypothesis of no difference can be obtained with the `testparm` command:

```
. testparm job#c.(female grades edufit workexp)
 ( 1)  2.job#c.female = 0
 ( 2)  3.job#c.female = 0
 ( 3)  2.job#c.grades = 0
 ( 4)  3.job#c.grades = 0
 ( 5)  2.job#c.edufit = 0
 ( 6)  3.job#c.edufit = 0
 ( 7)  2.job#c.workexp = 0
 ( 8)  3.job#c.workexp = 0

           chi2(  8) =   14.96
         Prob > chi2 =    0.0599
```

We find only mild evidence that employers look for different qualities in candidates according to the job for which they are being considered.

❑ Technical note

Allison (1999) stressed that the comparison between groups of the coefficients of logistic regression is problematic, especially in its latent-variable interpretation. In many common latent-variable models, only the regression coefficients divided by the scale of the latent variable are identified. Thus a comparison of logit regression coefficients between, say, men and women is meaningful only if one is willing to argue that the standard deviation of the latent residual does not differ between the sexes. The rank-ordered logit model is also affected by this problem. While we formulated the model with a scale-free residual, we can actually think of the model for the value of an alternative as being scaled by the standard deviation of the random term, representing other relevant attributes of alternatives. Again comparing attribute weights between jobs is meaningful to the extent that we are willing to defend the proposition that "all omitted attributes" are equally important for different kinds of jobs.

❑

Comparison of rologit and clogit

The rank-ordered logit model also has a sequential interpretation. A subject first chooses the best among the alternatives. Next he or she selects the best alternative among the remaining alternatives, etc. The decisions at each of the subsequent stages are described by a conditional logit model, and a subject is assumed to apply the same decision weights at each stage. Some authors have expressed concern that later choices may well be made more randomly than the first few decisions. A formalization of this idea is a heteroskedastic version of the rank-ordered logit model in which the scale of the random term increases with the number of decisions made (for example, Hausman and Ruud [1987]). This extended model is currently not supported by `rologit`. However, the hypothesis that the same decision weights are applied at the first stage and at later stages can be tested by applying a Hausman test.

First, we fit the rank-ordered logit model on the full ranking data for the first type of job,

```
. rologit pref age female edufit grades workexp boardexp if job==1, group(caseid)
> nolog
Rank-ordered logistic regression                Number of obs      =        80
Group variable: caseid                          Number of groups   =         8
No ties in data                                 Obs per group: min =        10
                                                               avg =     10.00
                                                               max =        10
                                                LR chi2(6)         =     54.13
Log likelihood = -68.34539                      Prob > chi2        =    0.0000

------------------------------------------------------------------------------
        pref |      Coef.   Std. Err.      z    P>|z|     [95% Conf. Interval]
-------------+----------------------------------------------------------------
         age |  -.0984926   .0820473    -1.20   0.230    -.2593024    .0623172
      female |  -.4487287   .3671307    -1.22   0.222    -1.168292    .2708343
      edufit |   .7658064   .3602366     2.13   0.034     .0597556    1.471857
      grades |   3.064534   .6148245     4.98   0.000       1.8595    4.269568
     workexp |   1.386427    .292553     4.74   0.000     .8130341    1.959821
    boardexp |   .6944377   .3762596     1.85   0.065    -.0430176    1.431893
------------------------------------------------------------------------------
```

and we save the estimates for later use with the `estimates` command.

```
. estimates store Ranking
```

To estimate the decision weights on the basis of the most preferred alternatives only, we create a variable, `best`, that is 1 for the best alternatives, and 0 otherwise. The `by` prefix is useful here.

```
. by caseid (pref), sort: gen best = pref == pref[_N] if job==1
(210 missing values generated)
```

By specifying `(pref)` with `by caseid`, we ensured that the data were sorted in increasing order on `pref` within `caseid`. Hence, the most preferred alternatives are last in the sort order. The expression `pref == pref[_N]` is true (1) for the most preferred alternatives, even if the alternative is not unique, and false (0) otherwise. If the most preferred alternatives were sometimes tied, we could still fit the model for the based-alternatives-only data via `rologit`, but `clogit` would yield different results because it deals with ties in a less appropriate way for continuous valuations. To ascertain whether there are ties in the selected data regarding applicants for research positions, we can combine `by` with `assert`:

```
. by caseid (pref), sort: assert pref[_N-1] != pref[_N] if job==1
```

There are no ties. We can now fit the model on the choice data by using either `clogit` or `rologit`.

```
. rologit best age edufit grades workexp boardexp if job==1, group(caseid) nolog
Rank-ordered logistic regression                Number of obs      =        80
Group variable: caseid                          Number of groups   =         8
No ties in data                                 Obs per group: min =        10
                                                               avg =     10.00
                                                               max =        10
                                                LR chi2(5)         =     17.27
Log likelihood = -9.783205                      Prob > chi2        =    0.0040

------------------------------------------------------------------------------
        best |      Coef.   Std. Err.      z    P>|z|     [95% Conf. Interval]
-------------+----------------------------------------------------------------
         age |  -.1048959   .2017068    -0.52   0.603    -.5002339    .2904421
      edufit |   .4558387   .9336775     0.49   0.625    -1.374136    2.285813
      grades |   3.443851   1.969002     1.75   0.080    -.4153223    7.303025
     workexp |   2.545648   1.099513     2.32   0.021     .3906422    4.700655
    boardexp |   1.765176   1.112763     1.59   0.113    -.4157988    3.946152
------------------------------------------------------------------------------
. estimates store Choice
```

The same results, though with a slightly different formatted header, would have been obtained by using `clogit` on these data.

```
. clogit best age edufit grades workexp boardexp if job==1, group(caseid) nolog
Conditional (fixed-effects) logistic regression   Number of obs   =        80
                                                  LR chi2(5)      =     17.27
                                                  Prob > chi2     =    0.0040
Log likelihood = -9.7832046                       Pseudo R2       =    0.4689

------------------------------------------------------------------------------
        best |      Coef.   Std. Err.      z    P>|z|     [95% Conf. Interval]
-------------+----------------------------------------------------------------
         age |  -.1048959   .2017068    -0.52   0.603    -.5002339    .2904421
      edufit |   .4558387   .9336775     0.49   0.625    -1.374136    2.285813
      grades |   3.443851   1.969002     1.75   0.080    -.4153223    7.303025
     workexp |   2.545648   1.099513     2.32   0.021     .3906422    4.700655
    boardexp |   1.765176   1.112763     1.59   0.113    -.4157988    3.946152
------------------------------------------------------------------------------
```

The parameters of the ranking and choice models look different, but the standard errors based on the choice data are much larger. Are we estimating parameters with the ranking data that are different from those with the choice data? A Hausman test compares two estimators of a parameter. One of the estimators should be efficient under the null hypothesis, namely, that choosing the second-best alternative is determined with the same decision weights as the best, etc. In our case, the efficient estimator of the decision weights uses the ranking information. The other estimator should be consistent, even if the null hypothesis is false. In our application, this is the estimator that uses the first-choice data only.

```
. hausman Choice Ranking
                 ---- Coefficients ----
             |      (b)          (B)            (b-B)     sqrt(diag(V_b-V_B))
             |     Choice       Ranking       Difference          S.E.
-------------+----------------------------------------------------------------
         age |   -.1048959    -.0984926      -.0064033         .1842657
      edufit |    .4558387     .7658064      -.3099676         .8613846
      grades |    3.443851     3.064534       .3793169         1.870551
     workexp |    2.545648     1.386427       1.159221         1.059878
   boardexp |    1.765176     .6944377       1.070739          1.04722
------------------------------------------------------------------------------
                         b = consistent under Ho and Ha; obtained from rologit
          B = inconsistent under Ha, efficient under Ho; obtained from rologit

    Test:  Ho:  difference in coefficients not systematic

                  chi2(5) = (b-B)'[(V_b-V_B)^(-1)](b-B)
                          =        3.05
                Prob>chi2 =      0.6918
```

We do not find evidence for misspecification. We have to be cautious, though, because Hausman-type tests are often not powerful, and the number of observations in our example is very small, which makes the quality of the method of the null distribution by a chi-squared test rather uncertain.

On reversals of rankings

The rank-ordered logit model has a property that you may find unexpected and even unfortunate. Compare two analyses with the rank-ordered logit model, one in which alternatives are ranked from "most attractive" to "least attractive", the other a reversed analysis in which these alternatives are ranked from "most unattractive" to "least unattractive". By unattractiveness, you probably mean just the opposite of attractiveness, and you expect that the weights of the attributes in predicting "attractiveness" to be minus the weights in predicting "unattractiveness". This is, however, *not* true for the rank-ordered logit model. The assumed distribution of the random residual takes the form $F(\epsilon) = 1 - \exp\{\exp(-\epsilon)\}$. This distribution is right-skewed. Therefore, slightly different models result from adding and subtracting the random residual, corresponding with high-to-low and low-to-high rankings. Thus the estimated coefficients will differ between the two specifications, though usually not in an important way. You may observe the difference by specifying the `reverse` option of `rologit`. Reversing the rank order makes rankings that are incomplete at the bottom become incomplete at the top. Only the first kind of incompleteness is supported by `rologit`. Thus, for this comparison, we exclude the alternatives that are not ranked, omitting the information that ranked alternatives are preferred over excluded ones.

```
. rologit pref grades edufit workexp boardexp if job==1 & pref!=0, group(caseid)
  (output omitted )
. estimates store Original
. rologit pref grades edufit workexp boardexp if job==1 & pref!=0, group(caseid)
> reverse
  (output omitted )
. estimates store Reversed
```

```
. estimates table Original Reversed, stats(aic bic)
```

Variable	Original	Reversed
grades	2.0032332	-1.0955335
edufit	-.13111006	-.05710681
workexp	1.2805373	-1.2096383
boardexp	.46213212	-.27200317
aic	96.750452	99.665642
bic	104.23526	107.15045

Thus, although the weights of the attributes for reversed rankings are indeed mostly of opposite signs, the magnitudes of the weights and their standard errors differ. Which one is more appropriate? We have no advice to offer here. The specific science of the problem will determine what is appropriate, though we would be surprised indeed if this helps here. Formal testing does not help much either, as the models for the original and reversed rankings are not nested. The model-selection indices, such as the AIC and BIC, however, suggest that you stick to the rank-ordered logit model applied to the original ranking rather than to the reversed ranking.

Saved results

`rologit` saves the following in `e()`:

Scalars

`e(N)`	number of observations
`e(ll_0)`	log likelihood of the null model ("all rankings are equiprobable")
`e(ll)`	log likelihood
`e(df_m)`	model degrees of freedom
`e(chi2)`	χ^2
`e(p)`	significance
`e(r2_p)`	pseudo-R^2
`e(N_g)`	number of groups
`e(g_min)`	minimum group size
`e(g_avg)`	average group size
`e(g_max)`	maximum group size
`e(code_inc)`	value for incomplete preferences
`e(N_clust)`	number of clusters
`e(rank)`	rank of `e(V)`

Macros

`e(cmd)`	`rologit`
`e(cmdline)`	command as typed
`e(depvar)`	name of dependent variable
`e(group)`	name of `group()` variable
`e(wtype)`	weight type
`e(wexp)`	weight expression
`e(title)`	title in estimation output
`e(clustvar)`	name of cluster variable
`e(offset)`	linear offset variable
`e(chi2type)`	`Wald` or `LR`; type of model χ^2 test
`e(reverse)`	`reverse`, if specified
`e(ties)`	`breslow`, `efron`, `exactm`
`e(vce)`	*vcetype* specified in `vce()`
`e(vcetype)`	title used to label Std. Err.
`e(properties)`	`b V`
`e(predict)`	program used to implement `predict`
`e(marginsok)`	predictions allowed by `margins`
`e(marginsnotok)`	predictions disallowed by `margins`
`e(asbalanced)`	factor variables `fvset` as `asbalanced`
`e(asobserved)`	factor variables `fvset` as `asobserved`

Matrices

`e(b)`	coefficient vector
`e(V)`	variance–covariance matrix of the estimators
`e(V_modelbased)`	model-based variance

Functions

`e(sample)`	marks estimation sample

Methods and formulas

`rologit` is implemented as an ado-file.

Allison and Christakis (1994) demonstrate that maximum likelihood estimates for the rank-ordered logit model can be obtained as the maximum partial-likelihood estimates of an appropriately specified Cox regression model for waiting time ([ST] **stcox**). In this analogy, a higher value for an alternative is formally equivalent to a higher hazard rate of failure. `rologit` uses `stcox` to fit the rank-ordered logit model based on such a specification of the data in Cox terms. A higher stated preference is represented by a shorter waiting time until failure. Incomplete rankings are dealt with via censoring. Moreover, decision situations (subjects) are to be treated as strata. Finally, as proposed by Allison and Christakis, ties in rankings are handled by the marginal-likelihood method, specifying that all strict preference orderings consistent with the stated weak preference ordering are equally likely. The marginal-likelihood estimator is available in stcox via the exactm option. The methods of the marginal likelihood due to Breslow and Efron are also appropriate for the analysis of rank-ordered logit models. Because in most applications the number of ranked alternatives by one subject will be fairly small (at most, say, 20), the number of ties is small as well, and so you rarely will need to turn to methods to restrict computer time. Because the marginal-likelihood estimator in stcox does not support the cluster adjustment or pweights, you should use the Efron method in such cases.

This command supports the clustered version of the Huber/White/sandwich estimator of the variance using vce(robust) and vce(cluster *clustvar*). See [P] **_robust**, particularly *Maximum likelihood estimators* and *Methods and formulas*. Specifying vce(robust) is equivalent to specifying vce(cluster *groupvar*), where *groupvar* is the identifier variable that links the alternatives.

Acknowledgment

The rologit command was written by Jeroen Weesie, Department of Sociology, Utrecht University, The Netherlands.

References

Allison, P. D. 1999. Comparing logit and probit coefficients across groups. *Sociological Methods and Research* 28: 186–208.

Allison, P. D., and N. Christakis. 1994. Logit models for sets of ranked items. In Vol. 24 of *Sociological Methodology*, ed. P. V. Marsden, 123–126. Oxford: Blackwell.

Beggs, S., S. Cardell, and J. A. Hausman. 1981. Assessing the potential demand for electric cars. *Journal of Econometrics* 17: 1–19.

de Wolf, I. 2000. *Opleidingsspecialisatie en arbeidsmarktsucces van sociale wetenschappers*. Amsterdam: ThelaThesis.

Hair, J. F., Jr., W. C. Black, B. J. Babin, and R. E. Anderson. 2010. *Multivariate Data Analysis*. 7th ed. Upper Saddle River, NJ: Pearson.

Hausman, J. A., and P. A. Ruud. 1987. Specifying and testing econometric models for rank-ordered data. *Journal of Econometrics* 34: 83–104.

Luce, R. D. 1959. *Individual Choice Behavior: A Theoretical Analysis*. New York: Dover.

Marden, J. I. 1995. *Analyzing and Modeling Rank Data*. London: Chapman & Hall.

McCullagh, P. 1993. Permutations and regression models. In *Probability Models and Statistical Analysis for Ranking Data*, ed. M. A. Fligner and J. S. Verducci, 196–215. New York: Springer.

Plackett, R. L. 1975. The analysis of permutations. *Applied Statistics* 24: 193–202.

Punj, G. N., and R. Staelin. 1978. The choice process for graduate business schools. *Journal of Marketing Research* 15: 588–598.

Thurstone, L. L. 1927. A law of comparative judgment. *Psychological Reviews* 34: 273–286.

Yellott, J. I., Jr. 1977. The relationship between Luce's choice axiom, Thurstone's theory of comparative judgment, and the double exponential distribution. *Journal of Mathematical Psychology* 15: 109–144.

Also see

[R] **rologit postestimation** — Postestimation tools for rologit

[R] **clogit** — Conditional (fixed-effects) logistic regression

[R] **logistic** — Logistic regression, reporting odds ratios

[R] **mlogit** — Multinomial (polytomous) logistic regression

[R] **nlogit** — Nested logit regression

[R] **slogit** — Stereotype logistic regression

[U] **20 Estimation and postestimation commands**

Title

rologit postestimation — Postestimation tools for rologit

Description

The following postestimation commands are available after `rologit`:

Command	Description
`contrast`	contrasts and ANOVA-style joint tests of estimates
`estat`	AIC, BIC, VCE, and estimation sample summary
`estimates`	cataloging estimation results
`hausman`	Hausman's specification test
`lincom`	point estimates, standard errors, testing, and inference for linear combinations of coefficients
`linktest`	link test for model specification
`lrtest`	likelihood-ratio test
`margins`[1]	marginal means, predictive margins, marginal effects, and average marginal effects
`marginsplot`	graph the results from margins (profile plots, interaction plots, etc.)
`nlcom`	point estimates, standard errors, testing, and inference for nonlinear combinations of coefficients
`predict`	predictions, residuals, influence statistics, and other diagnostic measures
`predictnl`	point estimates, standard errors, testing, and inference for generalized predictions
`pwcompare`	pairwise comparisons of estimates
`test`	Wald tests of simple and composite linear hypotheses
`testnl`	Wald tests of nonlinear hypotheses

[1] The default prediction statistic `pr` cannot be correctly handled by `margins`; however, `margins` can be used after `rologit` with the `predict(xb)` option.

See the corresponding entries in the *Base Reference Manual* for details.

Syntax for predict

`predict` [*type*] *newvar* [*if*] [*in*] [, *statistic* `nooffset`]

statistic	Description
Main	
`pr`	probability that alternatives are ranked first; the default
`xb`	linear prediction
`stdp`	standard error of the linear prediction

These statistics are available both in and out of sample; type `predict ... if e(sample) ...` if wanted only for the estimation sample.

Menu

Statistics > Postestimation > Predictions, residuals, etc.

Options for predict

Main

`pr`, the default, calculates the probability that alternatives are ranked first.

`xb` calculates the linear prediction.

`stdp` calculates the standard error of the linear prediction.

`nooffset` is relevant only if you specified `offset(`*varname*`)` for `rologit`. It modifies the calculations made by `predict` so that they ignore the offset variable; the linear prediction is treated as $\mathbf{x}_j\mathbf{b}$ rather than as $\mathbf{x}_j\mathbf{b} + \text{offset}_j$.

Remarks

See *Comparing respondents* and *Clustered choice data* in [R] **rologit** for examples of the use of `testparm`, an alternative to the `test` command.

See *Comparison of rologit and clogit* and *On reversals of rankings* in [R] **rologit** for examples of the use of `estimates`.

See *Comparison of rologit and clogit* in [R] **rologit** for an example of the use of `hausman`.

Methods and formulas

All postestimation commands listed above are implemented as ado-files.

Also see

[R] **rologit** — Rank-ordered logistic regression

[U] **20 Estimation and postestimation commands**

Title

rreg — Robust regression

Syntax

`rreg` *depvar* [*indepvars*] [*if*] [*in*] [, *options*]

options	Description
Model	
`tune(#)`	use # as the biweight tuning constant; default is `tune(7)`
Reporting	
`level(#)`	set confidence level; default is `level(95)`
`genwt(newvar)`	create *newvar* containing the weights assigned to each observation
display_options	control column formats, row spacing, line width, and display of omitted variables and base and empty cells
Optimization	
optimization_options	control the optimization process; seldom used
`graph`	graph weights during convergence
`coeflegend`	display legend instead of statistics

indepvars may contain factor variables; see [U] **11.4.3 Factor variables.**
depvar and *indepvars* may contain time-series operators; see [U] **11.4.4 Time-series varlists.**
`by`, `fracpoly`, `mfp`, `mi estimate`, `rolling`, and `statsby` are allowed; see [U] **11.1.10 Prefix commands.**
`coeflegend` does not appear in the dialog box.
See [U] **20 Estimation and postestimation commands** for more capabilities of estimation commands.

Menu

Statistics > Linear models and related > Other > Robust regression

Description

`rreg` performs one version of robust regression of *depvar* on *indepvars*.

Also see *Robust standard errors* in [R] **regress** for standard regression with robust variance estimates and [R] **qreg** for quantile (including median or least-absolute-residual) regression.

Options

Model

`tune(#)` is the biweight tuning constant. The default is 7, meaning seven times the median absolute deviation (MAD) from the median residual; see *Methods and formulas*. Lower tuning constants downweight outliers rapidly but may lead to unstable estimates (less than 6 is not recommended). Higher tuning constants produce milder downweighting.

Reporting

`level(#)`; see [R] **estimation options**.

`genwt(`*newvar*`)` creates the new variable *newvar* containing the weights assigned to each observation.

display_options: `noomitted`, `vsquish`, `noemptycells`, `baselevels`, `allbaselevels`, `cformat(%`*fmt*`)`, `pformat(%`*fmt*`)`, `sformat(%`*fmt*`)`, and `nolstretch`; see [R] **estimation options**.

Optimization

optimization_options: `iterate(#)`, `tolerance(#)`, [`no`]`log`. `iterate()` specifies the maximum number of iterations; iterations stop when the maximum change in weights drops below `tolerance()`; and `log`/`nolog` specifies whether to show the iteration log. These options are seldom used.

`graph` allows you to graphically watch the convergence of the iterative technique. The weights obtained from the most recent round of estimation are graphed against the weights obtained from the previous round.

The following option is available with `rreg` but is not shown in the dialog box:

`coeflegend`; see [R] **estimation options**.

Remarks

`rreg` first performs an initial screening based on Cook's distance > 1 to eliminate gross outliers before calculating starting values and then performs Huber iterations followed by biweight iterations, as suggested by Li (1985).

▷ Example 1

We wish to examine the relationship between mileage rating, weight, and location of manufacture for the 74 cars in our automobile data. As a point of comparison, we begin by fitting an ordinary regression:

```
. use http://www.stata-press.com/data/r12/auto
(1978 Automobile Data)

. regress mpg weight foreign

      Source |       SS       df       MS              Number of obs =      74
-------------+------------------------------           F(  2,    71) =   69.75
       Model |   1619.2877     2  809.643849           Prob > F      =  0.0000
    Residual |  824.171761    71   11.608053           R-squared     =  0.6627
-------------+------------------------------           Adj R-squared =  0.6532
       Total |  2443.45946    73  33.4720474           Root MSE      =  3.4071

------------------------------------------------------------------------------
         mpg |      Coef.   Std. Err.      t    P>|t|     [95% Conf. Interval]
-------------+----------------------------------------------------------------
      weight |  -.0065879   .0006371   -10.34   0.000    -.0078583   -.0053175
     foreign |  -1.650029   1.075994    -1.53   0.130      -3.7955    .4954422
       _cons |    41.6797   2.165547    19.25   0.000     37.36172    45.99768
------------------------------------------------------------------------------
```

We now compare this with the results from rreg:

```
. rreg mpg weight foreign
   Huber iteration 1:  maximum difference in weights = .80280176
   Huber iteration 2:  maximum difference in weights = .2915438
   Huber iteration 3:  maximum difference in weights = .08911171
   Huber iteration 4:  maximum difference in weights = .02697328
Biweight iteration 5:  maximum difference in weights = .29186818
Biweight iteration 6:  maximum difference in weights = .11988101
Biweight iteration 7:  maximum difference in weights = .03315872
Biweight iteration 8:  maximum difference in weights = .00721325

Robust regression                                      Number of obs =      74
                                                       F(  2,    71) =  168.32
                                                       Prob > F      =  0.0000
```

mpg	Coef.	Std. Err.	t	P>|t|	[95% Conf.	Interval]
weight	-.0063976	.0003718	-17.21	0.000	-.007139	-.0056562
foreign	-3.182639	.627964	-5.07	0.000	-4.434763	-1.930514
_cons	40.64022	1.263841	32.16	0.000	38.1202	43.16025

Note the large change in the foreign coefficient.

◁

❑ Technical note

It would have been better if we had fit the previous robust regression by typing rreg mpg weight foreign, genwt(w). The new variable, w, would then contain the estimated weights. Let's pretend that we did this:

```
. rreg mpg weight foreign, genwt(w)
  (output omitted)
. summarize w, detail
```

Robust Regression Weight

	Percentiles	Smallest		
1%	0	0		
5%	.0442957	0		
10%	.4674935	0	Obs	74
25%	.8894815	.0442957	Sum of Wgt.	74
50%	.9690193		Mean	.8509966
		Largest	Std. Dev.	.2746451
75%	.9949395	.9996715		
90%	.9989245	.9996953	Variance	.0754299
95%	.9996715	.9997343	Skewness	-2.287952
99%	.9998585	.9998585	Kurtosis	6.874605

We discover that 3 observations in our data were dropped altogether (they have weight 0). We could further explore our data:

```
. sort w
. list make mpg weight w if w<.467, sep(0)
```

	make	mpg	weight	w
1.	Datsun 210	35	2,020	0
2.	Subaru	35	2,050	0
3.	VW Diesel	41	2,040	0
4.	Plym. Arrow	28	3,260	.04429567
5.	Cad. Seville	21	4,290	.08241943
6.	Toyota Corolla	31	2,200	.10443129
7.	Olds 98	21	4,060	.28141296

Being familiar with the automobile data, we immediately spotted two things: the VW is the only diesel car in our data, and the weight recorded for the Plymouth Arrow is incorrect. ❑

▷ Example 2

If we specify no explanatory variables, **rreg** produces a robust estimate of the mean:

```
. rreg mpg
   Huber iteration 1:  maximum difference in weights = .64471879
   Huber iteration 2:  maximum difference in weights = .05098336
   Huber iteration 3:  maximum difference in weights = .0099887
Biweight iteration 4:  maximum difference in weights = .25197391
Biweight iteration 5:  maximum difference in weights = .00358606

Robust regression                                     Number of obs =      74
                                                      F(  0,    73) =    0.00
                                                      Prob > F      =       .
```

mpg	Coef.	Std. Err.	t	P>|t|	[95% Conf.	Interval]
_cons	20.68825	.641813	32.23	0.000	19.40912	21.96738

The estimate is given by the coefficient on _cons. The mean is 20.69 with an estimated standard error of 0.6418. The 95% confidence interval is [19.4, 22.0]. By comparison, ci (see [R] **ci**) gives us the standard calculation:

```
. ci mpg
```

Variable	Obs	Mean	Std. Err.	[95% Conf.	Interval]
mpg	74	21.2973	.6725511	19.9569	22.63769

◁

Saved results

`rreg` saves the following in `e()`:

Scalars

`e(N)`	number of observations
`e(mss)`	model sum of squares
`e(df_m)`	model degrees of freedom
`e(rss)`	residual sum of squares
`e(df_r)`	residual degrees of freedom
`e(r2)`	R-squared
`e(r2_a)`	adjusted R-squared
`e(F)`	F statistic
`e(rmse)`	root mean squared error
`e(rank)`	rank of `e(V)`

Macros

`e(cmd)`	`rreg`
`e(cmdline)`	command as typed
`e(depvar)`	name of dependent variable
`e(genwt)`	variable containing the weights
`e(title)`	title in estimation output
`e(model)`	`ols`
`e(vce)`	`ols`
`e(properties)`	`b V`
`e(predict)`	program used to implement `predict`
`e(marginsok)`	predictions allowed by `margins`
`e(asbalanced)`	factor variables `fvset` as `asbalanced`
`e(asobserved)`	factor variables `fvset` as `asobserved`

Matrices

`e(b)`	coefficient vector
`e(V)`	variance–covariance matrix of the estimators

Functions

`e(sample)`	marks estimation sample

Methods and formulas

`rreg` is implemented as an ado-file.

See Berk (1990), Goodall (1983), and Rousseeuw and Leroy (1987) for a general description of the issues and methods. Hamilton (1991a, 1992) provides a more detailed description of `rreg` and some Monte Carlo evaluations.

`rreg` begins by fitting the regression (see [R] **regress**), calculating Cook's D (see [R] **predict** and [R] **regress postestimation**), and excluding any observation for which $D > 1$.

Thereafter `rreg` works iteratively: it performs a regression, calculates case weights from absolute residuals, and regresses again using those weights. Iterations stop when the maximum change in weights drops below `tolerance()`. Weights derive from one of two weight functions, Huber weights and biweights. Huber weights (Huber 1964) are used until convergence, and then, from that result, biweights are used until convergence. The biweight was proposed by Beaton and Tukey (1974, 151–152) after the Princeton robustness study (Andrews et al. 1972) had compared various estimators. Both weighting functions are used because Huber weights have problems dealing with severe outliers, whereas biweights sometimes fail to converge or have multiple solutions. The initial Huber weighting should improve the behavior of the biweight estimator.

In Huber weighting, cases with small residuals receive weights of 1; cases with larger residuals receive gradually smaller weights. Let $e_i = y_i - \mathbf{X}_i\mathbf{b}$ represent the ith-case residual. The ith scaled residual $u_i = e_i/s$ is calculated, where $s = M/0.6745$ is the residual scale estimate and

$M = \text{med}(|e_i - \text{med}(e_i)|)$ is the median absolute deviation from the median residual. Huber estimation obtains case weights:

$$w_i = \begin{cases} 1 & \text{if } |u_i| \leq c_h \\ c_h/|u_i| & \text{otherwise} \end{cases}$$

rreg defines $c_h = 1.345$, so downweighting begins with cases whose absolute residual exceeds $(1.345/0.6745)M \approx 2M$.

With biweights, all cases with nonzero residuals receive some downweighting, according to the smoothly decreasing biweight function

$$w_i = \begin{cases} \{1 - (u_i/c_b)^2\}^2 & \text{if } |u_i| \leq c_b \\ 0 & \text{otherwise} \end{cases}$$

where $c_b = 4.685 \times \texttt{tune()}/7$. Thus when $\texttt{tune()} = 7$, cases with absolute residuals of $(4.685/0.6745)M \approx 7M$ or more are assigned 0 weight and thus are effectively dropped. Goodall (1983, 377) suggests using a value between 6 and 9, inclusive, for tune() in the biweight case and states that performance is good between 6 and 12, inclusive.

The tuning constants $c_h = 1.345$ and $c_b = 4.685$ (assuming tune() is set at the default 7) give rreg about 95% of the efficiency of OLS when applied to data with normally distributed errors (Hamilton 1991b). Lower tuning constants downweight outliers more drastically (but give up Gaussian efficiency); higher tuning constants make the estimator more like OLS.

Standard errors are calculated using the pseudovalues approach described in Street, Carroll, and Ruppert (1988).

Acknowledgment

The current version of rreg is due to the work of Lawrence Hamilton, Department of Sociology, University of New Hampshire.

References

Andrews, D. F., P. J. Bickel, F. R. Hampel, P. J. Huber, W. H. Rogers, and J. W. Tukey. 1972. *Robust Estimates of Location: Survey and Advances*. Princeton: Princeton University Press.

Beaton, A. E., and J. W. Tukey. 1974. The fitting of power series, meaning polynomials, illustrated on band-spectroscopic data. *Technometrics* 16: 147–185.

Berk, R. A. 1990. A primer on robust regression. In *Modern Methods of Data Analysis*, ed. J. Fox and J. S. Long, 292–324. Newbury Park, CA: Sage.

Goodall, C. 1983. M-estimators of location: An outline of the theory. In *Understanding Robust and Exploratory Data Analysis*, ed. D. C. Hoaglin, F. Mosteller, and J. W. Tukey, 339–431. New York: Wiley.

Gould, W. W., and W. H. Rogers. 1994. Quantile regression as an alternative to robust regression. In *1994 Proceedings of the Statistical Computing Section*. Alexandria, VA: American Statistical Association.

Hamilton, L. C. 1991a. srd1: How robust is robust regression? *Stata Technical Bulletin* 2: 21–26. Reprinted in *Stata Technical Bulletin Reprints*, vol. 1, pp. 169–175. College Station, TX: Stata Press.

——. 1991b. ssi2: Bootstrap programming. *Stata Technical Bulletin* 4: 18–27. Reprinted in *Stata Technical Bulletin Reprints*, vol. 1, pp. 208–220. College Station, TX: Stata Press.

——. 1992. *Regression with Graphics: A Second Course in Applied Statistics*. Belmont, CA: Duxbury.

——. 2009. *Statistics with Stata (Updated for Version 10)*. Belmont, CA: Brooks/Cole.

Huber, P. J. 1964. Robust estimation of a location parameter. *Annals of Mathematical Statistics* 35: 73–101.

Li, G. 1985. Robust regression. In *Exploring Data Tables, Trends, and Shapes*, ed. D. C. Hoaglin, F. Mosteller, and J. W. Tukey, 281–340. New York: Wiley.

Mosteller, F., and J. W. Tukey. 1977. *Data Analysis and Regression: A Second Course in Statistics*. Reading, MA: Addison–Wesley.

Relles, D. A., and W. H. Rogers. 1977. Statisticians are fairly robust estimators of location. *Journal of the American Statistical Association* 72: 107–111.

Rousseeuw, P. J., and A. M. Leroy. 1987. *Robust Regression and Outlier Detection*. New York: Wiley.

Street, J. O., R. J. Carroll, and D. Ruppert. 1988. A note on computing robust regression estimates via iteratively reweighted least squares. *American Statistician* 42: 152–154.

Verardi, V., and C. Croux. 2009. Robust regression in Stata. *Stata Journal* 9: 439–453.

Also see

[R] **rreg postestimation** — Postestimation tools for rreg

[R] **qreg** — Quantile regression

[R] **regress** — Linear regression

[MI] **estimation** — Estimation commands for use with mi estimate

[U] **20 Estimation and postestimation commands**

Title

rreg postestimation — Postestimation tools for rreg

Description

The following postestimation commands are available after `rreg`:

Command	Description
`contrast`	contrasts and ANOVA-style joint tests of estimates
`estat`	VCE and estimation sample summary
`estimates`	cataloging estimation results
`lincom`	point estimates, standard errors, testing, and inference for linear combinations of coefficients
`margins`	marginal means, predictive margins, marginal effects, and average marginal effects
`marginsplot`	graph the results from margins (profile plots, interaction plots, etc.)
`nlcom`	point estimates, standard errors, testing, and inference for nonlinear combinations of coefficients
`predict`	predictions, residuals, influence statistics, and other diagnostic measures
`predictnl`	point estimates, standard errors, testing, and inference for generalized predictions
`pwcompare`	pairwise comparisons of estimates
`test`	Wald tests of simple and composite linear hypotheses
`testnl`	Wald tests of nonlinear hypotheses

See the corresponding entries in the *Base Reference Manual* for details.

Syntax for predict

`predict` [*type*] *newvar* [*if*] [*in*] [, *statistic*]

statistic	Description
Main	
`xb`	linear prediction; the default
`stdp`	standard error of the linear prediction
`residuals`	residuals
`hat`	diagonal elements of the hat matrix

These statistics are available both in and out of sample; type `predict ... if e(sample) ...` if wanted only for the estimation sample.

Menu

Statistics > Postestimation > Predictions, residuals, etc.

Options for predict

Main

xb, the default, calculates the linear prediction.

stdp calculates the standard error of the linear prediction.

residuals calculates the residuals.

hat calculates the diagonal elements of the hat matrix. You must have run the rreg command with the genwt() option.

Methods and formulas

All postestimation commands listed above are implemented as ado-files.

Also see

[R] **rreg** — Robust regression

[U] **20 Estimation and postestimation commands**

Title

runtest — Test for random order

Syntax

`runtest` *varname* [*in*] [, *options*]

options	Description
`continuity`	continuity correction
`drop`	ignore values equal to the threshold
`split`	randomly split values equal to the threshold as above or below the threshold; default is to count as below
`mean`	use mean as threshold; default is median
`threshold(#)`	assign arbitrary threshold; default is median

Menu

Statistics > Nonparametric analysis > Tests of hypotheses > Test for random order

Description

`runtest` tests whether the observations of *varname* are serially independent—that is, whether they occur in a random order—by counting how many runs there are above and below a threshold. By default, the median is used as the threshold. A small number of runs indicates positive serial correlation; a large number indicates negative serial correlation.

Options

`continuity` specifies a continuity correction that may be helpful in small samples. If there are fewer than 10 observations either above or below the threshold, however, the tables in Swed and Eisenhart (1943) provide more reliable critical values. By default, no continuity correction is used.

`drop` directs `runtest` to ignore any values of *varname* that are equal to the threshold value when counting runs and tabulating observations. By default, `runtest` counts a value as being above the threshold when it is strictly above the threshold and as being below the threshold when it is less than or equal to the threshold.

`split` directs `runtest` to randomly split values of *varname* that are equal to the threshold. In other words, when *varname* is equal to threshold, a "coin" is flipped. If it comes up heads, the value is counted as above the threshold. If it comes up tails, the value is counted as below the threshold.

`mean` directs `runtest` to tabulate runs above and below the mean rather than the median.

`threshold(#)` specifies an arbitrary threshold to use in counting runs. For example, if *varname* has already been coded as a 0/1 variable, the median generally will not be a meaningful separating value.

Remarks

runtest performs a nonparametric test of the hypothesis that the observations of *varname* occur in a random order by counting how many runs there are above and below a threshold. If *varname* is positively serially correlated, it will tend to remain above or below its median for several observations in a row; that is, there will be relatively few runs. If, on the other hand, *varname* is negatively serially correlated, observations above the median will tend to be followed by observations below the median and vice versa; that is, there will be relatively many runs.

By default, runtest uses the median for the threshold, and this is not necessarily the best choice. If mean is specified, the mean is used instead of the median. If threshold(#) is specified, # is used. Because runtest divides the data into two states—above and below the threshold—it is appropriate for data that are already binary; for example, win or lose, live or die, rich or poor, etc. Such variables are often coded as 0 for one state and 1 for the other. Here you should specify threshold(0) because, by default, runtest separates the observations into those that are greater than the threshold and those that are less than *or equal* to the threshold.

As with most nonparametric procedures, the treatment of ties complicates the test. Observations equal to the threshold value are ties and can be treated in one of three ways. By default, they are treated as if they were below the threshold. If drop is specified, they are omitted from the calculation and the total number of observations is adjusted. If split is specified, each is randomly assigned to the above- and below-threshold groups. The random assignment is different each time the procedure is run unless you specify the random-number seed; see [R] **set seed**.

▷ Example 1

We can use runtest to check regression residuals for serial correlation.

```
. use http://www.stata-press.com/data/r12/run1
. scatter resid year, connect(l) yline(0) title(Regression residuals)
```

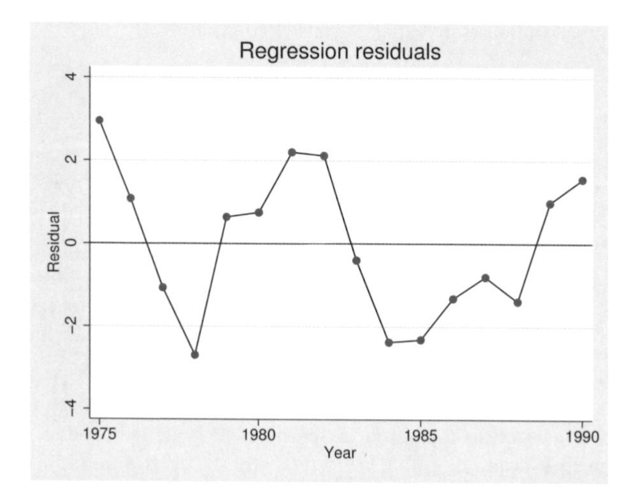

The graph gives the impression that these residuals are positively correlated. Excursions above or below zero—the natural threshold for regression residuals—tend to last for several observations. runtest can evaluate the statistical significance of this impression.

```
. runtest resid, thresh(0)
 N(resid <= 0) = 8
 N(resid >  0) = 8
           obs = 16
       N(runs) = 5
           z   = -2.07
     Prob>|z|  = .04
```

There are five runs in these 16 observations. Using the normal approximation to the true distribution of the number of runs, the five runs in this series are fewer than would be expected if the residuals were serially independent. The p-value is 0.04, indicating a two-sided significant result at the 5% level. If the alternative hypothesis is positive serial correlation, rather than any deviation from randomness, then the one-sided p-value is $0.04/2 = 0.015$. With so few observations, however, the normal approximation may be inaccurate. (Tables compiled by Swed and Eisenhart list five runs as the 5% critical value for a one-sided test.)

`runtest` is a nonparametric test. It ignores the magnitudes of the observations and notes only whether the values are above or below the threshold. We can demonstrate this feature by reducing the information about the regression residuals in this example to a 0/1 variable that indicates only whether a residual is positive or negative.

```
. generate byte sign = resid>0
. runtest sign, thresh(0)
 N(sign <= 0) = 8
 N(sign >  0) = 8
          obs = 16
      N(runs) = 5
          z   = -2.07
    Prob>|z|  = .04
```

As expected, `runtest` produces the same answer as before.

◁

❑ Technical note

The run test can also be used to test the null hypothesis that two samples are drawn from the same underlying distribution. The run test is sensitive to differences in the shapes, as well as the locations, of the empirical distributions.

Suppose, for example, that two different additives are added to the oil in 10 different cars during an oil change. The cars are run until a viscosity test determines that another oil change is needed, and the number of miles traveled between oil changes is recorded. The data are

```
. use http://www.stata-press.com/data/r12/additive, clear
. list
```

	additive	miles
1.	1	4024
2.	1	4756
3.	1	7993
4.	1	5025
5.	1	4188
6.	2	3007
7.	2	1988
8.	2	1051
9.	2	4478
10.	2	4232

To test whether the additives generate different distributions of miles between oil changes, we sort the data by `miles` and then use `runtest` to see whether the marker for each additive occurs in random order:

```
. sort miles
. runtest additive, thresh(1)
 N(additive <= 1) = 5
 N(additive >  1) = 5
              obs = 10
          N(runs) = 4
              z   = -1.34
        Prob>|z|  = .18
```

Here the additives do not produce statistically different results.

❑

❑ Technical note

A test that is related to the run test is the runs up-and-down test. In the latter test, the data are classified not by whether they lie above or below a threshold but by whether they are steadily increasing or decreasing. Thus an unbroken string of increases in the variable of interest is counted as one run, as is an unbroken string of decreases. According to Madansky (1988), the run test is superior to the runs up-and-down test for detecting trends in the data, but the runs up-and-down test is superior for detecting autocorrelation.

`runtest` can be used to perform a runs up-and-down test. Using the regression residuals from the example above, we can perform a `runtest` on their first differences:

```
. use http://www.stata-press.com/data/r12/run1
. generate resid_D = resid - resid[_n-1]
(1 missing value generated)
. runtest resid_D, thresh(0)
 N(resid_D <= 0) = 7
 N(resid_D >  0) = 8
             obs = 15
         N(runs) = 6
             z   = -1.33
       Prob>|z|  = .18
```

Edgington (1961) has compiled a table of the small sample distribution of the runs up-and-down statistic, and this table is reprinted in Madansky (1988). For large samples, the z statistic reported by runtest is incorrect for the runs up-and-down test. Let N be the number of observations (15 here), and let r be the number of runs (6). The expected number of runs in the runs up-and-down test is

$$\mu_r = \frac{2N-1}{3}$$

the variance is

$$\sigma_r^2 = \frac{16N-29}{90}$$

and the correct z statistic is

$$\widehat{z} = \frac{r-\mu_r}{\sigma_r}$$

❑

❑ Technical note

runtest will tolerate missing values at the beginning or end of a series, as occurred in the technical note above (generating first differences resulted in a missing value for the first observation). runtest, however, will issue an error message if there are any missing observations in the interior of the series (in the portion covered by the in *range* modifier). To perform the test anyway, simply drop the missing observations before using runtest.

❑

Saved results

runtest saves the following in r():

Scalars

r(N)	number of observations	r(p)	p-value of z
r(N_below)	number below the threshold	r(z)	z statistic
r(N_above)	number above the threshold	r(n_runs)	number of runs
r(mean)	expected number of runs	r(Var)	variance of the number of runs

Methods and formulas

runtest is implemented as an ado-file.

runtest begins by calculating the number of observations below the threshold, n_0; the number of observations above the threshold, n_1; the total number of observations, $N = n_0 + n_1$; and the number of runs, r. These statistics are always reported, so the exact tables of critical values in Swed and Eisenhart (1943) may be consulted if necessary.

The expected number of runs under the null is

$$\mu_r = \frac{2n_0 n_1}{N} + 1$$

the variance is

$$\sigma_r^2 = \frac{2n_0n_1\,(2n_0n_1 - N)}{N^2\,(N-1)}$$

and the normal approximation test statistic is

$$\widehat{z} = \frac{r - \mu_r}{\sigma_r}$$

Acknowledgment

`runtest` was written by Sean Becketti, a past editor of the *Stata Technical Bulletin*.

References

Edgington, E. S. 1961. Probability table for number of runs of signs of first differences in ordered series. *Journal of the American Statistical Association* 56: 156–159.

Madansky, A. 1988. *Prescriptions for Working Statisticians*. New York: Springer.

Swed, F. S., and C. Eisenhart. 1943. Tables for testing randomness of grouping in a sequence of alternatives. *Annals of Mathematical Statistics* 14: 66–87.